Applications of Genome Engineering in Plants

AF184403

Applications of Genome Engineering in Plants

Edited by

Santosh Kumar Upadhyay
Department of Botany
Panjab University
Chandigarh, India

WILEY

This edition first published 2024
© 2024 John Wiley & Sons Ltd

All rights reserved. No part of this publication may be reproduced, stored in a retrieval system, or transmitted, in any form or by any means, electronic, mechanical, photocopying, recording or otherwise, except as permitted by law. Advice on how to obtain permission to reuse material from this title is available at http://www.wiley.com/go/permissions.

The right of Santosh Kumar Upadhyay to be identified as the author of the editorial material in this work has been asserted in accordance with law.

Registered Office(s)
John Wiley & Sons, Inc., 111 River Street, Hoboken, NJ 07030, USA
John Wiley & Sons Ltd, The Atrium, Southern Gate, Chichester, West Sussex, PO19 8SQ, UK

For details of our global editorial offices, customer services, and more information about Wiley products visit us at www.wiley.com.

Wiley also publishes its books in a variety of electronic formats and by print-on-demand. Some content that appears in standard print versions of this book may not be available in other formats.

Trademarks: Wiley and the Wiley logo are trademarks or registered trademarks of John Wiley & Sons, Inc. and/or its affiliates in the United States and other countries and may not be used without written permission. All other trademarks are the property of their respective owners. John Wiley & Sons, Inc. is not associated with any product or vendor mentioned in this book.

Limit of Liability/Disclaimer of Warranty
While the publisher and authors have used their best efforts in preparing this work, they make no representations or warranties with respect to the accuracy or completeness of the contents of this work and specifically disclaim all warranties, including without limitation any implied warranties of merchantability or fitness for a particular purpose. No warranty may be created or extended by sales representatives, written sales materials or promotional statements for this work. This work is sold with the understanding that the publisher is not engaged in rendering professional services. The advice and strategies contained herein may not be suitable for your situation. You should consult with a specialist where appropriate. The fact that an organization, website, or product is referred to in this work as a citation and/or potential source of further information does not mean that the publisher and authors endorse the information or services the organization, website, or product may provide or recommendations it may make. Further, readers should be aware that websites listed in this work may have changed or disappeared between when this work was written and when it is read. Neither the publisher nor authors shall be liable for any loss of profit or any other commercial damages, including but not limited to special, incidental, consequential, or other damages.

Library of Congress Cataloging-in-Publication Data Applied for:

Hardback ISBN: 9781394183883

Cover Design: Wiley
Cover Image: © MARK GARLICK/SCIENCE PHOTO LIBRARY/Getty Images

Set in 9.5/12.5pt STIXTwoText by Straive, Pondicherry, India
Printed and bound by CPI Group (UK) Ltd, Croydon, CR0 4YY

C9781394183883_121223

Contents

List of Contributors

Mehtap AYDIN
Genetics and Bioengineering Department
Yeditepe University
Istanbul, Turkey

Uzma Badar
Department of Genetics
University of Karachi
Karachi, Pakistan

Purva Bhalothia
Birla Institute of Scientific Research (BISR)
Jaipur, Rajasthan, India

Mudasir Ahmad Bhat
School of Biosciences and Biotechnology
Baba Ghulam
Shah Badshah University, Rajouri
Jammu and Kashmir, India

Dipul Kumar Biswas
Department of Biology
Faculty of Arts and Science, Bolu Abant
İzzet Baysal Üniversitesi
Gölköy, Turkey

Amit Chauhan
Institute of Biochemistry and Biology
University of Potsdam, Potsdam-Golm
Germany

Max Planck Institute of Molecular
Plant Physiology
Potsdam-Golm, Germany

Prasanta K. Dash
ICAR-National Institute for Plant
Biotechnology, New Delhi
India

Tapasya Datta
CSIR-Central Institute of Medicinal and
Aromatic Plants (CSIR-CIMAP), Lucknow
Uttar Pradesh, India

Rupesh Deshmukh
Department of Biotechnology
Central University of Haryana
Haryana, India

Sameer Dixit
National Institute of Plant
Genome Research
New Delhi, India

Jesus Christian Grimaldi-Olivas
Ciencia y Tecnología de Alimentos
(Food Science and Technology)
Alimentos Funcionales y Nutracéuticos
(Nutraceuticals and Functional Foods)
Centro de Investigación en Alimentación y
Desarrollo, AC (CIAD), (Research Center
for Food and Development), Culiacán
Sinaloa, México

Buket Çakmak Güner
Department of Medical Biology
School of Medicine
Ankara University
Ankara, Turkey

Bhavuk Gupta
National Centre for Cell Science, Pune
Maharashtra, India

Payal Gupta
ICAR-National Institute for Plant
Biotechnology, New Delhi
India

Kathleen Hefferon
Department of Microbiology, Cornell
University, Ithaca
NY, USA

J. Basilio Heredia
Ciencia y Tecnología de Alimentos
(Food Science and Technology)
Alimentos Funcionales y Nutracéuticos
(Nutraceuticals and Functional Foods)
Centro de Investigación en Alimentación y
Desarrollo, AC (CIAD), (Research Center
for Food and Development), Culiacán
Sinaloa, México

Arif Tasleem Jan
School of Biosciences and Biotechnology
Baba Ghulam Shah Badshah
University, Rajouri
Jammu and Kashmir, India

Saima Jan
School of Biosciences and Biotechnology
Baba Ghulam Shah Badshah
University, Rajouri
Jammu and Kashmir, India

Pooja Jangir
Department of Botany
University of Rajasthan, Jaipur
Rajasthan, India

Luis Alfonso Jiménez-Ortega
Ciencia y Tecnología de Alimentos
(Food Science and Technology)
Alimentos Funcionales y Nutracéuticos
(Nutraceuticals and Functional Foods)
Centro de Investigación en Alimentación y
Desarrollo, AC (CIAD), (Research Center
for Food and Development), Culiacán
Sinaloa, México

Suhas G. Karkute
Division of Vegetable Improvement
ICAR-Indian Institute of Vegetable
Research, Varanasi
Uttar Pradesh, India

Sukhmandeep Kaur
National Agri-Food Biotechnology
Institute (NABI), Mohali
Punjab, India

Department of biotechnology
Panjab University, Chandigarh
India

Ayush Khandelwal
South Asian University
New Delhi, India

Purva Khandelwal
Department of Botany
University of Rajasthan, Jaipur
Rajasthan, India

Anil Kumar
National Institute of Plant Genomic
Research (NIPGR), Aruna Asaf Ali Marg
New Delhi, India

Brijesh Kumar
Patanjali Herbal Research Division
Patanjali Research Institute, Haridwar
Uttarakhand, India

Ravi Shankar Kumar
CSIR-Central Institute of Medicinal and Aromatic Plants (CSIR-CIMAP), Lucknow
Uttar Pradesh, India

Yuxin Li
School of Horticulture
Hainan University/Key Laboratory for Quality Regulation of Tropical Horticultural Crops of Hainan Province
Haikou, China

Wen Liu
School of Horticulture
Hainan University/Key Laboratory for Quality Regulation of Tropical Horticultural Crops of Hainan Province
Haikou, China

Nikhil Malhotra
ICAR-National Bureau of Plant Genetic Resources Regional Station, Shimla
Himachal Pradesh, India

Elena V. Mikhaylova
Institute of Biochemistry and Genetics
Ufa Federal Research Center RAS
Ufa, Russia

Brandon Estefano Morales-Merida
Ciencia y Tecnología de Alimentos (Food Science and Technology)
Alimentos Funcionales y Nutracéuticos (Nutraceuticals and Functional Foods)
Centro de Investigación en Alimentación y Desarrollo, AC (CIAD), (Research Center for Food and Development), Culiacán
Sinaloa, México

Mahendra Pawar
National Institute of Plant Genome Research
New Delhi, India

R. Prajapati
Department of Plant Pathology
Kansas State University
Manhattan, KS, USA

Hasthi Ram
National Institute of Plant Genome Research, New Delhi
Delhi, India

Sumreen Amin Shah
School of Biosciences and Biotechnology
Baba Ghulam Shah Badshah University, Rajouri
Jammu and Kashmir, India

Ashish Sharma
CSIR-Central Institute of Medicinal and Aromatic Plants (CSIR-CIMAP), Lucknow
Uttar Pradesh, India

Pooja Kanwar Shekhawat
Department of Botany
University of Rajasthan, Jaipur
Rajasthan, India

Erum Shoeb
Department of Genetics
University of Karachi
Karachi, Pakistan

Akanchha Shukla
Althea DRF Lifesciences
Rouse Avenue Institutional Area
New Delhi, India

Shumayla
Department of Botany
Kashyap Block
Panjab University, Chandigarh
India

Achuit K. Singh
Division of Vegetable Improvement
ICAR-Indian Institute of Vegetable Research, Varanasi
Uttar Pradesh, India

Hiteshwari Sinha
CSIR-National Botanical Research Institute, Council of Scientific and Industrial Research (CSIR-NBRI)
Rana Pratap Marg, Lucknow
Uttar Pradesh, India

Academy of Scientific and Innovative Research (AcSIR), Ghaziabad
Uttar Pradesh, India

Praveen Soni
Department of Botany
University of Rajasthan, Jaipur
Rajasthan, India

Prabodh Kumar Trivedi
CSIR-Central Institute of Medicinal and Aromatic Plants (CSIR-CIMAP), Lucknow
Uttar Pradesh, India

CSIR-National Botanical Research Institute, Council of Scientific and Industrial Research (CSIR-NBRI)
Rana Pratap Marg, Lucknow
Uttar Pradesh, India

Academy of Scientific and Innovative Research (AcSIR), Ghaziabad
Uttar Pradesh, India

K. Tyagi
Department of Plant Pathology
University of Minnesota, St Paul
Minneapolis, MN, USA

Santosh Kumar Upadhyay
Department of Botany, Kashyap Block
Panjab University, Chandigarh
India

Elif Karlik Urhan
Institution of Science
Department of Biotechnology
Istanbul University
Istanbul, Turkey

Jyothilakshmi Vadassery
National Institute of Plant Genome Research
New Delhi, India

Sanskriti Vats
National Agri-Food Biotechnology Institute (NABI), Mohali
Punjab, India

Regional Centre for Biotechnology, Faridabad
Haryana (NCR Delhi), India

Srividhya Venkataraman
Department of Cell and Systems Biology
University of Toronto
Toronto, Canada

Hengyi Xu
Departments of Molecular Biosciences and Oncology, Institute of Cellular and Molecular Biology
University of Texas at Austin
Austin, TX, USA

Yang Zhou
School of Horticulture
Hainan University/Key Laboratory for Quality Regulation of Tropical Horticultural Crops of Hainan Province
Haikou, China

Preface

Recent advances in the genome editing methods, especially the CRISPR-Cas system for the precise engineering of the target plant genes/genomes, enabled numerous applications of this technology for crop improvement programs and functional genomics studies. In the last few years, several advances have been done in the CRISPR-Cas tool kit including the identification of new Cas proteins having a diverse range of PAM specificity, base editing technologies and prime editing without any stand break, trans-gene free editing, nano-particle-mediated delivery of the CRISPR tool kit, development of various ready to use vectors for numerous purposes, etc. These advancements make this system more efficient and effective for genome engineering, both *in vivo* and *in vitro*.

The modified tool kit has been utilized in several agricultural and horticultural crop species including rice, tomato, banana, etc., either for nutritional improvement or for stress tolerance. Further, it has been demonstrated that the CRISPR tool kit can also be utilized in the non-transgenic mode for trans-gene-free editing, which has been recently deregulated by various regulatory bodies in several countries including India and USA. These announcements of deregulation have opened a new avenue and strongly boosted the research and developments by utilizing the genome engineering tools for numerous purposes in recent years.

Nutritional improvement and stress tolerance have been of prime importance in the last few decades due to increasing nutritional deficiency in the various regions and changes in climatic conditions. The use of genome engineering methods has been efficiently demonstrated for nutritional improvement in various agricultural and horticultural crops including rice, tomato, potato, banana, etc. Development of carotenoid- and flavonoid-rich cereals, tomato and fruit crops like banana, iron-enriched rice, maize, wheat, etc., are few examples of successful utilization of technology. Further, it is being utilized for abiotic and biotic stress tolerance, disease resistance, virus resistance, and ultimately for the improvement of yield in numerous important crop species.

In this book, the updated information related to the various technological advancements and applications for the nutritional improvements, disease resistance, abiotic stress tolerance, and yield improvement in various agricultural and horticultural crops by utilizing the genome engineering tool kits have been discussed.

Santosh Kumar Upadhyay,
Department of Botany, Panjab University,
Chandigarh, India.

About the Editor

Dr. Santosh Kumar Upadhyay is currently working as an Assistant Professor in the Department of Botany, Panjab University, Chandigarh, India. Prior to this, Dr. Upadhyay was DST-INSPIRE faculty at the National Agri-Food Biotechnology Institute, Mohali, Punjab, India. He did his doctoral work at the CSIR-National Botanical Research Institute, Lucknow, and received his PhD in Biotechnology from UP Technical University, Lucknow, India. He has been working in the field of Plant Biotechnology for more than 14 years. His present research focuses in the area of functional genomics. He is involved in the bio-prospecting and characterization of various insect toxic proteins from plant biodiversity and defence and stress signaling genes in bread wheat. His research group at PU has characterized numerous important gene families and long non-coding RNAs related to the abiotic and biotic stress tolerance and signaling in bread wheat. He has also established the method for genome editing in bread wheat using CRISPR-Cas system and developed a tool SSinder for CRISPR target site prediction. His research contribution led to the publication of more than 80 research papers in leading journals of international repute. Further, there are more than 5 national and international patents, 40 book chapters and eleven books in his credit. In recognition of his substantial research record, he has been awarded INSA Associate Fellow (2023), NAAS Young scientist award (2017–2018) and NAAS-Associate (2018) from the National Academy of Agricultural Sciences, India, INSA Medal for Young Scientist (2013) from the Indian National Science Academy, India, NASI- Young Scientist Platinum Jubilee Award (2012) from the National Academy of Sciences, India, and Altech Young Scientist Award (2011). He has also been the recipient of the prestigious DST-INSPIRE Faculty Fellowship (2012), and SERB-Early Career Research Award (2016) from the Ministry of Science and Technology, Government of India. Dr. Upadhyay also serves as a member of the editorial board and reviewer of several peer-reviewed international journals.

1

CRISPR/Cas-Mediated Genome Editing in Plants: A Historical Perspective

Anil Kumar[1#], *Shumayla*[2#], *and Santosh Kumar Upadhyay*[2]

[1] *National Institute of Plant Genomic Research (NIPGR), Aruna Asaf Ali Marg, New Delhi, India*
[2] *Department of Botany, Kashyap Block, Panjab University, Chandigarh, India*

1.1 Introduction

The production of food on a global scale is currently faced with enormous obstacles. By the year 2050, the population of the world is expected to reach 9.6 billion, with a corresponding rise in staple food crop demand by 60% (Tilman et al. 2011). There is a decline in the rate of increase in yield brought about by the "Green Revolution" and adverse climatic changes that are limiting plant production. Therefore, it is a prerequisite to generate cultivars with enhanced stress tolerance, improved quality, and increased yields. Moreover, the conventional methods of crop breeding are time-consuming, complicated, and laborious. There is a need for cost-effective and time-saving methods (Chen et al. 2019).

Nowadays, crop plant improvement and their sustainable production have been achieved by a newly emerged targeted genome engineering approach, an alternative to breeding and transgenic methods. Genome editing tool utilizes sequence-specific nucleases (SSNs) to introduce double-strand breaks (DSBs) at the desired loci within the genome (Wei et al. 2022). The introduced DSBs are immediately repaired either by a nonhomologous end joining (NHEJ) pathway or a homologous recombination (HR) mechanism. NHEJ is an error-prone pathway ideally recruited by most organisms, including higher plants that typically involves the ligation of broken ends that ultimately generates insertions and deletions (indels). NHEJ can also introduce double-stranded templates in the DSBs-generated overhangs (Li et al. 2019; Park and Beal 2019; Permyakova et al. 2022). Since the HR pathway uses homologous sequence as a template to repair the DSBs, this pathway can be utilized for the precise integration of the desired genes into the genome (Park and Beal 2019). Until recently, techniques such as zinc finger nucleases (ZFNs) and TAL effector nucleases (TALENS) were utilized by researchers. But due to their complex protein engineering processes, these are not widely adopted by the plant research community. However, due to

[#] Authors contributed equally to this work

Applications of Genome Engineering in Plants, First Edition. Edited by Santosh Kumar Upadhyay.
© 2024 John Wiley & Sons Ltd. Published 2024 by John Wiley & Sons Ltd.

advancements in sequencing technology and the availability of genomic information for a large number of plant species, genome editing with precision for crop improvement has become feasible.

CRISPR (clustered regularly interspaced palindromic repeats)/Cas (CRISPR-associated protein) has become the most widely used genome editing system due to its low cost, easy adaptability, and high specificity. It is an adaptive immune system from archaea and bacteria (Mali et al. 2013b; Yin et al. 2017). This technology mainly requires two major components: Cas9 enzyme and a single guide RNA (sgRNA). CRISPR/Cas is an RNA-guided machinery that exists naturally as a part of a prokaryotic defense system that cleaves foreign DNA in a sequence-specific manner. CRISPR repeats were first discovered as an unusual structure in the *Escherichia coli* IAP gene (Ishino et al. 1987), and later investigations revealed that the spacer sequences in CRISPR loci had similarities with phage and plasmid sequences (Bolotin et al. 2005; Mojica et al. 2005; Pourcel et al. 2005). The spacers are short fragments of the sequences integrated between the two adjacent repeats at the proximal end of a CRISPR locus. The CRISPR arrays, including the spacers, are transcribed and processed into small interfering CRISPR RNAs (crRNAs). These crRNAs further combine with the transactivating RNA (tracrRNA) to activate and guide a Cas9 nuclease to cleave complementary genomic target sequences, also known as protospacers (Barrangou 2007). However, the only requirement for the cleavage is the presence of a conserved protospacer-adjacent motif (PAM) downstream of the target DNA usually 5′-NGG-3′ (Gasiunas et al. 2012; Jinek et al. 2012) and less frequently 5′-NAG-3′ (Hsu et al. 2013). The CRISPR/Cas system of archaea and bacteria is classified into three major types (I–III) (Makarova et al. 2015), of which the simplest is the type II system from *Streptococcus pyogenes*. The type II system only requires individual expression of crRNA, tracrRNA, and Cas9. This is followed by the hybridization of crRNA with trans-activating crRNA (tracrRNA) and requires a host factor RNaseIII, which is important for crRNA maturation and recruitment of Cas9 to the target (Deltcheva et al. 2011; Jinek et al. 2012; Sushmita et al. 2021). However, the complexity of CRISPR/Cas has been made facile for genome editing by the fusion of crRNA and tracrRNA into a single-guide RNA (sgRNA/gRNA) (Jinek et al. 2012; Makarova and Koonin 2015).

Soon after its demonstration in bacteria, yeast, mammals, and model plants, CRISPR/Cas is being widely applied as a genome editing tool (Tripathi et al. 2021; Lu et al. 2022; Pant et al. 2022; Vicencio et al. 2022). The application appears to offer a significant promise for modifying crops to improve their nutritional content, resistance to pathogens, or metabolic pathways leading to important end products. This chapter of the book provides an outline of the evolution, discovery, classification, and application of CRISPR/Cas as well as its mechanism of action.

1.2 Historical Background

In 1985, a novel assembly of repetitive sequences in the 3′-end flanking regions of the IAP gene in *E. coli*, which plays a role in the isozyme conversion of alkaline phosphatase, was discovered for the first time. This was the beginning of the CRISPR framework. The structure consisted of direct repeats of five highly similar sequences, each of which contained 29 nucleotides, with 32 nucleotides serving as spacers between the repeats (Ishino et al. 1987). The sequencing of the bacterial genome in the following decade led to the discovery of many CRISPR loci. However, the biological function of the CRISPR array was not appreciated until 2005, when it was found that spacers contained sequences that were identical to those of viruses and plasmids (Bolotin et al. 2005; Mojica et al. 2005; Pourcel et al. 2005). In the latter part of the year 2007, it was discovered that the CRISPR array may be utilized by a strain of *Streptococcus thermophiles* to confer resistance against an invading bacteriophage by making use of the Cas genes (Barrangou 2007). It was confirmed that the spacer region in the CRISPR loci is a form of RNA-mediated immune system that is used by bacteria to defend themselves against nucleic acids that have invaded their territory. Similar observations were reported for other organisms. This technique assists bacteria in the process of acquiring short sequences into the CRISPR locus, which may then be utilized by transcription later on. CRISPR loci in *Streptococcus pyogenes* have been studied, and the results have shown that numerous Cas genes are involved in the recognition and cleavage of invading DNA (Heler et al. 2015). However, the role that Cas proteins play in CRISPR-based resistance is not entirely known as of yet.

Further research into the structure of the Cas protein uncovered similarities with endonuclease domains, which hints at a possible function in viral resistance via DSBs. Upon activation, spacers are translated into short crRNA, and Cas proteins then form an association with crRNA to create DSBs in the invading sequences (Garneau et al. 2010). Later on, CRISPR was split up into three major categories (I, II, and III) based on the gene conservation and locus arrangement of the different Cas genes that were discovered in bacteria and archaea. In 2005, the classification of the Cas genes was originally suggested by Haft et al. and then in 2011, Makarova et al. did a good job of expanding on that classification (Haft et al. 2005; Makarova et al. 2011). The structural functioning of Cas9 as an RNA-guided endonuclease was demonstrated by Jinek et al. (2012). Cas9 is comprised of two primary nuclease domains that primarily execute cleavage functions on both strands of DNA. This research was bolstered even further by the direct confirmation of the connection between the Cas9-guide RNA and target DNA complexes that was made possible by the imaging of their crystal structures (Nishimasu et al. 2014). Multiple authors demonstrated the use of CRISPR/Cas immune system based on RNA-mediated DNA targeting soon after understanding its biological role in the year 2007 (Marraffini and Sontheimer 2008; Garneau et al. 2010; Deltcheva et al. 2011).

The transition of the CRISPR/Cas from the immune system to the genome editing technique was not by fate, but it was designed and confirmed that editing could be performed by changing only 20 nt in the crRNA and further specificity of targeting by crRNA could be combined with tracrRNA into a chimeric single guide RNA (gRNA), thus reducing the system from three to two components (Jinek et al. 2012). Shortly thereafter, this two-component system was functionally demonstrated in eukaryotes such as humans, mice, and zebrafish, suggesting the other functions of the CRISPR locus gene were supported by eukaryotic endogenous enzymes (Cho et al. 2013; Cong et al. 2013; Hwang et al. 2013; Jinek et al. 2013; Mali et al. 2013b). Interestingly, it was shown that multiple gRNA could be used to achieve high-efficiency multiplex genome engineering simultaneously at different loci (Cong et al. 2013; Mali et al. 2013a). The first applications of CRISPR/Cas were reported in August 2013 by five research groups separately in the model plants *Arabidopsis*, tobacco, and rice using transient and stable transformation strategies (Feng et al. 2013; Li et al. 2013; Nekrasov et al. 2013; Shan et al. 2013; Xie and Yang 2013). Immediately, research on higher crop species including sorghum (Jiang et al. 2013), wheat (Upadhyay et al. 2013), and maize (Liang et al. 2014) were published. All these milestones in genome editing substantiate that CRISPR/Cas is a simple, versatile, and inexpensive tool for targeted genome editing in different organisms including plants.

1.3 Mechanism of CRISPR/Cas System

The CRISPR/Cas system recognizes and cleaves foreign DNA or RNA in a sequence-specific manner. The defense mechanism can be explained easily by first understanding the components of the CRISPR/Cas unit.

The CRISPR/Cas system can be broken down into two classes, Class I and Class II, according to the structure and functions of the Cas proteins. Class I includes types I, III, and IV, while Class II includes types II, V, and VI. The Class I systems are made up of Cas-protein complexes that contain multiple subunits, whereas the Class II systems only make use of a single Cas-protein. Because of type II CRISPR/Cas-9 simplicity, it has been the subject of extensive research and has seen considerable application in genetic engineering (Liu et al. 2020). Guide RNA, also known as gRNA, and CRISPR-associated proteins, also known as Cas-9 proteins, are the two parts that are necessary for the CRISPR/Cas-9 system. Cas9 is a multidomain endonuclease isolated from *Streptococcus pyogenes* (SpCas-9). It is known as a genetic scissor (Mei et al. 2016). It is responsible for cutting the target DNA to generate a double-stranded break in the DNA. There are three steps in the mechanism by which the type II CRISPR/Cas system (van der Oost et al. 2009; Garneau et al. 2010; Horvath and Barrangou 2010; Marraffini and Sontheimer 2010) recognizes and targets foreign genetic material.

1.3.1 Acquisition of Spacers

It is the first stage in CRISPR/Cas-mediated defense. At this point, the CRISPR/Cas system is modified to include the foreign genetic material, allowing the host to recognize and respond effectively to future invasions by the same phage strain. This stage is mediated by the proteins Cas1 and Cas2, which are found in all the CRISPR/Cas systems (Makarova et al. 2006). Foreign DNA is recognized and integrated into the CRISPR locus as a spacer during the acquisition process. Typically, the spacer is recognized by a short (2–5 bp) sequence of conserved nucleotides (PAMs) in the protospacer. The selection of the protospacers is determined by the presence of protospacer-adjacent short motifs, also known as PAMs, in proximity to the target sequence. PAMs are short sequences of two to five nucleotides that are unique to each CRISPR/Cas subtype and bacteria (Mojica et al. 2009). A single copy of 30 bp spacer is inserted at the leader side of the CRISPR array followed by its duplication (Garneau et al. 2010). The immunity against pathogen attack via CRISPR/Cas system can be suppressed by mutations in the PAM region of the viral genome (Garneau et al. 2010).

1.3.2 Biogenesis

During the expression phase of the CRISPR/Cas system functioning, the lengthy pre-crRNA is transcribed from the CRISPR locus and processed into crRNAs with the aid of Cas proteins (Cas1, Cas2, Cas9, and Cas4/Casn2). Karvelis et al. (2013) recently discovered that the tracrRNA also has a role in the synthesis of pre-crRNA in *S. pyogenes* (Karvelis et al. 2013). The pre-crRNA into crRNA processing is aided by the tracrRNA, which forms base-complementary pairings with the crRNA repeat region (Deltcheva et al. 2011). For defense against any foreign invader, the processed crRNA integrates into the CRISPR-associated complex and recognizes and base pairs with an exact target region of the foreign DNA (Deltcheva et al. 2011).

1.3.3 Interference with the Target

In the final stage, crRNA combines with Cas protein to form an effector complex that base pairs with the complementary strand in the invading DNA and induces DSB of the specific target region (Garneau et al. 2010). Hence, the proliferation and propagation of foreign pathogens are prevented by these effector ribonucleoprotein (RNP) complexes. The type II effector complexes function as RNA-guided endonucleases that recognize the target sequence by crRNA and guide the Cas protein for cleavage of DNA within the target protospacer (Garneau et al. 2010; Marraffini and Sontheimer 2010). Repairing this DSB is necessary for maintaining the host system. Now, if the break is repaired via NHEJ, the gene may become defective due to insertion or deletion mutations (INDELs) (Symington and Gautier 2011). Homologous direct repair (HDR) allows DNA insertion at the target gene, permitting gene insertion or substitution, if the repair mechanism is aided by another template DNA that has homologous sequence flanking to the target site. However, the implementation of a bacterial CRISPR/Cas system in any organism requires the assembly of a functional crRNA-tracrRNA-Cas9 unit.

1.4 Breakthrough Studies in CRISPR/Cas System

S. no	Breakthrough study	Approach	Species	References
1	Base editing	Cytosine base editor	*Oryza sativa, maize*	Li et al. (2020)
		Adenine base editor	*Triticum aestivum, O. sativa, rapeseed, and Arabidopsis thaliana*	Li et al. (2018), Kang et al. (2018), Hua et al. (2020)
		Cytosine-dependent DNA manipulation		
2	Prime editing	Nick in the unmodified DNA strand	*O. sativa and T. aestivum*	Lin et al. (2020b), Xu et al. (2020)
3	Multiplexing	Multiple gRNA delivery within single CRISPR/Cas complex	*T. aestivum, Oryza sativa, A. thaliana, and Bombyx mori*	Ma et al. (2014), Debbarma et al. (2019), Zhou et al. (2019)

(*Continued*)

(Continued)

S. no	Breakthrough study	Approach	Species	References
4	Chromosome engineering	18 kb stretch rearrangement via CRISPR/Cas; reciprocal chromosomal translocations	*A. thaliana*	Schmidt et al. (2019), Beying et al. (2020)
5	Homologous recombination	Gene replacement	*Nicotiana tobaccum*	Barakate et al. (2020)
		High-efficiency gene targeting with stable inheritance	*A. thaliana*	Hahn et al. (2018)
6	Epigenetic engineering	Histone modification	*A. thaliana*	Kuppu et al. (2020)
		DNA methylation editing	*Cucumis melo* L.	Giordano et al. (2022)
		Transposon silencing	*Hordeum vulgare*	Galli et al. (2022)
7	Conditional CRISPR/Cas systems	Efficient mutagenesis in specific cell types, tissues, or organs	*A. thaliana*	Decaestecker et al. (2019)
8	Crop improvement	Abiotic and biotic stress tolerant crops, biofortified crops, high yield crops, etc.	*O. sativa*, *Musa acuminata*, *T. aestivum*, etc.	Zhang et al. (2019), Lacchini et al. (2020), Kaur et al. (2020)

1.5 CRISPR Types

CRISPR/Cas system is categorized into two classes including Class I and Class II. The categorization is based on the number of subunits and the complexity of the system. Class-I CRISPR/Cas system has a multi-subunit effector protein and thus is more complex; however, Class-II system has a single subunit effector protein. Besides this classification, based on the type of effector Cas proteins and/or modules, both classes are further divided into their types and subtypes. Types I, III, and IV are part of the Class-I CRISPR system that has Cas3, Cas1, and CsfI as effector proteins, respectively. Types II, V, and VI are the types of Class-II with Cas9, Cas12, and Cas13 proteins as their effector modules, respectively. Moreover, based on their functional mechanisms, these types are further divided into their subtypes. The three stages of bacterial adaptation for the viral attack include the adaptation of spacer DNA, expression of CRISPR array, and then the action of effector complex called interference. The multi-subunit effector complex of Type-I and subtypes III-A and III-B CRISPR/Cas systems is called CRISPR-associated complex for antiviral defense (Cascade) complex including Cas3, Casm, and Cmr, respectively. These cascade complexes perform expression and maturation followed by interference. On the other hand, in Types II, III, and V, single multidomain proteins including Cas9, Cas10, and Cpf1, respectively, perform the expression, maturation, and interference stages.

1.6 Type of Cas Proteins

1.6.1 Cas1

Cas1 is a metal-dependent DNase and is one of the most conserved proteins that constitute the CRISPR-based bacterial immune system. Cas1 is the universal part of all types of CRISPR systems, and it has the capacity to bind to RNA in a non-sequence-specific manner. It is one of the important proteins that is requisite for the acquisition of spacer DNA (Makarova et al. 2011; Nuñez et al. 2014).

1.6.2 Cas2

Cas1 is a metal-dependent endoribonuclease and is essential for acquisition of spacer DNA. Cas2 is a universal protein of all the CRISPR/Cas system of bacteria. Cas1 and Cas2 function by forming a complex and recognize the CRISPR array. The symmetrical dimerized Cas2 binds to asymmetrical dimer of Cas1 and forms a complex that is involved in the identification of viral DNA invasion (Makarova et al. 2011; Nuñez et al. 2014).

1.6.3 Cas3

Cas3 is a translocase/helicase, and it is a part of type I CRISPR/Cas system and an important component of CRISPR interference for precise priming as well as deleting DNA into smaller fragments (Nuñez et al. 2014; He et al. 2020; Yoshimi et al. 2022). Cas3 has distinguished helicase and nuclease activities due to the presence of helicase domain known as Cas' and HD nuclease domain named as Cas' domain, respectively. It is involved in unidirectional and long-range deletion of genomic DNA upstream to the PAM region (Makarova et al. 2011; Morisaka et al. 2019).

1.6.4 Cas4

Cas4 is one of the important proteins that form the core of CRISPR/Cas system. It is a DNA exonuclease and can show 5′–3′ and 3′–5′ exonuclease activity (Zhang et al. 2012; Lemak et al. 2013). It has two domains including RecB nuclease domain responsible for nuclease activity and Fe-S cluster binding domain (Dixit et al. 2021). It is involved in the identification of 5′PAM sequence and shows a direct association with Cas1–Cas2 complex (Lee and Dhingra 2019).

1.6.5 Cas5

Cas5 is recognized as an endoribonuclease that is involved in the processing of pre-crRNA and the production of individual crRNA by binding to its 5′ handle. It is one of the components of CRISPR/Cas core complex that shows an interaction with the larger subunit of the cascade and Cas7 (Brendel et al. 2014).

1.6.6 Cas6

Cas6 is recognized as a metal-independent endoribonuclease. It is the part of CRISPR/Cas systems 1 and 3. It is recognized as an endoribonuclease and is found to be involved in the biogenesis of crRNA against the DNA of the invading viruses. It helps in stabilizing the crRNA and the processing of individual crRNA units (Carte et al. 2008; Makarova and Koonin 2015).

1.6.7 Cas7

These are the multicopy proteins that are the integral part of the Cascade complex. Cas7 is involved in the transcription process of pre-crRNA and also helps in the production of discrete crRNA (Makarova and Koonin 2015). In a recent study, Cas7–11 that comprised of Cas7.1–4, Cas11, INS, and CTE domains was reported to be used for programmable targeted RNA cleavage (Katoh et al. 2002; Kato et al. 2022).

1.6.8 Cas8

Cas8 is an integral member of CRISPR/Cas interference system that is required for the identification of PAM sequence in the invading bacteriophage DNA. Cas8 shows DNA binding by interacting with Cas5-Cas-cRNA complex. Cas8 is functionally analogous to the CasA module in type I CRISPR/Cas system (Cass et al. 2015; Kalwani et al. 2020).

1.6.9 Cas9

Cas9 is one of the extensively studied and utilized protein in precise genome editing. This is the signature protein of type II CRISPR/Cas system. Cas9 has the nuclease activities driven by RuvC and HNH domains for double-stranded break (Jinek et al. 2012). The most studied Cas9 protein was isolated from *Streptococcus pyogenes* that requires a PAM sequence of -NGG at the 3′ end of the target site. Several modifications have been done in Cas9 protein to design new variants with different functions and utilities. Cas9 and its variants have been used extensively for CRISPR/Cas9-mediated targeted genome editing (Liu et al. 2020).

1.6.10 Cas10

Cas10 is the signature effector protein of Type-III CRISPR/Cas system that is required for the crRNA processing and interference. Cas10 can create a non-PAM-based single-stranded break both in RNA and DNA. Three different domains have been identified in Cas10 including a nuclease domain, a palm domain responsible for the production of secondary messengers, and a Zn ribbon that shows homology with the cascade module (Lin et al. 2020a).

1.6.11 Cas11

Cas 11 is the integral member of type III and two type I systems including 1A and 1E CRISPT/Cas systems. Previous studies suggest that certain type I systems that do not have

Cas11 subunits as effector proteins produce Cas11d small subunit variant from the Cas10d ORFs (Schunder et al. 2013; McBride et al. 2020).

1.6.12 Cas12

Cas12 is also known as Cpf1 or Cas12a (class 2 type V-A system). This is one of the most important and studied Cas proteins after Cas9. Cas12 does not require tracrRNA and can effectively cleave the target DNA. Cas12 has RNase activity, and hence it is involved in the processing of self-pre-crRNA processing. Due to its smaller size, it has a significant importance as genome editing protein (Zetsche et al. 2015). Cas12b (C2c1) is the recently introduced genome editing effector protein that has a T-rich preferred PAM. This protein creates staggered ends at double-stranded DNA breaks. Among Cas9, Cas12a, and Cas12b, which are major genome editing effector proteins, Cas12b has a smaller size (Ming et al. 2020).

1.6.13 Cas13

Cas13 is an RNA-guided RNA endonuclease that cleaves single-stranded RNA at the precise target site. Cas13 has two domains including higher eukaryotic-prokaryotic nucleotide (HEPN) binding domain that is responsible for its RNase activity. It has gRNA maturation activity that is potentially provided by HEPN2 and Halical-1 domains. Cas13 has four small subtypes including Cas13a (C2c2) (Type VI-A), Cas13b (C2c4) (Type VI-B), Cas13c (C2c7) (Type VI-C), Cas13d, Cas13X (Type VI-X), and Cas13Y (Type VI-Y) (Abudayyeh et al. 2017; O'Connell 2019).

1.6.14 Cas14

Cas14 proteins size ranged from 40–70 kDa that act as effector proteins and cleave only single-stranded DNA without the requirement PAM sequence. Cas14 is highly specific to the middle nucleotides of its target sites, and hence this property can be utilized to mark a specific locus at genomic level by fusing any fluorophore to the protein. Moreover, this can be used to detect SNPs (Savage 2019).

1.7 CRISPR/Cas Modification

After the discovery of CRISPR/Cas9 system as a genome editing tool, several researchers engineered Cas proteins for their versatile utilization.

1.7.1 Nickase

Cas9 nickases are the modified versions of Cas9 that create a nick in either of the strands of double-stranded DNA. For instance, two variants of Cas9 nickases are generated by creating a D10A mutation in RuvC domain that cleaves target strand and a H840A mutation in HNH domain that cleaves nontarget strands (Jinek et al. 2012). These variants can be used in pairs to minimize the off-target efficiency of CRISPR/Cas9 system. Moreover, these can further be modified by making fusion proteins and generating new variants such as base editors.

1.7.2 Dead Cas9 (dCas9)

This variant of Cas9 was created by the inactivation of both the nuclease domains of Cas9. dCas9 does not have nuclease activity; however, this could make a complex with sgRNA that further binds to the DNA at target site (Jinek et al. 2012). These variants are further modified by making a fusion with other proteins such as reporter proteins.

1.7.3 Base Editors

Taking the advantages of dCas9 and nickases, base editors were generated by fusing these proteins with base editing proteins. There are primarily two variants of base editors including cytosine base editors (CBEs) and adenine base editors (ABEs). In CBEs, cytidine deaminase is fused to dCas9 to create cytidine (C) to thymidine (T) conversion; however, in ABEs, the TadA (tRNA-specific adenosine deaminase) from *E. coli* is fused to dCas9 that creates the conversion of adenosine (A) to inosine (I), which is translated as guanosine (G) resulting in A to G conversion. The primary reason to use TadA domain is that adenine deaminase base editors arenatively absent in nature (Wolf et al. 2002). Studies have shown base editing in plants such as wheat and rice (Li et al. 2018).

1.7.4 Prime Editors

The prime editors are the fusion proteins of Cas9 nickase (H840A) and a modified reverse transcriptase. The system also requires a guide RNA that is extended by addition modified sequences serving as template for reverse transcription that would lead to a mismatch at the target site. This mismatch can either be retained by the host system or repaired to its original sequence that would serve as another chance for editing (Scholefield and Harrison 2021).

1.8 CRISPR/Cas as a Genome Editing Tool and Its Application

After the discovery of this system, several researchers have used it for various molecular biology research (Upadhyay 2021). The system was initially used for precise genome editing for gene knockout; however, the further advancements paved new ways for its utilization in gene knockin, localization, fine tuning of genes, and RNA editing. Some of the primary applications are discussed further.

1.8.1 Gene Knockout

As described in the section above, CRISPR/Cas9 system of *Streptococcus pyogenes* is the extensively studied system. Researchers further modified this system to be used as a precise genome engineering tool. The target site is first fused with a gRNA scaffold; the fused product is known as single-guide RNA (sgRNA). This cassette is cloned into a binary vector system that also harbors the Cas9 open reading frame (ORF) driven by an appropriate constitutive and/or inducible promoter depending upon the nature of the work. Because Cas9 is originally a bacterial protein, the coding sequence of Cas9 is first codon optimized according to the host such as plants and animals. The expression constructs are transformed into the host system. For instance, in plants, the expression constructs are transferred extensively by agrobacterium-mediated plant transformation and particle bombardment methods. However, the transformation is not limited to these methods. In successful transformants, Cas9 will bind to the target sites based on the homology to sgRNA to the target site; however, the cleavage of DNA will be done based on the PAM sequence present adjacent to the target site. Cas9 cleaves the double-stranded DNA at the precise position guided by sgRNA. These breaks are repaired by either of the host DNA repair mechanisms including non-homologous end joining and homologous direct repair that are often error prone, leading the indel mutation at the cleavage site that in turn disrupt the gene function either by frame shift mutation or stop codon gain, etc. Cas9, Cas9 nickase, Cas12a, and Cas12b can be utilized for gene knockout (Liang et al. 2017; Ming et al. 2020).

1.8.2 DNA Insertion

Taking advantage of homologous direct repair, it is possible to insert DNA precisely at the targeted site on the host genome. To perform the gene insertion, a donor DNA template is provided to the cell and this switches the repair mechanism to HDR instead of NHEJ (Jasin and Haber 2016; Gallagher and Haber 2018; Nambiar et al. 2019). PCR-amplified DNA template, single-stranded oligonucleotides, and plasmids are being used for donor templates. The donor template is designed in such a way that it harbors the homologous sequences known as the homologous arms from the upstream and downstream regions of the target site, and in between these arms, the desired DNA with mutation is present. To avoid the cleavage of the donor template, the PAM sequence can be altered in the donor template with missense mutation. The double-stranded breaks are created by Cas9 at the target site, and the break is repaired by the insertion of donor DNA template. Although HDR mechanism can intentionally be switched, however, it has a very low efficiency, and therefore double-stranded DNA can also be repaired by NHEJ. The desired cells with mutation can further be screened. CRISPR/Cas-based DNA insertion has been reported in different studies (Dahan-Meir et al. 2018; Lu et al. 2020).

1.8.3 Base Editing

It is possible to utilize the CRISPR/Cas system for genome editing without creating a double-stranded break and create a precise mutation at a single nucleotide instead (Zhang et al. 2019). The binary vectors that harbor a fusion protein of dCas9 and base editors' as effector proteins are used in such studies. The complex of sgRNA and base editors is guided to its target loci at the genome and the effector proteins (CBE/ABE) make the targeted single base editing. The complementary to the converted base is repaired by the host systems that lead to permanent base conversion at the target loci. Several modifications such as incorporation of DNA glycosylase inhibitor (UGI) or bacteriophage Mu-GAM protein have been done in the fusion protein to enhance the expression and hence the efficiency of base editing (Komor et al. 2017). Moreover, base editing of RNA is the further advancement of the tool. The fusion protein of dCas13b and adenosine deaminase is an effector protein for RNA base editing that converts A–I. Because of the functional equivalency of I–G, the final modification is A–G in RNA.

1.8.4 Gene Activation and/or Repression

The ability of dCas9 to bind to the DNA at the targeted site and not showing any nuclease activity created several utilizations of the protein such as the activation and/or repression of genes. dCas9 can be fused to different activators and repressors, and the promoter region of the genes can be targeted for gene regulation. The fusion of dCas9 with one or more activators will result in the activation of the downstream genes known as CRISPRa and fusion with one or more repressors will repress the expression of downstream genes known as CRISPRi (KREB-dCAS9) (Alerasool et al. 2020). Several plasmid vectors are available that can be utilized for CRISPR/Cas-mediated gene regulation in different host systems; however, there are still several new opportunities for the further advancement of this system to design precise, specialized, and gene-specific regulatory tools. Some of the initial CRISPRa systems include dCas9-VP64 (Lowder et al. 2015; Piatek et al. 2015). More efficient CRISPRa systems are now being designed such as dCas9-SunTag, dCas9TV, dCasEV2.1, and CRISPR-ACT3.0 (Gilbert et al. 2014; Chavez et al. 2015; Li et al. 2017; Papikian et al. 2019; Selma et al. 2019; Pan et al. 2021; Xiong et al. 2021).

1.8.5 Epigenetic Modifications

Several methods have been established to regulate the epigenetic marks at a specific locus, thereby repressing or activating gene expression. The DNA binding domain and epigenetic effector domain are the prime components of epigenetic editing tools. The CRISPR/Cas9 system, which can be easily programmed, is currently the primary choice for targeted genome editing. dCas9 having the ability to bind to DNA at precise and guided target loci is fused to epigenetic modifiers including p300, LSD1, MQ1, and TET1 (Pulecio et al. 2017).

1.8.6 Localization

dCas9 is fused with reporter genes or fluorophores and then transformed into the host system. The complex shows a guided and targeted binding at the genomic loci, and further reporter assays reveal loci of interest. The early version of this was CASFISH, in which fluorescent *in situ* hybridization can be performed to label and identify locus of the gene in live cells (Deng et al. 2015). Due to the multiplexing of the target site, several genomic locations can be mapped. Moreover, there are different modules of dCas9 fused to different colored fluorophores that give a multi-colored imaging for simultaneous marking of the genomic loci.

1.8.7 RNA Editing

With the discovery of Cas13 and Cas14 proteins, it is possible to target RNA for interference such as against different plant viruses, mRNA degradation, editing, alternative splicing modulation, RNA localization. Cas13 and its other subtypes can be modified and fused to base editor such as adenosine deaminase acting on RNA (ADAR) (Kavuri et al. 2022). Moreover, the nuclease activity of these proteins can effectively be used as RNA editing tool. Although CRISPR/Cas-based RNA editing is yet to be explored, however, this can potentially be developed into much more efficient tool than RNAi.

1.9 Conclusion

The discovery of CRISPR/Cas system has emerged enormous possibilities for precise genome editing in different host systems. The system has been developed as an excellent functional genomics tool to decipher the functional role of several new genes in plants. Targeted single and simultaneous multiplexed genome targeting makes CRISPR/Cas system the preferred genome editing tool over other available functional tools. To date, this tool has been used for the genetic modulation in plants for several trait development such as high yielding, stress-tolerant crops, biofortified crops with enhanced metabolite production. Further advancements and multidisciplinary potential of CRISPR/Cas system have also generated new opportunities for the plant biologists especially in the field of agricultural science. Moreover, besides the availability of new tools in CRISPR/Cas system, the improved methods of plant transformation have also reduced the laborious and time-consuming processes that have paced up the tailback in the field of plant genetic engineering. Furthermore, the possibility of the production of transgene-free and/or non-transgenic genome-edited plants has paved new ways for the acceptance of genetically modified plants as mainstream crops. However, certain challenges, such as reducing the off-targets of CRISPR/Cas systems and improving efficiency and reproducibility of the methods, require a continuous attention to utilize this tool to its full potential.

References

Abudayyeh, O.O., Gootenberg, J.S., Essletzbichler, P. et al. (2017). RNA targeting with CRISPR-Cas13. *Nature* 550: 280–284.

Alerasool, N., Segal, D., Lee, H., and Taipale, M. (2020). An efficient KRAB domain for CRISPRi applications in human cells. *Nature Methods* 17 (11): 1093–1096.

Barakate, A., Keir, E., Oakey, H., and Halpin, C. (2020). Stimulation of homologous recombination in plants expressing heterologous recombinases. *BMC Plant Biology* 20: 1–10.

Barrangou, R. (2007). CRISPR provides acquired resistance against viruses in prokaryotes. *Science* 315: 1709–1712.

Beying, N., Schmidt, C., Pacher, M. et al. (2020). CRISPR–Cas9-mediated induction of heritable chromosomal translocations in *Arabidopsis*. *Nature Plants* 6 (6): 638–645.

Bolotin, A., Quinquis, B., Sorokin, A., and Dusko, E.S. (2005). Clustered regularly interspaced short palindrome repeats (CRISPRs) have spacers of extrachromosomal origin. *Microbiology* 151: 2551–2561.

Brendel, J., Stoll, B., Lange, S.J. et al. (2014). A complex of Cas proteins 5, 6, and 7 is required for the biogenesis and stability of clustered regularly interspaced short palindromic repeats (crispr)-derived rnas (crrnas) in *Haloferax volcanii*. *The Journal of Biological Chemistry* 289: 7164–7177.

Carte, J., Wang, R., Li, H. et al. (2008). Cas6 is an endoribonuclease that generates guide RNAs for invader defense in prokaryotes. *Genes & Development* 22: 3489–3496.

Cass, S.D.B., Haas, K.A., Stoll, B. et al. (2015). The role of Cas8 in type I CRISPR interference. *Bioscience Reports* 35: 197.

Chavez, A., Scheiman, J., Vora, S. et al. (2015). Highly efficient Cas9-mediated transcriptional programming. *Nature Methods* 12: 326–328.

Chen, K., Wang, Y., Zhang, R. et al. (2019). CRISPR/Cas genome editing and precision plant breeding in agriculture. *Annual Review of Plant Biology* 70: 667–697.

Cho, S.W., Kim, S., Kim, J.M., and Kim, J.S. (2013). Targeted genome engineering in human cells with the Cas9 RNA-guided endonuclease. *Nature Biotechnology* 31 (3): 230–232.

Cong, L., Ran, F.A., Cox, D. et al. (2013). Multiplex genome engineering using CRISPR/Cas systems. *Science* 339: 819–823.

Dahan-Meir, T., Filler-Hayut, S., Melamed-Bessudo, C. et al. (2018). Efficient in planta gene targeting in tomato using geminiviral replicons and the CRISPR/Cas9 system. *The Plant Journal* 95: 5–16.

Debbarma, J., Sarki, Y.N., Saikia, B. et al. (2019). Ethylene response factor (ERF) family proteins in abiotic stresses and CRISPR–Cas9 genome editing of ERFs for multiple abiotic stress tolerance in crop plants: a review. *Molecular Biotechnology* 61 (2): 153–172.

Decaestecker, W., Buono, R.A., Pfeiffer, M.L. et al. (2019). CRISPR-TSKO: a technique for efficient mutagenesis in specific cell types, tissues, or organs in *Arabidopsis*. *The Plant Cell* 31: 2868–2887.

Deltcheva, E., Chylinski, K., Sharma, C.M. et al. (2011). CRISPR RNA maturation by trans-encoded small RNA and host factor RNase III. *Nature* 471 (7340): 602–607.

Deng, W., Shi, X., Tjian, R. et al. (2015). CASFISH: CRISPR/Cas9-mediated in situ labeling of genomic loci in fixed cells. *Proceedings of the National Academy of Sciences of the United States of America* 112: 11870–11875.

Dixit, B., Anand, V., Hussain, M.S., and Kumar, M. (2021). The CRISPR-associated Cas4 protein from Leptospira interrogans demonstrate versatile nuclease activity. *Current Research in Microbial Sciences* 2: 100040.

Feng, Z., Zhang, B., Ding, W. et al. (2013). Efficient genome editing in plants using a CRISPR/Cas system. *Cell Research* 23: 1229–1232.

Gallagher, D.N. and Haber, J.E. (2018). Repair of a site-specific DNA cleavage: old-school lessons for cas9-mediated gene editing. *ACS Chemical Biology* 13: 397–405.

Galli, M., Martiny, E., Imani, J. et al. (2022). CRISPR/SpCas9-mediated double knockout of barley Microrchidia MORC1 and MORC6a reveals their strong involvement in plant immunity, transcriptional gene silencing and plant growth. *Plant Biotechnology Journal* 20: 89–102.

Garneau, J.E., Dupuis, M.È., Villion, M. et al. (2010). The CRISPR/Cas bacterial immune system cleaves bacteriophage and plasmid DNA. *Nature* 468 (7320): 67–71.

Gasiunas, G., Barrangou, R., Horvath, P., and Siksnys, V. (2012). Cas9-crRNA ribonucleoprotein complex mediates specific DNA cleavage for adaptive immunity in bacteria. *Proceedings of the National Academy of Sciences of the United States of America* 109: E2579–E2586.

Gilbert, L.A., Horlbeck, M.A., Adamson, B. et al. (2014). Genome-scale CRISPR-mediated control of gene repression and activation. *Cell* 159: 647–661.

Giordano, A., Santo Domingo, M., Quadrana, L. et al. (2022). CRISPR/Cas9 gene editing uncovers the roles of CONSTITUTIVE TRIPLE RESPONSE 1 and REPRESSOR OF SILENCING 1 in melon fruit ripening and epigenetic regulation. *Journal of Experimental Botany* 73: 4022–4033.

Haft, D.H., Selengut, J., Mongodin, E.F., and Nelson, K.E. (2005). A guild of 45 CRISPR-associated (Cas) protein families and multiple CRISPR/Cas subtypes exist in prokaryotic genomes. *PLoS Computational Biology* 1: e60.

Hahn, F., Eisenhut, M., Mantegazza, O., and Weber, A.P.M. (2018). Homology-directed repair of a defective glabrous gene in *arabidopsis* with Cas9-based gene targeting. *Frontiers in Plant Science* 9: 424.

He, L., James, M.S.J., Radovcic, M. et al. (2020). Cas3 protein-A review of a multi-tasking machine. *Genes* 11: 208.

Heler, R., Samai, P., Modell, J.W. et al. (2015). Cas9 specifies functional viral targets during CRISPR–Cas adaptation. *Nature* 519 (7542): 199–202.

Horvath, P. and Barrangou, R. (2010). CRISPR/Cas, the immune system of bacteria and archaea. *Science* 327: 167–170.

Hsu, P.D., Scott, D.A., Weinstein, J.A. et al. (2013). DNA targeting specificity of RNA-guided Cas9 nucleases. *Nature Biotechnology* 31 (9): 827–832.

Hua, K., Tao, X., Liang, W. et al. (2020). Simplified adenine base editors improve adenine base editing efficiency in rice. *Plant Biotechnology Journal* 18: 770–778.

Hwang, W.Y., Fu, Y., Reyon, D. et al. (2013). Efficient genome editing in zebrafish using a CRISPR-Cas system. *Nature Biotechnology* 31 (3): 227–229.

Ishino, Y., Shinagawa, H., Makino, K. et al. (1987). Nucleotide sequence of the iap gene, responsible for alkaline phosphatase isoenzyme conversion in *Escherichia coli*, and identification of the gene product. *Journal of Bacteriology* 169: 5429–5433.

Jasin, M. and Haber, J.E. (2016). The democratization of gene editing: insights from site-specific cleavage and double-strand break repair. *DNA Repair (Amst)* 44: 6–16.

Jiang, W., Zhou, H., Bi, H. et al. (2013). Demonstration of CRISPR/Cas9/sgRNA-mediated targeted gene modification in *Arabidopsis*, tobacco, sorghum and rice. *Nucleic Acids Research* 41.

Jinek, M., Chylinski, K., Fonfara, I. et al. (2012). A programmable dual-RNA-guided DNA endonuclease in adaptive bacterial immunity. *Science* 337: 816–821.

Jinek, M., East, A., Cheng, A. et al. (2013). RNA-programmed genome editing in human cells. *eLife* e00471. https://doi.org/10.7554/eLife.00471.

Kalwani, P., Rath, D., and Ballal, A. (2020). Novel molecular aspects of the CRISPR backbone protein 'Cas7' from cyanobacteria. *The Biochemical Journal* 477: 971–983.

Kang, B.C., Yun, J.Y., Kim, S.T. et al. (2018). Precision genome engineering through adenine base editing in plants. *Nature Plants* 4 (7): 427–431.

Karvelis, T., Gasiunas, G., Miksys, A. et al. (2013). crRNA and tracrRNA guide Cas9-mediated DNA interference in *Streptococcus thermophilus*. *RNA Biology* 10: 841–851.

Kato, K., Zhou, W., Okazaki, S. et al. (2022). Structure and engineering of the type III-E CRISPR-Cas7-11 effector complex. *Cell* 185: 2324–2337. e16.

Katoh, K., Misawa, K., Kuma, K.I., and Miyata, T. (2002). MAFFT: a novel method for rapid multiple sequence alignment based on fast fourier transform. *Nucleic Acids Research* 30: 3059–3066.

Kaur, N., Alok, A., Kumar, P. et al. (2020). CRISPR/Cas9 directed editing of lycopene epsilon-cyclase modulates metabolic flux for $β$-carotene biosynthesis in banana fruit. *Metabolic Engineering* 59: 76–86.

Kavuri, N.R., Ramasamy, M., Qi, Y., and Mandadi, K. (2022). Applications of CRISPR/Cas13-based RNA editing in plants. *Cells* 11: 2665.

Komor, A.C., Zhao, K.T., Packer, M.S. et al. (2017). Improved base excision repair inhibition and bacteriophage Mu Gam protein yields C:G-to-T: a base editors with higher efficiency and product purity. *Science Advances* 3: eaao4774.

Kuppu, S., Ron, M., Marimuthu, M.P.A. et al. (2020). A variety of changes, including CRISPR/Cas9-mediated deletions, in CENH3 lead to haploid induction on outcrossing. *Plant Biotechnology Journal* 18: 2068–2080.

Lacchini, E., Kiegle, E., Castellani, M. et al. (2020). CRISPR-mediated accelerated domestication of African rice landraces. *PLoS One* 15: e0229782.

Lee, H., Dhingra, Y., and Sashital, D.G. (2019). The Cas4-Cas1-Cas2 complex mediates precise prespacer processing during CRISPR adaptation. *eLife* 8: e44248.

Lemak, S., Beloglazova, N., Nocek, B. et al. (2013). Toroidal structure and DNA cleavage by the CRISPR-associated [4Fe-4S] cluster containing Cas4 nuclease SSO0001 from sulfolobus solfataricus. *Journal of the American Chemical Society* 135: 17476–17487.

Li, J.F., Norville, J.E., and Aach, J. (2013). Multiplex and homologous recombination-mediated genome editing in *Arabidopsis* and *Nicotiana benthamiana* using guide RNA and Cas9. *Nature Biotechnology* 31: 688–691.

Li, Z., Zhang, D., Xiong, X. et al. (2017). A potent Cas9-derived gene activator for plant and mammalian cells. *Nature Plants* 3: 930–936.

Li, C., Zong, Y., Wang, Y. et al. (2018). Expanded base editing in rice and wheat using a Cas9-adenosine deaminase fusion. *Genome Biology* 19: 1–9.

Li, F., Sun, Q., Liu, K. et al. (2019). The deubiquitinase OTUD5 regulates Ku80 stability and non-homologous end joining. *Cellular and Molecular Life Sciences* 76: 3861–3873.

Li, C., Zhang, R., Meng, X. et al. (2020). Targeted, random mutagenesis of plant genes with dual cytosine and adenine base editors. *Nature Biotechnology* 38: 875–882.

Liang, Z., Zhang, K., Chen, K., and Gao, C. (2014). Targeted mutagenesis in Zea mays using TALENs and the CRISPR/Cas system. *Journal of Genetics and Genomics* 41: 63–68.

Liang, Z., Chen, K., Li, T. et al. (2017). Efficient DNA-free genome editing of bread wheat using CRISPR/Cas9 ribonucleoprotein complexes. *Nature Communications* 8: 14261.

Lin, J., Feng, M., Zhang, H., and She, Q. (2020a). Characterization of a novel type III CRISPR-Cas effector provides new insights into the allosteric activation and suppression of the Cas10 DNase. *Cell Discovery* 6 (1): 1–16.

Lin, Q., Zong, Y., Xue, C. et al. (2020b). Prime genome editing in rice and wheat. *Nature Biotechnology* 38: 582–585.

Liu, Z., Dong, H., Cui, Y. et al. (2020). Application of different types of CRISPR/Cas-based systems in bacteria. *Microbial Cell Factories* 19 (1): 1–14.

Lowder, L.G., Zhang, D., Baltes, N.J. et al. (2015). A CRISPR/Cas9 toolbox for multiplexed plant genome editing and transcriptional regulation. *Plant Physiology* 169: 971–985.

Lu, Y., Tian, Y., Shen, R. et al. (2020). Targeted, efficient sequence insertion and replacement in rice. *Nature Biotechnology* 38 (12): 1402–1407.

Lu, Y., Yang, H., Bai, J. et al. (2022). CRISPR-Cas based molecular diagnostics for foodborne pathogens. *Critical Reviews in Food Science and Nutrition*. https://doi.org/10.1080/10408398.2022.2153792.

Ma, S., Chang, J., Wang, X. et al. (2014). CRISPR/Cas9 mediated multiplex genome editing and heritable mutagenesis of BmKu70 in Bombyx mori. *Scientific Reports* 4 (1): 1–6.

Makarova, K.S. and Koonin, E.V. (2015). Annotation and classification of CRISPR-Cas systems. *Methods in Molecular Biology* 1311: 47–75.

Makarova, K.S., Grishin, N.V., Shabalina, S.A. et al. (2006). A putative RNA-interference-based immune system in prokaryotes: computational analysis of the predicted enzymatic machinery, functional analogies with eukaryotic RNAi, and hypothetical mechanisms of action. *Biology Direct* 1: 1–26.

Makarova, K.S., Haft, D.H., Barrangou, R. et al. (2011). Evolution and classification of the CRISPR-Cas systems. *Nature Reviews Microbiology* 9: 467–477.

Makarova, K.S., Wolf, Y.I., Alkhnbashi, O.S. et al. (2015). An updated evolutionary classification of CRISPR-Cas systems. *Nature Reviews Microbiology* 13: 722–736.

Mali, P., Aach, J., Stranges, P.B. et al. (2013a). CAS9 transcriptional activators for target specificity screening and paired nickases for cooperative genome engineering. *Nature Biotechnology* 31: 833–838.

Mali, P., Yang, L., Esvelt, K.M. et al. (2013b). RNA-guided human genome engineering via Cas9. *Science* 339: 823–826.

Marraffini, L.A. and Sontheimer, E.J. (2008). CRISPR interference limits horizontal gene transfer in staphylococci by targeting DNA. *Science* 322: 1843–1845.

Marraffini, L.A. and Sontheimer, E.J. (2010). CRISPR interference: RNA-directed adaptive immunity in bacteria and archaea. *Nature Reviews Genetics* 11: 181–190.

McBride, T.M., Schwartz, E.A., Kumar, A. et al. (2020). Diverse CRISPR-Cas complexes require independent translation of small and large subunits from a single gene. *Molecular Cell* 80: 971–979.e7.

Mei, Y., Wang, Y., Chen, H. et al. (2016). Recent progress in CRISPR/Cas9 technology. *Journal of Genetics and Genomics* 43: 63–75.

Ming, M., Ren, Q., Pan, C. et al. (2020). CRISPR–Cas12b enables efficient plant genome engineering. *Nature Plants* 6 (3): 202–208.

Mojica, F.J.M., Díez-Villaseñor, C., García-Martínez, J., and Soria, E. (2005). Intervening sequences of regularly spaced prokaryotic repeats derive from foreign genetic elements. *Journal of Molecular Evolution* 60: 174–182.

Mojica, F.J.M., Díez-Villaseñor, C., García-Martínez, J., and Almendros, C. (2009). Short motif sequences determine the targets of the prokaryotic CRISPR defence system. *Microbiology* 155: 733–740.

Morisaka, H., Yoshimi, K., Okuzaki, Y. et al. (2019). CRISPR-Cas3 induces broad and unidirectional genome editing in human cells. *Nature Communications* 10 (1): 1–13.

Nambiar, T.S., Billon, P., Diedenhofen, G. et al. (2019). Stimulation of CRISPR-mediated homology-directed repair by an engineered RAD18 variant. *Nature Communications* 10 (1): 1–13.

Nekrasov, V., Staskawicz, B., Weigel, D. et al. (2013). Targeted mutagenesis in the model plant *Nicotiana benthamiana* using Cas9 RNA-guided endonuclease. *Nature Biotechnology* 31: 691–693.

Nishimasu, H., Ran, F.A., Hsu, P.D. et al. (2014). Crystal structure of Cas9 in complex with guide RNA and target DNA. *Cell* 156: 935–949.

Nuñez, J.K., Kranzusch, P.J., Noeske, J. et al. (2014). Cas1–Cas2 complex formation mediates spacer acquisition during CRISPR–Cas adaptive immunity. *Nature Structural & Molecular Biology* 21: 528.

O'Connell, M.R. (2019). Molecular mechanisms of RNA targeting by Cas13-containing type VI CRISPR-Cas systems. *Journal of Molecular Biology* 431: 66–87.

van der Oost, J., Jore, M.M., Westra, E.R. et al. (2009). CRISPR-based adaptive and heritable immunity in prokaryotes. *Trends in Biochemical Sciences* 34: 401–407.

Pan, C., Wu, X., Markel, K. et al. (2021). CRISPR–Act3.0 for highly efficient multiplexed gene activation in plants. *Nature Plants* 7 (7): 942–953.

Pant, S., Ritika, N.P., Ghati, A. et al. (2022). Employment of the CRISPR/Cas9 system to improve cellulase production in *Trichoderma reesei*. *Biotechnology Advances* 60: 108022.

Papikian, A., Liu, W., Gallego-Bartolomé, J., and Jacobsen, S.E. (2019). Site-specific manipulation of *Arabidopsis* loci using CRISPR-Cas9 suntag systems. *Nature Communications* 10: 729.

Park, S. and Beal, P.A. (2019). Off-target editing by CRISPR-guided DNA base editors. *Biochemistry* 58: 3727–3734.

Permyakova, N.V., Marenkova, T.V., Belavin, P.A. et al. (2022). CRISPR/Cas9-mediated targeted DNA integration: rearrangements at the junction of plant and plasmid DNA. *International Journal of Molecular Sciences* 23: 8636.

Piatek, A., Ali, Z., Baazim, H. et al. (2015). RNA-guided transcriptional regulation in planta via synthetic dCas9-based transcription factors. *Plant Biotechnology Journal* 13: 578–589.

Pourcel, C., Salvignol, G., and Vergnaud, G. (2005). CRISPR elements in *Yersinia pestis* acquire new repeats by preferential uptake of bacteriophage DNA, and provide additional tools for evolutionary studies. *Microbiology* 151: 653–663.

Pulecio, J., Verma, N., Mejía-Ramírez, E. et al. (2017). CRISPR/Cas9-based engineering of the epigenome. *Cell Stem Cell* 21: 431.

Savage, D.F. (2019). Viewpoint – Cas14: big advances from small CRISPR proteins. *Biochemistry* 58: 1024.

Schmidt, C., Pacher, M., and Puchta, H. (2019). Efficient induction of heritable inversions in plant genomes using the CRISPR/Cas system. *The Plant Journal* 98: 577–589.

Scholefield, J. and Harrison, P.T. (2021). Prime editing – an update on the field. *Gene Therapy* 28 (7): 396–401.

Schunder, E., Rydzewski, K., Grunow, R., and Heuner, K. (2013). First indication for a functional CRISPR/Cas system in *Francisella tularensis*. *International Journal of Medical Microbiology* 303: 51–60.

Selma, S., Bernabé-Orts, J.M., Vazquez-Vilar, M. et al. (2019). Strong gene activation in plants with genome-wide specificity using a new orthogonal CRISPR/Cas9-based programmable transcriptional activator. *Plant Biotechnology Journal* 17: 1703–1705.

Shan, Q., Wang, Y., Li, J. et al. (2013). Targeted genome modification of crop plants using a CRISPR-Cas system. *Nature Biotechnology* 31: 686–688.

Sushmita, Kaur, G., Upadhyay, S.K., and Verma, P.C. (2021). An overview of genome-engineering methods. *Wiley*. https://doi.org/10.1002/9781119672425.ch1.

Symington, L.S. and Gautier, J. (2011). Double-strand break end resection and repair pathway choice. *Annual Review of Genetics* 45: 247–271.

Tilman, D., Balzer, C., Hill, J., and Befort, B.L. (2011). Global food demand and the sustainable intensification of agriculture. *Proceedings of the National Academy of Sciences of the United States of America* 108: 20260–20264.

Tripathi, J.N., Ntui, V.O., Shah, T., and Tripathi, L. (2021). CRISPR/Cas9-mediated editing of DMR6 orthologue in banana (Musa spp.) confers enhanced resistance to bacterial disease. *Plant Biotechnology Journal* 19: 1291–1293.

Upadhyay, S.K., Kumar, J., Alok, A., and Tuli, R. (2013). RNA-guided genome editing for target gene mutations in wheat. *G3: Genes, Genomes, Genetics* 3: 2233–2238.

Upadhyay, S.K. (2021). Genome engineering for crop improvement. *Wiley*. https://doi.org/10.1002/9781119672425.

Vicencio, J., Sánchez-Bolaños, C., Moreno-Sánchez, I. et al. (2022). Genome editing in animals with minimal PAM CRISPR-Cas9 enzymes. *Nature Communications* 13 (1): 1–13.

Wei, X., Pu, A., Liu, Q. et al. (2022). The bibliometric landscape of gene editing innovation and regulation in the worldwide. *Cell* 11: 2682.

Wolf, J., Gerber, A.P., and Keller, W. (2002). tadA, an essential tRNA-specific adenosine deaminase from *Escherichia coli. The EMBO Journal* 21: 3841.

Xie, K. and Yang, Y. (2013). RNA-guided genome editing in plants using a CRISPR-Cas system. *Molecular Plant* 6: 1975–1983.

Xiong, X., Liang, J., Li, Z. et al. (2021). Multiplex and optimization of dCas9-TV-mediated gene activation in plants. *Journal of Integrative Plant Biology* 63: 634–645.

Xu, W., Zhang, C., Yang, Y. et al. (2020). Versatile nucleotides substitution in plant using an improved prime editing system. *Molecular Plant* 13: 675–678.

Yin, K., Gao, C., and Qiu, J.L. (2017). Progress and prospects in plant genome editing. *Nature Plants* 3 (8): 1–6.

Yoshimi, K., Takeshita, K., Kodera, N. et al. (2022). Dynamic mechanisms of CRISPR interference by *Escherichia coli* CRISPR-Cas3. *Nature Communications* 13: 4917.

Zetsche, B., Gootenberg, J.S., Abudayyeh, O.O. et al. (2015). Cpf1 is a single RNA-guided endonuclease of a class 2 CRISPR-Cas system. *Cell* 163: 759–771.

Zhang, J., Kasciukovic, T., and White, M.F. (2012). The CRISPR associated protein Cas4 Is a 5′ to 3′ DNA exonuclease with an iron-sulfur cluster. *PLoS One* 7: e47232.

Zhang, R., Liu, J., Chai, Z. et al. (2019). Generation of herbicide tolerance traits and a new selectable marker in wheat using base editing. *Nature Plants* 5 (5): 480–485.

Zhou, J., Xin, X., He, Y. et al. (2019). Multiplex QTL editing of grain-related genes improves yield in elite rice varieties. *Plant Cell Reports* 38: 475–485.

2

CRISPR/Cas-Mediated Multiplex Genome Editing in Plants and Applications

R. Prajapati[1] *and K. Tyagi*[2]

[1] *Department of Plant Pathology, Kansas State University, Manhattan, KS, USA*
[2] *Department of Plant Pathology, University of Minnesota, St Paul, Minneapolis, MN, USA*

2.1 Introduction

Researchers have used traditional plant breeding technology for a long time to select value-added traits by combining hundreds or thousands of plant genotypes that have unique qualities such as improved nutrition and increased shelf-life and yield. However, traditional breeding technology is an expensive and time-consuming process due to longer breeding period and lower chances of required mutation (Rani et al. 2020). Advance transgenic technology is comparatively a more versatile tool that can be used to overcome many of these limitations. However, these transgenic methods also have their own limitations such as presence of marker genes that raises global consequences. During last few years, many new genetic modification tools have emerged rapidly. These technologies depend on specific engineered endonucleases or EENs, which cut DNA at some specific locations by recognizing specific nucleotide sequences (Xiong et al. 2015; Sushmita et al. 2021). The Zinc finger nuclease (ZFN) belongs to the first-generation EENs and has been used successfully for DNA alteration in various plants including *Arabidopsis*, maize, and tobacco (Beerli and Barbas 2002; Xiong et al. 2015). *T*ranscription *a*ctivator-*l*ike *e*ffector *n*ucleases (TALENs), another alternative to ZFNs, is a comparatively easier process; however, it involves binding of nucleotide sequences to transcription activator-like effector (TALE), which requires customization for every new target (Klug 2010). In recent times, another powerful genome editing tool has been used known as clustered regularly interspaced short palindromic repeats or CRISPR. Main principle behind CRISPR/Cas9 technology was originated from type-II prokaryotic adaptive immune system. It was first reported in bacteria in 1987 (Ishino et al. 1987) and was first characterized in 2007 (Barrangou et al. 2007). CRISPR is involved in bacterial defense mechanism and present in almost about 90% of archaea and 40% of bacterial species (Makarova et al. 2011). CRISPR/Cas9 can identify a protospacer adjacent motif (PAM) on a specific target nucleotide sequence and binds with the help of small guide RNA (gRNA). The gRNA recruits Cas9 for

Applications of Genome Engineering in Plants, First Edition. Edited by Santosh Kumar Upadhyay.
© 2024 John Wiley & Sons Ltd. Published 2024 by John Wiley & Sons Ltd.

specific cleavage and produce double-stranded breaks (DSBs). This induces nonhomologous end joining (NHEJ)-mediated DNA repair mechanism and leads to genomic alterations (Arora and Narula 2017; Upadhyay 2021). CRISPR/Cas technology has been robustly used to alter plant genes; however, with emergence of Cas9 variants and need to target multiple genes at a time, multiplex CRISPR/Cas technique has taken the foreground. Multiple gRNAs are cloned into a single CRISPR construct to facilitate editing of several sites simultaneously, leading to wide application of this technology (Cong et al. 2013; Bhalothia et al. 2020). Here, we discuss the role of multiplex CRISPR/Cas technology in modifying plant genome and crop improvement. The multiplex CRISPR/Cas gene editing requires efficient construct designing and delivery to effectively alter the plant genome (Figure 2.1).

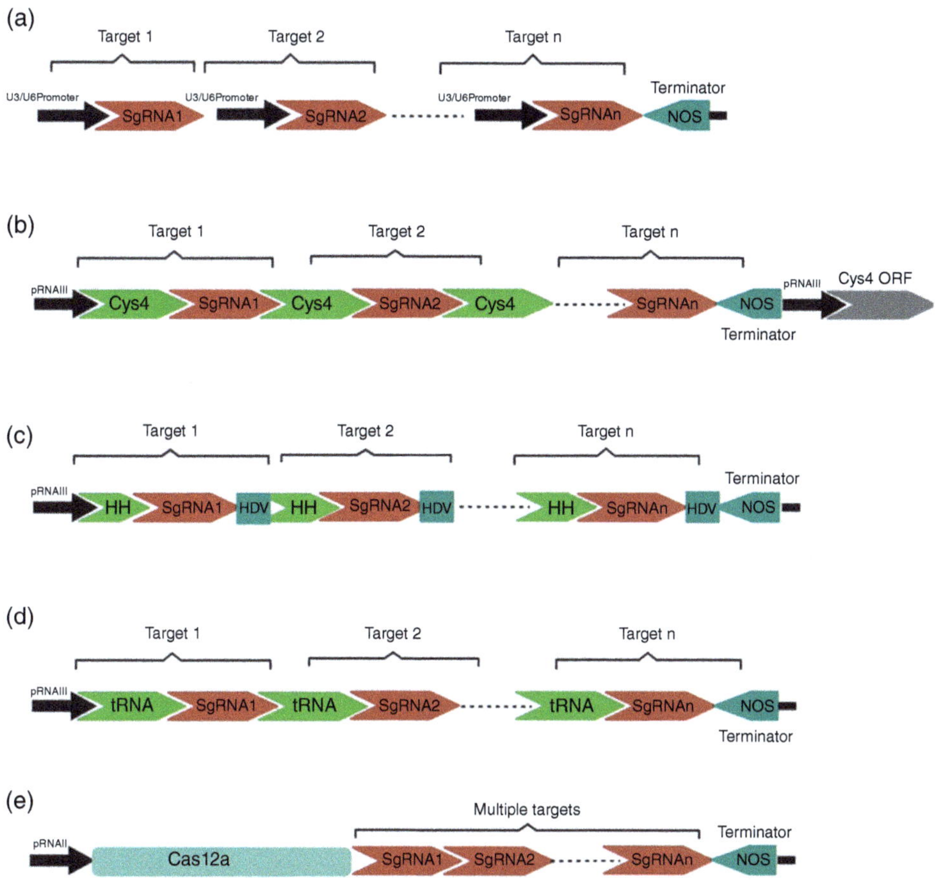

Figure 2.1 A graphical representation of different multiplex gRNA expression system of plant. (a) Various gRNA cloned under U3/U6 promoter to be expressed by RNA polymerase III system. (b–d) Various gRNAs are cloned after RNA III polymerase promoter that generates a big transcript composed of several gRNAs, which then get separated into individual gRNAs by post-transcriptional modification machinery. Separation of each gRNAs is regulated by Csy4 and tRNAs. (e) Various gRNAs cloned under RNA polymerase II promoter. All the gRNAs are generated as single transcript that will be cleaved individually by Cas12a system. *Source:* The figure is adapted and modified from Abdelrahman et al. (2021).

2.2 Construct Design for Multiplex CRISPR/Cas Genome Editing

The three main factors that are essential for construct preparation for CRISPR/Cas multiplex genome editing are 1) Cas protein, 2) specificity of sgRNAs, and 3) promoters or gene regulatory elements regulating the Cas and gRNAs (Xing et al. 2014; Alok et al. 2021).

1) **Cas protein**: Apart from Cas9, CRISPR from *Prevotella* and *Francisella* 1 (Cpf1) has been used for efficient genome editing in plants. Cpf1, which is now termed as Cas12a, is a class II type V endonuclease that has various advantages over Cas9 (Zaidi et al. 2017). The major advantage of Cas12a is its relatively smaller size than Cas9. Moreover, it requires a short crRNA (42 nt) and does not require multiple regulatory elements. A single pol III promoter can transcribe several crRNAs altogether (Zetsche et al. 2017). Other orthologues of Cas12 such as Cas12b (also known as C2c1) have been also used for genome editing. Cas12b is a class 2 type V-B endonuclease and requires both crRNA and tracrRNA for target identification. It recognizes 5′-T-rich regions as PAM sequence and makes DSBs with 5′-overhangs (Shmakov et al. 2015). CasX (or Cas12e) and Cas Y (or Cas12d), which can produce DSBs, have also been identified (Burstein et al. 2017; Liu et al. 2019b). Cas14 is another nuclease that does not require PAM sequence and can cleave single-stranded DNA (Harrington et al. 2018).
2) **sgRNAs**: Cas9 nucleases interact with CRISPR RNA (crRNA) and transactivating CRISPR RNA (tracrRNA), which are fused together to form a single gRNA (sgRNA). The CRISPR/Cas9 system requires an RNA designing with a short-guide sequence (sgRNA) that directs a Cas9 nuclease for cleaving any target sequence. Various models and algorithms that could predict gRNAs with minimum off targets are required for designing the high-efficiency gRNAs. Knowledge of genomic features including GC content, epigenetic factors, and specific nucleotide position has been utilized to develop models for designing sgRNA for CRISPR/Cas applications. Several platforms that use these models to predict efficient sgRNA include E-CRISP, CCTOP CHOPCHOP, and CRISPOR (Uniyal et al. 2019). The basic idea behind successful multiplex genome editing is to design and insert multiple sgRNAs efficiently into one binary vector. Multiple cloning steps can be utilized to insert different sgRNAs expression cassettes targeting different sequences into a single vector or an advanced Golden Gate cloning method can be used to insert multiple sgRNAs by generating sticky ends in the target DNA. This technique facilitates the cloning of different gRNAs into a single binary vector. Till date 2–24 gRNAs have been used for multiplexed genome editing in plants (Upadhyay et al. 2013; Zhou et al. 2014; Yan et al. 2015; Stuttmann et al. 2021).
3) **Gene regulatory elements or promoters**: The multiplex CRISPR systems can be divided into two broad categories as far as expression of construct is concerned (Montecillo et al. 2020).
   i) **Single transcriptional cassette system**: In this system, expression of both Cas9 and sgRNAs is under a single promoter and terminator. RNA Pol II promoters are most frequently used for this purpose as they are capable of transcribing long transcripts. This approach has been useful where tissue-specific editing or promoter-inducible application is required. However, it has one limitation as the promoter

may not be able to transcribe the whole construct if number of gRNAs are more (Abdelrahman et al. 2021).

ii) **Multiple transcriptional cassette system**: This can be further divided into mixed dual promoter system that uses two different kinds of promoters for Cas gene and gRNAs and dual promoter system that uses same promoter with Cas gene and gRNAs individually (Hassan et al. 2021).

2.3 Strategies for Processing Multiple-Guide RNAs

Various strategies have been used for post-transcriptional processing of multiple gRNAs from a transcript using different enzymes (Figure 2.1).

1) **Ribozyme-mediated self-cleavage**: One of the strategy uses natural self-cleaving ribozyme enzyme for making multiple gRNAs. This method was first reported in yeast for single gRNA expression but was later used for multiplex genome editing other organisms. The ribozyme method uses either Pol II or Pol III promoter for generation of multiple gRNAs. A transcript consisting of two ribozyme-gRNA-ribozyme gene (RGR) units was designed for rice with either *U6* (*OsU6*) promoter or *Actin1* promoter (a Pol II promoter) (Feng and Yang 2020).
2) **Polycistronic tRNA–gRNA (PTG)**: This produces multiple guide RNAs from a single transcript by using tRNAs and endogenous RNases, which recognize and excise the distinct tRNA structure. The PTG recognizes shorter sequence of ~80 nt tRNA and is expressed with Pol II and Pol III promoters. It has been used in various plant systems including rice (Dong et al. 2017).
3) **Cys 4 based system**: Another method utilizes Cas family endonucleases, such as Csy4, to excise gRNAs that are expressed under Pol II or Pol III promoters. Csy4 recognizes a very short 28-nt sequence that flanks gRNAs; 10–12 gRNAs can be expressed with a single Pol II promoter and processed by Csy4 (McCarty et al. 2020). The Csy4-based system was successfully used for multiplex genome editing in plants when the transcript was expressed under Pol II promoter of *Cestrum yellow leaf curling virus* (CmYLCV) (Čermák et al. 2017). Although highly efficient, a major disadvantage of Csy4-based system is high cytotoxicity.
4) **Cpf1based excision**: As mentioned earlier, Cpf1 or Cas12a is a class II CRISPR nucleases; however, Cpf1 is smaller in size than Cas9 and recognizes AT-rich sequence sites. Moreover, it does not require tracrRNA and the cassette is generally expressed using a crRNA array with both Pol II and Pol III promoters. It recognizes hairpin structure within the spacer sequence and cleaves pre-crRNA to produce mature crRNAs (Zetsche et al. 2017). Cpf1 has been efficiently used for multiplexed gene editing of four *OsBEL* genes in rice (Wang et al. 2017b).
5) **Endogenous RNase system**: In this system, endogenous tRNA-processing machinery consisting of endogenous RNases such as RNase III and RNase T1 is used to process multiple gRNAs that are flanked by 77-nt long pre-tRNA genes. A SpCas9 and gRNA cassette with a single Pol II promoter was used to transform and edit rice genes efficiently (Mikami et al. 2017).

2.4 Delivery of CRISPR/Cas Construct into Plant Cells

The mode of delivery of the CRISPR/Cas construct or components is crucial for generating edited plants. Various delivery methods including agrobacterium-mediated, virus-mediated delivery, particle bombardment, nanoparticle-mediated delivery, and PEG-mediated transformation have been used to achieve multiplex genome editing in plants. Some of the commonly used methods have been discussed here:

2.4.1 Agrobacterium-Mediated Delivery

Agrobacterium tumefaciens that harbor genetically engineered constructs based on binary vectors have been widely used for genomic editing. *Agrobacterium tumefaciens* have been previously utilized to deliver CRISPR/Cas components to various plant species including *Arabidopsis thaliana*, *Nicotiana tabacum*, and *Oryza sativa*. Transgenic plants can be sexually propagated to obtain homozygous population and positive transformants can be screened for the presence of antibiotic resistance (Kuluev et al. 2019; Sandhya et al. 2020). The technique can be used to transform the plants through their leaves, callus formation, and floral structures. This technique has high editing efficiency and is very robust and simple to use (Mao et al. 2019). Multiplex editing of few rice genes was successfully achieved using Agrobacterium harboring pCAMBIA-based binary vectors with two guide RNAs and Cpf1 nucleases (Wang et al. 2017a). Another bacterium, *A. rhizogenes*, has also been used as a delivery tool for CRISPR/Cas genome editing. This bacterium produces hairy root upon successful delivery of genetically engineered vectors, hence providing an easy way for confirming the transformants. This method has been used to deliver CRISPR/Cas components to various plants such as soybean (Cai et al. 2015).

2.4.2 Virus-Mediated Delivery

Viral vectors have the ability to replicate within various hosts and hence can be used as vectors for delivery of CRISPR/Cas constructs. Several engineered plant RNA and DNA viruses are utilized to either generate gene knockouts or insert DNA (Scholthof et al. 1996). Most used RNA viruses are tobacco rattle virus (TRV), tobacco mosaic virus (TMV), barley stripe mosaic virus (BSMV), and beet necrotic yellow vein virus (BNYVV), which have been used to efficiently deliver gRNAs to plants including *Arabidopsis*, Nicotiana, wheat, and barley. Geminiviruses that are DNA-based viruses including yellow dwarf virus (BeYDV), cabbage leaf curl (CaLCuV), wheat dwarf virus (WDV), and tomato leaf curl virus (ToLCV) have been used in various studies to deliver CRISPR components (Gil-Humanes et al. 2017; Montecillo et al. 2020). Few of the advantages of virus-based delivery system are higher transcript levels of sgRNAs due to viral replication, reduced time due to lack of transformation and regeneration process, and production of transgene-free edited plants (Mahas et al. 2019). Efficient multiplexed CRISPR/Cas-based genome editing has been achieved in *N. benthamiana* using TRV- and TMV-based delivery vectors (Ali et al. 2015; Cody et al. 2017). The main limitation of using a viral-based delivery system can be limited cargo size of viruses. Large sequences may not be able to integrate into virus efficiently (Yin et al. 2015).

2.4.3 Particle Bombardment-Based Delivery

The particle bombardment method of transformation requires a "gene gun or biolistic gun" and carrier particles such as gold, silver, and tungsten for delivery of Cas9 and gRNA proteins. The method requires optimal pressure conditions generated by helium to deliver CRISPR components to the explants used. The advantage of this method is delivery of multiple copies of desired construct to the host increasing the chance of editing (Montecillo et al. 2020). Successful transformants have been reported in various plants including potato, maize, sorghum, and brassica (Svitashev et al. 2015; Murovec et al. 2018; Liu et al. 2019a; Sandhya et al. 2020).

2.5 Broader Implications of CRISPR/Cas Multiplex Gene Editing

CRISPR/Cas multiplex gene editing finds its use in various aspects of crop improvement. Some of these are listed below (Figure 2.2)

2.5.1 Simultaneous Knockout of Multiple Genes

Previously multigene knockout in plants were generated by either T-DNA insertion mutagenesis or chemical mutagenesis. The mutant lines for individual gene knockouts were screened and crossed to obtain homozygous double, triple, or higher mutants. The process is time consuming, tedious, and may sometime produces weaker phenotypes. CRISPR/Cas9 multiplex editing enables multigene knockouts with a single transformation, thus reducing time and effort. This may also lead to the discovery of newer and stronger traits (Minkenberg et al. 2017).

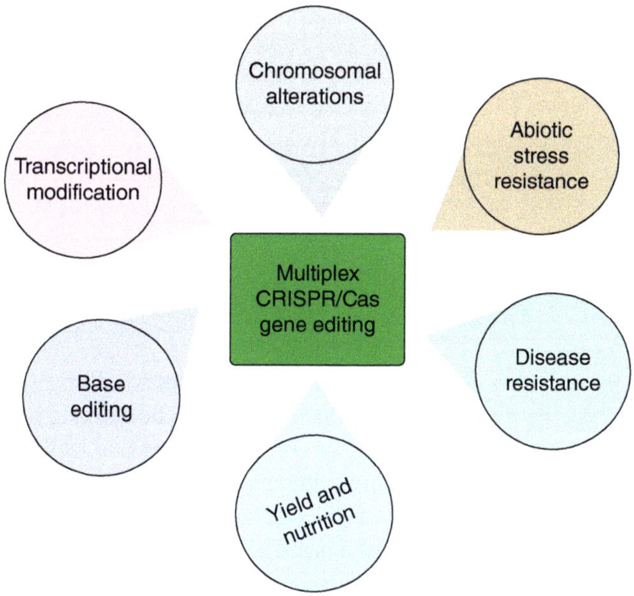

Figure 2.2 Overview of applications of multiplex CRISPR/Cas gene editing in plants.

2.5.2 Targeted Chromosomal Deletions

To study the functions of group of genes, deletions of bigger fragments within the chromosome would be more efficient compared to smaller modifications using a single gRNA. Targeted chromosomal deletions using multiplex CRISPR/Cas editing would greatly facilitate functional characterization of clustered genes or the cis regulatory elements which transcriptionally regulate the genes (Hsu et al. 2013; Minkenberg et al. 2017). This could also help target protein-coding genes simultaneously to evaluate their role in plants. Moreover, deletions in non-coding regions of genome could also help understand them better. A previous study reported deletion of 450 bp intronic region from *Arabidopsis* gene *AGAMOUS* (*AG*), which is involved in its development (Yan et al. 2016). Further, this technique could also have implications in studying agronomic traits of various crops and deletion of loci contributing to an unwanted trait could improve quality of crops. Reports suggest chromosomal deletion of up to 245 and 2 kb has been generated using multiplex CRISPR/Cas9 editing in rice and Nicotiana, respectively (Zhou et al. 2014; Lowder et al. 2017).

2.5.3 Transcriptional Activation or Repression of Genes

Multiplex genome editing can be used to enhance the efficacy of gene activation or repression as several transcription factors could be modulated simultaneously (Gilbert et al. 2013). Multiplexed CRISPR-based transcriptional repression of genes mostly recruits an inactive mutant of Cas9, nuclease dead Cas9 (dCas9) with suppressed endonuclease activity (Moradpour and Abdulah 2020). For efficient activation or repression, the dCas9 complex should be guided toward the promoter region. dCas9–sgRNA complex interferes with binding of transcriptional factors or RNA polymerase or by inhibiting elongation (Qi et al. 2013). A previous study successfully deployed CRISPR-based multiplex editing by simultaneously activating three independent endogenous genes, Os11g35410, Os03g01240, and Os04g39780, in rice (Lowder et al. 2017). Efficient transcriptional activation was achieved by designing three gRNAs each to target *Arabidopsis Production of Anthocyanin Pigment1* (*AtPAP1*; encodes a transcription factor) and *miR319* (Lowder et al. 2015). A 2- to 7-fold activation in AtPAP1 expression and 3- and 7.5-fold activation in the expression of miR319 was observed. Moreover, expression of *Arabidopsis Cleavage Stimulating Factor64* (*AtCSTF64*) gene was repressed by using dCAS9-based multiplex genome editing complex.

2.5.4 Base Editing

Base editing through CRISPR/Cas multiplex editing has emerged as powerful tool for single-base mutation (Bharat et al. 2020). The cytidine base editor (CBE) and adenine base editor (ABE) are the two types of base editors that facilitate either deamination of cytosine (C) to uracil (U) or deamination of adenosine (A) to inosine (I) (Komor et al. 2016; Nishida et al. 2016). SpCas9, dCas9, or Cpf1 is most widely used for this purpose. Multiplex base-editing generates various desired SNPs at multiple loci in different plant genes, which may be beneficial for crop improvement. Adenine base editor (rBE14) fused with nCas9 (D10A) converted A·T to G·C in *OsMPK6*, *OsSERK2*, and *OsWRKY45* genes of rice (Yan et al. 2018). Moreover, using CBE and Cas9 fusion, a herbicide-resistant tomato containing multiple point mutations in the acetolactate synthase (ALS) gene was generated (Shimatani et al. 2017). Overall, the base editing tools provide a platform for precision editing in various plant species, thus enhancing the scope of crop improvement.

2.6 Application of CRISPR/Cas Multiplex Gene Editing in Generating Disease Resistant Plants

Global crop production is threatened by viral, fungal or bacterial diseases leading to 20–40% loss in agricultural yield. A major application of genome editing approaches in plants is to develop disease-resistant varieties. This can be achieved by modifying 1) host susceptibility genes (S-genes), 2) genes regulating plant hormones, 3) negative regulators of plant immunity, and 4) targeting pathogen/pest genes (Kamburova et al. 2017; Singh et al. 2022). Due to the ability to precisely manipulate plant genomes, CRISPR/Cas has emerged as a powerful tool to generate disease-resistant crops. With advancement in CRISPR/Cas technology, several plant or pathogen genes can be targeted simultaneously to enhance disease tolerance.

2.6.1 Disease Resistance Against Viruses

CRISPR/Cas-mediated multiplex gene editing has been explored to develop effective ways for inducing resistance against plant viruses. This can be done in two ways: either by targeting host genes that facilitate virus replication or by modifying the viral genome and hence disrupting replication of the virus. In a previous study, multiplex gene editing was deployed to target isoflavonoid pathway in soybean. A single CRISPR/Cas-based construct was used to target three soybean genes flavanone-3-hydroxylase 1 and 2 and flavone synthase (*GmF3H1*, *GmF3H2* and *GmFNSII-1*). The editing of these three genes enhances the production of isoflavone genistein by inhibiting the production of other isoflavones including daidzein and glycitein that may compete with genistein. Improvement in the production of genistein leads to resistance against *Soybean mosaic virus* in transgenic soybean plants.

CRISPR/Cas-based targeted mutagenesis of cassava eukaryotic translation initiation factor 4E (*eIF4E)* gene isoforms novel cap-binding protein *nCBP-1* and *nCBP-2* provides resistance against *Cassava brown streak virus* (CBSV), the causative agent of Cassava brown streak disease (CBSD). The CBP double mutants showed delayed symptoms, less viral load, and reduced disease severity. Further multiplex CRISPR/Cas-based genome editing has been used to target eIF gene family and its paralogues in various other plants to provide resistance against viruses including *Rice tungro spherical virus* (Macovei et al. 2018), *Zucchini yellow mosaic virus*, *Cucumber vein yellowing virus* (*Ipomovirus*), and *Papaya ring spot mosaic virus-W* (Chandrasekharan et al. 2016; Tripathi et al. 2020).

Apart from host genes, diseases resistance has been achieved by targeting three or more genes of *Cotton Leaf Curl Viruses* (CLCuVs) that cause cotton leaf curl disease by expressing multiple guide RNAs in *Nicotiana benthamiana* plants. The simultaneous targeting of multiple intergenic regions (IR) of DNA geminiviruses such as c*otton leaf curl Kokhran virus* (CLCuKoV) and *tomato yellow leaf curl virus* (TYLCV) impairs their ability to replicate by interfering with DNA repair system (NHEJ) within the host *N. benthiamiana,* thus leading to the detection of lower viral load in leaves (Ali et al. 2016). In a similar study, CRISPR/Cas9 constructs harbouring multiple guide RNAs were used to target various IRs of chilli leaf curl virus (ChiLCV), a dominant begomovirus in chilli (*Capsicum annuum*). The mutants generated were incapable to produce severe symptoms and viral titer in model plant *N. benthiamiana* due to disruption of viral rep and coat protein (Roy et al. 2019). Moreover, multiplex CRISPR/Cas approach was employed to target four different coding sequence of Rep proteins in *wheat dwarf virus* (WDV) that infects both wheat and barley (Kis et al. 2019). The barley plants infected with mutated WDV showed disease resistance and fewer symptoms. Overall, these studies suggest that CRISPR/Cas-mediated multiplex genome editing is a powerful tool to provide resistance against viruses.

2.6.2 Disease Resistance Against Fungi

Fungal pathogens are most devastating pathogens causing up to 30% yield losses every year. Thus, newer strategies are needed to provide fungal disease tolerance in plants and CRISPR/Cas technology has proved beneficial in this area. A previous study depicted that targeting two miRNAs, miR482b and miR482c, simultaneously through multiplex CRISPR/Cas gene editing approach enhances tolerance against *Phytophthora infestans*, the causal agent of leaf blight in tomato (Hong et al. 2021). CRISPR/Cas9 was used to mutate three homeo-alleles of *mildew resistance locus O* (*MLO*), i.e. *TaMLO-A1, TaMLO-B1, TaMLO-D1* to develop resistance against *Blumeria graminis* f. sp. *tritici*, which cause powdery mildew disease in wheat (Tyagi et al. 2021). Furthermore, apple DIPM genes (*DIPM-1, DIPM-2*, and *DIPM-4*) were targeted using CRISPR/Cas9-based gene editing to provide tolerance against fire blight disease caused by *Erwinia amylovora* (Malnoy et al. 2016). DIPMs are leucine-rich repeat receptor-like kinase receptors which act as susceptibility factors by directly interacting with the disease-specific (dsp) gene of *Erwinia*. A recent study used CRISPR/Cas9 technique to induce chromosomal rearrangement in two soybean NLR gene clusters, Rpp1-like and Rps1, by targeting DSBs in these clusters. The rearrangement of clusters could induce immunity in soybean against the fungi *Phakopsora pachyrhizi* and *Phytophthora sojae* (Nagy et al. 2021; Paul et al. 2021). A CRISPR/Cas9-based targeted mutagenesis of the *Brassica napus* WRKY genes, i.e. *BnWRKY11* and *BnWRKY70*, with multiple sgRNA cassettes led to enhanced resistance against *Sclerotinia sclerotiorum* (Sun et al. 2018).

CRISPR/Cas9-mediated multiplex gene ediding was efficiently used for fungal gene mutagenesis. Two genes of *Alternaria alternata pksA* and *brm2* that are involved in melanin biosynthesis pathway were mutated (Wenderoth et al. 2017). Melanin biosynthesis has been associated with conidia formation in various filamentous fungi including Alternaria (Thomma 2003). Several genes of *Ustilaginoidea virens,* the causal agent of rice false smut, were efficiently mutated through multiplexed CRISPR/Cas9 system (Liang et al. 2018). The mutants generated were defective in USTA ustiloxin and UvSLT2 MAP kinase genes including UA01, UA13, and UA21. The mutants demonstrated lower cell wall integrity and were sensitive to cytoplasmic stresses. CRISPR/Cas9 was used to enhance resistance of *Brassica napus* against necrotrophic fungal pathogens *Sclerotinia sclerotiorum* and *Botrytis cinerea* by targeting multiple copies *BnQCR8* gene (Zaynab et al. 2020; Zhang et al. 2021). *BnQCR8* is a subunit of the cytochrome b-c1 complex, which interacts with an effector-like protein (*SsSSVP1*) in *Sclerotinia* and its homologue in *Botrytis*. Together these studies have suggested that mutagenesis of multiple host and fungal pathogen genes simultaneously could increase the chance of improved resistance against the diseases.

2.6.3 Disease Resistance Against Bacteria

During bacterial infection, bacteria deploy type III apparatus to secrete effectors into the host. These effectors modulate host's defense by targeting susceptibility factors (Zaidi et al. 2017). These *S* genes and other negative regulators of plant immunity are potential targets for CRISPR/Cas9-mediated gene editing to improve defense against bacterial pathogens. CRISPR/Cas-based genome editing has been widely used to target rice susceptible genes, specifically SWEET genes, against the bacterial pathogen *Xanthomonas oryzae pv oryzae (Xoo)* causative agent of bacterial blight. Five mutations were generated simultaneously in the promoter region of *SWEET13* and *SWEET14* genes of rice using multiplex CRISPR/Cas9 genome editing technique. The rice mutants thus generated showed resistance to most of the strains of Xoo (Oliva et al. 2019; Singh et al. 2022). It is well established that *Xoo* hijacks the *SWEET* family genes to infect the host through its endogenous transcription activator-like effectors (TALE). TALEs interact via effector binding elements (EBEs) which are present in the promoter region of *OsSWEET* genes (Chen et al. 2012). The CRISPR/Cas9-mediated mutagenesis in four such EBEs present in the promoter region of the *SWEET14* of super basmati rice provided resistance against Xoo by inhibiting interaction between EBEs and TALEs including AvrXa7, PthXo3, and TalF (Zafar et al. 2020; Singh et al. 2022). A triple mutant of rice was generated by knocking out three S genes *Pi21* (encodes a proline-rich protein), *Bsr-d1* (encodes a C2H2-type transcription factor), and *Xa5* (a bacterial blight resistance gene) using CRISPR/Cas9 multiplex genome editing technology (Tao et al. 2021). These mutant plants showed increased resistance to *Magnaporthe oryzae* and *Xanthomonas oryzae pv. oryzae (Xoo)*. Mutations were induced into the TMS5 (encodes RNase ZS1), Pi21, and Xa13 genes of rice elite variety Pinzhan using the CRISPR/Cas9 multiplex genome editing, and the triple tms5/pi21/xa13 mutants were found resistant against bacterial blight (Li et al. 2019).

2.7 Application of CRISPR/Cas Multiplex Gene Editing in Abiotic Stress-Tolerant Crop Production

Over the past few years, earth's atmosphere changed significantly, which caused global warming. Due to increased global warming, rain and water level have depleted severely, which are responsible for abiotic stress-mediated crop loss. The major abiotic stresses caused by global warming includes salinity, drought, and extreme high or low temperature conditions. These abiotic stresses combined or independently affect growth and development of plants followed by decreased crop yield and quality (Zhang et al. 2021). Over the time, many transcription factors, structural elements, non-coding RNAs, and regulatory factors have been identified to have direct or indirect role in plant response against various abiotic stresses. These genes or elements can be used as putative targets for crop improvement against various abiotic stresses by using traditional transgenic approach or by using modern CRISPR/Cas gene editing technique (Zafar et al. 2020; Upadhyay 2021).

2.7.1 Drought Tolerance

Wheat is one of the most commonly used crop plants all over the world. Similar to other crop plants, wheat also needs lot of water and in drought condition its production declines significantly. Recently multiplex CRISPR/Cas9 genome editing has been used significantly in many plant species including wheat. In wheat, *SAL1* gene has been already identified to work in drought tolerance by encoding two enzymes: 3′(2′), 5′-bisphosphate nucleosidase and inositol polyphosphate 1-phosphatase. Wheat genome consists of six homologues of *TaSAL1* genes (4A, 7A, 5A, 5B, 5D, and 7D), and an additional gene was found as tandem duplication (4A1 and 4A2) on chromosome 4A. By using multiplex CRISPR/Cas9 technique, gRNAs were designed for all seven *TaSAL1* genes to generate *TaSAL1* knockout mutant of wheat plant. *TaSAL1* mutant plants exhibited rolled leaf and bended stem phenotype in young plants. These CRISPR-generated wheat plant showed significantly higher germination rate and drought-tolerant phenotype. Moreover, it has been known that ethylene plays an important role in plant survival during high temperature and drought stress condition. Over the time, several studies have shown that plant sensitivity toward ethylene or depleting the biosynthesis of ethylene during drought stress condition can increase the fruit yield in maize plant. A previous study has identified *AGROS8* as a negative regulator of ethylene response in maize plant. Shi et al. has used advanced CRISPR/Cas9 tool to generate *AGROS8* overexpressing maize plants by promoter modification and promoter swapping. It was found that these CRISPR-edited plants gave high yield under drought stress as compared to the WT plants (Abdallah et al. 2022).

2.7.2 Salinity Tolerance

Salinity stress is caused by the presence of higher amount of salts in soil. Salinity stress affects cell osmosis, stomatal closure, and reduced water uptake that eventually affects photosynthetic efficiency. Multiplex CRISPR/Cas9-mediated genome editing has also been used in the production of salinity stress-tolerant plants. Hybrid proline-rich proteins (*HyPRPs*) have been recently targeted to produce salinity tolerance variety. *SlHyPRP* belongs to the subgroup of putative cell wall glycoproteins and harbors PRD (proline-rich domain) at their N-terminal domain and consists of a conserved 8CM (eight cystein motifs) at their C-terminal domain. *HyPRPs* of tobacco and pepper plants have already shown to have role in plant cell death and work as positive and negative regulators of biotic stress (Yeom et al. 2012). In contrast, tomato hybrid PRPs (*SlHyPRPs*) have been identified as a positive regulator of salinity and dehydration (Li et al. 2016a). Multiplex genome editing tool has been efficiently used in tomato plants to generate high salinity-tolerant plants by modifying multiple functional domain (s) of *SlHyPRP1* genes (Tran et al. 2021).

2.7.3 Herbicide Resistance

Glyphosate is one of the major and rapidly adopted herbicide used in fields of soybean, maize, chili pepper, and sugar bean. Over the time, many genes were identified in crop plants that can be targeted to generate herbicide-resistant plants (Sony et al. 2023). In plant cells phosphoenole pyruvate (PEP) is used for the production of aromatic compounds; 5-enolpyruvylshikimate 3-phosphate synthase (EPSPS) is a key enzyme that participates in above mentioned reaction. Herbicide glyphosate inhibits EPSPS enzyme by attaching to it that eventually inhibits PEP binding to EPSPS. Glyphosate-dependent inhibition of EPSPS affects the production of aromatic compound production and eventually leads to plant death. In a previous study, multiplex genome editing was adopted to replace second exon with another TIPS substitution containing exon in rice. Multiple sgRNAs were designed to target intron 1 and intron 2 of OsEPSPS gene (Sony et al. 2023). Chlorsufuron and bispyribac sodium are another major herbicide used in farming. These herbicides target branched chain amino acid biosynthesis by affecting a major enzyme acetolactate synthase 1 (ALS1) involved in this process. Previous reports have shown that mutation in either P178S or P165 position of *ALS1* using CRISPR has generated only few chorosulfuron-resistant plants. A previous study by Sun et al. (2016) revealed that multiplex CRISPR technology can be used to generate chorosulfuron-resistant rice plants by targeting multiple amino acids in *ALS1* gene.

2.8 Application of CRISPR/Cas Multiplex Gene Editing in Enhancing Crop Yield, Nutrition, and Related Traits

It has been known that suppression of starch branching enzymes (SBEs) is responsible for higher amount of resistant starch (RS) in cereal crops (Carciofi et al. 2012). RS is beneficial for health as it is not retained in the stomach and pass out of the body leading to lower diabetes risk. CRISPR/Cas9 multiplex genome editing technique has been used to target four *SBE* genes in U.S. rice cultivar Presidio to enhance its quality and RS amount. Besides this, multiplex genome editing has also been used in various studies to modulate and understand the role of different rice genes including ideal plant architecture1 (*IPA1*), grain number 1a (*Gn1a*), dense and erect panicle (*DEP1*) grain size (*GS3*), and amino acid transporter (*OsAAP*) in enhancing rice grain yield, number, architecture, and weight (Li et al. 2016b; Xu et al. 2016; Lu et al. 2018). Moreover, five key genes in the γ-aminobutyric acid (GABA) shunt pathway including GABA transaminases (*GABA-TP1*, *GABA-TP2,* and *GABA-TP3*), cationic amino acid transporter 9 (*CAT9*), and succinate semialdehyde dehydrogenase (*SSADH)* were targeted through CRIPSR/Cas multiplex editing to enhance

GABA accumulation in different tissues of tomato (Li et al. 2018b). The study was useful to anticipate role of GABA homeostasis in tomato growth and development. Some of the proteins such as α-amylase/trypsin inhibitors (ATI) are involved in causing wheat allergies including Baker's asthma (Mansueto et al. 2019). A previous study targeted ATI subunits WTAI-CM3 and WTAI-CM16 in Italian durum wheat cultivar Svevo using CRISPR/Cas9 multiplex genome editing (Camerlengo et al. 2020). This may lead to reduced number of allergens in wheat. In a previous, CRISPR/Cpf1-mediated multiplex editing was adopted to achieve improved oil configuration by targeting fatty acid desaturase genes *FAD2-1A* and *FAD2-1B* in soybean (Kim et al. 2017; Hussain et al. 2018). These genes are known to convert oleic acid into linolenic acid. Further, to enhance the lycopene content in fruit, CRISPR/Cas9 was used to modulate genes including stay-green 1 (*SGR1*) and lycopene β-cyclase family genes that are involved in carotenoid metabolic pathway of tomato (Li et al. 2018a). Overall, CRISPR/Cas9 multiplex gene editing strategy has provided platform to target various crop or fruit genes simultaneously to enhance their quantitative/qualitative traits (Dixit et al. 2021).

2.9 Conclusion

Over the time, multiplex genome editing (as summarized in Figure 2.3) has advanced the production of variety of crops by producing multiple mutations in different alleles, by targeting multiple genes of a single gene family, by targeting unrelated genes that have common function, or by targeting different genes to generate variety of traits. There has been tremendous advancement in CRISPR/Cas gene editing technology in order to become one of the most powerful and practical methods. Different varients of CRISPR/Cas have been developed, which are used rigorously in several research laboratories. Even though a single mutation in a single gene is also enough to achieve desired phenotype; however, as discussed here, sometimes researchers target more than one gene or more than one target site in single gene to enhance the phenotypic effect. Conventional methods in such cases consume more time and are laborious. Here, the multiplex CRISPR/Cas gene editing technique provides a platform to achieve the precise editing and the desired phenotype. Moreover, with discovery of more and more efficient Cas proteins and transgene-free delivery systems, multiplex gene editing will be used more frequently in agricultural research.

Acknowledgments

We acknowledge the authors whose work has not been cited in this book chapter due to space limitations. RP is supported by Post-doctoral Research Fellowship from the Department of Plant Pathology, Kansas State University. KT is supported by the Post-doctoral Research Fellowship from the Department of plant pathology, University of Minnesota, USA.

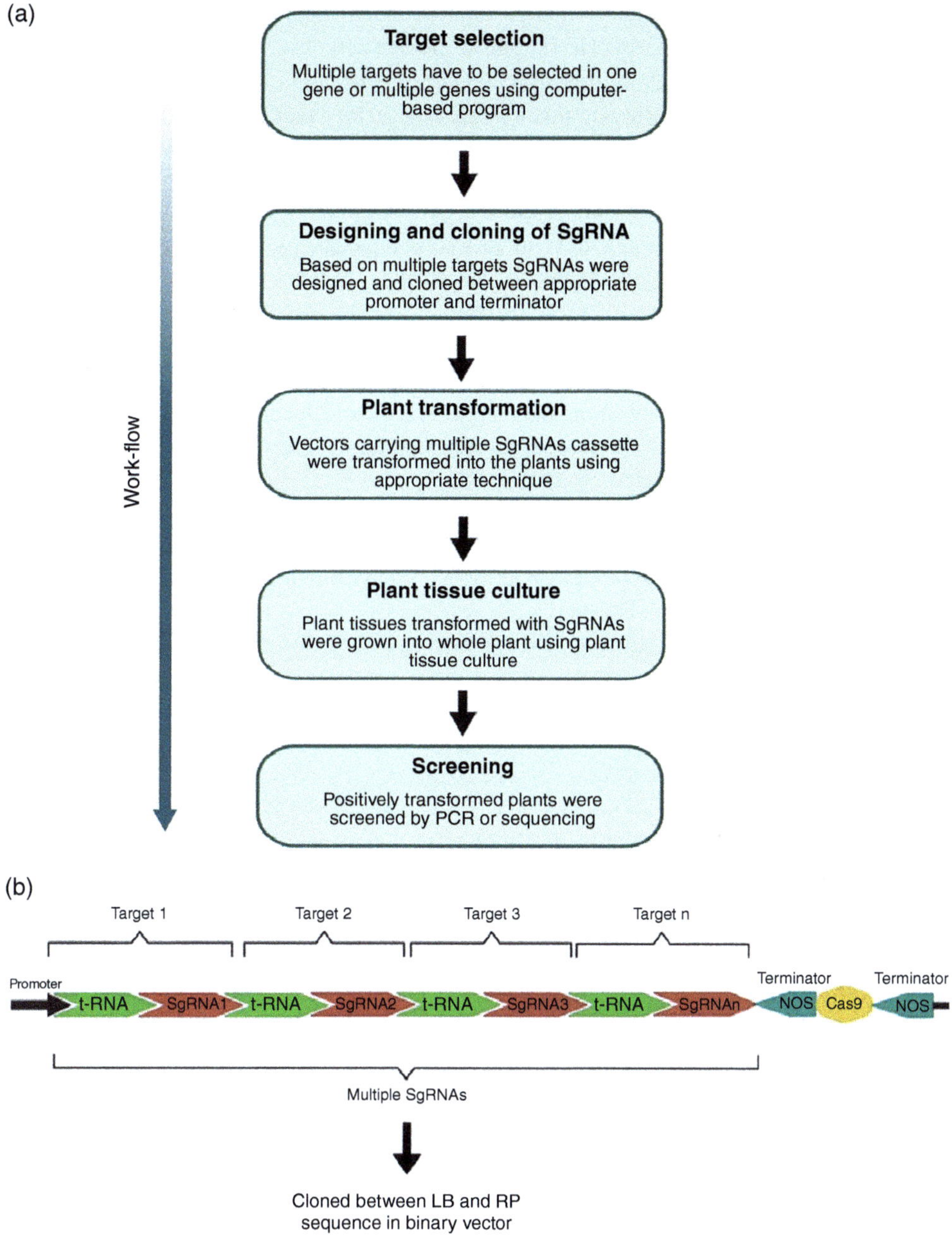

Figure 2.3 Strategy for multiplex CRISPR/Cas genome editing. (a) Flow chart showing steps involved in multiplex CRISPR/Cas gene editing in plants. (b) Assembly of multiplex CRISPR/Cas9 cassette.

References

Abdallah, N.A., Elsharawy, H., Abulela, H.A. et al. (2022). Multiplex CRISPR/Cas9-mediated genome editing to address drought tolerance in wheat. *GM Crops & Food* 6: 1–17.

Abdelrahman, M., Wei, Z., Rohila, J.S., and Zhao, K. (2021). Multiplex genome-editing technologies for revolutionizing plant biology and crop improvement. *Frontiers in Plant Science* 12: 721203.

Ali, Z., Abul-Faraj, A., Li, L. et al. (2015). Efficient virus-mediated genome editing in plants using the CRISPR/Cas9 system. *Molecular Plant* 8 (8): 1288–1291.

Ali, Z., Ali, S., Tashkandi, M. et al. (2016). CRISPR/Cas9-mediated immunity to geminiviruses: differential interference and evasion. *Scientific Reports* 6 (1): 1–13.

Alok, A., Chauhan, H., Upadhyay, S.K. et al. (2021). Compendium of plant-specific CRISPR vectors and their technical advantages. *Life* 11: 1021. https://doi.org/10.3390/life11101021.

Arora, L. and Narula, A. (2017). Gene editing and crop improvement using CRISPR-Cas9 system. *Frontiers in Plant Science* 8: 1932.

Barrangou, R., Fremaux, C., Deveau, H. et al. (2007). CRISPR provides acquired resistance against viruses in prokaryotes. *Science* 315 (5819): 1709–1712.

Beerli, R.R. and Barbas, C.F. (2002). Engineering polydactyl zinc-finger transcription factors. *Nature Biotechnology* 20 (2): 135–141.

Bhalothia, P., Yajnik, K., Alok, A., and Upadhayay, S.K. (2020). The current progress of CRISPR/Cas9 development in plants. In: *Genome Engineering via CRISPR-Cas9 Systems* (ed. S. Dhar). Elsevier. https://doi.org/10.1016/B978-0-12-818140-9.00011-8.

Bharat, S.S., Li, S., Li, J. et al. (2020). Base editing in plants: current status and challenges. *The Crop Journal* 8 (3): 384–395.

Burstein, D., Harrington, L.B., Strutt, S.C. et al. (2017). New CRISPR–Cas systems from uncultivated microbes. *Nature* 542 (7640): 237–241.

Cai, Y., Chen, L., Liu, X. et al. (2015). CRISPR/Cas9-mediated genome editing in soybean hairy roots. *PLoS One* 10 (8): e0136064.

Camerlengo, F., Frittelli, A., Sparks, C. et al. (2020). CRISPR-Cas9 multiplex editing of the α-amylase/trypsin inhibitor genes to reduce allergen proteins in durum wheat. *Frontiers in Sustainable Food Systems* 4: 104.

Carciofi, M., Blennow, A., Jensen, S.L. et al. (2012). Concerted suppression of all starch branching enzyme genes in barley produces amylose-only starch granules. *BMC Plant Biology* 12 (1): 1–16.

Čermák, T., Curtin, S.J., Gil-Humanes, J. et al. (2017). A multipurpose toolkit to enable advanced genome engineering in plants. *The Plant Cell* 29 (6): 1196–1217.

Chandrasekaran, J., Brumin, M., Wolf, D. et al. (2016). Development of broad virus resistance in non-transgenic cucumber using CRISPR/Cas9 technology. *Molecular Plant Pathology* 17 (7): 1140–1153.

Chen, L.Q., Qu, X.Q., Hou, B.H. et al. (2012). Sucrose efflux mediated by SWEET proteins as a key step for phloem transport. *Science* 335 (6065): 207–211.

Cody, W.B., Scholthof, H.B., and Mirkov, T.E. (2017). Multiplexed gene editing and protein overexpression using a tobacco mosaic virus viral vector. *Plant Physiology* 175 (1): 23–35.

Cong, L., Ran, F.A., Cox, D. et al. (2013). Multiplex genome engineering using CRISPR/Cas systems. *Science* 339 (6121): 819–823.

Dixit, S., Shukla, A., Singh, V. et al. (2021). Engineering of plant metabolic pathway for nutritional improvement. *Wiley*. https://doi.org/10.1002/9781119672425.ch20.

Dong, F., Xie, K., Chen, Y. et al. (2017). Polycistronic tRNA and CRISPR guide-RNA enables highly efficient multiplexed genome engineering in human cells. *Biochemical and Biophysical Research Communications* 482 (4): 889–895.

Gilbert, L.A., Larson, M.H., Morsut, L. et al. (2013). CRISPR-mediated modular RNA-guided regulation of transcription in eukaryotes. *Cell* 154 (2): 442–451.

Gil-Humanes, J., Wang, Y., Liang, Z. et al. (2017). High-efficiency gene targeting in hexaploid wheat using DNA replicons and CRISPR/Cas9. *The Plant Journal* 89 (6): 1251–1262.

Harrington, L.B., Burstein, D., Chen, J.S. et al. (2018). Programmed DNA destruction by miniature CRISPR-Cas14 enzymes. *Science* 362 (6416): 839–842.

Hassan, M.M., Zhang, Y., Yuan, G. et al. (2021). Construct design for CRISPR/Cas-based genome editing in plants. *Trends in Plant Science* 26 (11): 1133–1152.

Hong, Y., Meng, J., He, X. et al. (2021). Editing miR482b and miR482c simultaneously by CRISPR/Cas9 enhanced tomato resistance to Phytophthora infestans. *Phytopathology* 111 (6): 1008–1016.

Hsieh-Feng, V. and Yang, Y. (2020). Efficient expression of multiple guide RNAs for CRISPR/Cas genome editing. *Abiotech* 1 (2): 123–134.

Hsu, P.D., Scott, D.A., Weinstein, J.A. et al. (2013). DNA targeting specificity of RNA-guided Cas9 nucleases. *Nature Biotechnology* 31 (9): 827–832.

Hussain, B., Lucas, S.J., and Budak, H. (2018). CRISPR/Cas9 in plants: at play in the genome and at work for crop improvement. *Briefings in Functional Genomics* 17 (5): 319–328.

Ishino, Y., Shinagawa, H., Makino, K. et al. (1987). Nucleotide sequence of the iap gene, responsible for alkaline phosphatase isozyme conversion in *Escherichia coli*, and identification of the gene product. *Journal of Bacteriology* 169 (12): 5429–5433.

Kamburova, V.S., Nikitina, E.V., Shermatov, S.E. et al. (2017). Genome editing in plants: an overview of tools and applications. *International Journal of Agronomy* 131: 12–21.

Kim, H., Kim, S.T., Ryu, J. et al. (2017). CRISPR/Cpf1-mediated DNA-free plant genome editing. *Nature Communications* 8 (1): 1–7.

Kis, A., Hamar, É., Tholt, G. et al. (2019). Creating highly efficient resistance against wheat dwarf virus in barley by employing CRISPR/Cas9 system. *Plant Biotechnology Journal* 17 (6): 1004.

Klug, A. (2010). The discovery of zinc fingers and their applications in gene regulation and genome manipulation. *Annual Review of Biochemistry* 79: 213–231.

Komor, A.C., Kim, Y.B., Packer, M.S. et al. (2016). Programmable editing of a target base in genomic DNA without double-stranded DNA cleavage. *Nature* 533 (7603): 420–424.

Kuluev, B.R., Gumerova, G.R., Mikhaylova, E.V. et al. (2019). Delivery of CRISPR/Cas components into higher plant cells for genome editing. *Russian Journal of Plant Physiology* 66 (5): 694–706.

Li, J., Ouyang, B., Wang, T. et al. (2016a). HyPRP1 gene suppressed by multiple stresses plays a negative role in abiotic stress tolerance in tomato. *Frontiers in Plant Science* 7: 967.

Li, M., Li, X., Zhou, Z. et al. (2016b). Reassessment of the four yield-related genes Gn1a, DEP1, GS3, and IPA1 in rice using a CRISPR/Cas9 system. *Frontiers in Plant Science* 7: 377.

Li, R., Li, R., Li, X. et al. (2018a). Multiplexed CRISPR/Cas9-mediated metabolic engineering of γ-aminobutyric acid levels in *Solanum lycopersicum*. *Plant Biotechnology Journal* 16 (2): 415–427.

Li, X., Wang, Y., Chen, S. et al. (2018b). Lycopene is enriched in tomato fruit by CRISPR/Cas9-mediated multiplex genome editing. *Frontiers in Plant Science* 9: 559.

Li, S., Shen, L., Hu, P. et al. (2019). Developing disease-resistant thermosensitive male sterile rice by multiplex gene editing. *Journal of Integrative Plant Biology* 61 (12): 1201–1205.

Liang, Y., Han, Y., Wang, C. et al. (2018). Targeted deletion of the USTA and UvSLT2 genes efficiently in *Ustilaginoidea virens* with the CRISPR-Cas9 system. *Frontiers in Plant Science* 9: 699.

Liu, G., Li, J., and Godwin, I.D. (2019a). Genome editing by CRISPR/Cas9 in sorghum through biolistic bombardment. In: *Sorghum* (ed. Z.-Y. Zhao), 169–183. New York: Humana Press.

Liu, J.J., Orlova, N., Oakes, B.L. et al. (2019b). CasX enzymes comprise a distinct family of RNA-guided genome editors. *Nature* 566 (7743): 218–223.

Lowder, L.G., Zhang, D., Baltes, N.J. et al. (2015). A CRISPR/Cas9 toolbox for multiplexed plant genome editing and transcriptional regulation. *Plant Physiology* 169 (2): 971–985.

Lowder, L.G., Paul, J.W., and Qi, Y. (2017). Multiplexed transcriptional activation or repression in plants using CRISPR-dCas9-based systems. In: *Plant Gene Regulatory Networks* (ed. K. Kaufmann), 167–184. New York: Humana Press.

Lu, K., Wu, B., Wang, J. et al. (2018). Blocking amino acid transporter Os AAP 3 improves grain yield by promoting outgrowth buds and increasing tiller number in rice. *Plant Biotechnology Journal* 16 (10): 1710–1722.

Macovei, A., Sevilla, N.R., Cantos, C. et al. (2018). Novel alleles of rice eIF4G generated by CRISPR/Cas9-targeted mutagenesis confer resistance to *Rice tungro spherical virus*. *Plant Biotechnology Journal* 16 (11): 1918–1927.

Mahas, A., Ali, Z., Tashkandi, M., and Mahfouz, M.M. (2019). Virus-mediated genome editing in plants using the CRISPR/Cas9 system. In: *Plant Genome Editing with CRISPR Systems* (ed. Y. Qi), 311–326. New York: Humana Press.

Makarova, K.S., Haft, D.H., Barrangou, R. et al. (2011). Evolution and classification of the CRISPR–Cas systems. *Nature Reviews Microbiology* 9 (6): 467–477.

Malnoy, M., Viola, R., Jung, M.H. et al. (2016). DNA-free genetically edited grapevine and apple protoplast using CRISPR/Cas9 ribonucleoproteins. *Frontiers in Plant Science* 7: 1904.

Mansueto, P., Soresi, M., Iacobucci, R. et al. (2019). Non-celiac wheat sensitivity: a search for the pathogenesis of a self-reported condition. *Italian Journal of Medicine* 13 (1): 15–23.

Mao, Y., Botella, J.R., Liu, Y., and Zhu, J.K. (2019). Gene editing in plants: progress and challenges. *National Science Review* 6 (3): 421–437.

McCarty, N.S., Graham, A.E., Studená, L., and Ledesma-Amaro, R. (2020). Multiplexed CRISPR technologies for gene editing and transcriptional regulation. *Nature Communications* 11 (1): 1281.

Mikami, M., Toki, S., and Endo, M. (2017). In planta processing of the SpCas9–gRNA complex. *Plant and Cell Physiology* 58 (11): 1857–1867.

Minkenberg, B., Wheatley, M., and Yang, Y. (2017). CRISPR/Cas9-enabled multiplex genome editing and its application. *Progress in Molecular Biology and Translational Science* 149: 111–132.

Montecillo, J.A.V., Chu, L.L., and Bae, H. (2020). CRISPR-Cas9 system for plant genome editing: current approaches and emerging developments. *Agronomy* 10 (7): 1033.

Moradpour, M. and Abdulah, S.N.A. (2020). CRISPR/dC as9 platforms in plants: strategies and applications beyond genome editing. *Plant Biotechnology Journal* 18 (1): 32–44.

Murovec, J., Guček, K., Bohanec, B. et al. (2018). DNA-free genome editing of *Brassica oleracea* and B. rapa protoplasts using CRISPR-Cas9 ribonucleoprotein complexes. *Frontiers in Plant Science* 9: 1594.

Nagy, E.D., Stevens, J.L., Yu, N. et al. (2021). Novel disease resistance gene paralogs created by CRISPR/Cas9 in soy. *Plant Cell Reports* 40 (6): 1047–1058.

Nishida, K., Arazoe, T., Yachie, N. et al. (2016). Targeted nucleotide editing using hybrid prokaryotic and vertebrate adaptive immune systems. *Science* 353 (6305): aaf8729.

Oliva, R., Ji, C., Atienza-Grande, G. et al. (2019). Broad-spectrum resistance to bacterial blight in rice using genome editing. *Nature Biotechnology* 37 (11): 1344–1350.

Paul, N.C., Park, S.W., Liu, H. et al. (2021). Plant and fungal genome editing to enhance plant disease resistance using the CRISPR/Cas9 system. *Frontiers in Plant Science* 12: 1534.

Qi, L.S., Larson, M.H., Gilbert, L.A. et al. (2013). Repurposing CRISPR as an RNA-guided platform for sequence-specific control of gene expression. *Cell* 152 (5): 1173–1183.

Rani, M., Tyagi, K., and Jha, G. (2020). Advancements in plant disease control strategies. In: *Advancement in Crop Improvement Techniques* (ed. N. Tuteja), 141–157. Woodhead Publishing.

Roy, A., Zhai, Y., Ortiz, J. et al. (2019). Multiplexed editing of a begomovirus genome restricts escape mutant formation and disease development. *PLoS One* 14 (10): e0223765.

Sandhya, D., Jogam, P., Allini, V.R. et al. (2020). The present and potential future methods for delivering CRISPR/Cas9 components in plants. *Journal of Genetic Engineering and Biotechnology* 18 (1): 1–11.

Scholthof, H.B., Scholthof, K.B.G., and Jackson, A.O. (1996). Plant virus gene vectors for transient expression of foreign proteins in plants. *Annual Review of Phytopathology* 34 (1): 299–323.

Shimatani, Z., Kashojiya, S., Takayama, M. et al. (2017). Targeted base editing in rice and tomato using a CRISPR-Cas9 cytidine deaminase fusion. *Nature Biotechnology* 35 (5): 441–443.

Shmakov, S., Abudayyeh, O.O., Makarova, K.S. et al. (2015). Discovery and functional characterization of diverse class 2 CRISPR-Cas systems. *Molecular Cell* 60 (3): 385–397.

Singh, J., Sharma, D., Brar, G.S. et al. (2022). CRISPR/Cas tool designs for multiplex genome editing and its applications in developing biotic and abiotic stress-resistant crop plants. *Molecular Biology Reports* 49 (12): 1–25.

Sony, S.K., Kaul, T., Motelb, K.F.A. et al. (2023). CRISPR/Cas9-mediated homology donor repair base editing confers glyphosate resistance to rice (*Oryza sativa* L.). *Frontiers in Plant Science* 14: 1122926.

Stuttmann, J., Barthel, K., Martin, P. et al. (2021). Highly efficient multiplex editing: one-shot generation of 8× Nicotiana benthamiana and 12× *Arabidopsis* mutants. *The Plant Journal* 106 (1): 8–22.

Sun, Y., Zhang, X., Wu, C. et al. (2016). Engineering herbicide-resistant rice plants through CRISPR/Cas9-mediated homologous recombination of acetolactate synthase. *Molecular Plant* 9 (4): 628–631.

Sun, Q., Lin, L., Liu, D. et al. (2018). CRISPR/Cas9-mediated multiplex genome editing of the BnWRKY11 and BnWRKY70 genes in *Brassica napus* L. *International Journal of Molecular Sciences* 19 (9): 2716.

Sushmita, K.G., Upadhyay, S.K., and Verma, P.C. (2021). An overview of genome-engineering methods. In: *Genome Engineering for Crop Improvement* (ed. S.K. Upadhyay). Wiley https://doi.org/10.1002/9781119672425.ch1.

Svitashev, S., Young, J.K., Schwartz, C. et al. (2015). Targeted mutagenesis, precise gene editing, and site-specific gene insertion in maize using Cas9 and guide RNA. *Plant Physiology* 169 (2): 931–945.

Tao, H., Shi, X., He, F. et al. (2021). Engineering broad-spectrum disease-resistant rice by editing multiple susceptibility genes. *Journal of Integrative Plant Biology* 63 (9): 1639–1648.

Thomma, B.P. (2003). *Alternaria* spp.: from general saprophyte to specific parasite. *Molecular Plant Pathology* 4 (4): 225–236.

Tran, M.T., Doan, D.T.H., Kim, J. et al. (2021). CRISPR/Cas9-based precise excision of SlHyPRP1 domain (s) to obtain salt stress-tolerant tomato. *Plant Cell Reports* 40 (6): 999–1011.

Tripathi, L., Ntui, V.O., and Tripathi, J.N. (2020). CRISPR/Cas9-based genome editing of banana for disease resistance. *Current Opinion in Plant Biology* 56: 118–126.

Tyagi, S., Kumar, R., Kumar, V. et al. (2021). Engineering disease resistant plants through CRISPR-Cas9 technology. *GM Crops & Food* 12 (1): 125–144.

Uniyal, A.P., Mansotra, K., Yadav, S.K., and Kumar, V. (2019). An overview of designing and selection of sgRNAs for precise genome editing by the CRISPR-Cas9 system in plants. *3 Biotech* 9 (6): 1–19.

Upadhyay, S.K. (2021). *Genome Engineering for Crop Improvement*. Wiley https://doi.org/10.1002/9781119672425.

Upadhyay, S.K., Kumar, J., Alok, A., and Tuli, R. (2013). RNA guided genome editing for multiple target gene mutations in wheat. *G3- Genes|Genomes|Genetics* 3 (12): 2233–2238.

Wang, M., Lu, Y., Botella, J.R. et al. (2017a). Gene targeting by homology-directed repair in rice using a geminivirus-based CRISPR/Cas9 system. *Molecular Plant* 10 (7): 1007–1010.

Wang, M., Mao, Y., Lu, Y. et al. (2017b). Multiplex gene editing in rice using the CRISPR-Cpf1 system. *Molecular Plant* 10 (7): 1011–1013.

Wenderoth, M., Pinecker, C., Voß, B., and Fischer, R. (2017). Establishment of CRISPR/Cas9 in *Alternaria alternata*. *Fungal Genetics and Biology* 101: 55–60.

Xing, H.L., Dong, L., Wang, Z.P. et al. (2014). A CRISPR/Cas9 toolkit for multiplex genome editing in plants. *BMC Plant Biology* 14 (1): 1–12.

Xiong, J.S., Ding, J., and Li, Y. (2015). Genome-editing technologies and their potential application in horticultural crop breeding. *Horticulture Research* 2: 15019.

Xu, R., Yang, Y., Qin, R. et al. (2016). Rapid improvement of grain weight via highly efficient CRISPR/Cas9-mediated multiplex genome editing in rice. *Journal of Genetics and Genomics= Yi chuan xue bao* 43 (8): 529–532.

Yan, L., Wei, S., Wu, Y. et al. (2015). High-efficiency genome editing in *Arabidopsis* using YAO promoter-driven CRISPR/Cas9 system. *Molecular Plant* 8 (12): 1820–1823.

Yan, W., Chen, D., and Kaufmann, K. (2016). Efficient multiplex mutagenesis by RNA-guided Cas9 and its use in the characterization of regulatory elements in the AGAMOUS gene. *Plant Methods* 12 (1): 1–9.

Yan, F., Kuang, Y., Ren, B. et al. (2018). Highly efficient A·T to G·C base editing by Cas9n-guided tRNA adenosine deaminase in rice. *Molecular Plant* 11 (4): 631–634.

Yeom, S.I., Seo, E., Oh, S.K. et al. (2012). A common plant cell-wall protein HyPRP1 has dual roles as a positive regulator of cell death and a negative regulator of basal defense against pathogens. *The Plant Journal* 69 (5): 755–768.

Yin, K., Han, T., Liu, G. et al. (2015). A geminivirus-based guide RNA delivery system for CRISPR/Cas9 mediated plant genome editing. *Scientific Reports* 5: 14926.

Zafar, K., Khan, M.Z., Amin, I. et al. (2020). Precise CRISPR-Cas9 mediated genome editing in super basmati rice for resistance against bacterial blight by targeting the major susceptibility gene. *Frontiers in Plant Science* 11: 575.

Zaidi, S.S.E.A., Mahfouz, M.M., and Mansoor, S. (2017). CRISPR-Cpf1: a new tool for plant genome editing. *Trends in Plant Science* 22 (7): 550–553.

Zaynab, M., Sharif, Y., Fatima, M. et al. (2020). CRISPR/Cas9 to generate plant immunity against pathogen. *Microbial Pathogenesis* 141: 103996.

Zetsche, B., Heidenreich, M., Mohanraju, P. et al. (2017). Multiplex gene editing by CRISPR–Cpf1 using a single crRNA array. *Nature Biotechnology* 35 (1): 31–34.

Zhang, X., Cheng, J., Lin, Y. et al. (2021). Editing homologous copies of an essential gene affords crop resistance against two cosmopolitan necrotrophic pathogens. *Plant Biotechnology Journal* 19 (11): 2349.

Zhou, H., Liu, B., Weeks, D.P. et al. (2014). Large chromosomal deletions and heritable small genetic changes induced by CRISPR/Cas9 in rice. *Nucleic Acids Research* 42 (17): 10903–10914.

3

Cas Variants Increased the Dimension of the CRISPR Tool Kit

Sameer Dixit[1], Akanchha Shukla[2], Mahendra Pawar[1], and Jyothilakshmi Vadassery[1]

[1] *National Institute of Plant Genome Research, New Delhi, India*
[2] *Althea DRF Lifesciences, Rouse Avenue Institutional Area, New Delhi, India*

3.1 Introduction

Genome architecture of eukaryotic organisms is highly complex and composed of billions of nucleotides (Gibbs 2020). This genomic information is the basis of every aspect of an organism's life. Therefore, the ability to modify these nucleotide bases at precisely predetermined positions or genome editing has enormous application in the field of medical biotechnology. Several techniques such as I-SceI endonuclease, Fok I endonuclease, zinc finger nuclease, and transcription activator-like effector nucleases have been identified in past, but the discovery of CRISPR has revolutionized the field due to its robust editing efficiency (Rouet et al. 1994; Kim et al. 1996; Bibikova et al. 2001; Boch et al. 2009; Sushmita et al. 2021; Upadhyay 2021). Clustered regularly interspaced short palindromic repeats (CRISPR) is an acquired immune system found in the genomes of bacteria and archaea (Adli 2018). CRISPR repeat clusters are similar to other typical tandem repeats in the genome, except they are separated by non-repeating DNA sequences known as spacers and also found adjunct to many highly conserved genes called CRISPR-associated (Cas) genes (Adli 2018).

CRISPR elements were initially identified in 1987 by Japanese researchers from Dr. Nakata's group, while working on the identification of the gene responsible for the conversion of alkaline phosphatase isozyme in *Escherichia coli,* and later the term CRISPR was coined by Spanish microbiologist, Mojica (Ishino et al. 1987; Jansen et al. 2002). Since long, the function of CRISPR-Cas system was poorly understood. CRISPR-Cas system was linked with bacteriophage resistance, due to the similarity of spacer sequence with the bacteriophage genome. This link was experimentally proven by Barrangou et al. demonstrating that new spacer sequences originating from the virus was inserted into the host genome after viral infection (Barrangou et al. 2007). Moreover, any modulation in these inserted spacer sequences changed the viral resistance phenotype of the host cell (Barrangou et al. 2007). Thereafter many key discoveries such as CRISPR RNAs (crRNAs)-guided activity of Cas enzymes, presence of protospacer-adjacent motifs (PAMs), and DNA catalytic activity of Cas9 came into light (Brouns et al. 2008; Deveau et al. 2008; Garneau

Applications of Genome Engineering in Plants, First Edition. Edited by Santosh Kumar Upadhyay.
© 2024 John Wiley & Sons Ltd. Published 2024 by John Wiley & Sons Ltd.

et al. 2010). However, from an application point of view, a major breakthrough was achieved by Sapranauskas et al. when they showed that CRISPR/Cas system from *Streptococcus thermophilus* provides immunity in *E. coli* (Sapranauskas et al. 2011). This was the first evidence that CRISPR system is transferable from one organism to another organism. Soon after in 2012, two important studies were published, which independently demonstrated that Cas9–crRNA ribonucleoprotein complex performed specific DNA cleavage for adaptive immunity in bacteria (Gasiunas et al. 2012; Jinek et al. 2012) Even Jinek et al. showed that a single chimeric RNA of tracrRNA and crRNA, called a single guide RNA (sgRNA), is sufficient to make ribonucleoprotein complex with Cas. This led to the start of an application era of CRISPR-Cas system. In 2013, three independent reports demonstrated that CRISPR-Cas system can be used for *in vivo* genome editing in eukaryotic cells (Cong et al. 2013; Jinek et al. 2013; Mali et al. 2013). Since then, continuous research has improved the dimensions and applicability of the CRISPR-Cas system.

3.2 General Architecture and Mechanism of CRISPR-Cas System

Due to the continuous evolutionary pressure by the environmental mobile genetic elements such as phages, bacteria developed various CRISPR-type immune responses for survival. Naturally CRISPR-Cas immune responses involve three main phases, i.e. adaptation, expression, and interference. During the adaptation stage, target DNA is recognized by a short motif known as a protospacer-adjacent motif (PAM) that binds with a distinct complex of Cas proteins. After binding by Cas complex, a part of the target DNA, termed as protospacer, is cleaved. Protospacer then duplicates and inserts as a spacer at the 5′ end of the CRISPR array. In the expression stage, entire CRISPR array is transcribed as a single transcript, termed called as pre-crRNA, which is further processed either by single or multiple Cas proteins or by host RNases into mature crRNA. Interference is the last stage of CRISPR-type immune responses in which the Cas-crRNA ribonucleoprotein complex recognizes and binds to the sequence of invading genetic element, which is similar to the spacer sequence and eventually gets cleaved by a Cas nuclease or other recruited nucleases.

CRISPR-Cas system has been discovered in the 40% and 90% genomes of bacteria and archaea, respectively. Mostly the frequency of CRISPR locus is one per genome but some prokaryotes have up to eight CRISPR loci in their genome. Structurally CRISPR loci consist of three key domains, i.e. CRISPR array, leader sequence, and genes encoding Cas proteins. CRISPR array comprises 21–47 nucleotides, long direct repeats that are surrounded by 20–72 nucleotides long spacer sequence (Karginov and Hannon 2010). The conservation of direct repeats varies in prokaryotes; however, the sequence of spacer is highly variable (Karginov and Hannon 2010). The main function of leader sequence present near to CRISPR array is to provide adaptation and transcription of CRISPR array during CRISPR-type immune response. Alkhnbashi et al. showed that CRISPR loci are inactive due to the lack of leader sequence during adaptation phase; however, these loci still contribute to interference phase (Alkhnbashi et al. 2016). The last part of CRISPR loci is genes encoding Cas protein, which are required for the acquisition and destruction of invading genetic material. Cas genes present as a cluster of multiple individual genes whose order and orientation vary in a different type of CRISPR systems.

3.3 Classification of CRISPR-Cas System

The lack of any conserved gene shared among all types of CRISPR-Cas systems makes it difficult for its inclusive, straightforward phylogenetic classification. Therefore, researchers used multiple computational-based approaches that mostly involved the presence and the orientation of signatures genes, organization of CRISPR loci, and sequence-based clustering for its classification. Three major attempts have been made by Makarova et al. to classify CRISPR-Cas systems (Makarova et al. 2011, 2015, 2020). According to recent classification, CRISPR-Cas systems are divided into 2 classes, 6 types, and 33 subtypes (Makarova et al. 2020) (Table 3.1).

Table 3.1 Recent classification of CRISPR-Cas systems.

Class 1					
Type	**Subtype**	**Variant**	**Native target**	**Effector complex**	**Origin**
I	I-A		DNA	*cas11(csa5)-cas7-cas5-cas8a1*	*Archaeoglobus fulgidus*
	I-B		DNA	*cas8b1-cas7-cas5*	*Clostridium kluyveri*
	I-C		DNA	*cas5-cas8c-cas7*	*Bacillus halodurans*
	I-D		DNA	*cas3″-cas10d-cas7(csc2)-cas5(csc1)*	*Cyanothece sp. 8802*
	I-E		DNA	*cas8e(cse1)-cas11(cse2)-cas7-cas5-cas6*	*Escherichia coli K12*
	I-F	F1	DNA	*cas8f1(csy1)-cas5f1(csy2)-cas7f1(csy3)-cas6f*	*Yersinia pseudo-tuberculosis*
		F2	DNA	*cas7f2(PBPRB1993)-cas5f2(PBPRB1992)-cas6f*	*Shewanella putrefaciens CN-32*
		F3		*cas8f3/cas5f3-cas7f3-cas6f*	*Vibrio crassostreae J5 20*
	I-G		DNA	*cas8u2-cas7-cas5-cas6*	*Geobacter sulfurreducens*
III	III-A		DNA/RNA	*cas10-cas11(csm2)-cas7(csm3)-cas5(csm4)-cas7(csm5)*	*Staphylococcus epidermidis*
	III-B		DNA/RNA	*cas7(cmr1)-cas10-cas5(cmr3)—cas7(cmr4)-cas11(cmr5)-cas7(cmr6)*	*Pyrococcus furiosus*
	III-C		DNA/RNA	*cas7(cmr1)-cas7-cas10-cas7(cmr4)-cas11(cmr5)-cas5(cmr3)*	*Methanothermobacter thermautotrophicus*
	III-D		RNA?	*cas10- cas7-cas5(csx10)-cas11-cas7-cas7-csx19-cas7*	*Synechocystis sp. 6803*
	III-E		RNA?	*cas7-cas7-cas7-cas11*	*Candidatus Scalindua brodae*
	III-F		DNA?	*cas10-cas5-cas11-cas7*	*Thermotoga lettingae TMO*

Table 3.1 (Continued)

Type	Subtype	Variant	Native target	Effector complex	Origin
IV	IV-A			*cas8like(csf1)-cas7 (csf2)-cas5(csf3)*	*Thioalkalivibrio sp. K90mix*
	IV-B			*cas8like(csf1)-cas11-cas7(csf2)-cas5(csf3)*	*Rhodococcus jostii RHA1*
	IV-C		DNA?	*LS-cas11-cas7(csf2)-cas5*	*Thermoflexia bacterium*
				Class 2	
II	II-A		DNA	*cas9*	*Streptococcus thermophilus*
	II-B		DNA	*cas9*	*Legionella pneumophila str. Paris*
	II-C	C1	DNA	*cas9*	*Neisseria lactamica 020-06*
		C2	DNA	*cas9*	*Micrarchaeum acidiphilum ARMAN-1*
V	V-A		DNA	*cas12a*	*Francisella cf. Novicida Fx1*
	V-B	B1	DNA	*cas12b1*	*Alicyclobacillus acidoterrestris*
		B2	DNA	*cas12b2*	*Planctomycetes bacterium RGB_13_46_10*
	V-C		DNA	*cas12c*	*Oleiphilus sp.*
	V-D		DNA	*cas12d*	*Bacterium CG09_39_24*
	V-E		DNA	*cas12e*	*Deltaproteobacteria bacterium*
	V-F	F1	DNA	*cas12f1(cas14a)*	*Uncultured archaeon*
		F1(U3)	DNA	*c2c10*	*Bacillus thuringiensis HD-771*
		F2	DNA	*cas12f2(cas14b)*	*Uncultured archaeon*
		F3		*cas12f3(cas14c)*	*Candidatus Micrarchaeota archaeon*
		U1		*c2c4*	*Gordonia otitidis*
		U2		*c2c8*	*Cyanothece sp. PCC 8801*
		U4		*c2c9*	*Rothia dentrocariosa M567*
	V-G		RNA	*cas12g*	*Hot springs metagenome*
	V-H			*cas12h*	*Hypersaline lake sediment metagenome (JGI)*
	V-I		DNA	*cas12i*	*Freshwater metagenome (JGI)*
	V-K	U5		*cas12k*	*Cyanothece sp. PCC 8801*
VI	VI-A		RNA	*cas13a*	*Leptotrichia shahii*
	VI-B	B1	RNA	*cas13b1*	*Prevotella buccae*
		B2	RNA	*cas13b2*	*Bergeyella zoohelcum*
	VI-C		RNA?	*cas13c*	*Fusobacterium perfoetens*
	VI-D		RNA	*cas13d*	*Ruminococcus bicirculans*

Source: Adapted from (Makarova et al. 2020).

3.3.1 Class 1 CRISPR-Cas System

Class 1 CRISPR-Cas systems are characterized by the occurrence of multiple subunits in crRNA-effector complex. This class is subdivided into three major types I, III, and IV.

Type I system is further subdivided into seven subtypes from I-A to I-G. Typically in most subtypes of type I, the whole CRISPR-Cas system is encoded by a single operon except type I-A and I-B, which have two or more operons. All subtypes of type I CRISPR-Cas system consist of either Cas3 or its variant Cas3' gene in its effector complex; this encodes a helicase that is required to unzip double-stranded DNA or RNA–DNA duplexes. Apart from helicase domain, Cas3 proteins also contain N-terminal HD family endonuclease domain for the cleavage of target DNA. In subtype I-G (formerly known as I-U) and some subtype I-A systems, this domain present at the carboxyl terminus. In some systems, like I-F, Cas3 is also fused with Cas2. Subtype I-B consist of gene arrangement, i.e. *cas1-cas2-cas3-cas4-cas5-cas6-cas7-cas8* and considered as an ancestral type I system. Subtypes I-A and I-C have also preserved same gene arrangement with rearranged order. Subtype I-D is composed of a distinct variant of Cas3 and has Cas10d in the place of Cas8. Cas8 is the most diverged protein in all subtypes of type 1 system.

Type III system has more compositional and organizational diversity than type I systems. Type III system was earlier subdivided in to four subtypes (subtypes III A–D), but recently two more subtypes, i.e. III-E and III-F have been discovered from *Candidatus Scalindua brodae* and *Thermotoga lettingae TMO,* respectively. Some subtypes (III-A, -B, -C) of type III contains reverse transcriptase in the adaptation module, thus it has ability to co-transcriptionally target DNA and RNA. Phylogenetically, III-A, III-D, III-E, and III-F are closer to each other as they consist of *csm* genes in their effector module, whereas III-D and III-E make a separate clade and have *cmr* genes. All types III systems (except III-E) contain extensively diverse *cas10* as a signature gene, which translates the largest protein of this system. Cas10 protein is a multidomain protein consisting of Palm domain, a variant of the RNA recognition motif, along with HD family nuclease domain (distinct from type I Cas3 HD domains I). Cas10 is expected to have cyclases and polymerases activity. Recently identified type III-E system has more resemblance with class 2 systems as its effector module is constructed with single multidomain protein that consists of fusion of multiple Cas7 proteins and a putative Cas11 (Csm2-like) for crRNA binding. However, due to the sequence analysis, domain composition and orientation positioned it inside type III of class 1. Further effector module of type III has conserved aspartate residues, which is involved in RNA cleavage in its homolog Csm4. Thus, III-E has predicted pre-crRNA and RNA cleavage activity. Subtype III-E system also comprise putative ancillary gene which translated caspase family protease. These proteases usually performed caspase-mediated cell death by using CHAT domain that fuses with tetratricopeptide repeats. The presence of these ancillary genes suggests the potential involvement of type III-E system in multifaceted defense pathways. Subtype III-F recently classified as a separate subtype and have Cas7 and Cas5 subunits in effector module like other type III systems. In contrast, III-F only possess single Cas7-like protein encoding gene. Cas10 conserves all catalytic residues of HD-domain but acquired by amino acid substitutions in the catalytic site of cyclase or polymerase domain.

As per recent Classification, **Type IV** has 3 subtypes (IV-A, -B, -C) and mostly discovered in the plasmids of several bacteria. Type IV systems are not able to cleave invading foreign genomes as they lack Cas proteins in adaptation module. Pre-crRNA processing is performed by distinct Cas6 protein, while effector module consists of Cas5, Cas7, and Csf1 proteins. This system is mostly present in thermophilic microorganisms.

3.3.2 Class 2 CRISPR-Cas System

Single large multidomain protein in the effector complexes is the distinguishable feature of Class 2 CRISPR-Cas systems. Recent classification shows a burst in class 2 CRISPR-Cas systems, earlier including only 2 types and 4 subtypes is now expanded to 3 types (types II, V, and VI) and 17 subtypes.

Type II system is very simple in gene architecture and includes three subtypes, i.e. II-A, II-B, and II-C. All type II systems have signature *cas9* gene, encoding large multidomain protein that works in adaptation as well as effector module, including the cleavage of foreign DNA. Cas9 has characteristic Arg-rich cluster along with two nuclease domains. Along with *cas9,* all characterized type II systems also contain *cas1* and *cas2* genes. However, subtype II-A also consists of *csn2* that is replaced by *cas4* in subtype II-B and subtype II-C (Only in C1 variant). Some type II loci also transcribed a tracrRNAs, which shares partial complementation to the repeats of same CRISPR array. Subtype II-C system is the most frequent type II CRISPR-Cas system found in bacteria.

Type V system emerges as a highly expanded system in recent classification and consists of 10 subtypes (V-A to V-K) including several variants. In type V system, Cas12 (formerly known as Cpf1) is present in effector complex. Cas12 is fundamentally differ with Cas9 by the domain architecture. In contrast of two nuclease domains of Cas9, which individually sliced one strand of the target DNA, Cas12 has RuvC-like domain for double stand DNA cut. Mostly type V system targets DNA molecule; however, some subtypes of type V, i.e. type V-G identified in *Hot springs metagenome,* also able to target RNA molecule. Most type V system also consist of genes, which is either partially (V-C and V-D) or completely missing (V-I, V-H, and V-G) in some subtypes. Subtype V-K has *tnsB, tnsC, and tniQ* in the place of *cas1, cas2, and cas4* genes.

Type VI is considered as a successor of HEPN domain containing toxin–antitoxin modules, as the signature effector protein of this system, Cas13, consists of two individual HEPN domains. However, it is very interesting that N-terminal HEPN domains of Cas13 are phylogenetically distinct from C-terminal HEPN domains of Cas13 suggesting its evolution from two common ancestors. Currently type VI system is less diverse than other systems, but the possibility for the discovery of new subtypes remains open. Due to the presence of HERN RNase domain protein in effector complex, type VI exclusively cleaved invading foreign RNA.

3.4 Different Application-Based CRISPR-Cas System

Different CRISPR-Cas systems have distinct properties that allow them to target different types of substrates (ssRNA, ssDNA, and dsDNA) and have effector proteins of varying sizes. Factors such as target sequence length, cleavage pattern, PAM/PFS sequences, and component of the guide RNA differ between systems, making each system uniquely suited for specific applications.

3.4.1 Cas9

Cas9 was first discovered from *Streptococcus pyogenes* bacteria that utilized the CRISPR-Cas9 system against foreign bacteriophages. The CRISPR-Cas9 system consists of a four-component system that includes small molecules crRNA and tracrRNA (Deltcheva et al. 2011; Barrangou 2015). Emmanuelle Charpentier and Jennifer Doudna further modified the CRISPR-Cas9 system by combining crRNA and tracrRNA to form guide RNA that could cut the DNA at specified target with the help of Cas9. This breakthrough was awarded Nobel prize in chemistry in 2020 (Jinek et al. 2012). Cas9 has different variants: eSpCas9 (1.1)7, SpCas9-HF1, HypaCas9, evoCas9, Sniper-Cas9, VQR12, VRER12, VRQR8, QQR1, xCas9, SpCas9-NG, and nCas9 (Kim et al. 2020; Alok et al. 2021; Lee et al. 2023). The CRISPR-Cas9 system has an effector protein size in the range of 2.9–4.9 kb and targets dsDNA substrates and requires a 3′-GG PAM sequence. The target sequence length is 20–24 nt and the cleavage pattern is a staggered 5′-overhang. Cas9 has two nuclease domains, HNH and RuvC, which cleaves the target and nontarget DNA strands, respectively. The modified CRISPR-Cas9 system is successfully used as genome editing tool in different organisms including *Danio rerio* (Hwang et al. 2013), *Drosophila melanogaster* (Gratz et al. 2013), *Caenorhabditis elegans* (Friedland et al. 2013), *Arabidopsis thaliana* (Jiang et al. 2013), mice (Wang et al. 2013), and monkeys (Guo and Li 2015).

3.4.2 Cas12

Cas12 was characterized from *Francisella novicida* (Zetsche et al. 2015). Cas12 has different variants naming sequentially from Cas12a to Cas12f and are found in different archaea and bacteria (Paul and Montoya 2020). CRISPR-Cas12 system targets dsDNA substrates with an effector protein size of 3.6–3.9 kb. Cas12 requires only crRNA for targeting as opposed to Cas9 that requires both crRNA and tracrRNA. Cas12 requires a 5′-TT(T) PFS sequence with target sequence length of 20–24 nucleotides. The cleavage pattern is a staggered 5′-overhang with collateral ssDNA cleavage. Cas12 poses a single RuvC domain with endoribonuclease activity for DNA cleavage. CRISPR-Cas12 system has found applications in multiplexed genome editing, gene regulation, and diagnostics (Kim et al. 2017).

3.4.3 Cas14

Cas14 was discovered from Jennifer Doudna's lab. Cas14 is found in the superphylum of extremophile archaea. Cas14 is present only in archaea, and thus it is considered to be elementary as compared to the other *cas* genes from bacteria (Aquino-Jarquin 2019). There are 24 variants of cas14 gene reported so far and they are classified into three subgroups: cas14a, cas14b, and cas14c. The effector protein used by the CRISPR-Cas14 system has a size range of 1.2–2.1 kb and targets ssDNA substrates. It needs a 5′-TT(T) PFS sequence and uses crRNA as guide RNA. The length of the target sequence is 20–25 nucleotides, and the pattern of cleavage is gRNA-dependent ssDNA and dsDNA cuts with collateral ssDNA cleavage (Savage 2019).

3.4.4 Cas13

Cas13 was discovered from *Leptotrichia shahii* bacteria in 2016. CRISPR-Cas13 system is a single component system that includes a small molecule, crRNA. Cas13 is an RNA-guided RNA endonuclease, and hence, unlike the other CRIPSR-Cas systems that target DNA, Cas13 cleaves single-stranded RNA (Abudayyeh et al. 2016). Cas13 family consists of different subtypes including Cas13a, Cas13b, Cas13c and Cas13d etc. Cas13 system consists of an effector protein size of 4.1 kb requiring a 3′-H PAM sequence with target sequence of 20 nucleotides, and cleavage pattern is gRNA-dependent RNA cut with collateral RNA cleavage (Gootenberg et al. 2017). Cas13 contains two higher eukaryote and prokaryote nucleotide-binding (HEPN) RNase domains for target cleavage. Mutations in these domains result in inactive Cas13 protein. CRISPR-Cas13 system is used as a tool for RNA knockdown and modification in mammalian cells (Moon et al. 2019).

3.4.5 Cas3

Cas3 was initially identified as COG1203 from an *in silico* analysis of different thermophilic Archaea and bacteria in 2002 (Makarova et al. 2002). Cas3 has seven variants: 1-A, 1-B, 1-C, 1-D, 1-E, 1-F, and 1-U (Makarova et al. 2018). CRISPR-Cas3 system requires protein cascade for its activity. Cas3 forms complex with cascade and targets ssDNA with help of crRNA. Cas3 uses HD-nuclease domain along with two RecA-like domains for ATP-dependent target degradation (Sinkunas et al. 2011). CRSIPR-Cas3 system is re-engineered to generate large deletions in chromosomes up to 30–100 kb in human cells (Makarova et al. 2002; Dolan et al. 2019; Morisaka et al. 2019.)

3.5 Advancement and Reengineering of CRISPR-Cas System

The CRISPR-Cas system is revolutionary in the field of genetic engineering. It is now easily accessible to users, relatively high efficiency, simplicity of use, compatibility with genetic screening, having all these merits allow its universal adaptability including basic science and commercial applications (Shivram et al. 2021). However, like any other technology, it also has several shortcomings that need to be addressed for its full-fledged use. Currently researchers following two directions:1) discovery of new Cas proteins, 2) reengineering of preestablished CRISPR-Cas9 system to increase the dimension of CRISPR-Cas technology (Moon et al. 2019). One of the major concerns with the CRISPR-Cas system is off-target effects. Several modifications have been done in traditional CRISPR-spCas9 system to overcome this problem, such as the development of mutated spCas9, i.e. dCas9, nickase Cas9, for precise gene editing (Komor et al. 2016; Moon et al. 2019; Bhalothia et al. 2020; Sushmita et al. 2021). Additionally, the selection of specific delivery method further reduces the off target, e.g. local delivery of sgRNA-Cas9 as a ribonucleotide protein complex provides transient Cas9 activity, which resulted fewer off targets. Reengineering of sgRNA further adds another layer of specific Cas activity, recently researcher shows that addition of cytosine stretches to the 5′-end of conventional single-guide RNAs constrained genome-editing activity of Cas9 (Kawamata et al. 2023). In addition to all, development of fusion

Table 3.2 Different modes of nucleotide modification using CRISPR system.

Modifications	Available CRISPR-Cas systems	Mechanism	Advantage
Double strand breakage	Cas9, Cas12a, Cas12b, Cas12e, Cas14	DSB repair pathway	Gene knockout Gene correction
DNA base modifier	Cas9- Deaminase, Cas12a- Deaminase,	Single nucleotide substitution	C > T conversion A > G conversion
RNA base modifier	Cas13a- Deaminase	Single nucleotide substitution	C > U conversion A > G conversion
DNA integration	Cas12k + TnsB/TnsC/TniQ	Transposon based integration	Site-specific integration
Transcriptional regulation	Cas9 + Repressor/Activator	Promoter binding	Gene regulation
Epigenetic regulation	Cas9 + TET/ DNMT	DNA methylation	Gene regulation
Histone modification	Cas9 + HDM/HMT/ HDAC/HAT	Histone modification	Gene regulation

Cas protein provides completely new dimension to CRISPR technology, fusion of uracil glycosylase inhibitor and APOBEC1 enzyme to nickase Cas9 specifically converts cytosines into uracil (U or T) between the 4th and 8th bases of the nonbinding strand of the sgRNA without requiring an exogenous DNA template or causing double-strand DNA breaks, this lead to the invention of CRISPR-STOP technology by introducing early STOP codons in target genes (Kuscu et al. 2017).

Conventional CRISPR-Cas9 system can only target certain types of DNA sequences, which limits the CRISPR applicability. To overcome this problem, scientists discover new variant of Cas proteins such as Cas13 that can also bind to RNA molecule. Currently, CRISPR technology is not only limited to develop gene KO line but efficiently perform DNA integration, base modification, gene regulation and many more (Table 3.2). Further scientists are also exploring new delivery methods, such as viral vectors, nanoparticles, to improve the efficiency and specificity of this technology.

Overall, the CRISPR-Cas system has the potential to revolutionize medicine, agriculture, and biotechnology. Ongoing research and development will continue to improve and expand the capabilities of this powerful tool.

3.6 Conclusions

CRISPR-Cas gene editing technology is a breakthrough in molecular biology that provides a flexible approach for genome alteration/regulation and transcriptional regulation. Although only a few types of CRISPR-Cas systems are being extensively used, constant discovery and development of CRISPR systems from prokaryotic species is continuously expanding the potential of this technology to revolutionize medicine, agriculture, and biotechnology.

CRISPR-Cas technology has extensive prospects. However, non-specific delivery, leading to off-target changes to the genome sequences, are the major setbacks of this technology that may cause immediate or delayed adverse effects. Thus, improved methods to develop safe, stable, efficient systems, with broader adaptations will be important for future advancement.

Acknowledgments

We wish to apologize to all colleagues whose work, because of lack of space, could not be cited. S.D. is also thankful to DBT, India for receiving M.K.-Bhan YRFP grant. We thank to all my colleagues for helpful discussions.

References

Abudayyeh, O.O., Gootenberg, J.S., Konermann, S. et al. (2016). C2c2 is a single-component programmable RNA-guided RNA-targeting CRISPR effector. *Science* 353 (6299): aaf5573.

Adli, M. (2018). The CRISPR tool kit for genome editing and beyond. *Nature Communications* 9 (1): 1911.

Alkhnbashi, O.S., Shah, S.A., Garrett, R.A. et al. (2016). Characterizing leader sequences of CRISPR loci. *Bioinformatics* 32 (17): i576–i585.

Alok, A., Chauhan, H., Upadhyay, S.K. et al. (2021). Compendium of plant-specific CRISPR vectors and their technical advantages. *Life* 11: 1021. https://doi.org/10.3390/life11101021.

Aquino-Jarquin, G. (2019). CRISPR-Cas14 is now part of the artillery for gene editing and molecular diagnostic. *Nanomedicine: Nanotechnology, Biology and Medicine* 18: 428–431.

Barrangou, R. (2015). Diversity of CRISPR-Cas immune systems and molecular machines. *Genome Biology* 16 (1): 1.

Barrangou, R., Fremaux, C., Deveau, H. et al. (2007). CRISPR provides acquired resistance against viruses in prokaryotes. *Science* 315 (5819): 1709–1712.

Bhalothia, P., Yajnik, K., Alok, A., and Upadhyay, S.K. (2020). The current progress of CRISPR/Cas9 development in plants. *Elsevier*. https://doi.org/10.1016/B978-0-12-818140-9.00011-8.

Bibikova, M., Carroll, D., Segal, D.J. et al. (2001). Stimulation of homologous recombination through targeted cleavage by chimeric nucleases. *Molecular and Cellular Biology* 21 (1): 289–297.

Boch, J., Scholze, H., Schornack, S. et al. (2009). Breaking the code of DNA binding specificity of TAL-type III effectors. *Science* 326 (5959): 1509–1512.

Brouns, S.J., Jore, M.M., Lundgren, M. et al. (2008). Small CRISPR RNAs guide antiviral defense in prokaryotes. *Science* 321 (5891): 960–964.

Cong, L., Ran, F.A., Cox, D. et al. (2013). Multiplex genome engineering using CRISPR/Cas systems. *Science* 339 (6121): 819–823.

Deltcheva, E., Chylinski, K., Sharma, C.M. et al. (2011). CRISPR RNA maturation by trans-encoded small RNA and host factor RNase III. *Nature* 471 (7340): 602–607.

Deveau, H., Barrangou, R., Garneau, J.E. et al. (2008). Phage response to CRISPR-encoded resistance in *Streptococcus thermophilus*. *Journal of Bacteriology* 190 (4): 1390–1400.

Dolan, A.E., Hou, Z., Xiao, Y. et al. (2019). Introducing a spectrum of long-range genomic deletions in human embryonic stem cells using type I CRISPR-Cas. *Molecular Cell* 74 (5): 936–950. e5.

Friedland, A.E., Tzur, Y.B., Esvelt, K.M. et al. (2013). Heritable genome editing in *C. elegans* via a CRISPR-Cas9 system. *Nature Methods* 10 (8): 741–743. https://doi.org/10.1038/nmeth.2532.

Garneau, J.E., Dupuis, M.È., Villion, M. et al. (2010). The CRISPR/Cas bacterial immune system cleaves bacteriophage and plasmid DNA. *Nature* 468 (7320): 67–71.

Gasiunas, G., Barrangou, R., Horvath, P., and Siksnys, V. (2012). Cas9–crRNA ribonucleoprotein complex mediates specific DNA cleavage for adaptive immunity in bacteria. *Proceedings of the National Academy of Sciences* 109 (39): E2579–E2586.

Gibbs, R.A. (2020). The human genome project changed everything. *Nature Reviews Genetics* 21 (10): 575–576.

Gootenberg, J.S., Abudayyeh, O.O., Lee, J.W. et al. (2017). Nucleic acid detection with CRISPR-Cas13a/C2c2. *Science* 356 (6336): 438–442.

Gratz, S.J., Cummings, A.M., Nguyen, J.N. et al. (2013). Genome engineering of *Drosophila* with the CRISPR RNA-guided Cas9 nuclease. *Genetics* 194 (4): 1029–1035.

Guo, X. and Li, X.-J. (2015). Targeted genome editing in primate embryos. *Cell Research* 25 (7): 767–768. https://doi.org/10.1038/cr.2015.64.

Hwang, W.Y., Fu, Y., Reyon, D. et al. (2013). Efficient genome editing in zebrafish using a CRISPR-Cas system. *Nature Biotechnology* 31 (3): 227–229. https://doi.org/10.1038/nbt.2501.

Ishino, Y., Shinagawa, H., Makino, K. et al. (1987). Nucleotide sequence of the iap gene, responsible for alkaline phosphatase isozyme conversion in *Escherichia coli*, and identification of the gene product. *Journal of Bacteriology* 169 (12): 5429–5433.

Jansen, R., Embden, J.D.V., Gaastra, W., and Schouls, L.M. (2002). Identification of genes that are associated with DNA repeats in prokaryotes. *Molecular Microbiology* 43 (6): 1565–1575.

Jiang, W., Zhou, H., Bi, H. et al. (2013). Demonstration of CRISPR/Cas9/sgRNA-mediated targeted gene modification in *Arabidopsis*, tobacco, sorghum and rice. *Nucleic Acids Research* 41 (20): e188–e188. https://doi.org/10.1093/nar/gkt780.

Jinek, M., Chylinski, K., Fonfara, I. et al. (2012). A programmable dual-RNA–guided DNA endonuclease in adaptive bacterial immunity. *Science* 337 (6096): 816–821.

Jinek, M., East, A., Cheng, A. et al. (2013). RNA-programmed genome editing in human cells. *eLife* 2: e00471.

Karginov, F.V. and Hannon, G.J. (2010). The CRISPR system: small RNA-guided defense in bacteria and archaea. *Molecular Cell* 37 (1): 7–19.

Kawamata, M., Suzuki, H.I., Kimura, R., and Suzuki, A. (2023). Optimization of Cas9 activity through the addition of cytosine extensions to single-guide RNAs. *Nature Biomedical Engineering* 1–20.

Kim, Y.G., Cha, J., and Chandrasegaran, S. (1996). Hybrid restriction enzymes: zinc finger fusions to Fok I cleavage domain. *Proceedings of the National Academy of Sciences* 93 (3): 1156–1160.

Kim, H., Kim, S.-T., Ryu, J. et al. (2017). CRISPR/Cpf1-mediated DNA-free plant genome editing. *Nature Communications* 8 (1): 14406.

Kim, N., Kim, H.K., Lee, S. et al. (2020). Prediction of the sequence-specific cleavage activity of Cas9 variants. *Nature Biotechnology* 38 (11): 1328–1336.

Komor, A.C., Kim, Y.B., Packer, M.S. et al. (2016). Programmable editing of a target base in genomic DNA without double-stranded DNA cleavage. *Nature* 533 (7603): 420–424.

Kuscu, C., Parlak, M., Tufan, T. et al. (2017). CRISPR-STOP: gene silencing through base-editing-induced nonsense mutations. *Nature Methods* 14 (7): 710–712.

Lee, J., Lim, K., Kim, A. et al. (2023). Prime editing with genuine Cas9 nickases minimizes unwanted indels. *Nature Communications* 14 (1).

Makarova, K.S., Aravind, L., Grishin, N.V. et al. (2002). A DNA repair system specific for thermophilic archaea and bacteria predicted by genomic context analysis. *Nucleic Acids Research* 30 (2): 482–496.

Makarova, K.S., Haft, D.H., Barrangou, R. et al. (2011). Evolution and classification of the CRISPR–Cas systems. *Nature Reviews Microbiology* 9 (6): 467–477.

Makarova, K.S., Wolf, Y.I., Alkhnbashi, O.S. et al. (2015). An updated evolutionary classification of CRISPR–Cas systems. *Nature Reviews Microbiology* 13 (11): 722–736.

Makarova, K.S., Wolf, Y.I., and Koonin, E.V. (2018). Classification and nomenclature of CRISPR-Cas systems: where from here? *The CRISPR Journal* 1 (5): 325–336.

Makarova, K.S., Wolf, Y.I., Iranzo, J. et al. (2020). Evolutionary classification of CRISPR–Cas systems: a burst of class 2 and derived variants. *Nature Reviews Microbiology* 18 (2): 67–83.

Mali, P., Yang, L., Esvelt, K.M. et al. (2013). RNA-guided human genome engineering via Cas9. *Science* 339 (6121): 823–826.

Moon, S.B., Kim, D.Y., Ko, J.H., and Kim, Y.S. (2019). Recent advances in the CRISPR genome editing tool set. *Experimental & Molecular Medicine* 51 (11): 1–11.

Morisaka, H., Yoshimi, K., Okuzaki, Y. et al. (2019). CRISPR-Cas3 induces broad and unidirectional genome editing in human cells. *Nature Communications* 10 (1): 5302.

Paul, B. and Montoya, G. (2020). CRISPR-Cas12a: functional overview and applications. *Biomedical Journal* 43 (1): 8–17.

Rouet, P., Smih, F., and Jasin, M. (1994). Introduction of double-strand breaks into the genome of mouse cells by expression of a rare-cutting endonuclease. *Molecular and Cellular Biology* 14 (12): 8096–8106.

Sapranauskas, R., Gasiunas, G., Fremaux, C. et al. (2011). The *Streptococcus thermophilus* CRISPR/Cas system provides immunity in *Escherichia coli*. *Nucleic Acids Research* 39 (21): 9275–9282.

Savage, D.F. (2019). Cas14: big advances from small CRISPR proteins. *Biochemistry* 58 (8): 1024–1025.

Shivram, H., Cress, B.F., Knott, G.J., and Doudna, J.A. (2021). Controlling and enhancing CRISPR systems. *Nature Chemical Biology* 17 (1): 10–19.

Sinkunas, T., Gasiunas, G., Fremaux, C. et al. (2011). Cas3 is a single-stranded DNA nuclease and ATP-dependent helicase in the CRISPR/Cas immune system. *The EMBO Journal* 30 (7): 1335–1342.

Sushmita, Kaur, G., Upadhyay, S.K., and Verma, P.C. (2021). An overview of genome-engineering methods. *Wiley*. https://doi.org/10.1002/9781119672425.ch14.

Upadhyay, S.K. (2021). Genome engineering for crop improvement. Wiley. https://doi.org/10.1002/9781119672425.

Wang, H., Yang, H., Shivalila, C.S. et al. (2013). One-step generation of mice carrying mutations in multiple genes by CRISPR/Cas-mediated genome engineering. *Cell* 153 (4): 910–918.

Zetsche, B., Gootenberg, S.J., Abudayyeh, O.O. et al. (2015). Cpf1 is a single RNA-guided endonuclease of a class 2 CRISPR-Cas system. *Cell* 163 (3): 759–771. https://doi.org/10.1016/j.cell.2015.09.038.

4

Advancement in Delivery Systems and Vector Selection for CRISPR/Cas-Mediated Genome Editing in Plants

Sanskriti Vats[1,2], Sukhmandeep Kaur[1,3], Amit Chauhan[4,5], Dipul Kumar Biswas[6], and Rupesh Deshmukh[7]

[1] *Agri-Biotechnology Department, National Agri-Food Biotechnology Institute (NABI), Mohali, Punjab, India*
[2] *Regional Centre for Biotechnology, Faridabad, Haryana (NCR Delhi), India*
[3] *Department of Biotechnology, Panjab University, Chandigarh, Panjab, India*
[4] *Institute of Biochemistry and Biology, University of Potsdam, Potsdam-Golm, Germany*
[5] *Max Planck Institute of Molecular Plant Physiology, Potsdam-Golm, Germany*
[6] *Department of Biology, Faculty of Arts and Science, Bolu Abant İzzet Baysal Üniversitesi, Gölköy, Turkey*
[7] *Department of Biotechnology, Central University of Haryana, Mahendragarh, Haryana, India*

4.1 Introduction

Currently we face a global challenge of producing enough food for the growing population with unstable climatic conditions. The investment in sequencing, assembling, and annotating plant genome has resulted in the discovery of novel genes and associated functions. This generated information can be used for genome editing is a set of techniques that provide scientists the ability to modify the genetic material in a controlled manner. These gene editing technologies can deliberately insert, remove, or modify DNA in living cells. Zinc finger nucleases (ZFNs) and transcriptional activator-like effector nucleases (TALENs) were developed first, leveraging the DNA binding domains of proteins to generate targeted double-strand breaks (DSBs) that encourage DNA repair (Gaj et al. 2013; Manghwar et al. 2019). Since 2002–2011, although ZFNs and TALENs have been widely used for GE in humans, animals, and plant cells, there are still some limitations that prevent their effective use (Tang et al. 2017; Manghwar et al. 2019). CRISPR/Cas technique on the other hand is facile, easy to apply, and cheap with higher efficiencies. CRISPRs in bacterial DNA were identified in 1987 (Ishino et al. 1987; Vats et al. 2019; Upadhyay 2021). Nevertheless, their role in bacteria was not understood until 2005 (Ishino et al. 1987; Bolotin et al. 2005), and in 2007, CRISPRs were proven to confer immunity in combination with Cas proteins (Barrangou et al. 2007; Vats et al. 2019). Doudna and colleagues were the first group to carry out gene editing using CRISPR/Cas system (Jinek et al. 2012; Waddington et al. 2016), and shortly thereafter many independent research groups edited genes in animal and plant systems (Waddington et al. 2016; Vats et al. 2019; Singh et al. 2022).

Applications of Genome Engineering in Plants, First Edition. Edited by Santosh Kumar Upadhyay.
© 2024 John Wiley & Sons Ltd. Published 2024 by John Wiley & Sons Ltd.

The designed single guide RNA (sgRNA) guides the Cas protein through its 5′ CRISPR RNA (crRNA) base pairing component that creates DSBs 3 base pairs upstream of the protospacer adjacent motifs (PAMs) (Ceasar et al. 2016) (Figure 4.1). Among the different Cas orthologs and paralogs available, Cas9 from *Streptococcus pyogenes* is the most widely used endonuclease that recognizes the 5′-NGG-3′ PAM sequence (N can be any nucleotide base) (Asmamaw and Zawdie 2021). Precise genomic modifications have been achieved by the development of multiple CRISPR-based base and prime editors targeting various traits in vital crops (Adli 2018; Haque et al. 2018; Zhu et al. 2020; Alok et al. 2021). However, their widespread application is limited to crops with efficient transformation protocols, making the transfer of editing reagents to target cells a bottleneck.

This chapter deals with technological advances in the reagent delivery pipelines, vector selection, stable transformation, and single cell regeneration. Regeneration is the major bottleneck here, even more than stable integration of foreign DNA into plant cells. Genome editing has resulted in an increased demand for plant transformation. The reagents can either be transiently delivered into protoplasts or stably transformed as a transgene. Here, primary challenges are long tissue culture times, low efficiency, and low titres of DNA repair template, required for homology directed repair, delivered by *Agrobacterium*-mediated transformation. Each challenge offers a possibility for a suite of engineering enhancements that are discussed here.

4.2 Advancement in Delivery Systems and Vector Selection for CRISPR/Cas-Mediated Genome Editing in Plants

4.2.1 Vector Selection Based on Application and Availability in Plants

First of all CRISPR/Cas approach was established in human cells and thereafter emergence of this excellent tool into plant systems (Mali et al. 2013). Although this approach was tested successfully in plant system but possess comparatively lower range of genome editing efficiency (Shan et al. 2013; Ma et al. 2016). After this, several attempts were carried out for development of good vector system, which ensure good genome editing in plants in terms of knockout, knockin, knockdown, and prompter alterations. All these events can have positive attributes to plants in terms of quality, quantity, and tolerance to stresses (Ma et al. 2016). As we know with the advent in plants, biotechnology makes the possibility of emergence of new genome editing tools, which leads to precise genome editing in plants and becomes better replacement for the traditional approaches like gamma radiation and EMS mutagenesis. *Agrobacterium*-mediated or biolistic mode of plant transformation are the two most common plant genome editing approaches that facilitate the entry of CRISPR/Cas 9 into plant system (Razzaq et al. 2019). These advent genetic CRISPR tool and vectors are originated or evolved from IIA CRISPR/Cas derived from bacteria *Streptococcus pyogenes* (Doudna and Charpentier 2014). These new emerging CRISPR tools and vectors pave the way for green revolution through alteration in the genome and will be helpful to tailor-made new cultivars having noval traits. In future, progress in plant transformations can make its applicability in wider range of plant species (Altpeter et al. 2016) (Figure 4.2).

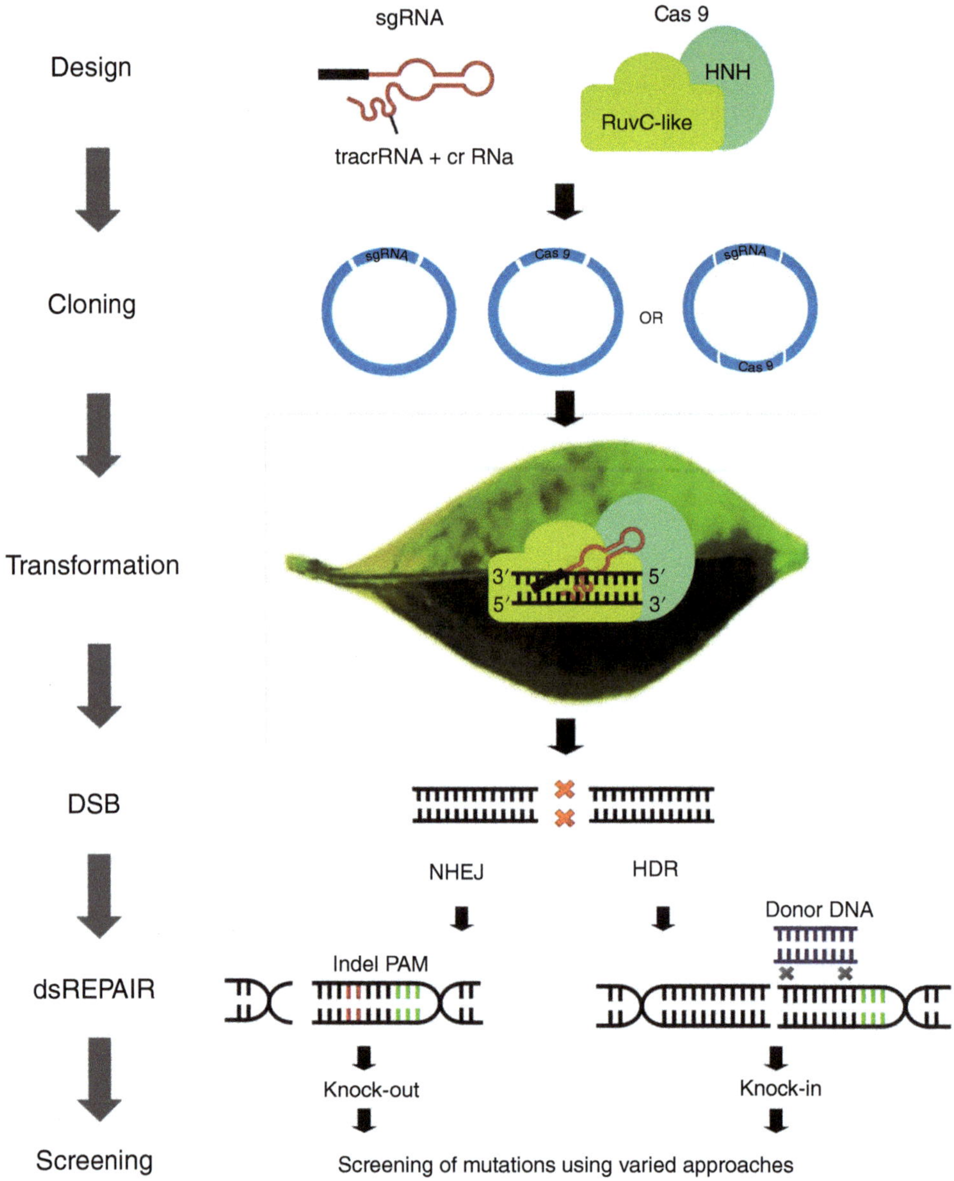

Figure 4.1 Schematic representation of the genome editing process in plants through CRISPR/Cas gene editing tools. First, the sgRNA includes a spacer (black), crRNA, and tracrRNA (both shown here in red), and Cas9 is composed of two domains: the HNH and the RuvC-like domain. HNH domain cuts the DNA strand complementary to the sgRNA, and RuvC-like domain cleaves the other DNA strand. Coding sequences of Cas and sgRNA are cloned together or individually into a vector (blue), which is transformed into plant cells. The sgRNA and Cas9 are expressed in the plant, which then lead to DSB, resulting in the activation of DNA repair machinery leading to the modification of DNA sequence and subsequently in the protein coded by sequences and conclusively in the phenotype. The final step is the screening of mutations, which is usually done by PCR and sequencing. *Source:* Sanskriti Vats et al. 2019/MDPI/CC BY 4.0.

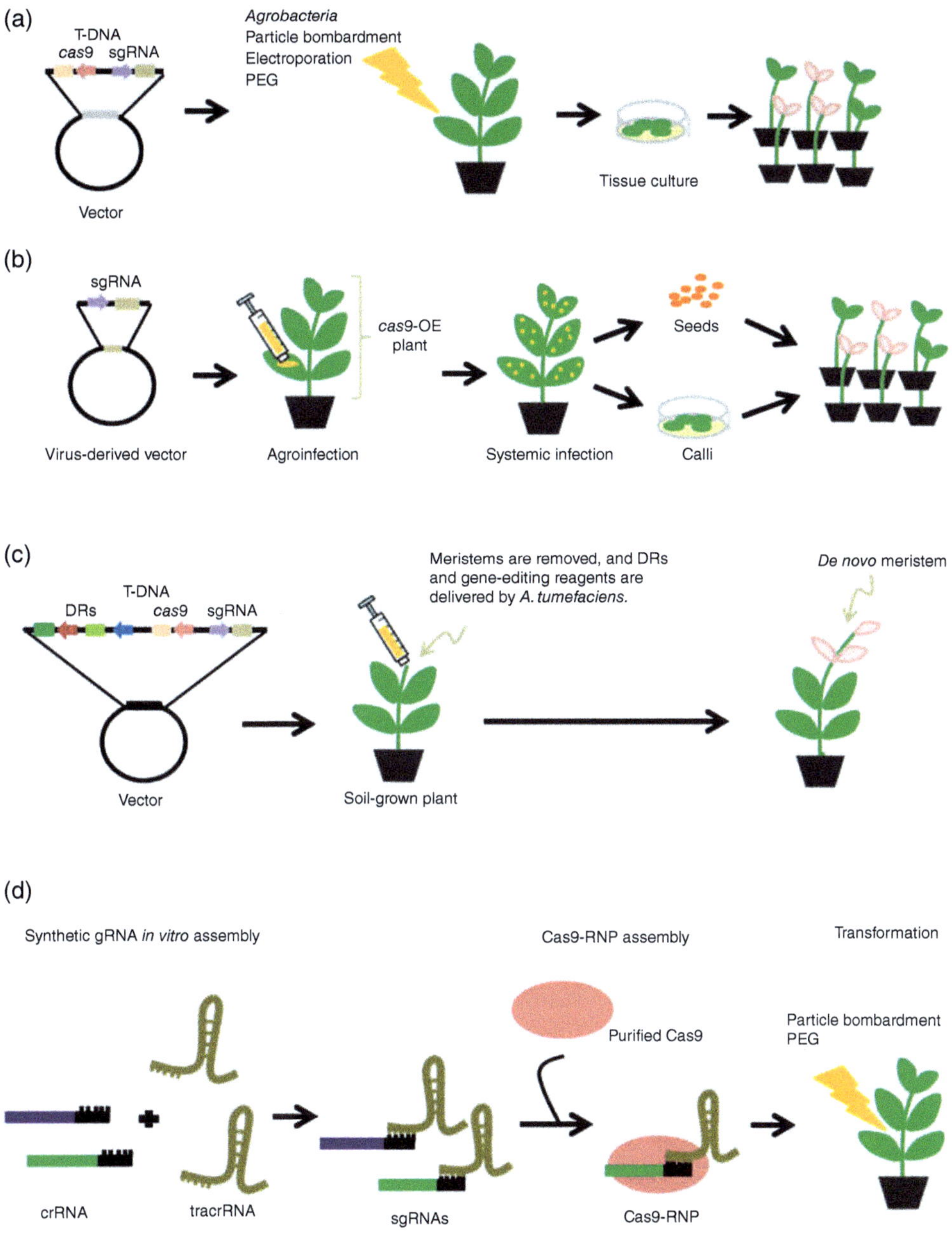

Figure 4.2 Schematic representation of the different kinds of plant transformation approaches via genome editing tools. (a) Various plant transformation method. (b) Virus-mediated sgRNA delivery. Virus-derived vectors are used to deliver sgRNAs in plants. (c) *De novo* meristem induction. (d) Cas9-sgRNA ribonucleoproteins (RNPs) transformation. *Source:* Picture adapted from Montecillo et al. (2020).

4.2.2 Plant Transformation Methodologies

There are various genetic transformation systems such as *Agrobacterium*-mediated transformation, direct transformation of plant protoplasts via polyethylene glycol or electroporation method, and biolistic transformation of plant tissues, which have been created for genetic transformation in plants (Rakoczy-Trojanowska 2002). All of these transformation systems have been successfully adapted in plants, but in some instances, transformation is difficult because their efficiencies differ and depend mostly on the plant species (Montecillo et al. 2020). Genetic engineering of staple crops has become one of the most effective solutions to address this problem. The common transformation methods for plant transgenesis using these transformation methods to transport CRISPR-Cas9 constructs into plants and to successfully deliver editing systems to different plant species with different transformation efficiencies have been proven (Borrelli et al. 2018). It was found that *Agrobacterium tumefaciens*-mediated transformation has edge over biolistic approach and leads to more positive transformation for barley Golden promise cultivar transformation experiment.

Some important plants like wheat have low transformation efficiency and this acts as a constraint for the crop improvement using CRISPR/Cas9. In order to enhance the transformations in such type of species, several steps must be taken into consideration, which includes types of wheat genotype, tissue culture media composition, immature embryo quality, strains of *A. tumefaciens,* and transformation, and some factors must be taken into account such as the choice of genotype, quality of immature embryos, composition of the media, *A. tumefaciens* strain, pre-treatment of embryos, and tissue handling. With proper optimization of these factors, Agrobacterium-mediated transformation resulted in as high as 90% transformation (Ishida et al. 2015).

The CRISPR-Cas9 system, an excellent plant genetic engineering tool has the precision in cutting plant genomic sequences and recent published data showed plant codon optimized CRISPR-Cas9 possesses excellence in genome editing in *Arabidopsis*, tobacco, wheat, and rice (Belhaj et al. 2013; Li et al. 2013; Shan et al. 2013; Bhalothia et al. 2020). In past, researchers have utilized this CRISPR-Cas9 system in various applications like protoplast transfection, hairy root engineering, biolistic bombardment of callus, and leaf agro-infiltration, and these applications help in studying genomic alterations caused due to CRISPR-Cas9 system (Alok et al. 2018; Upadhyay 2021). These genomic alterations can be insertion, deletion, or frame shift. Apart from these studies, successful inheritable germline mutation was carried out using CRISPR-Cas9-mediated Agrobacterium germline transformation in *Arabidopsis thaliana* (Feng et al. 2014; Jiang et al. 2014) and rice (Zhang et al. 2014; Zhou et al. 2014). More relevant studies in this direction was also found in sorghum (Jiang et al. 2013), wheat (Upadhyay et al. 2013), maize (Liang et al. 2014), sweet orange (Jia and Wang 2014), tomato (Brooks et al. 2014), potato (Butler et al. 2015), liverwort *Marchantia polymorpha L.* (Sugano et al. 2014), barley and *Brassica oleracea* (Lawrenson et al. 2015), soybean (Jacobs et al. 2015), melon (Tzuri et al. 2015), and poplar (Fan et al. 2015). In addition to this, transient expression of CRISPR-Cas9 DNA in wheat calli and immature embryos were achieved using biolistic method and lead to formation of transgene free homozygous desired transformants (Zhang et al. 2016). So far, we have discussed the brilliance of CRISPR-Cas9 mediated genome editing in different plants like rice, barley, wheat, and maize. However, the genotype of a crop species is also a contributing factor in genome editing mediated by CRISPR-Cas9. Apart from these, plant genotype is also an important factor governing the genome editing efficiency. For performing CRISPR-Cas9-mediated gene replacement using co-delivering the DNA templates with CRISPR-Cas9 reagents, particle bombardment transformation methods are preferred over the Agrobacterium-mediated transformation method (Montecillo et al. 2020). Co-delivering multiple copies of the donor DNA template into the plant host through biolistic methods increases the efficiency of gene replacement via homologous recombination (Svitashev et al. 2015; Baysal et al. 2016). Other important aspects like PEG-mediated protoplast transformation with CRISPR-Cas9 RNPs has also looks promising in the direction of plant genome editing and had shown excellent genome editing efficiency in apples and grapes (Malnoy et al. 2016). In future section of this chapter, we will look all these methods in detail. However, these exciting plant transformation protocol have great potential to pay path for green revolution and assistance of CRISPR/Cas9 system make it more promising. In future more biotechnological and computational advancement can speed up these attempts.

4.3 Emerging Advanced CRISPR/Cas Systems and the Increased Demand for Quick Transformation Protocols

The use of CRISPR/Cas systems for genome editing has made major advancements by providing ease and efficacy in the process of adding sequences of interest and deleting undesirable sequences from the genome (Ran et al. 2013; Manghwar et al. 2019). However, this system has its own limitations that have made it difficult to use it as an ideal DNA editing and delivery method. Some problems with this system include the challenge of packing CRISPR/Cas9 system in a viral vector for delivery due to its large size; the ability to only recognize the target site if NGG PAM site is present near it; addition of nonspecific and off-target mutations; and the constraint related to time, expenses, and resources (Manghwar et al. 2019). To perform efficient genome editing and to solve the abovementioned shortcomings of CRISPR/Cas9 system, various CRISPR systems modified from the traditional system have been developed in recent years.

A variant and orthologue of Cas9, which is derived from *Campylobacter jejuni* (CjCas9), is much smaller than Cas9 derived from *Staphylococcus aureus* (SaCas9) and *Streptococcus pyogenes* (SpCas9) and therefore is a more efficient candidate for genome editing (Kim et al. 2017a). This 2.95 kbp long variant Cas9 cuts a small number of specific sites in the genome and has been observed as an efficient means of inducing mutations at specific targets like Hif1a or Vegfa gene in retinal pigment epithelium (RPE) cells or mouse muscle cells through delivery by an adeno-associated virus (AAV) vector (Kim et al. 2017a). A SpCas9 variant called xCas9 was generated using phase-assisted continuous evolution that has the ability to recognize a wider range of protospacer adjacent motif (PAM) sequences such as GAA, GAT, and NG (Hu et al. 2018). This variant is more efficient in editing than SpCas9 in terms of its DNA recognition and specificity ability, and lower chances of off-target activity at both NGG and non-NGG PAM target sites. Another variant of SpCas9 known as SpCas9-NG created by Nishimasu et al. (2018) has the ability to recognize sequences apart from NGG sequence, thus increasing the target range of the CRISPR system. The efficacy of Cas9-NG variant was also evaluated in another independent study conducted by Ren et al. (2019). This study demonstrated that Cas9-NG can recognize, target, and edit all NGA, NGC, NGG, and NGT sites apart from NG PAM in plants, proving the potential of this efficient variant in targeting a broader range of sites and expanding the scope of various genome-editing tools in plants. The gene editing efficacies of these two SpCas9 variants, xCas9 and SpCas9-NG, was evaluated in rice (Hua et al. 2019). It was observed that xCas9-induced mutations at target sites comprising NG and GAT PAM sequences with great efficacy, while SpCas9-NG robustly performed editing of various NG PAMs containing target sites. This demonstrated the greater specificity and efficacy of xCas9 and SpCas9-NG than SpCas9, from which they are derived. These variants have the ability to expand the target scope of various genome-editing tools in a plethora of plants.

An RNA-guided class 2 CRIPSR system commonly known as Cpf1 (CRIPSR/Cas12a) performs DNA editing mechanism in a different and more efficient manner than its analogue Cas9. This CRISPR system depends on T-rich protospacer-adjacent motif and creates DSB with staggered ends at the target site improving the chances of knockin processes and efficiency of nonhomologous end-joining (NHEJ)-based DNA insertion mechanisms (Zetsche et al. 2015; Bin Moon et al. 2018). Cpf1-based CRISPR system requires a 42-nt CRISPR RNA (crRNA), which is much shorter than the ~100-nt gRNA utilized by Cas9, making Cpf1 a much better, cost effective, and efficient genome editing tool (Zaidi et al. 2017). Variants of Cas12a have also been engineered to further increase the target scope of Cas12a-based CRISPR system. Kleinstiver et al. (2019) have generated an enhanced *Acidaminococcus* sp. Cas12a variant (enAsCas12a) with a significantly higher GE activity due to its ability to recognize and target a wide range of PAM sequences that were beyond reach earlier. A recently isolated compact protein, Cas14a, which can target single-stranded DNA (ssDNA) independent of sequence and cleave in an unrestricted manner, thus proving to be an important CRISPR system to induce resistance against ssDNA viruses in various plants (Khan et al. 2019). Also, this CRISPR system is more efficient than CRISPR/Cas9 system due to small size of Cas14a and its ability to target sites without the requirement of flanking PAM sequences (Harrington et al. 2018). Thus, Cas14a nuclease seems to be a promising emerging tool for genome editing, especially in plants.

Apart from DNA editing, some CRISPR systems with RNase activity and ability to target RNAs with greater specificity and efficiency have also been identified (Abudayyeh et al. 2017; Wolter and Puchta 2018). This RNA-targeting and RNA-guided Cas13a nuclease (previously referred as C2c2) system has the ability to target and cleave explicit sequences in single-stranded RNA molecules, proving its capacity in RNA manipulation procedures in a similar manner as is the capacity of Cas9 and Cas12 in DNA manipulation (Wolter and Puchta 2018). The efficiency of this CRISPR/Cas13a system was proved in an experiment conducted by Aman et al. (2018), where they used RNA interference to develop resistance against Turnip Mosaic Virus (TuMV, an RNA virus) in plants. This proved the efficacy of Cas13a in inducing RNA interference and potential to employ this CRISPR system for RNA manipulations and inducing resistance against certain viruses in plants.

A different system for genome editing that is currently emerging and advancing is base editing (BE), which recruits cytidine deaminases to introduce changes in nucleotides without the need for double strand breaks (DSBs) and donor DNA templates, and has lesser rate of off-target mutations, making this system more efficient than homology-directed repair (HDR) mechanism (Hess et al. 2017; Yang et al. 2017). The various types of BE system recruit cytidine deaminases through the use of Cas9 or Cpf1 variants that employ DNA mismatch repair pathways to induce required C to T alterations (Komor et al. 2017; Li et al. 2018; Zong et al. 2018). Billon et al. (2017) were able to employ base-editing CRISPR tools to suppress the activity of genes by inducing C to T alterations using selected sgRNAs to convert codons CAG, TGG, CAA, and CAG into stop codons. The efficiency of CRISPR-mediated base editing has been significantly demonstrated in plants (Shimatani et al. 2017; Cai et al. 2020; Choi et al. 2021). To further enhance the applicability of CRISPR-dependent base editing, an intein-mediated split-nCas9 system has been developed, which can induce effective genome editing at target sites and also overcome the limitation of packaging due to large size of Cas9 (Yuan et al. 2022). This emerging genome editing tool that employs base editing strategies along with CRISPR system has shown great potential in introducing predictable changes in DNA and enhancing sequence diversity.

These emerging genome editing systems play a major role in optimization of plant transformation protocols and changing the course of crop genetics toward positive and valuable results (Altpeter et al. 2016). There are highly efficient protocols available for genome editing and transformation in plants; however, a protocol that is efficient as well as rapid and quick is the milestone that scientists are trying to reach (Veillet et al. 2019; Rodrigues et al. 2021; Zhang et al. 2021a). The optimization of Agrobacterium-mediated transformation protocols with the CRISPR-gene editing system have proved to have enhanced regeneration efficiency in *tropical japonica* rice varieties (Molina-Risco et al. 2021). Therefore, the coupling of the modified CRISPR systems described above with the transformation protocols especially Agrobacterium-mediated transformation system may result in being a game-changer in the field of plant biotechnology. Also, the development and further modifications in these CRISPR systems may help in increasing the potential of genome editing tools and their scope in advancing the efficiency and effectiveness of transformation mechanisms in plants.

4.4 Advancements in *Agrobacterium*-Meditated Stable Transformation of Plants

The Agrobacterium-mediated transformation is utilized due to the ability of Agrobacterium strains to cause infection in plants through attachment and transfer of T-DNA and virulence (vir) effector proteins, which later enter the nucleus of plant cells and T-DNA integrates to the host genome and lead to regeneration of transgenic plants (Figure 4.3). However, lower target range and insertion of the gene of interest into a specific and planned location in the complex plant genome is the limiting and challenging factor for Agrobacterium-mediated transformation (Altpeter et al. 2016; Peterson et al. 2021). The quality of HDR-based gene insertion process has been improved by increasing the efficiency of Agrobacterium-mediated transformation and designing of novel and improved vectors for delivery in plants (Peterson et al. 2021). In this study, three vectors were designed, two of which were almost similar with *Wus* and *Bbm* genes, while the third one included *Wus2* instead of *Wus* gene, and all the vectors were engineered to contain SpCas9 and gRNA. This advancement

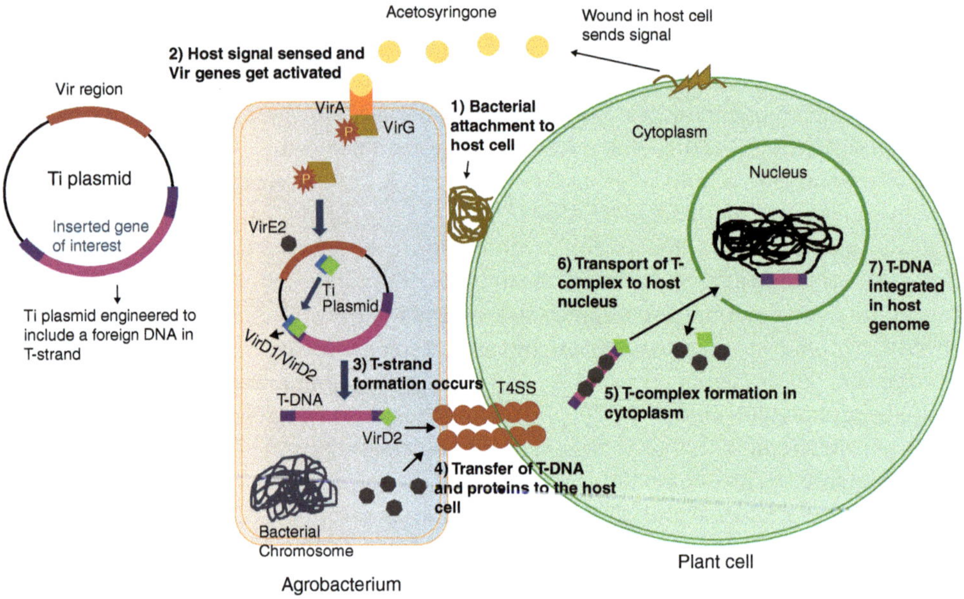

Figure 4.3 An overview of *Agrobacterium*-mediated transformation for the generation of stable inheritable mutations in plants.

in vector construction improved the transformation of maize embryo cells through *Agrobacterium*-mediated infection. The transformation susceptibility can also be downregulated by some factors and proteins, such as an *Arabidopsis* MYB transcription factor (MTF1) whose overexpression decreases *Agrobacterium*-mediated transformation of plants (Sardesai et al. 2013, 2014). However, when MTF1 expression was decreased using RNAi strategies, the transformation susceptibility increased (Sardesai et al. 2014). Therefore, the downregulation of these genes that act against *Agrobacterium* infection through priming of the plants can lead to increased efficacy of *Agrobacterium*-mediated transformation.

The development of various modifications and variants of Cas9, which are highly efficient with enhanced target range, lower off-target mutations, and increased capacity to introduce mutations in DNA with base editing, has also led to advancements in stable and effective *Agrobacterium*-mediated transformation in plants. The primary difficulty in employing genome-editing mechanisms in plants is the presence of large and more complex genomic material and the rigid cell walls around the plant cells (Laforest and Nadakuduti 2022). Thus, the delivery of genome modifying reagents has been a strenuous process. However, the emerging CRISPR-mediated genome editing system and the recent advancements in this system have made it significantly easier and efficient to employ CRISPR-dependent genome editing and *Agrobacterium*-mediated transformation in plants (Nadakuduti and Enciso-Rodríguez 2021). The optimization of Agrobacterium-mediated transformation and CRISPR system and comparison of conventional CRISPR systems with the variants are some of the methods for improving the transformation efficiency in plants. The optimized protocols involving induction of callus from mature seeds and evaluation of novel knockout mutations to obtain results that can be visibly evaluated for the efficiency of gene editing may lead to improved transformation mediated by *Agrobacterium* in plant varieties that are difficult to transform and regenerate (Molina-Risco et al. 2021). The comparison of SpCas9, LbCpf1, and xCas9 CRISPR systems was undertaken by Liu et al. (2020) in their study, where they evaluated the efficiencies of these systems independently with *OsU6a*, *TaU3*, and *TaU6* promoters that regulate sgRNA, revealing *TaU3* being more efficient in haploid induction in wheat.

Prime genome editing mechanism that employs CRISPR–Cas9 nickase (H840A)–reverse transcriptase fusions in combination with the prime editing guide RNAs (pegRNAs) has significantly improved the transformation of plants via *Agrobacterium* infection as this process can introduce mutations without the requirement of a donor DNA and DSBs (Lin et al. 2020). The CRISPR-dependent base editing system can also be used to introduce mutations in the genome of *Agrobacterium* species to develop strains that have higher capacity of transformation and lower off-target and nonspecific mutations (Rodrigues et al. 2021). A Super-*Agrobacterium* ver. 4 with increased transformation efficiency is a much more effective strain for gene transfer in a quick and less laborious manner than other conventional strains (Nonaka et al. 2019). The *A. tumefaciens* strain GV2260 was used for this study, and this strain was updated to ver. 4 through the introduction of GABA transaminase (*gabT*) and ACC deaminase (*acdS*) genes (Nonaka et al. 2019). In the future, the development of advanced modifications in the CRISPR system and the *Agrobacterium* strains can further help in enhancing the capacity of Agrobacterium-mediated transformation to efficiently deliver the gene targeting reagents and regenerate transformed plants.

4.5 Improvement of *Agrobacterium*-Mediated Transformation System by Developmental Regulators and Modular *Agrobacterium* Strains

Developmental regulators (DRs) comprise a number of modulators that form a complex regulatory network, which derives the regeneration machinery and plays a role in directing and switching the cell fate in plants by governing pluripotency capacity. These developmental regulators (DRs) are found to be involved in plant regeneration and can significantly enhance callus induction and formation (Lian et al. 2022). It has been shown previously by Lowe et al. (2016) that overexpression of certain DRs like *Baby boom (Bbm)* and *Wuschel2 (Wus2)* maize genes increased the *Agrobacterium*-mediated transformation efficiency in maize inbred lines that are not cooperative to *Agrobacterium* infection and also in plants like sorghum, sugarcane, and rice. The analysis of positive effect of various DRs, such as *Plethora (Plt5), Wound Induced Dedifferentiation 1 (WIND1), Wuschel (Wus),* and *Enhanced Shoot Regeneration (ESR1)* on the transformation process through *A. tumefaciens* infection has been observed in multiple plants (Lian et al. 2022). This study also used a fused form of two DRs, *Wus* and *Baby-Boom (WUS-P2A-BBM)*, and demonstrated the potential of the above mentioned DRs in promotion and formation of callus; regeneration of shoot; and improvement of genetic transformation efficiency of *Agrobacterium*-dependent system in tomato, sweet pepper, snapdragon (*Capsicum annum*), and two Brassica cabbage varieties (*Brassica rapa*, varieties, “Bok choy” and “Pei Tsai”).

The connected delivery of developmental regulars (DRs) and the reagents used for genome editing through *Agrobacterium* infection that can be introduced either through simultaneous culture of seedlings with *Agrobacterium* in a liquid culture or by injection of *Agrobacterium* to plants has made the generation of transgenic plants easier and faster as this process does not require the time-consuming conventional tissue culture step (Maher et al. 2020). Furthermore, a fusion protein consisting of Growth-Regulating Factor 4 (GRF4) and its cofactor GRF-Interacting Factor 1 (GIF1), *GRF4–GIF1*, has been found to have positive impact on transformation and regeneration efficiency in both monocotyledonous and dicotyledonous plants (Debernardi et al. 2020). The strategy of combining GRF4-GIF1 with CRISPR/Cas9 genome editing machinery was also employed, which improved the genome-editing frequency in plants (Debernardi et al. 2020). The lone expression of DRs or in combination with CRISPR system through *Agrobacterium* infection has proved to be extremely efficient and beneficial for DNA manipulation in a plethora of plants that are otherwise unresponsive to the *Agrobacterium*-mediated transformation.

The presence of type IV secretion system (T4SS) in Agrobacterium that facilitates the transfer of various effectors to plant cells to cause infection could be advantageous to transfer other types of proteins and nucleic acid molecules into plants (Alvarez-Martinez and Christie 2009). Apart from T4SS, another secretion system, namely, T6SS, can also be optimized for co-delivery of various effectors, genes of interest that do not integrate into host genome, proteins, and RNA molecules into plants cells via *Agrobacterium* infection, thus creating more efficient *Agrobacterium* strains (Wu et al. 2012; Lin et al. 2013). A simplified and modified method of transformation by *Agrobacterium*, AgarTrap, which is faster and less expensive than the conventional *Agrobacterium*-mediated transformation, has been shown to have good transformation efficiency and ability to improve genome editing procedures in plants that are difficult to transform (Tsuboyama et al. 2018). Moreover, engineering the *Agrobacterium* to form novel and improved strains by including modifications that can enhance the transformation efficiency in plants has been focus of researchers to increase the scope of genome editing in plants. GV2260 *A. tumefaciens* strain was engineered to generate a *Super-Agrobacterium* ver. 4 strain, which included the genes like ACC deaminase (*acdS*) and GABA transaminase (*gabT*) that are responsible for degradation of ACC, the ethylene precursor, and GABA (Nonaka et al. 2019). This updated strain demonstrated a quick, less laborious, and an increased ability to incur stable transformation by 3.6 fold as compared to the original strain. Therefore, the development of strategies that can improve and enhance the efficacy of *Agrobacterium*-mediated transformation may help in increasing the scope of various genome editing tools and employing these combined methods to generate transgenic plants with desired characteristics.

4.6 Non-*Agrobacterium* Systems for Plant Transformation

The transformation process is typically connected with Agrobacterium infection, which is most studied machinery for transformation; however, there are a number of organisms that have been reported to perform this similar transformation method, but at lower frequency. These organisms might have gained the transformation function through the Ti-plasmid that may have been transferred to them via conjugation with other bacterial species like *A. tumefaciens* in the natural habitats (Lacroix and Citovsky 2019). A number of different *Rhizobium* and other diverse species of bacteria, such as *Sinorhizobium meliloti*, have the ability to transfer genetic products at a lesser efficient rate than *Agrobacterium* into various plants (Broothaerts et al. 2005). *Ensifer adhaerens* (OV14)-mediated transformation (EMT) has been performed in a variety of plants and the efficacy of this system has proven to be equivalent to that of Agrobacterium in terms of transformation and integration output (Wendt et al. 2012; Zuniga-Soto et al. 2015). However, all these non-Agrobacterium agents do not comprise the exact homologs of *vir* genes, therefore have lower transformation efficiency. A *Rhizobium* species, *Rhizobium etli* CFN42, has recently been identified with capability of driving the transfer and integration of gene of interest into the genome of host plants due to presence of an efficient and self-machinery of whole set of proteins that are very close homologs of proteins encoded by *vir* genes involved in transformation process (Lacroix and Citovsky 2016). Furthermore, these transformation methods may be refined additionally, which may result in construction of enhanced and evolved versions of non-*Agrobacterium* strains having characteristics more beneficial and superior than those constituted in *Agrobacterium*, consequently, giving an avenue to expand the scope of genome editing tools in plants.

4.7 Viral Vectors for Delivery of CRISPR Reagents and Increasing Donor Titer

Cell wall rigidity, genomic rearrangements, and polyploidy are some of major hurdles that deteriorate the entry of any foreign biomolecules in to plant cells (Laforest and Nadakuduti 2022). Plasmid DNA is the most common genome editing reagent, transformed by *Agrobacterium*-mediated stable or transient methods and particle bombardment (Kujur et al. 2021). It is necessary to stably express the Cas9 and gRNA in transgenic plants so that progeny with targeted modifications in the genome will be heritable to the next generations. Despite of having multitude of advancements in the genome editing system, the efficient delivery of gene editing reagents still limits its vast application. In the direction to fill this gap in the plant genome editing, many studies are currently going on. Among these, viral vectors look promising.

In the recent years, both RNA and DNA viral vectors have been reported to successfully deliver these genome editing reagents (Ali et al. 2015; Ellison et al. 2020; Ma et al. 2020; Kujur et al. 2021; Laforest and Nadakuduti 2022). Of these, the DNA viruses belong to geminivirus group. Upon entry into the host cells, host DNA polymerase comes into action and converts the single-stranded genome of virus to the double-stranded genome. As a second option, single-stranded virus genome may be encapsulated by coat protein (CP) to produce virions, which move across different cells with the help of plasmodesmata. Geminiviruses depend on replication initiator protein (Rep), which limits their application in gene editing because of its interactions with several core protein members of host cells involved in different cell regulatory process-like cell cycle, DNA methylation machinery, DNA replication, and DNA repair (Ali et al. 2015; Zaidi and Mansoor 2017; Ellison et al. 2020; Ma et al. 2020; Kujur et al. 2021).

RNA viruses involved in genome editing process include both positive-strand viruses (PSVs) and negative-strand viruses (NSVs). Most widely used PSVs are Tobacco rattle virus (TRV), barley stripe mosaic virus (BSMV), tobacco mosaic virus (TMV), foxtail mosaic virus (FMV), and these reported the host genome mutation frequency of 30–56, 40–78, 70, 85, and 3–91%, respectively (Wang et al. 2020; Kujur et al. 2021; Laforest and Nadakuduti 2022). Despite its benefits, genome editing via PSVs has a significant drawback. Due to PSVs' limited cargo capacity, which is roughly 1 kb, they were unable to carry CRISPR/Cas9 nucleases of greater size, which typically range from 3 to 4.2 kb (approximately 160 kDa) (Kujur et al. 2021). Researchers used constitutively expressed Cas9 transgenic plants to carry out virus-mediated genome editing to get around this limitation. Among the PSVs, Tobacco rattle virus (TRV) is a bipartite positive-strand RNA and was successfully used to deliver sgRNAs into Cas9 overexpressing transgenic *Nicotiana* transgenic plants but have the limitation of low genome editing efficiency into next progenies. In order to address this hereditary issue, the endogenous mobile RNA Flowering Locus T (FT) has been coupled to sgRNA to improve and facilitate systemic transmission within the plant to reach the meristem cells (Laforest and Nadakuduti 2022). Barley stripe mosaic virus (BSMV) is a tripartite positive-strand RNA virus reported to deliver sgRNA in wheat, maize, and Cas9 transgenic *Nicotiana benthamiana* (Hu et al. 2019; Li et al. 2021). Because of the absence of tissue culture and regeneration requirements, it was highly helpful in wheat. Wheat M1 generations demonstrated 100% mutation rates were virus-free and further demonstrated a greater heritability of edits in M2 offspring. Additionally, multiplex edited wheat mutants were created by co-transforming a number of BMSV vectors (Li et al. 2021). Rhabdoviruses have shown promise in subsequent attempts to resolve the low cargo capacity-related issue. To do this end, the recombinant plant cytorhabdovirus barley yellow striate mosaic virus (BYSMV) was effectively employed to express Cas9 proteins that target the green fluorescent protein in the leaves of transgenic tobacco line 16c (Gao et al. 2019). Rhabdoviruses still have one disadvantage; they cannot reach meristem or germline cells, which prevents them from editing the genes of plants grown from the seeds of initially mutated plants (Kujur et al. 2021; Laforest and Nadakuduti 2022). In a recent study, the cargo capacity restriction was successfully circumvented by the potato virus X (PVX) vector. This virus's filamentous genome was encapsulated by its coat protein, which makes it the perfect vector system for delivering both sgRNA and Cas9 into tobacco leaves at once. Despite the benefits, PVX has several limitations, including a host range that is exclusive to Solanaceae species and the need for aseptic tissue regeneration to produce mutant offspring (Ariga et al. 2020; Kujur et al. 2021; Laforest and Nadakuduti 2022).

Negative-strand viruses (NSVs) are also used to deliver CRISPR reagents and have some advantages and disadvantages. Because NSVs' genomes are encased, RNAase not able to digest these nucleocapsids. Sonchus yellow net rhabdovirus (SYNV), which successfully delivered both Cas9 and sgRNA to cause the desired mutation in the host plant, was recently discovered (Ma et al. 2020). The SYNV tgtRNA (tRNA-gRNA-tRNA)-Cas9 cassette, which facilitated the precise processing of this transcript to release legitimate sgRNA, was created via an endogenous tRNA-processing method. As a result of this tRNA-based multiplex editing strategy, somatic cells showed high levels of mutagenesis, ranging from 40% to 90% for distinct genes. As we previously explained, plants grown from seeds of originally mutated plants would not carry the rhabdovirus-caused mutations. This is the primary drawback of SYNV-derived genome editing. Another drawback of SYNV is its limited host range, which prevents it from being administered to a variety of crop species. Another obstacle of using rhabdovirus as a viral vector is the difficulty in finding and producing infected clones (Kujur et al. 2021; Laforest and Nadakuduti 2022).

In conclusion, altering these viral vectors will improve the success rate of genome editing in plants and result in the development of new crop varieties with a broader spectrum of qualitative and quantitative traits. Some previous accomplishments have resulted in improvements in viral vector-based genome editing, such as tRNA-based multiplex genome editing utilizing SYNV tgtRNA and genome editing in meristem or germline tissue using coupling of Flowering locus T with sgRNA in TRV vectors. The previous research clears up some problems and paves the way for new suggested solutions. As far we are understood, the SYNV system is a rhabdovirus that does not have the capacity to cause germline or meristem tissue mutation and does not necessitate any CRISPR/Cas9 transgenic lines to function. We can introduce mutations in meristem or germline tissues by integrating mobile RNA sequences into sgRNAs in the SYNV viral system. Another possibility is to utilize newly discovered hypercompact Casϕ in the place of Cas9 in viral vectors (Laforest and Nadakuduti 2022). PSVs with hypercompact Casϕ and sgRNA combined with Flowering locus T have a better chance of overcoming the cargo capacity issue and boosting the efficiency of heritability of mutations to progeny. The possibility for gene editing in plants would be further increased by the establishment of the BYSMV system for SSN-mediated gene targeting in more species and a PVX-like system for non-solanaceous plants. The potential for exciting and novel research in this field has been opened by the advent of virus vector systems and breakthroughs in genome editing technology.

4.8 *De novo* Meristem Induction

Stem cells are vital for the plant body because they facilitate morphogenesis and tissue regeneration in the case of injury or aging. These stem cells are anchored in multicellular structures called meristems. Meristems are necessary for the plastic expansion of the highly branched root and shoot systems. Due to totipotent nature of plant cells, they can differentiate into a variety of cell types (Montecillo et al. 2020; Nicolas et al. 2022). By employing tissue culture, somatic cells are normally allowed to proliferate and expand undifferentiated as calli. Next, totipotency is induced in these calli, causing them to differentiate further into shoots and roots. The process of promoting *de novo* meristem induction and inducing totipotency is not straightforward; it has certain requirements that vary with plant species. Among these requirements appropriate media composition and hormone regime are critical parameters for meristem induction. Significant advances in tissue culture has been made possible by the use of developmental regulators (DR) that are geared toward *de novo* meristem induction (Pierre-Jerome et al. 2018; Montecillo et al. 2020).

The process of incorporating changes into plants through tissue culture is complex and takes months to years. This pushed scientists to think about efficient strategies for introducing transgenes into meristems or egg cells directly. These attempts are ineffective and restricted to *A. thaliana* and its relatives. Phytohormones, like as auxin and cytokinin, are essential for triggering organogenesis in somatic cells, although they coordinate with a number of developmental regulators. The majority of these developmental regulators (DRs) are transcription factors that promote and induce somatic cell totipotency (Pierre-Jerome et al. 2018; El-Mounadi et al. 2020; Montecillo et al. 2020; Nasti and Voytas 2021; Nicolas et al. 2022).

Recent findings have demonstrated that ectopic expression of the DRs Wuschel (WUS), shoot meristemless (STM), and a variant of monopteros (MP) BABYBOOM induces meristem activity in *Arabidopsis* leaf cells (Gallois et al. 2002; Ckurshumova et al. 2014). In a similar study, Lowe et al. (2016, 2018) showed DRs BABYBOOM- and WUSCHEL-mediated somatic embryogenesis in maize somatic cells. The same type of remark was also published about sorghum (Mookkan et al. 2017). Co-delivering DRs and genome editing agents to seedlings or soil-grown plants using *Agrobacterium* transformation assisted in promoting meristem activity and induced the formation of shoots. The recovered shoots were either genetically altered or transgenic. Next-generation seeds acquired the edits after fertilization. This strategy was tested in *Nicotiana benthamiana* and seemed to be most promising (Maher et al. 2020; Nasti and Voytas 2021). By eliminating the process of tissue culture, the preceding method will be extremely beneficial in speeding genome editing in shoots. Therefore, in comparison to previous methods, this strategy is effective for producing gene-edited plants rapidly. This strategy will be used in crops species in the future to ensure improvement in quantity and quality features.

4.9 Biolistics and Protoplast Systems for CRISPR-Based Genome Editing

Apart from *Agrobacterium*-mediated transformation, there are several other methods stimulating the delivery of CRISPR reagents into plants. Among these techniques, biolistic and protoplast transformation are two major approaches, and these have some advantages and disadvantages.

4.9.1 Biolistic Approach

The biolistic method, also known as particle bombardment or the "gene gun," involves coating nucleic acids on gold and tungsten microprojectiles, which are then shot into plant cells at a high speed. This causes the plant cell to break down, allowing the entry of foreign nucleic acids into the host cells (Ismagul et al. 2018). Biolistics have diverse cargo capacity including plasmid DNA, ssDNA, recombinant proteins, and RNA or ribonucleic proteins (RNPs) assembled from IVTs. Despite having the capacity to carry diverse cargo, biolistic has several significant disadvantages because of the random genomic integration of the cargo DNA at various places. RNPs cargo have enhanced transformation efficiency and precision in comparison to DNA cargo, which has been proved in various cereal crops where RNPs-mediated biolistic transformation produced transgene-free, gene-edited cereal plants (Svitashev et al. 2016; Liang et al. 2017; Banakar et al. 2019, 2020; Zhang et al. 2021b). RNPs are able to overcome the issue of off-target editing owing to their short shelf life, demonstrating its superiority to plasmid DNA delivery method (Liang et al. 2017). RNPs not only prevent off-target editing, but they also promote chromosomal rearrangements. This was demonstrated in a research in maize where RNPs cooperated with guide RNAs flanking the junctions of the desired inversion to promote a 75.5 Mb inversion on chromosome two (Schwartz et al. 2020). Such type of genetic engineering is valuable to design tailor-made crops. The use of SAM as explants for bombardment brings about further advancement in biolistic-mediated gene editing and eliminates the need for human labor in the preparation of calli or immature embryos. Next-generation seeds receive the edits from targeted SAM (Hamada et al. 2017; Hamada et al. 2018; Imai et al. 2020). Furthermore, as according to earlier reports, pollen was used as explant tissue for bombardment in *Nicotiana benthamiana* and retained fertility even after particle bombardment, which then delivered the cargo into the ovule during fertilization (Nagahara et al. 2021). Despite these achievements, some technical improvements are required to limit non-targeted insertions and control the variability between different bombardments. In such an attempt, a double-barreled gene gun works in tandem with cell counting software that has an internal standard, reducing the standard deviation among bombardments by 50% (Miller et al. 2021).

4.9.2 Protoplast Approach

Protoplasts can be isolated by enzymatic digestion from plants (Jeong et al. 2021). The first report of protoplast isolation was discovered in bacteria and fungi, and protoplast isolation was then implanted in plants. Protoplasts have the ability to dedifferentiate, and cultured protoplasts have the propensity to reform cell walls and can also undergo cell division, suggesting that protoplasts have the ability to regenerate entire plants. Because protoplasts lack cell walls, they are excellent transient systems for evaluating genome editing efficiency before moving on to the actual labor-intensive transformation protocol and avoiding unnecessary wastage in the transformation process (Gilliard et al. 2021; Jeong et al. 2021).

The transformation of protoplasts is accomplished through two methods: (PEG)-mediated transfection and electro-transfection. These methods facilitated the delivery of DNA, IVTs, or RNPs into protoplasts, resulting in transient CRISPR cassette expression. Cargo plasmid DNA has the potential for random integration, whereas IVTs and RNPs have the potential for DNA-free genome editing due to their ability to bypass transcription and translation machinery and rapidly degrade following CRISPR activity (Andersson et al. 2018; Liang et al. 2018; González et al. 2020; Gilliard et al. 2021; Zhang et al. 2021b; Laforest and Nadakuduti 2022).

Despite many achievements related to plant regeneration via protoplast transformation, there are many drawbacks to this approach that raise questions for the scientific community. Plant regeneration from protoplasts is limited to some species, challenging scientists to establish this in recalcitrant species, including important crop and medicinal species. Following successful establishment in important plant species, it can be combined with modern technology such as flow cytometry and omics approaches to significantly improve genome editing (Klimek-Chodacka et al. 2021; Nicolia et al. 2021; Yu et al. 2021; Zhang et al. 2021b). Multiplex genome editing and coupled transformation of three Prime Repair exonuclease 2 (TREX2) and CRISPR/Cas9 reagents have improved this (Weiss et al. 2020).

4.10 Generation of Transgene-Free CRISPR-Edited Lines

After successful establishment of CRIPSR/Cas9-mediated edits in plants, the researchers are now aiming to set up approaches for quickly generating genome-edited plants that are devoid of any transgene. The removal of Cas9 is crucial to get rid of any unwanted mutation in genome. With regards to this, several tools and approaches are established, which leads to reduction in off-target mutations and precise the genome editing. Removal of transgene alleviate the safety concern about genome-edited plants. Although unnecessary mutation in plants can be eliminated by crossing in successive generations. Several methods have been developed for generating null segregates in plants, which include mendelian segregation analysis, programmed self-elimination method, transient expression of cassette expressing CRISPR reagents, and ribonucleoprotein (RNP)-based genome editing (Wada et al. 2020). These are briefly described below.

4.10.1 Mendelian Segregation Analysis

According to mendal's segregation law, alleles segregate in future progenies. This is the common method for removing transgene. First CRISPR/Cas9 cassettes integration is confirmed by genomic PCR and antibiotic selection, and thereafter from the selected positive plants, edited plants are selected. In the next generation, mendalian segregation analysis is carried out to keep only genome-edited plants that are devoid of the transgene. However, this method is very laborious and time consuming (Gao et al. 2016; Lu et al. 2017). To compensate for this, Gao et al. (2016) developed fluorescence-based CRISPR vectors where T2 line *Arabidopsis* seeds emit mCherry fluorescence under microscope if Cas9 is still present in plants.

4.10.2 Programmed Self-Elimination Method

Finding null segregants from mendalian segregation analysis is costly, time consuming, and laborious method. To get rid of these, programmed self-elimination approach is a good alternative. This strategy was established first time in rice by utilizing suicide genes (BARNASE and CMS) (He et al. 2018). BARNASE is a toxic nuclease enzyme and CMS is a rice male gametophyte-specific lethal protein. According to their concept, plants that contain Cas9 as transgene and two suicidal genes will die off but null segregants will survive in the T1 generation (He et al. 2018). This approach is restricted to plants that can be transformed and produce seeds, but not to asexually propagated crops (Wada et al. 2020).

4.10.3 Transient Expression of CRISPR/Cas9 Cassette

Other methods for generating null segregants is the transient expression of CRISPR/Cas9 DNA or mRNA. In this PEG-mediated transformation of CRISPR/Cas9 vectors in plant protoplast is carried out, and whole plants generated thereof. As per previous reports, various monocot and dicot protoplasts were successfully transformed using PEG. Although this has proved to be an efficient editing approach in a few important crops like rice, potato, *Brassica*, *Nicotiana tobacum*, and tomato, its role in other economically important crops has been difficult to establish (Andersson et al. 2017; Lin et al. 2018). To overcome this limitation, *Agrobacterium*-mediated transient expression was carried out for tobacco leaf discs, which resulted in 17.2% null segregants among the all genome-edited plants (Chen et al. 2018). Despite of this, still there are chances of transgene integration into the host genome. So, it is necessary to develop a method that should be efficient and promote easy screening of null segregants.

RNP-based DNA-free genome editing is another approach for generating plants devoid of any transgene. In this method, *in vitro* RNP formed from Cas9 protein and gRNA are transformed into plant protoplast, which further resulted in targeted mutation in plants. Transferred RNP complex is devoid of any donor DNA template, that is why edited progenies are devoid of any transgene (Wada et al. 2020). Woo et al. (2015) established this approach in rice, *Arabidopsis*, tobacco, and lettuce where they transformed the preassembled Cas9-gRNA RNP complex into the respective protoplast. From results, they observed genome editing efficiency of 8.4–44% and stable transmission of mutations in future progenies. Apart from this study, null segregants obtained from RNP-mediated genome editing is also identified in other plant species like apple, Chinese cabbage, grapevine, and maize (Malnoy et al. 2016; Liang et al. 2017; Murovec et al. 2018).

Zhang et al. (2016) tried to optimize the transient transformation methods; in his study, they compared the transient expression of both Cas9 and *in vitro* transcripts (IVT) of Cas9-coding sequences. They performed particle bombardment transformation on calli from two wheat varieties named as hexaploid bread wheat (*Triticum aestivum* L.) and tetraploid durum wheat (*T. turgidum* L.). Here, two lines of hexaploid bread wheat contained mutations in all six alleles. He identified 43.8–86.6% of the T0 mutants devoid of transgene in Cas9 gene-mediated transformation. In comparison to this, IVT-mediated transformation caused all transgene-free plants. Thus, considering these results, the IVT-based transient transformation in protoplast seems to be promising approach to generate null segregants. Various methods for transformation of RNP complex into plant material are reported. Here, wheat and maize immature embryos were transformed using biolistic method. In maize, they obtained genome editing efficiency of 47% in the presence of selection marker and 2.4–9.7% in absence of selection marker, whereas in wheat 4.4% of genome editing efficiency was found (Svitashev et al. 2016; Liang et al. 2017). In a different study, RNP complex from Cas12a and gRNA resulted in mutation in transformed soybean and tobacco (Kim et al. 2017b).

As a conclusion from all previous studies, in comparison to DNA-based approach, RNP-based genome editing has shown the remarkable reduction in transgene integration in plants. Thus, RNP-mediated genome editing looks superior over all available methods with a point of view of quick recovery of transgene-free plants and it will further expand in efficiency or potentiality.

References

Abudayyeh, O.O., Gootenberg, J.S., Essletzbichler, P. et al. (2017). RNA targeting with CRISPR–Cas13. *Nature* 550 (7675): 280–284.

Adli, M. (2018). The CRISPR tool kit for genome editing and beyond. *Nature Communications* 9 (1): 1911.

Ali, Z., Abul-Faraj, A., Li, L. et al. (2015). Efficient virus-mediated genome editing in plants using the CRISPR/Cas9 system. *Molecular Plant* 8 (8): 1288–1291.

Alok, A., Kumar, J., and Upadhyay, S.K. (2018). Engineering in hairy roots using CRISPR/Cas9-mediated editing. In: *Hairy Roots* (ed. V. Srivastava, S. Mehrotra, and S. Mishra). Singapore: Springer. https://doi.org/10.1007/978-981-13-2562-5_15.

Alok, A., Chauhan, H., Upadhyay, S.K. et al. (2021). Compendium of plant-specific CRISPR vectors and their technical advantages. *Life* 11: 1021. https://doi.org/10.3390/life11101021.

Altpeter, F., Springer, N.M., Bartley, L.E. et al. (2016). Advancing crop transformation in the era of genome editing. *The Plant Cell* 28 (7): 1510–1520.

Alvarez-Martinez, C.E. and Christie, P.J. (2009). Biological diversity of prokaryotic type IV secretion systems. *Microbiology and Molecular Biology Reviews* 73 (4): 775–808.

Aman, R., Ali, Z., Butt, H. et al. (2018). RNA virus interference via CRISPR/Cas13a system in plants. *Genome Biology* 19 (1): 1–9.

Andersson, M., Turesson, H., Nicolia, A. et al. (2017). Efficient targeted multiallelic mutagenesis in tetraploid potato (*Solanum tuberosum*) by transient CRISPR-Cas9 expression in protoplasts. *Plant Cell Reports* 36: 117–128.

Andersson, M., Turesson, H., Olsson, N. et al. (2018). Genome editing in potato via CRISPR-Cas9 ribonucleoprotein delivery. *Physiologia Plantarum* 164 (4): 378–384.

Ariga, H., Toki, S., Ishibashi, K.J.P., and Physiology, C. (2020). Potato virus X vector-mediated DNA-free genome editing in plants. *Plant and Cell Physiology* 61 (11): 1946–1953.

Asmamaw, M. and Zawdie, B. (2021). Mechanism and applications of CRISPR/Cas-9-mediated genome editing. *Biologics: Targets and Therapy* 353–361.

Banakar, R., Eggenberger, A.L., Lee, K. et al. (2019). High-frequency random DNA insertions upon co-delivery of CRISPR-Cas9 ribonucleoprotein and selectable marker plasmid in rice. *Scientific Reports* 9 (1): 19902.

Banakar, R., Schubert, M., Collingwood, M. et al. (2020). Comparison of CRISPR-Cas9/Cas12a ribonucleoprotein complexes for genome editing efficiency in the rice phytoene desaturase (OsPDS) gene. *Rice* 13 (1): 1–7.

Barrangou, R., Fremaux, C., Deveau, H. et al. (2007). CRISPR provides acquired resistance against viruses in prokaryotes. *Science* 315 (5819): 1709–1712.

Baysal, C., Bortesi, L., Zhu, C. et al. (2016). CRISPR/Cas9 activity in the rice OsBEIIb gene does not induce off-target effects in the closely related paralog OsBEIIa. *Molecular Breeding* 36: 1–11.

Belhaj, K., Chaparro-Garcia, A., Kamoun, S., and Nekrasov, V. (2013). Plant genome editing made easy: targeted mutagenesis in model and crop plants using the CRISPR/Cas system. *Plant Methods* 9 (1): 1–10.

Bhalothia, P., Yajnik, K., Alok, A., and Upadhayay, S.K. (2020). The current progress of CRISPR/Cas9 development in plants. In: *Genome Engineering via CRISPR-Cas9 Systems* (ed. S. Dhar). Elsevier. https://doi.org/10.1016/B978-0-12-818140-9.00011-8.

Billon, P., Bryant, E.E., Joseph, S.A. et al. (2017). CRISPR-mediated base editing enables efficient disruption of eukaryotic genes through induction of STOP codons. *Molecular Cell* 67 (6): 1068-1079. e1064.

Bin Moon, S., Lee, J.M., Kang, J.G. et al. (2018). Highly efficient genome editing by CRISPR-Cpf1 using CRISPR RNA with a uridinylate-rich 3′-overhang. *Nature Communications* 9 (1): 1–11.

Bolotin, A., Quinquis, B., Sorokin, A., and Ehrlich, S.D. (2005). Clustered regularly interspaced short palindrome repeats (CRISPRs) have spacers of extrachromosomal origin. *Microbiology* 151 (8): 2551–2561.

Borrelli, V.M., Brambilla, V., Rogowsky, P. et al. (2018). The enhancement of plant disease resistance using CRISPR/Cas9 technology. *Frontiers in Plant Science* 9: 1245.

Brooks, C., Nekrasov, V., Lippman, Z.B., and Van Eck, J. (2014). Efficient gene editing in tomato in the first generation using the clustered regularly interspaced short palindromic repeats/CRISPR-associated9 system. *Plant Physiology* 166 (3): 1292–1297.

Broothaerts, W., Mitchell, H.J., Weir, B. et al. (2005). Gene transfer to plants by diverse species of bacteria. *Nature* 433 (7026): 629–633.

Butler, N.M., Atkins, P.A., Voytas, D.F., and Douches, D.S. (2015). Generation and inheritance of targeted mutations in potato (*Solanum tuberosum* L.) using the CRISPR/Cas system. *PLoS One* 10 (12): e0144591.

Cai, Y., Chen, L., Zhang, Y. et al. (2020, 1996). Target base editing in soybean using a modified CRISPR/Cas9 system. *Plant Biotechnology Journal* 18 (10).

Ceasar, S.A., Rajan, V., Prykhozhij, S.V. et al. (2016). Insert, remove or replace: a highly advanced genome editing system using CRISPR/Cas9. *Biochimica et Biophysica Acta (BBA)-Molecular Cell Research* 1863 (9): 2333–2344.

Chen, L., Li, W., Katin-Grazzini, L. et al. (2018). A method for the production and expedient screening of CRISPR/Cas9-mediated non-transgenic mutant plants. *Horticulture Research* 5.

Choi, M., Yun, J.-Y., Kim, J.-H. et al. (2021). The efficacy of CRISPR-mediated cytosine base editing with the RPS5a promoter in *Arabidopsis thaliana*. *Scientific Reports* 11 (1): 1–9.

Ckurshumova, W., Smirnova, T., Marcos, D. et al. (2014). Irrepressible MONOPTEROS/ARF 5 promotes de novo shoot formation. *New Phytologist* 204 (3): 556–566.

Debernardi, J.M., Tricoli, D.M., Ercoli, M.F. et al. (2020). A GRF–GIF chimeric protein improves the regeneration efficiency of transgenic plants. *Nature Biotechnology* 38 (11): 1274–1279.

Doudna, J.A. and Charpentier, E. (2014). The new frontier of genome engineering with CRISPR-Cas9. *Science* 346 (6213): 1258096.

Ellison, E.E., Nagalakshmi, U., Gamo, M.E. et al. (2020). Multiplexed heritable gene editing using RNA viruses and mobile single guide RNAs. *Nature Plants* 6 (6): 620–624.

El-Mounadi, K., Morales-Floriano, M.L., and Garcia-Ruiz, H.J.F.I.P.S. (2020). Principles, applications, and biosafety of plant genome editing using CRISPR-Cas9. *Frontiers in Plant Science* 11: 56.

Fan, D., Liu, T., Li, C. et al. (2015). Efficient CRISPR/Cas9-mediated targeted mutagenesis in populus in the first generation. *Scientific Reports* 5 (1): 12217.

Feng, Z., Mao, Y., Xu, N. et al. (2014). Multigeneration analysis reveals the inheritance, specificity, and patterns of CRISPR/Cas-induced gene modifications in *Arabidopsis*. *Proceedings of the National Academy of Sciences* 111 (12): 4632–4637.

Gaj, T., Gersbach, C.A., and Barbas, C.F. (2013). ZFN, TALEN, and CRISPR/Cas-based methods for genome engineering. *Trends in Biotechnology* 31 (7): 397–405.

Gallois, J.-L., Woodward, C., Reddy, G.V., and Sablowski, R. (2002). Combined SHOOT MERISTEMLESS and WUSCHEL trigger ectopic organogenesis in *Arabidopsis*. *Development* (29): 3207–3217.

Gao, X., Chen, J., Dai, X. et al. (2016). An effective strategy for reliably isolating heritable and Cas9-free *Arabidopsis* mutants generated by CRISPR/Cas9-mediated genome editing. *Plant Physiology* 171 (3): 1794–1800.

Gao, Q., Xu, W.Y., Yan, T. et al. (2019). Rescue of a plant cytorhabdovirus as versatile expression platforms for planthopper and cereal genomic studies. *New Phytologist* 223 (4): 2120–2133.

Gilliard, G., Huby, E., Cordelier, S. et al. (2021). Protoplast: a valuable toolbox to investigate plant stress perception and response. *Frontiers in Plant Science* 12: 749581.

González, M.N., Massa, G.A., Andersson, M. et al. (2020). Reduced enzymatic browning in potato tubers by specific editing of a polyphenol oxidase gene via ribonucleoprotein complexes delivery of the CRISPR/Cas9 system. *Frontiers in Plant Science* 10: 1649.

Hamada, H., Linghu, Q., Nagira, Y. et al. (2017). An in planta biolistic method for stable wheat transformation. *Scientific Reports* 7 (1): 11443.

Hamada, H., Liu, Y., Nagira, Y. et al. (2018). Biolistic-delivery-based transient CRISPR/Cas9 expression enables in planta genome editing in wheat. *Scientific Reports* 8 (1): 1–7.

Haque, E., Taniguchi, H., Hassan, M.M. et al. (2018). Application of CRISPR/Cas9 genome editing technology for the improvement of crops cultivated in tropical climates: recent progress, prospects, and challenges. *Frontiers in Plant Science* 9: 617.

Harrington, L.B., Burstein, D., Chen, J.S. et al. (2018). Programmed DNA destruction by miniature CRISPR-Cas14 enzymes. *Science* 362 (6416): 839–842.

He, Y., Zhu, M., Wang, L. et al. (2018). Programmed self-elimination of the CRISPR/Cas9 construct greatly accelerates the isolation of edited and transgene-free rice plants. *Molecular Plant* 11 (9): 1210–1213.

Hess, G.T., Tycko, J., Yao, D., and Bassik, M.C. (2017). Methods and applications of CRISPR-mediated base editing in eukaryotic genomes. *Molecular Cell* 68 (1): 26–43.

Hu, J.H., Miller, S.M., Geurts, M.H. et al. (2018). Evolved Cas9 variants with broad PAM compatibility and high DNA specificity. *Nature* 556 (7699): 57–63.

Hu, J., Li, S., Li, Z. et al. (2019). A barley stripe mosaic virus-based guide RNA delivery system for targeted mutagenesis in wheat and maize. *Molecular Plant Pathology* 20 (10): 1463–1474.

Hua, K., Tao, X., Han, P. et al. (2019). Genome engineering in rice using Cas9 variants that recognize NG PAM sequences. *Molecular Plant* 12 (7): 1003–1014.

Imai, R., Hamada, H., Liu, Y. et al. (2020). In planta particle bombardment (iPB): a new method for plant transformation and genome editing. *Plant Biotechnology* 37 (2): 171–176.

Ishida, Y., Tsunashima, M., Hiei, Y., and Komari, T. (2015). Wheat (*Triticum aestivum* L.) transformation using immature embryos. *Agrobacterium Protocols* 1: 189–198.

Ishino, Y., Shinagawa, H., Makino, K. et al. (1987). Nucleotide sequence of the iap gene, responsible for alkaline phosphatase isozyme conversion in *Escherichia coli*, and identification of the gene product. *Journal of Bacteriology* 169 (12): 5429–5433.

Ismagul, A., Yang, N., Maltseva, E. et al. (2018). A biolistic method for high-throughput production of transgenic wheat plants with single gene insertions. *BMC Plant Biology* 18 (1): 1–8.

Jacobs, T.B., LaFayette, P.R., Schmitz, R.J., and Parrott, W.A. (2015). Targeted genome modifications in soybean with CRISPR/Cas9. *BMC Biotechnology* 15: 1–10.

Jeong, Y.Y., Lee, H.-Y., Kim, S.W. et al. (2021). Optimization of protoplast regeneration in the model plant *Arabidopsis thaliana*. *Plant Methods* 17 (1): 1–16.

Jia, H. and Wang, N. (2014). Targeted genome editing of sweet orange using Cas9/sgRNA. *PLoS One* 9 (4): e93806.

Jiang, W., Zhou, H., Bi, H. et al. (2013). Demonstration of CRISPR/Cas9/sgRNA-mediated targeted gene modification in *Arabidopsis*, tobacco, sorghum and rice. *Nucleic Acids Research* 41 (20): e188–e188.

Jiang, W., Yang, B., and Weeks, D.P. (2014). Efficient CRISPR/Cas9-mediated gene editing in *Arabidopsis thaliana* and inheritance of modified genes in the T2 and T3 generations. *PLoS One* 9 (6): e99225.

Jinek, M., Chylinski, K., Fonfara, I. et al. (2012). A programmable dual-RNA–guided DNA endonuclease in adaptive bacterial immunity. *Science* 337 (6096): 816–821.

Khan, M.Z., Haider, S., Mansoor, S., and Amin, I. (2019). Targeting plant ssDNA viruses with engineered miniature CRISPR-Cas14a. *Trends in Biotechnology* 37 (8): 800–804.

Kim, E., Koo, T., Park, S.W. et al. (2017a). In vivo genome editing with a small Cas9 orthologue derived from *Campylobacter jejuni*. *Nature Communications* 8 (1): 1–12.

Kim, H., Kim, S.-T., Ryu, J. et al. (2017b). CRISPR/Cpf1-mediated DNA-free plant genome editing. *Nature Communications* 8 (1): 14406.

Kleinstiver, B.P., Sousa, A.A., Walton, R.T. et al. (2019). Engineered CRISPR–Cas12a variants with increased activities and improved targeting ranges for gene, epigenetic and base editing. *Nature Biotechnology* 37 (3): 276–282.

Klimek-Chodacka, M., Gieniec, M., and Baranski, R.J.I.J.M.S. (2021). Multiplex site-directed gene editing using polyethylene glycol-mediated delivery of CRISPR gRNA: Cas9 ribonucleoprotein (RNP). Complexes to Carrot Protoplasts. *International Journal of Molecular Sciences* 22 (19): 10740.

Komor, A.C., Zhao, K.T., Packer, M.S. et al. (2017). Improved base excision repair inhibition and bacteriophage Mu Gam protein yields C: G-to-T: a base editors with higher efficiency and product purity. *Science Advances* 3 (8): eaao4774.

Kujur, S., Senthil-Kumar, M., and Kumar, R.J.P.C.R. (2021). Plant viral vectors: expanding the possibilities of precise gene editing in plant genomes. *Plant Cell Reports* 40: 931–934.

Lacroix, B. and Citovsky, V. (2016). A functional bacterium-to-plant DNA transfer machinery of *Rhizobium etli*. *PLoS Pathogens* 12 (3): e1005502.

Lacroix, B. and Citovsky, V. (2019). Pathways of DNA transfer to plants from *Agrobacterium tumefaciens* and related bacterial species. *Annual Review of Phytopathology* 57: 231.

Laforest, L.C. and Nadakuduti, S.S. (2022). Advances in delivery mechanisms of CRISPR gene-editing reagents in plants. *Frontiers in Genome Editing* 4: 1.

Lawrenson, T., Shorinola, O., Stacey, N. et al. (2015). Induction of targeted, heritable mutations in barley and *Brassica oleracea* using RNA-guided Cas9 nuclease. *Genome Biology* 16: 1–13.

Li, J.-F., Norville, J.E., Aach, J. et al. (2013). Multiplex and homologous recombination–mediated genome editing in *Arabidopsis* and *Nicotiana benthamiana* using guide RNA and Cas9. *Nature Biotechnology* 31 (8): 688–691.

Li, X., Wang, Y., Liu, Y. et al. (2018). Base editing with a Cpf1–cytidine deaminase fusion. *Nature Biotechnology* 36 (4): 324–327.

Li, T., Hu, J., Sun, Y. et al. (2021). Highly efficient heritable genome editing in wheat using an RNA virus and bypassing tissue culture. *Molecular Plant* 14 (11): 1787–1798.

Lian, Z., Nguyen, C.D., Liu, L. et al. (2022). Application of developmental regulators to improve in planta or in vitro transformation in plants. *Plant Biotechnology Journal* 20 (8): 1622.

Liang, Z., Zhang, K., Chen, K., and Gao, C. (2014). Targeted mutagenesis in *Zea mays* using TALENs and the CRISPR/Cas system. *Journal of Genetics and Genomics* 41 (2): 63–68.

Liang, Z., Chen, K., Li, T. et al. (2017). Efficient DNA-free genome editing of bread wheat using CRISPR/Cas9 ribonucleoprotein complexes. *Nature Communications* 8 (1): 14261.

Liang, Z., Chen, K., Zhang, Y. et al. (2018). Genome editing of bread wheat using biolistic delivery of CRISPR/Cas9 in vitro transcripts or ribonucleoproteins. *Nature Protocols* 13 (3): 413–430.

Lin, J.-S., Ma, L.-S., and Lai, E.-M. (2013). Systematic dissection of the agrobacterium type VI secretion system reveals machinery and secreted components for subcomplex formation. *PLoS One* 8 (7): e67647.

Lin, C.S., Hsu, C.T., Yang, L.H. et al. (2018). Application of protoplast technology to CRISPR/Cas9 mutagenesis: from single-cell mutation detection to mutant plant regeneration. *Plant Biotechnology Journal* 16 (7): 1295–1310.

Lin, Q., Zong, Y., Xue, C. et al. (2020). Prime genome editing in rice and wheat. *Nature Biotechnology* 38 (5): 582–585.

Liu, H., Wang, K., Jia, Z. et al. (2020). Efficient induction of haploid plants in wheat by editing of TaMTL using an optimized agrobacterium-mediated CRISPR system. *Journal of Experimental Botany* 71 (4): 1337–1349.

Lowe, K., Wu, E., Wang, N. et al. (2016). Morphogenic regulators baby boom and Wuschel improve monocot transformation. *The Plant Cell* 28 (9): 1998–2015.

Lowe, K., La Rota, M., Hoerster, G. et al. (2018). Rapid genotype "independent" *Zea mays* L.(maize) transformation via direct somatic embryogenesis. *In Vitro Cellular & Developmental Biology-Plant* 54: 240–252.

Lu, H.P., Liu, S.M., Xu, S.L. et al. (2017). CRISPR-S: an active interference element for a rapid and inexpensive selection of genome-edited, transgene-free rice plants. *Plant Biotechnology Journal* 15 (11): 1371.

Ma, X., Zhu, Q., Chen, Y., and Liu, Y.-G. (2016). CRISPR/Cas9 platforms for genome editing in plants: developments and applications. *Molecular Plant* 9 (7): 961–974.

Ma, X., Zhang, X., Liu, H., and Li, Z.J.N.P. (2020). Highly efficient DNA-free plant genome editing using virally delivered CRISPR–Cas9. *Nature Plants* 6 (7): 773–779.

Maher, M.F., Nasti, R.A., Vollbrecht, M. et al. (2020). Plant gene editing through de novo induction of meristems. *Nature Biotechnology* 38 (1): 84–89.

Mali, P., Esvelt, K.M., and Church, G.M. (2013). Cas9 as a versatile tool for engineering biology. *Nature Methods* 10 (10): 957–963.

Malnoy, M., Viola, R., Jung, M.-H. et al. (2016). DNA-free genetically edited grapevine and apple protoplast using CRISPR/Cas9 ribonucleoproteins. *Frontiers in Plant Science* 7: 1904.

Manghwar, H., Lindsey, K., Zhang, X., and Jin, S. (2019). CRISPR/Cas system: recent advances and future prospects for genome editing. *Trends in Plant Science* 24 (12): 1102–1125.

Miller, K., Eggenberger, A.L., Lee, K. et al. (2021). An improved biolistic delivery and analysis method for evaluation of DNA and CRISPR-Cas delivery efficacy in plant tissue. *Scientific Reports* 11 (1): 7695.

Molina-Risco, M., Ibarra, O., Faion-Molina, M. et al. (2021). Optimizing agrobacterium-mediated transformation and CRISPR-Cas9 gene editing in the tropical japonica rice variety presidio. *International Journal of Molecular Sciences* 22 (20): 10909.

Montecillo, J.A.V., Chu, L.L., and Bae, H. (2020). CRISPR-Cas9 system for plant genome editing: current approaches and emerging developments. *Agronomy* 10 (7): 1033.

Mookkan, M., Nelson-Vasilchik, K., Hague, J. et al. (2017). Selectable marker independent transformation of recalcitrant maize inbred B73 and sorghum P898012 mediated by morphogenic regulators BABY BOOM and WUSCHEL2. *Plant Cell Reports* 36: 1477–1491.

Murovec, J., Guček, K., Bohanec, B. et al. (2018). DNA-free genome editing of *Brassica oleracea* and B. rapa protoplasts using CRISPR-Cas9 ribonucleoprotein complexes. *Frontiers in Plant Science* 9: 1594.

Nadakuduti, S.S. and Enciso-Rodríguez, F. (2021). Advances in genome editing with CRISPR systems and transformation technologies for plant DNA manipulation. *Frontiers in Plant Science* 11: 637159.

Nagahara, S., Higashiyama, T., and Mizuta, Y.J.P.R. (2021). Detection of a biolistic delivery of fluorescent markers and CRISPR/Cas9 to the pollen tube. *Plant Reproduction* 34 (3): 191–205.

Nasti, R.A. and Voytas, D.F.J.P.O.T.N.A.O.S. (2021). Attaining the promise of plant gene editing at scale. *Proceedings of the National Academy of Sciences* 118 (22): e2004846117.

Nicolas, A., Maugarny-Calès, A., Adroher, B. et al. (2022). De novo stem cell establishment in meristems requires repression of organ boundary cell fate. *The Plant Cell* 34 (12): 4738–4759.

Nicolia, A., Andersson, M., Hofvander, P. et al. (2021). Tomato protoplasts as cell target for ribonucleoprotein (RNP)-mediated multiplexed genome editing. *Plant Cell, Tissue and Organ Culture (PCTOC)* 144 (2): 463–467.

Nishimasu, H., Shi, X., Ishiguro, S. et al. (2018). Engineered CRISPR-Cas9 nuclease with expanded targeting space. *Science* 361 (6408): 1259–1262.

Nonaka, S., Someya, T., Kadota, Y. et al. (2019). Super-agrobacterium ver. 4: improving the transformation frequencies and genetic engineering possibilities for crop plants. *Frontiers in Plant Science* 10: 1204.

Peterson, D., Barone, P., Lenderts, B. et al. (2021). Advances in *Agrobacterium* transformation and vector design result in high-frequency targeted gene insertion in maize. *Plant Biotechnology Journal* 19 (10): 2000–2010.

Pierre-Jerome, E., Drapek, C., and Benfey, P.N.J.A.R.O.C. (2018). Regulation of division and differentiation of plant stem cells. *Annual Review of Cell and Developmental Biology* 34: 289–310.

Rakoczy-Trojanowska, M. (2002). Alternative methods of plant transformation-a short review. *Cellular and Molecular Biology Letters* 7 (3): 849–858.

Ran, F., Hsu, P.D., Wright, J. et al. (2013). Genome engineering using the CRISPR-Cas9 system. *Nature Protocols* 8 (11): 2281–2308.

Razzaq, A., Saleem, F., Kanwal, M. et al. (2019). Modern trends in plant genome editing: an inclusive review of the CRISPR/Cas9 toolbox. *International Journal of Molecular Sciences* 20 (16): 4045.

Ren, B., Liu, L., Li, S. et al. (2019). Cas9-NG greatly expands the targeting scope of the genome-editing toolkit by recognizing NG and other atypical PAMs in rice. *Molecular Plant* 12 (7): 1015–1026.

Rodrigues, S.D., Karimi, M., Impens, L. et al. (2021). Efficient CRISPR-mediated base editing in *Agrobacterium* spp. *Proceedings of the National Academy of Sciences* 118 (2): e2013338118.

Sardesai, N., Lee, L.-Y., Chen, H. et al. (2013). Cytokinins secreted by *Agrobacterium* promote transformation by repressing a plant myb transcription factor. *Science Signaling* 6 (302): ra100-ra100.

Sardesai, N., Laluk, K., Mengiste, T., and Gelvin, S.B. (2014). The *Arabidopsis* Myb transcription factor MTF1 is a unidirectional regulator of susceptibility to *Agrobacterium*. *Plant Signaling & Behavior* 9 (6): ra100.

Schwartz, C., Lenderts, B., Feigenbutz, L. et al. (2020). CRISPR–Cas9-mediated 75.5-Mb inversion in maize. *Nature Plants* 6 (12): 1427–1431.

Shan, Q., Wang, Y., Li, J. et al. (2013). Targeted genome modification of crop plants using a CRISPR-Cas system. *Nature Biotechnology* 31 (8): 686–688.

Shimatani, Z., Kashojiya, S., Takayama, M. et al. (2017). Targeted base editing in rice and tomato using a CRISPR-Cas9 cytidine deaminase fusion. *Nature Biotechnology* 35 (5): 441–443.

Singh, S., Rahangdale, S., Pandita, S. et al. (2022). CRISPR/Cas9 for insect pests management: a comprehensive review of advances and applications. *Agriculture* 12: 1896. https://doi.org/10.3390/agriculture12111896.

Sugano, S.S., Shirakawa, M., Takagi, J. et al. (2014). CRISPR/Cas9-mediated targeted mutagenesis in the liverwort *Marchantia polymorpha* L. *Plant and Cell Physiology* 55 (3): 475–481.

Svitashev, S., Young, J.K., Schwartz, C. et al. (2015). Targeted mutagenesis, precise gene editing, and site-specific gene insertion in maize using Cas9 and guide RNA. *Plant Physiology* 169 (2): 931–945.

Svitashev, S., Schwartz, C., Lenderts, B. et al. (2016). Genome editing in maize directed by CRISPR–Cas9 ribonucleoprotein complexes. *Nature Communications* 7 (1): 13274.

Tang, X., Lowder, L.G., Zhang, T. et al. (2017). A CRISPR–Cpf1 system for efficient genome editing and transcriptional repression in plants. *Nature Plants* 3 (3): 1–5.

Tsuboyama, S., Nonaka, S., Ezura, H., and Kodama, Y. (2018). Improved G-AgarTrap: a highly efficient transformation method for intact gemmalings of the liverwort *Marchantia polymorpha*. *Scientific Reports* 8 (1): 1–10.

Tzuri, G., Zhou, X., Chayut, N. et al. (2015). A 'golden'SNP in CmOr governs the fruit flesh color of melon (C ucumis Melo). *The Plant Journal* 82 (2): 267–279.

Upadhyay, S.K. (2021). *Genome Engineering for Crop Improvement*. Wiley Ltd. https://doi.org/10.1002/9781119672425.

Upadhyay, S.K., Kumar, J., Alok, A., and Tuli, R. (2013). RNA-guided genome editing for target gene mutations in wheat. *G3: Genes, Genomes, Genetics* 3 (12): 2233–2238.

Vats, S., Kumawat, S., Kumar, V. et al. (2019). Genome editing in plants: exploration of technological advancements and challenges. *Cells* 8 (11): 1386.

Veillet, F., Perrot, L., Chauvin, L. et al. (2019). Transgene-free genome editing in tomato and potato plants using agrobacterium-mediated delivery of a CRISPR/Cas9 cytidine base editor. *International Journal of Molecular Sciences* 20 (2): 402.

Wada, N., Ueta, R., Osakabe, Y., and Osakabe, K.J.B.P.B. (2020). Precision genome editing in plants: state-of-the-art in CRISPR/Cas9-based genome engineering. *BMC Plant Biology* 20: 1–12.

Waddington, S.N., Privolizzi, R., Karda, R., and O'Neill, H.C. (2016). A broad overview and review of CRISPR-Cas technology and stem cells. *Current Stem Cell Reports* 2: 9–20.

Wang, M., Gao, S., Zeng, W. et al. (2020). Plant virology delivers diverse toolsets for biotechnology. *Viruses* 12 (11): 1338.

Weiss, T., Wang, C., Kang, X. et al. (2020). Optimization of multiplexed CRISPR/Cas9 system for highly efficient genome editing in *Setaria viridis*. *The Plant Journal* 104 (3): 828–838.

Wendt, T., Doohan, F., and Mullins, E. (2012). Production of phytophthora infestans-resistant potato (*Solanum tuberosum*) utilising ensifer adhaerens OV14. *Transgenic Research* 21 (3): 567–578.

Wolter, F. and Puchta, H. (2018). The CRISPR/Cas revolution reaches the RNA world: Cas13, a new swiss army knife for plant biologists. *The Plant Journal* 94 (5): 767–775.

Woo, J.W., Kim, J., Kwon, S.I. et al. (2015). DNA-free genome editing in plants with preassembled CRISPR-Cas9 ribonucleoproteins. *Nature Biotechnology* 33 (11): 1162–1164.

Wu, C.-F., Lin, J.-S., Shaw, G.-C., and Lai, E.-M. (2012). Acid-induced type VI secretion system is regulated by ExoR-ChvG/ChvI signaling cascade in *Agrobacterium tumefaciens. PLoS Pathogens.* https://doi.org/10.1371/journal.ppat.1002938.

Yang, B., Li, X., Lei, L., and Chen, J. (2017). APOBEC: from mutator to editor. *Journal of Genetics and Genomics* 44 (9): 423–437.

Yu, J., Tu, L., Subburaj, S. et al. (2021). Simultaneous targeting of duplicated genes in *Petunia protoplasts* for flower color modification via CRISPR-Cas9 ribonucleoproteins. *Plant Cell Reports* 40: 1037–1045.

Yuan, G., Lu, H., De, K. et al. (2022). An intein-mediated split–nCas9 system for base editing in plants. *ACS Synthetic Biology* 11 (7): 2513–2517.

Zaidi, S.S.-E.-A. and Mansoor, S.J.F.I.P.S. (2017). Viral vectors for plant genome engineering. *Frontiers in Plant Science* 8: 539.

Zaidi, SS.-e.-A., Mahfouz, M.M., and Mansoor, S. (2017). CRISPR-Cpf1: a new tool for plant genome editing. *Trends in Plant Science* 22 (7): 550–553.

Zetsche, B., Gootenberg, J.S., Abudayyeh, O.O. et al. (2015). Cpf1 is a single RNA-guided endonuclease of a class 2 CRISPR-Cas system. *Cell* 163 (3): 759–771.

Zhang, H., Zhang, J., Wei, P. et al. (2014). The CRISPR/C as9 system produces specific and homozygous targeted gene editing in rice in one generation. *Plant Biotechnology Journal* 12 (6): 797–807.

Zhang, Y., Liang, Z., Zong, Y. et al. (2016). Efficient and transgene-free genome editing in wheat through transient expression of CRISPR/Cas9 DNA or RNA. *Nature Communications* 7 (1): 12617.

Zhang, X., Xu, G., Cheng, C. et al. (2021a). Establishment of an Agrobacterium-mediated genetic transformation and CRISPR/Cas9-mediated targeted mutagenesis in Hemp (*Cannabis Sativa* L.). *Plant Biotechnology Journal* 19 (10): 1979–1987.

Zhang, Y., Iaffaldano, B., and Qi, Y.J.P.C. (2021b). CRISPR ribonucleoprotein-mediated genetic engineering in plants. *Plant Communications* 2 (2): 100168.

Zhou, H., Liu, B., Weeks, D.P. et al. (2014). Large chromosomal deletions and heritable small genetic changes induced by CRISPR/Cas9 in rice. *Nucleic Acids Research* 42 (17): 10903–10914.

Zhu, H., Li, C., and Gao, C. (2020). Applications of CRISPR–Cas in agriculture and plant biotechnology. *Nature Reviews Molecular Cell Biology* 21 (11): 661–677.

Zong, Y., Song, Q., Li, C. et al. (2018). Efficient C-to-T base editing in plants using a fusion of nCas9 and human APOBEC3A. *Nature Biotechnology* 36 (10): 950–953.

Zuniga-Soto, E., Mullins, E., and Dedicova, B. (2015). Ensifer-mediated transformation: an efficient non-agrobacterium protocol for the genetic modification of rice. *Springerplus* 4 (1): 1–10.

5

Role of Nanotechnology in the Advancement in Genome Editing in Plants

Mehtap AYDIN

Genetics and Bioengineering Department, Yeditepe University, Istanbul, Turkey

5.1 An Overview of Plant Genome Editing

The first studies on genome editing consist of the mutagenesis process. These studies contain genetic modifications in the form of single nucleotide polymorphisms (SNPs) to increase the cereals that plant and animal breeders give to their animals or rise milk yield in dairy cows. The impressive knowledge of genome editing within plant groups focuses on applying genome technology to plant genetic modification (Consortium 2008; Songstad et al. 2017). Traditional plant breeding has contributed significantly to the world's nutrition over the years, and various breeding programs have improved high-yielding crops resistant to environmental conditions. Looking at plant breeding programs of the past centuries relied on genetic variations of natural and mutant origin to select suitable genetic combinations. These breeding programs based on these chemical compounds and irradiation methods had some disadvantages. Interspecies crossovers are not always targeted in mutagenesis, and a considerable portion of the genome is transferred rather than a single gene. When one nucleotide needs to be transported, thousands of nucleotides might mutate. Transgenic breeding projects evolved at the end of the twentieth century to address many of the issues that arose due to this (Mohanta et al. 2017).

Genome editing (GE) has revolutionized biology, with significant changes in both plant and animal genomes (Singh et al. 2022; Ahmar et al. 2021; Upadhyay 2021). GE is broadly classified into three parts, meganuclease (MegaN) and zinc finger nucleases (ZFNs), and transcription activator-like nucleases (TALENs). TALENs are regarded as tools of the second generation, while the CRISPR/CRISPR-associated protein 9 (Cas9) nuclease system is regarded as a means of the third generation (Manghwar et al. 2019; Lin et al. 2020; Shushmita et al. 2021).

Applications of Genome Engineering in Plants, First Edition. Edited by Santosh Kumar Upadhyay.
© 2024 John Wiley & Sons Ltd. Published 2024 by John Wiley & Sons Ltd.

5.1.1 Meganuclease

Meganucleases are specific restriction enzymes that produce homologous recombination in both plants and animals and have a long recognition site. Meganucleases are also called target endonucleases because of their high-efficiency endonuclease activity. MNs produce double-strand breaks (DSBs) at a particular spot, which are repaired by homology-directed repair (HDR) or nonhomologous end joining (NHEJ), leading to gene knockouts or knock-ins in *Arabidopsis* and tobacco (Vats et al. 2022). More research is needed to redesign MNs. Therefore, systems such as ZFN, TALEN, and CRISPR-Cas9 are being addressed in many studies as they are more efficient and more straightforward (Vats et al. 2022).

5.1.2 Zinc Finger Nucleases

Zinc finger nucleases (ZFNs) are DNA fragmentation proteins that are specially designed and engineered to cut specific DNA regions. ZFNs facilitate targeted gene editing by the formation of DSBs in DNA. Each ZFN has a DNA-binding field with a chain of two-finger modules that identifies a unique 6-bp hexamer in the DNA sequence, and DNA-cleaving domain comprised of a FokI nuclease domain (Figure 5.1). A zinc finger protein is formed by combining these domains (ZFP). The DNA-binding and DNA-cleaving domains combine to generate a highly specialized genomic scissor (Carroll 2011; Gupta et al. 2012; Mohanta et al. 2017).

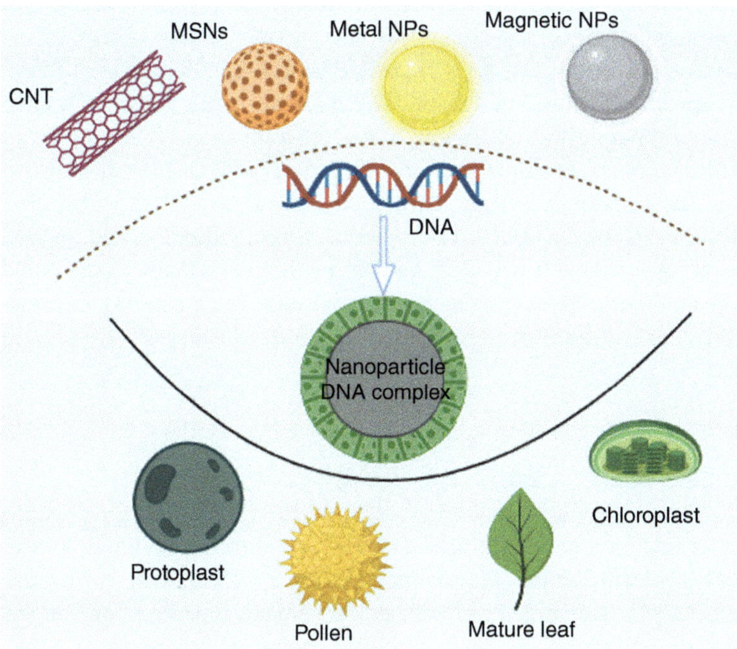

Figure 5.1 Nanoparticles of different shapes are used for genome editing. To build the most stable bioconjugate complex, the shape, and size can be adjusted to bind particular biomolecules. *Source:* Adapted from Ahmar et al. (2021).

5.1.3 Transcription Activator-Like Effectors Nucleases

Transcription activator-like effector nucleases (TALENs), which are genome editing tools in plants, have emerged in addition to ZFNs. Xanthomonas bacteria that produce TALEs contain TALE proteins and alter gene transcription in host plants (Boch and Bonas 2010; Mohanta et al. 2017). The way TALENs use DSBs is similar to ZFNs. ZFN FokI domains, on the other hand, bind to particular DNA-binding domains of highly conserved duplications generated from transcription activator-like effectors (TALEs) (Boch and Bonas 2010). Each monomer at the DNA binding site contains 34 amino acids (AA) conserved tandem repeats. Each of the monomers binds a single nucleotide in the target DNA. These specific nucleotides are identified by variable residues 12 and 13 of the monomer (Vats et al. 2022). The binding specificity of TALEs is easier than ZFNs, and TALEs have broader applications in life sciences. However, when looking at plant genome editing, there are few studies, and neither system is rich in performance in genome editing (Zhang et al. 2019).

5.1.4 CRISPR/Cas9 Based Genome Editing

Identification of CRISPRs was described in bacterial DNA in 1987, but their function was understood in 2005 (Ishino et al. 1987; Bolotin et al. 2005). CRISPR/Cas9 plant gene editing was first performed in 2013 in *Arabidopsis*, Nicotiana, and rice as model plants (Upadhyay et al. 2013; Li et al. 2013a, b). The CRISPR/Cas9 technique is cost-effective and acceptable; it is used effectively, especially in cereal products, due to its short duration. It is expected to eliminate the problems associated with transgenic technology in genome editing and replace technological approaches for commercial release (Ansari et al. 2020). In research so far constructs have two classes, six types, and 33 CRISPR/Cas9 subtypes based on different mechanisms for target-specific cutting. The reason why these six types differ from each other is that the effector components are different. Type 2 effector uses the multiprotein nature of caspases, while type 2 uses single protein effectors. Type 2 caspases consist of Cas9 (Alok et al. 2018, 2021; Makarova et al. 2018, 2020; Ahmar et al. 2020; Bhalothia et al. 2020). The application of the CRISPR/Cas9 system in crop production and applied research provides several advantages. It is widely used in healing, especially against cold and heat stress of plants and biotic stress (caused by fungi, viruses, and bacteria). In previous studies, the CRISPR/Cas9 system increased the size and weight of rice, wheat, and corn (Chen et al. 2019).

5.2 Nanoparticles used as Genome Editing Tools in Plants

Nanoparticles (NPs) can come from both natural and man-made sources. Natural processes that produce NPs can be volcanic eruptions, plants, animals, or photochemical events. Recently, plants and animals have emerged as intermediate biological sources for the production of nanoparticles and constitute the green source of nanoparticles (Singh et al. 2016). Genome editing of biomolecules by means of nanoparticles provides a great advantage. Due to their physiological properties, NPs can cross plant cells directly without a mechanical force (Bhat et al. 2021). The uptake of NPs differs in monocot and dicot plants, resulting in the plant's growth of roots and branches. The uptake of NPs to the cell membrane is either directly de-energized or endocytosis is energy dependent. Cationic NPs interact faster than anionic NPs due to the negative cell membrane (Vats et al. 2022).

Nanotechnology is a modern science that makes significant use of research in molecular biology. The use of nanotechnology in the production of genetically modified (GM) organisms (GMOs) is a powerful technique, including the use of NPs as nanocarriers by creating a binding complex with modifier molecules (CRISPR/Cas system) and transport into plant cells (Demirer et al. 2021). Today, the application of nanotechnology can support the removal of all barriers to producing genetically modified plants and biotransformation in addition to molecular developments. Nanomaterials engineering is also the latest emerging technology to develop sustainable agriculture systems (Singh et al. 2016). The improving of nanodevices can maximize the use of limited resources while minimizing the impact of pressures on energy production (Giraldo et al. 2019).

Traditional biomolecule delivery systems in plants have significant limitations, including low effectiveness of gene transmission, a limited species range for application, limited cargo kinds, and tissue damage. The ever-evolving advances in nanotechnology have opened up new possibilities to overcome traditional methods. Nanoparticles (NPs) have opened up new opportunities to deliver RNA, DNA, and proteins (Niazian et al. 2017; Demirer et al. 2019, 2020). There are many transformation procedures. One of these methods is the two primary genetic transformation methods used in plants, and this method is a genotype-specific method for gene transmission. Agrobacterium-mediated transformation (AMT) is the first genetic transformation method. It is the general method for introducing nontarget DNA into the nuclear genome, and this method is available for a limited number of plants. The AMT approach causes random DNA integration, altering endogenous plant genes and causing variance in gene expression due to the inserted locations (Cunningham et al. 2018; Ahmar et al. 2021). The most important contribution of nanotechnology to plant genetics is its effective delivery to different plant varieties. Delivery is a significant challenge in plants, as abiotic transfection methods (lipid-mediated delivery, heat shock, and electroporation) are ineffective in intact plants. Targeted delivery of specific nanoparticles allows visualization of processes such as cargo conduction and release (Demirer et al. 2021). DNA plasmids encoding CRISPR-Cas markers are widely used nanoparticles. Due to the different plasmid sizes, suitable reagents should be determined to load CRISPR DNA plasmids into nanoparticles in studies using nanomaterials (Demirer et al. 2021).

Minimizing the need for renewal of plants from tissue culture for plant seed transformation can be realized thanks to the developed nanomaterial technologies and can produce directly arranged offspring. To avoid regeneration, pollen transformation (the male gametophyte) is a potential strategy, as modified plants may be formed immediately by pollination with edited pollen. Due to the mechanical properties of nanoparticles, they can be used for the CRISPR editing of pollen. In addition, optimization is required to maximize the uptake of nanoparticles by pollen (Demirer et al. 2021). Editing germline cells in intact flowers containing male and female eggs is a different approach to pollen transformation. Applying CRISPR reagents to flowers via nanoparticles could regulate pollen, ovaries, and embryo cells. The shoot apical meristem that tissue culture-engineered plants can produce is another promising tissue target for nanomaterials (Demirer et al. 2021).

Nanoparticles are one of the most often employed methods of gene transfer. Inorganic NPs, carbon-based NPs, silicon-based NPs, and polymeric NPs are among them. Inorganic NPs (gold, silver, iron oxide) carry solely DNA as genetic cargo inside plants (Vats et al. 2022).

Plant leaves and roots are the most prevalent sites for nanomaterial absorption. The waxy cuticles in the leaves, which have tiny pores, decrease the time nanomaterials spend on the leaf surface and prevent their entry from the leaf. Instead, nanoparticles are more likely to enter leaves through stomata because the pores are more present. The location and number of stomata on the leaves vary according to the plant species, and environmental conditions affect the size of the stomata. (Su et al. 2019).

5.2.1 Mesoporous Silica Nanoparticles

Mesoporous silica nanoparticles (MSNs) have lately been of interest to many studies due to their large surface area, large structure, good thermal structure, and ability to bind to surfaces. The pores of silica particles are homogeneous in size and flexible in structure (Vats et al. 2022). In a study conducted in 2007, it was stated that MSN showed high transfection efficiency in *Nicotiana tabacum* plant and efficiency when the amount of DNA was less. Chemical inducers were loaded onto MSN particles and coated with gold NPs (100–200 nm). MSN covered with NPs was conjugated with GFP plasmid and transfected into *N. tabacuma* (Torney et al. 2007). Compared to other NPs, gold nanoparticles (AuNP) show exceptional physicochemical stability and biocompatibility. CRISPR-Cas9 modes can be produced via gold-based nanomaterials (Vats et al. 2022).

5.2.2 Carbon Nanotubes Carbon

Carbon nanotubes (CNTs) have optical and electrical properties and are structurally excellent mechanical allotropes. CNTs differ in their diameter and length, as well as the number of layers in the structure. CNTs can be classified single-walled carbon nanotubes (SWCNTs) and multi-walled carbon nanotubes (MWCNTs) (Aqel et al. 2012). In 2015, Burlaka et al. used CNTs as passive delivery materials in plants *in vivo* and *in vitro*. They used SWCNTs that provide direct delivery of DNA to protoplasts (Burlaka et al. 2015). In addition, a study conducted in *Triticum aestivum* demonstrated high-efficiency GFP expression with passive CNT delivery (Vats et al. 2022). The polymerization of vinyl monomers can synthesize nanocapsules by free radicals. Multilayer nanoparticles might be employed using polymers to transport CRISPR/Cas9 components to cell types. In recent studies, the nucleolin aptamer and MUC1-specific aptamer have been used to transport vectors for enhanced gene regulation and endocytosis in cancer cells (He et al. 2019). CRISPR/Cas9 tools can be transmitted using lipid-like molecules called lipidoids. Aliphatic tails in the structure of lipidoids increase stability and facilitate the initiation of the transport process (Vats et al. 2022).

Recent research has indicated that nanoparticles, especially carbon nanotubes (NCTs), can be used to independently acquire data on certain plant species and tissues. Entities of nanoparticles smaller than 500 nm contain genetic cargo chambers and foci where the cell walls are transported to study for detection of biomolecules (Kwak et al. 2019; Sharma and Lew 2022).

5.2.3 Lipid-Based Nanoparticles

Lipid-based NPs are advantageous because of their high carrying capacity. These NPs have advantages such as easy formation, increased carrying capacity, and controllable physicochemical properties. Having these advantages, lipid-based NPs are the preferred variety for gene editing. Duan et al. noted that lipid-like nanoparticles are used to accurately deliver CRISPR/Cas9 delivery and edit the correct genome. If the surface of the liposome is cationic, it can adsorb hydrophilic charged molecules such as nucleic acids and RNPs. If the surface of the liposome is cationic, it can adsorb hydrophilic charged molecules such as nucleic acids and RNPs. Ionizable or cationic lipids, sgRNA, and Spy Cas9 mRNA may all be used to functionalize LNPs (Duan et al. 2021; Vats et al. 2022). An environmentally and easy method of transfection, the magnetoreception condenses residues containing nucleic acids using magnetic fields. Transfection methods based on biochemical and physical processes can be combined in this system (Li et al. 2018). The thin layer in the pollen wall of the cotton plant allows the permeability of exogenous genes to the interior of the pollen. Magnetic nanoparticles can be employed as DNA carriers in pollen magnetofection, and these nanoparticles have magnetic field orientation ability (Vats et al. 2022).

5.2.4 Polymer-Based Nanoparticles

Polymer-based nanoparticles are widely used NPs for nucleic acids. They tend to bond quickly with genetic material negatively charged by electrical interaction. Chitosan and polyethyleneimine, cationic polymers, are widely used for CRISPR/Cas9 delivery (Zuris et al. 2015). The high target specificity of RNP transmissions is more advantageous when compared to the CRISPR system. Nevertheless, the cell wall restricts the plant's passive transport of genetic materials. For this reason, many researchers are required to use nanocarriers such as protein and RNA, DNA that can pass directly through the cell wall of the plant and can be loaded (Vats et al. 2022).

5.3 Point of View: The Nanotechnology and Plant Genome Editing

The development of genetically modified crops is accomplished by using a bacterial pathogen or by inserting foreign DNA into the plant host. These enhancements are accomplished by using plant transformation technologies, which are both costly and time-consuming. Many applications, such as overcoming the stress faced by plants, various genetic arrangements, and transferring small molecules, are among the applications of nanotechnology in agriculture (Landry and Mitter 2019). It has revolutionized implementing the CRISPR/Cas9 system since its introduction. However, many studies have experienced problems with the CRISPR application. In this case, improving cargo transportation and germline transformation efficiency is necessary (Ahmed et al. 2013; Vats et al. 2022).

Exploration of any advanced genome editing technologies is heavily reliant on biological knowledge of the target attribute and a well-defined issue. Various proteomic, transcriptomic, and genomic investigations should be prioritized to better understand and resolve agricultural efficiency (Chaudhary et al. 2019). Scientists doing genetics research are making various research discoveries to develop high-yield products. Increasing the nutrient ratio and continuing nanotechnology production against population growth and changing climatic conditions with each passing day are essential.

5.4 The Approach to Transferring Biomolecules to Plants and Its Limitations

Genetic transformation in plants includes two basic steps. The first is the transport of genetic cargo, whereas the second is the regeneration of changed plants. The regeneration ability varies according to how the biomolecule is transferred and whether a stable transformation is desired. Gene gun mediated and Agrobacterium-mediated transformation (AMT) are standard methods applied for plant genetic transformation. These practices have been embraced for soybean, maize, wheat, rice, and many different plant varieties (Cunningham et al. 2018; Ahmar et al. 2021). Although transformation and generation studies have been carried out for years, transformation and generation problems continue in many plant types. For example, AMT efficiency depends on the rice type. This situation also depends on the fact that the plant is monocot because monocots are challenging for AMT (Van Eck 2018; Ahmar et al. 2021).

5.5 Role of Nanotechnology in Agriculture

Nanotechnology offers some solutions to minimize the environment's adverse effects and effects of climate changes on food production (Hofmann et al. 2020). Plant nanotechnology is the field of functional analysis using NMs. Due to their distribution and displacement within the plant cell, NMs become suitable for use in plants (Kah et al. 2019). Many NMs, including nanofertilizers, nanopesticides, and nanobiosensors, have been studied in the studies carried out so far. Nanofertilizers are nanoparticles coated on plant nutrients, while nanopesticides are pest control components and trap the active substance. Nanobiosensors are sensitive parts that are compact and coupled together. Nanobiosensors and proper management of pesticides, crop yield, can be increased (Usman et al. 2020). Nanobiosensors can increase crop productivity with appropriate soil, fertilizer, and pesticide selection and provide plant protection by communicating with farmers. Nanobiosensors have a long shelf life compared to the old generation sensors due to their fast electron kinetics and high stability (Ahmar et al. 2021).

By using different NMs, it is possible to increase both crop production and soil quality. Elements such as zinc, copper, silver, and manganese can be used for transmission in the plant, thus increasing the efficiency of the transfer of fertilizers to the plant. NP silver doses substantially increased the seed germination rate in watermelon and squash (Wang et al. 2020).

Successes that can revolutionize agricultural technology can be achieved by using the smart NPs given in Table 5.1. NPs have been used in fruits and many plant species, including olive, blackberry, sativa, and pear. When we look at the studies conducted in recent years, MNPs have been applied as particles with antifungal properties in fruit-bearing trees such

Table 5.1 Nanomaterials (NMs) are thought to be advantageous to some of agricultural crops.

NM	Plant	Implementation	Effect
CdSe/ZnS	*Arabidopsis*, onion	Plant interactions with nanoparticles, imaging, fluorescence detection, and nanobiosensors are all areas of research	Biosensors aid in pathogen detection, the formation of reactive oxygen species (ROS), and the viability of cell and root development
ZnO	*Arabidopsis*, cucumber, lettuce, rice, radish, maize	Micronutrient delivery with nanopesticides	Increased seed germination, decreased onion production, and suppression of root development in rye, radish, maize, and lettuce
$CaCO_3$	Peanut	Nutrient solution	Increased plant biomass and yield Improved
Mg	Black eyed pea	Nanofertilizers	Enhanced chlorophyll content and stability of plasma membrane and yield
Cu	Cucumber, wheat, sorghum, and lettuce	NP interactions in plants	Accumulation of NP in cucumber's root, increasing total biomass, toxicity, and total nitrogen, shoot, and soot in wheat and mung bean
Au	Barley, rice, flame lily, *arabidopsis*, tomato	NP interactions in plants	Tomato and barley are not harmed, while flame lily germination development are boosted, as are NP accumulations in the roots
SiO_2	*Arabidopsis*	Plant–NP interaction	Decreased flowering period in onion plant, decrease in yield and increase in growth, increase in seed germination, stopping of root growth in plants such as lettuce, cucumber, and radish
CuO	Rice, cucumber, wheat, *arabidopsis*	Genetic engineering in plants	Cu boosted plant growth critical nutrients, increased ROS generation, and decreased shoot and root length
Chitosan	Tea, wheat	Nanoherbicides, and genetically modified plants	Plant growth stimulation, biocompatible and biodegradable material
$Ca_5(PO_4)OH$	Soybean	Solution for nutrition	increased biomass, yield, and growth
Fe_3O_4	Wheat, soybean, maize	NP interactions in plants	Increased chlorophyll content in soybean, increased plant height and leaf area in wheat, and reduced the visibility of brown patches on maize leaves

(Continued)

Table 5.1 (Continued)

NM	Plant	Implementation	Effect
Liposomes	Tomato, benth	Nutrient and DNA delivery	Enhanced DNA transport and cell targeting, as well as better nucleic acid protection
Mn	Chickpea	NP interactions in plants	Improved chlorophyll content, as well as increased shoot and root length
Ag	Maize, watermelon, rice, brown mustard, summer squash	NP interactions in plants	Summer squash and watermelons grew faster, maize root development was inhibited, and radishes seedling growth was inhibited

Source: Adapted from Ahmar et al. (2021).

as grapes and pears (Ahmar et al. 2021). In addition, MNPs can be applied as biosensor particles for the detection of biochemical degradation in both humans and plants (Thakur et al. 2020). NPs are not limited to specific products; they have a wide range of uses, from industrial products and medicinal products to woody trees (Ahmar et al. 2021).

5.6 Conclusion

The knowledge deficiency in understanding the ecotoxicity, exposure routes, and lifecycle effect estimates of nanoparticles may make it difficult to negotiate regulatory bodies' environmental and safety analyses. In addition, economic, ethical, commercial, and legal difficulties related to the use of nanoparticles complicate the studies. For these reasons, the problems regarding the safe use of nanoparticles for plant gene editing remain unresolved (Sharma and Lew 2022). There is no doubt that much research can be done with nanotechnology, but it is necessary to consider some difficulties while doing the work. The size, concentration, and interaction of the nanomaterial are very significant. If the concentration of NP to be used is too high, oxidative stress can occur due to the accumulation of these materials in the cell. It is crucial whether the NPs are transferred to the cell by active energy or by simple diffusion. In addition, after the molecules are transferred, the material's compatibility and the nanoparticle's functionality should be examined (Vats et al. 2022).

References

Ahmar, S., Saeed, S., Khan, M.H.U. et al. (2020). A revolution toward gene-editing technology and its application to crop improvement. *International Journal of Molecular Sciences* 21 (16): 1–28. https://doi.org/10.3390/ijms21165665.

Ahmar, S., Mahmood, T., Fiaz, S. et al. (2021). Advantage of nanotechnology-based genome editing system and its application in crop improvement. *Frontiers in Plant Science* 12 (May): https://doi.org/10.3389/fpls.2021.663849.

Ahmed, F., Arshi, N., Kumar, S., and Gill, S.S. (2013). Crop improvement under adverse conditions. In: *Crop Improvement Under Adverse Conditions* (ed. N. Tuteja and S.S. Gill), 1–394. Springer https://doi.org/10.1007/978-1-4614-4633-0.

Alok, A., Kumar, J., and Upadhyay, S.K. (2018). Engineering in hairy roots using CRISPR/Cas9-mediated editing. In: *Hairy Roots* (ed. V. Srivastava, S. Mehrotra, and S. Mishra). Singapore: Springer. https://doi.org/10.1007/978-981-13-2562-5_15.

Alok, A., Chauhan, H., Upadhyay, S.K. et al. (2021). Compendium of plant-specific CRISPR vectors and their technical advantages. *Life* 11: 1021. https://doi.org/10.3390/life11101021.

Ansari, W.A., Chandanshive, S.U., Bhatt, V. et al. (2020). Genome editing in cereals: approaches, applications and challenges. *International Journal of Molecular Sciences* 21: https://doi.org/10.3390/ijms21114040.

Aqel, A., El-nour, K.M.M.A., Ammar, R.A.A., and Al-warthan, A. (2012). Carbon nanotubes, science and technology part (I) structure, synthesis and characterisation. *Arabian Journal of Chemistry* 5 (1): 1–23. https://doi.org/10.1016/j.arabjc.2010.08.022.

Bhalothia, P., Yajnik, K., Alok, A., and Upadhayay, S.K. (2020). The current progress of CRISPR/Cas9 development in plants. In: *Genome Engineering via CRISPR-Cas9 Systems* (ed. S. Dhar). Elsevier. https://doi.org/10.1016/B978-0-12-818140-9.00011-8.

Bhat, J.A., Rajora, N., Raturi, G. et al. (2021). Silicon nanoparticles (SiNPs) in sustainable agriculture: major emphasis on the practicality, efficacy and concerns. *Nanoscale Advances* 3 (14): 4019–4028. https://doi.org/10.1039/d1na00233c.

Boch, J. and Bonas, U. (2010). Xanthomonas AvrBs3 family-type III effectors: discovery and function. *Annual Review of Phytopathology* 48: 419–436. https://doi.org/10.1146/annurev-phyto-080508-081936.

Bolotin, A., Quinquis, B., Sorokin, A., and Dusko Ehrlich, S. (2005). Clustered regularly interspaced short palindrome repeats (CRISPRs) have spacers of extrachromosomal origin. *Microbiology* 151 (8): 2551–2561. https://doi.org/10.1099/mic.0.28048-0.

Burlaka, O.M., Pirko, Y.V., Yemets, A.I., and Blume, Y.B. (2015). Plant genetic transformation using carbon nanotubes for DNA delivery. *Cytology and Genetics* 49 (6): 349–357. https://doi.org/10.3103/S009545271506002X.

Carroll, D. (2011). Genome engineering with zinc-finger nucleases. *Genetics* 188 (4): 773–782. https://doi.org/10.1534/genetics.111.131433.

Chaudhary, J., Khatri, P., Singla, P. et al. (2019). Advances in omics approaches for abiotic stress tolerance in tomato. *Biology* 8 (4): 1–19. https://doi.org/10.3390/biology8040090.

Chen, K., Wang, Y., Zhang, R. et al. (2019). CRISPR/Cas genome editing and precision plant breeding in agriculture. *Annual Review of Plant Biology* 70: 667–697. https://doi.org/10.1146/annurev-arplant-050718-100049.

Consortium, B.H. (2008). Genome-wide survey of SNP variation uncovers the genetic structure of cattle breeds A B admixed breeds principal component 1 downloaded from. *Science* 324 (5926): 528–532. www.sciencemag.org/cgi/content/full/324/5926/522/DC1http://science.sciencemag.org/.

Cunningham, F.J., Goh, N.S., Demirer, G.S. et al. (2018). Nanoparticle-mediated delivery towards advancing plant genetic engineering. *Trends in Biotechnology* 36 (9): 882–897. https://doi.org/10.1016/j.tibtech.2018.03.009.

Demirer, G.S., Zhang, H., Matos, J.L. et al. (2019). High aspect ratio nanomaterials enable delivery of functional genetic material without DNA integration in mature plants. *Nature Nanotechnology* 14 (5): 456–464. https://doi.org/10.1038/s41565-019-0382-5.

Demirer, G.S., Zhang, H., Goh, N.S. et al. (2020). Carbon nanocarriers deliver siRNA to intact plant cells for efficient gene knockdown. *Science Advances* 6 (26): https://doi.org/10.1126/sciadv.aaz0495.

Demirer, G.S., Silva, T.N., Jackson, C.T. et al. (2021). Nanotechnology to advance CRISPR–Cas genetic engineering of plants. *Nature Nanotechnology* 16 (3): 243–250. https://doi.org/10.1038/s41565-021-00854-y.

Duan, L., Ouyang, K., Xu, X. et al. (2021). Nanoparticle delivery of CRISPR/Cas9 for genome editing. *Frontiers in Genetics* 12 (May): https://doi.org/10.3389/fgene.2021.673286.

Giraldo, J.P., Wu, H., Newkirk, G.M., and Kruss, S. (2019). Nanobiotechnology approaches for engineering smart plant sensors. *Nature Nanotechnology* 14 (6): 541–553. https://doi.org/10.1038/s41565-019-0470-6.

Gupta, A., Christensen, R.G., Rayla, A.L. et al. (2012). An optimized two-finger archive for ZFN-mediated gene targeting. *Nature Methods* 9 (6): 588–590. https://doi.org/10.1038/nmeth.1994.

He, X.Y., Liu, B.Y., Peng, Y. et al. (2019). Multifunctional vector for delivery of genome editing plasmid targeting β-Catenin to remodulate cancer cell properties. *ACS Applied Materials and Interfaces* 11 (1): 226–237. https://doi.org/10.1021/acsami.8b17481.

Hofmann, T., Lowry, G.V., Ghoshal, S. et al. (2020). Technology readiness and overcoming barriers to sustainably implement nanotechnology-enabled plant agriculture. *Nature Food* 1 (7): 416–425. https://doi.org/10.1038/s43016-020-0110-1.

Ishino, Y., Shinagawa, H., Makino, K. et al. (1987). Nucleotide sequence of the iap gene, responsible for alkaline phosphatase isoenzyme conversion in *Escherichia coli*, and identification of the gene product. *Journal of Bacteriology* 169 (12): 5429–5433. https://doi.org/10.1128/jb.169.12.5429-5433.1987.

Kah, M., Tufenkji, N., and White, J.C. (2019). Nano-enabled strategies to enhance crop nutrition and protection. *Nature Nanotechnology* 14 (6): 532–540. https://doi.org/10.1038/s41565-019-0439-5.

Kwak, S.Y., Lew, T.T.S., Sweeney, C.J. et al. (2019). Chloroplast-selective gene delivery and expression in planta using chitosan-complexed single-walled carbon nanotube carriers. *Nature Nanotechnology* 14 (5): 447–455. https://doi.org/10.1038/s41565-019-0375-4.

Landry, M.P. and Mitter, N. (2019). How nanocarriers delivering cargos in plants can change the GMO landscape. *Nature Nanotechnology* 14 (6): 512–514. https://doi.org/10.1038/s41565-019-0463-5.

Li, J.F., Norville, J.E., Aach, J. et al. (2013a). Multiplex and homologous recombination-mediated genome editing in *Arabidopsis* and *Nicotiana benthamiana* using guide RNA and Cas9. *Nature Biotechnology* 31 (8): 688–691. https://doi.org/10.1038/nbt.2654.

Li, W., Teng, F., Li, T., and Zhou, Q. (2013b). Simultaneous generation and germline transmission of multiple gene mutations in rat using CRISPR-Cas systems. *Nature Biotechnology* 31 (8): 684–686. https://doi.org/10.1038/nbt.2652.

Li, L., Hu, S., and Chen, X. (2018). Non-viral delivery systems for CRISPR/Cas9-based genome editing: challenges and opportunities. *Biomaterials* 171: 207–218. https://doi.org/10.1016/j.biomaterials.2018.04.031.

Lin, Q., Zong, Y., Xue, C. et al. (2020). Prime genome editing in rice and wheat. *Nature Biotechnology* 38 (5): 582–585. https://doi.org/10.1038/s41587-020-0455-x.

Makarova, K.S., Wolf, Y.I., and Koonin, E.V. (2018). Classification and nomenclature of CRISPR-Cas systems: where from here? *The CRISPR Journal* 1 (5): 325–336. https://doi.org/10.1089/crispr.2018.0033.

Makarova, K.S., Wolf, Y.I., Iranzo, J. et al. (2020). Evolutionary classification of CRISPR–Cas systems: a burst of class 2 and derived variants. *Nature Reviews Microbiology* 18 (2): 67–83. https://doi.org/10.1038/s41579-019-0299-x.

Manghwar, H., Lindsey, K., Zhang, X., and Jin, S. (2019). CRISPR/Cas system: recent advances and future prospects for genome editing. *Trends in Plant Science* 24 (12): 1102–1125. https://doi.org/10.1016/j.tplants.2019.09.006.

Mohanta, T.K., Bashir, T., Hashem, A. et al. (2017). Genome editing tools in plants. *Genes* 8 (12): https://doi.org/10.3390/genes8120399.

Niazian, M., Sadatnoori, S.A., Galuszka, P., and Mortazavian, S.M.M. (2017). Tissue culture-based agrobacterium-mediated and in planta transformation methods. *Czech Journal of Genetics and Plant Breeding* 53 (4): 133–143. https://doi.org/10.17221/177/2016-CJGPB.

Sharma, P. and Lew, T.T.S. (2022). Principles of nanoparticle design for genome editing in plants. *Frontiers in Genome Editing* 4 (March): 1–8. https://doi.org/10.3389/fgeed.2022.846624.

Singh, S., Rahangdale, S., Pandita, S. et al. (2022). CRISPR/Cas9 for insect pests management: a comprehensive review of advances and applications. *Agriculture* 12: 1896. https://doi.org/10.3390/agriculture12111896.

Singh, D.P., Singh, H.B., and Prabha, R. (2016). *Microbial Inoculants in Sustainable Agricultural Productivity: Vol. 2: Functional Applications*, 1–308. Springer. https://doi.org/10.1007/978-81-322-2644-4.

Songstad, D.D., Petolino, J.F., Voytas, D.F., and Reichert, N.A. (2017). Genome editing of plants. *Critical Reviews in Plant Sciences* 36 (1): 1–23. https://doi.org/10.1080/07352689.2017.1281663.

Su, Y., Ashworth, V., Kim, C. et al. (2019). Delivery, uptake, fate, and transport of engineered nanoparticles in plants: a critical review and data analysis. *Environmental Science Nano analysis* https://doi.org/10.1039/c9en00461k.

Sushmita, K.G., Upadhyay, S.K., and Verma, P.C. (2021). An overview of genome-engineering methods. In: *Genome Engineering for Crop Improvement* (ed. S.K. Upadhyay). Wiley. https://doi.org/10.1002/9781119672425.ch1.

Thakur, A., Sharma, N., Bhatti, M. et al. (2020). Synthesis of barium ferrite nanoparticles using rhizome extract of *Acorus Calamus*: characterization and its efficacy against different plant phytopathogenic fungi. *Nano-Structures and Nano-Objects* 24: 100599. https://doi.org/10.1016/j.nanoso.2020.100599.

Torney, O.I.S., Trewyn, B.G., Lin, V.S., and Wang, K.A.N. (2007). Mesoporous silica nanoparticles deliver DNA and chemicals into plants. *Nature Nanotechnology* 295–300. https://doi.org/10.1038/nnano.2007.108.

Upadhyay, S.K. (2021). *Genome Engineering for Crop Improvement*. Wiley. https://doi.org/10.1002/9781119672425.

Upadhyay, S.K., Kumar, J., Alok, A. et al. (2013). RNA guided genome editing for multiple target gene mutations in wheat. *G3: Genes|Genomes|Genetics.* 3 (12): 2233–2238.

Usman, M., Farooq, M., Wakeel, A. et al. (2020). Nanotechnology in agriculture: current status, challenges and future opportunities. *Science of the Total Environment* 721: 137778. https://doi.org/10.1016/j.scitotenv.2020.137778.

Van Eck, J. (2018). The status of setaria viridis transformation: agrobacterium-mediated to floral dip. *Frontiers in Plant Science* 9 (May): 1–5. https://doi.org/10.3389/fpls.2018.00652.

Vats, S., Kumawat, S., Brar, J. et al. (2022). Opportunity and challenges for nanotechnology application for genome editing in plants. *Plant Nano Biology* 1 (March): 100001. https://doi.org/10.1016/j.plana.2022.100001.

Wang, L., Sun, S., Wu, T. et al. (2020). Natural variation and CRISPR/Cas9-mediated mutation in GmPRR37 affect photoperiodic flowering and contribute to regional adaptation of soybean. *Plant Biotechnology Journal* 18 (9): 1869–1881. https://doi.org/10.1111/pbi.13346.

Zhang, H.X., Zhang, Y., and Yin, H. (2019). Genome editing with mRNA encoding ZFN, TALEN, and Cas9. *Molecular Therapy* 27 (4): 735–746. https://doi.org/10.1016/j.ymthe.2019.01.014.

Zuris, J.A., Thompson, D.B., Shu, Y. et al. (2015). Cationic lipid-mediated delivery of proteins enables efficient protein-based genome editing in vitro and in vivo. *Nature Biotechnology* 33 (1): 73–80. https://doi.org/10.1038/nbt.3081.

6

Genome Editing for Crop Biofortification

Erum Shoeb[1], Srividhya Venkataraman[2], Uzma Badar[1], and Kathleen Hefferon[3]

[1] *Department of Genetics, University of Karachi, Karachi, Pakistan*
[2] *Department of Cell and Systems Biology, University of Toronto, Toronto, Canada*
[3] *Department of Microbiology, Cornell University, Ithaca, NY, USA*

6.1 Introduction

Hidden hunger describes the deficiency of essential micronutrient in humans, which occurs due to regular intake of diet without sufficient nutrients. Billions of people worldwide are affected with the deficiencies of iron, zinc, iodine, or vitamins A and E, predominantly in developing countries. In order to address the issues related to food within the developing countries, a cost-effective and sustainable strategy must be developed (Lowe 2021).

A scientific solution to the problem is 'biofortification' of crops. It is the process to increase nutritional content, such as essential vitamins and minerals, through conventional breeding or genetic modifications of crops that are the primary food source of the population. The basic aim of biofortification is to improve the health outcomes of the people occurring due to lack of micronutrient in their diet, such as iron, zinc, and vitamin A. Micronutrient deficiencies can cause various health issues, ranging from fatigue and weakness to anemia, blindness, stunted growth, impaired development in children, and increased susceptibility to infectious diseases (Foley et al. 2021; Kiran et al. 2022).

Crop modification has been revolutionized by the powerful tools of genome editing. Technologies such as, ZFNs, TALENs, and CRISPR/Cas9 offer precise and desirable improvement of DNA sequences for crop biofortification. Researchers have developed modified crops with increased nutritional value using genome editing tools targeting specific genes involved in biosynthesis and uptake of essential micronutrients in crops (Wang et al. 2019; Zaidi et al. 2020; Gao 2021; Upadhyay 2021).

It is important to implement sustainable programs to address the issue of hidden hunger at the population level. In this chapter, current research trends are being discussed, which are undertaken to biofortify and improve the nutritional quality of the crops.

Applications of Genome Engineering in Plants, First Edition. Edited by Santosh Kumar Upadhyay.
© 2024 John Wiley & Sons Ltd. Published 2024 by John Wiley & Sons Ltd.

6.2 Current Global Status of Micronutrient Malnutrition

Global population is expected to surpass 9 billion in the next 25 years. Improved varieties of crops with higher yield and better nutritional value are required to meet the global requirements (Simkin 2019). It is not favorable to simply increase the yield by cultivating double the amount of crops as we are growing these days considering the negative environmental impacts and the limitations of cultivatable land, water, and fertilizers (Godfray 2014). To minimize the environmental damage and maximize the yield, genetic manipulation offers a solution for the development of new varieties of staple crops with significantly improved nutritional quality.

Billions of people throughout the world are affected with micronutrient malnutrition, predominately in low- and middle-income populations (Ahmed et al. 2016). Micronutrient deficiencies account for higher mortality rate among young children and childbearing mothers (Stevens et al. 2022). Deficiencies in iron, zinc, folate, vitamin A, vitamin D, vitamin B12, and iodine can cause susceptibility against infections, blindness, stunted growth, cognitive impairment, and even death (Black et al. 2013).

Countries in Southeast Asia and sub-Saharan Africa reported to have high frequency of children under the age of five suffering from issues like stunting and underweight (Akombi et al. 2017; Ssentongo et al. 2021). Food insecurity is defined by the Food and Agriculture Organization of the United Nations as the "situation when people lack secure access to sufficient amounts of safe and nutritious food for normal growth and development and an active and healthy life." In the recent past, 9.2% of the population worldwide were found to be severely food insecure, more than 50% of those living in Asia; around 40% in Africa, and the rest of 10% split between rest of the world (Roser and Ritchie 2019). During COVID-19 pandemic in 2020, world hunger increased by more than 1% in one year, making it even more challenging to achieve the target of Zero Hunger by 2030 (FAO et al. 2020).

6.3 Importance of Biofortification in Ensuring Food Security

Economically struggling countries need to improve nutritional quality of diet taken by the people, especially young children and expecting mothers, to provide essential micronutrients. To overcome the deficiency of micronutrients, it was suggested around 100 years ago to supplement the food with micronutrients, called as fortification (Kumar et al. 2019). It was started with fortification of table salt with iodine and fluoride in Europe and USA (Leung et al. 2012). Later, food fortification was practiced for vitamins in milk, butter, and sugar in other parts of the world (Mishra et al. 2022). Moreover, in South Asia vegetable oil was supplemented with vitamin A and D as a mandatory measure for food fortification (Liu et al. 2014). In the present decade, food-fortified wheat with folic acid is still being used (Lockyer et al. 2018). However, contrary to its benefits, fortification could not sufficiently nourish people living in poverty-stricken regions of the world as supplementation require funds to access the market and to purchase the nutrients (UNICEF 2007). Furthermore, communities living in poor countries mostly depend on their own farming products instead of purchasing fortified products available in the market (Kumar et al. 2019). The solution was "biofortification," i.e. to biologically improve the nutritional value of the main staple crops by virtue of either agronomic interventions, conventional breeding, or modern genetic manipulation tools (Listman et al. 2019).

Biofortification has been proposed to be the most applicable and feasible solution to malnutrition and hidden hunger. Effects of iron and zinc biofortified crops were studied in different countries and found to improve both physical and cognitive conditions of micronutrient-deficient populations (Okwuonu et al. 2021). For instance, in Rwanda, consumption of iron-biofortified beans within five months resulted into considerable improvement in hemoglobin, ferritin, total body iron as well as behavioral and cognitive performance of women of reproductive age (Haas et al. 2016). Same results were observed when biofortified pearl millet was consumed by young children in India (Scott et al. 2018). Zinc biofortified crops have the ability to maintain active immune system and enhance resistance against infectious diseases (Wessels et al. 2017; Read et al. 2019). Biofortified crops introduced recently have impacted positively, such as orange-fleshed sweet potato, yellow cassava, and provitamin A maize that have been found to contain more vitamin A than traditional crops (Omari et al. 2019).

Biofortification has two significant advantages: cost-effectiveness and adaptability. Once developed nutritionally improved biofortified crops can be grown by farmers at virtually no extra cost. Secondly, micronutrient-rich crops can be adapted to all the environments owing to the selection of indigenous cultivars for modifications. Biofortified crops are developed keeping in mind the dietary needs of the population by identifying the target crop and level of required micronutrient. Biofortification strengthens the farmers by increasing the micronutrient level of the crops of their choice making it reachable to rural and urban areas (Bouis and Saltzman 2017).

6.4 Strategies for Biofortification

The main focus of biofortification has been iron, zinc, vitamins, and essential amino acids (Hirschi 2009). To improve the micronutrient value of staple crops, several strategies have been explored such as agronomic interventions including use of enriched fertilizers, breeding techniques by crossing and selecting the desired varieties, and most importantly, DNA manipulation approaches to produce new varieties with a higher nutrient value (Dixit et al. 2021; Gupta et al. 2022).

About two-third of the global population consumes rice as main food source (Rhowell Jr et al. 2021). Rice is naturally devoid of vitamins A, C, D, and B12 (Ghosh et al. 2019b). Even the nutrients that are present are mostly found in bran and very minute levels in the endosperm (de Pee 2014). Several strategies have been adopted in the past for the biofortification of rice. Agronomic approaches were not successful due to the sensitivity level of vitamins (Mozafar 1993). Harsh environmental conditions were acceptable by essential minerals, but the required content was not added to the endosperm through agronomic biofortification (Masunaga and Fong 2018). Since in original cultivars of rice there was no significant variation in terms of level of vitamin availability, conventional breeding techniques were not applicable (Tiozon Jr et al. 2021). Multiple genome editing strategies have successfully resulted in elevating the content of vitamins in rice.

World famous Golden Rice (GR) is a genetically modified rice variety that produces provitamin A, β-carotene, in the rice endosperm (Burkhardt et al. 1997). The entire biosynthetic pathway of β-carotene was inserted into the rice endosperm through Agrobacterium-mediated transformation to significantly increase total carotenoids level in milled rice (Ye et al. 2000; Datta et al. 2006).

The most widely grown cereal grain and staple food for 35% of the world's population is wheat. Wheat supplies more calories and protein to the diet of human population than any other crop. Owing to the earlier efforts of increasing yield, most of the commercial wheat varieties grown today are low in minerals and protein due to the suboptimal levels of micronutrients worldwide (Gupta et al. 2022). For the development of wheat varieties with higher nutrient levels, naturally occurring wild varieties with high concentrations of the micronutrients have to be explored. In 2003, International Maize and Wheat Improvement Center (CIMMYT) initiated a program in this direction and contributed few biofortified cultivars of wheat with higher concentrations of Zn and Fe (Singh et al. 2017).

Maize is a major source of food security worldwide and is a major staple food in sub-Saharan Africa. In addition, maize contributes in the cycle of livestock-to-meat around the world (Tanumihardjo et al. 2020). Maize is a source of micronutrients but the commonly used maize varieties do not have sufficient nutritional impact due to the low concentrations of most micronutrients (Prasanna et al. 2020). Fortunately, large genetic variations in maize allows biofortification of maize through the development of modified and improved cultivars with higher levels of important micronutrients (Bouis and Saltzman 2017). Scientists have been developing biofortified cultivars with high nutritional value through conventional breeding: for instance, lysine and tryptophan-rich quality protein maize (Twumasi-Afriyie et al. 2011); provitamin A carotenoids-rich orange maize (Pixley et al. 2013); and enhanced kernel Zn maize (Andersson et al. 2017).

6.4.1 Chloroplast Metabolic Engineering for Developing Nutrient-Dense Food Crops

Chloroplasts are organelles present in plants and have self-sufficient genomes and perform crucial functions in numerous vital metabolic activities like photosynthesis, amino acid, and fatty acid biosynthesis besides in the manufacture of various secondary metabolites. The favorable traits of plastid transformation in crops ensure the hyperexpression of foreign protein and deliver the prospect of many favorable agronomic characteristics without causing considerable environmental hazards (Bansal and Saha 2012). Plastid transformation has a vast perspective for the induction of significant agronomic attributes in plants. A conventional photosynthetically operational plant cell comprises of nearly 100 chloroplasts, whereas every single chloroplast includes approximately 100 replicates of DNA (Řepková 2010). Thus, each cell comprises numerous duplicates of the chloroplast genome resulting in better concentrations of protein accumulation obtained through transgene insertions in the chloroplast genome in contrast to the nuclear genome. The elimination of the plant cell wall by enzymes provides the induction of foreign DNA, RNA, and protein in protoplasts by polyethylene glycol (PEG) management and electroporation (Reed and Bargmann 2021). Chloroplast genetic engineering extends numerous exceptional benefits, involving multi-gene engineering and high-level transgene expression in a specific transformation occurrence and transgene restrained by maternal inheritance, along with lack of gene silencing, unwanted foreign DNA, position and pleiotropic effects (Daniell et al. 2005). According to Christou and Twyman (2004), durability and yields could be enhanced by redirecting the biomass from stalks and stems to palatable portions with the help of alteration of plant architecture. By the phenomenon of phytochrome engineering, the transgenic potato plants having an *Arabidopsis* phytochrome cDNA exhibited a lowered shade reaction by regulating the stem height even in overcrowded situations ultimately switching extra biomass to the tubers. Chloroplast biotechnology is in general a better way to make available crop plants with better nutritional contents, improved photosynthetic functioning, improved agronomic characteristics, and manufacture of significant fatty acids. The plastid transformation was originally determined in flowering plants about 30 years back (Khan et al. 2019). Lau et al. (2014) suggested that initiation of intensity of nutrient components may vary the flavor of a plant, possibly affecting its palatability. In future, by supplementing the beta-carotene route in rice plants for the production of rice with elevated amounts of vitamin A (yielding the Golden rice strain) can be established as one of the acceptable alterations to nutrient composition practicable by metabolic engineering.

Hussein et al. (2007) showed that chloroplast transgenic tobacco plants expressing mercuric ion reductase (*merA*) and organomercuric lyase (*merB*) can store mercury in roots to intensities exceeding the concentration of soil, till 200 μg/g, exclusive of any injurious outcome possibly acquire 100 times more Hg in leaves in comparison to the plants that are not transformed. Wurbs et al. (2007) described about fourfold more provitamin A yield (b-carotene) by the induction of lycopene b-cyclase into the chloroplasts from 2 distinctive heterologous resources (the bacteria *Erwinia herbicola* and the fungus *Phycomyces blakesleeanus*) in comparison to wild type tobacco. Hasunuma et al. (2008) effectively produced higher than 0.5% (dry weight) astaxanthin (higher than 70% of entire caroteniods) in the leaves of transpastomic tobacco plants. In addition, tobacco transformants also produced new carotenoid, 4-ketoantheraxanthin. By the initial addition of lycopene β-cyclase genes from (*Narcissus pseudonarcissus*) Daffodil and *Erwinia herbicola* into plastid genome of tomato Apel and Bock (2009) described a considerably elevated deposition of the entire carotenoid content with a rise of above 50%, and largely composed of the lycopene transformed to β-carotene with provitamin A levels achieving 1 mg/gm dry weight in fruits of tomatoes. Andrea and Rodriguez-Concepcion (2019) demonstrated the critical function of chloroplast protein (plastidail Hsp70 chaperones) quality control elements for management of carotenoid biosynthesis. It plays an important role in harmonizing tasks for each carotenoid biosynthesis and chromoplast differentiation throughout natural fruit ripening in tomato. Approaches to develop CO_2 fixation to resolve the catalytic inadequacies of Rubisco, both by raising CO_2 within the chloroplast or by direct enzyme engineering have been established. The occurrence of plastid transformation concerns the initial development of acquiring the plant with Rubisco L-subunit in tobacco protoplast. Probably bifunctional catalysis of Rubisco is the most important challenge in higher plants because oxygenation of RuBP controls photorespiration that is highly inefficient for the plant. Modification of rubisco catalysis is critical to address future climatic conditions, such as excessive heat and/or water scarcity along with the required enhancements in fiber yields and grains (Sharwood 2017). Harada et al. (2014) successfully constructed reproducible chloroplast-transformed lettuce with 94.5% of ketocarotenoids, with astaxanthin fatty acid esters. The transplastomic tool is widely operated (Bock 2014) to introduce resistance genes in the plastid genome to develop plants resistant to insect pests or tolerant to herbicides to direct recombinant proteins for engineering metabolic pathways and molecular farming (e.g. vaccines). Potato transplastomic plants were regenerated by Valkov et al. (2021) expressing recombinant GFP protein into the chloroplasts and amyloplasts or long double-stranded RNAs for pest and insect regulation.

The exploitation of transient transformation of protoplasts can bypass transgenesis, thus producing stable chloroplast-transformed (transplastomic) plants *Arabidopsis thaliana* (Ruf et al. 2021), plastid transformation provides an alternate source in plant transformation (Yu et al. 2020). Wang et al. (2018) revealed the development of homoplasmotic rice should be beneficial to transform plastid genome of monocotyledonous crops with unmanageable nature. Macro-chloroplasts in potato (Occhialini et al. 2020) were efficiently transformed by biolistic DNA delivery and proficiently generated in homoplasmic transplasmotic lines with similar amount of heterologous protein of its wild types providing increasing surface area of organelle of potential targets. De-La-Pena et al. (2022) revealed that the chloroplast transformation has numerous benefits over the traditional genomic transformation like transgene safety, high expression and recognition of superior proficiency regulatory components are crucial for obtaining and enhancing effective chloroplast transformation in different species. Chloroplast transgenics provide the capability to enable multigene metabolic engineering to produce novel biofortified crops. Plastid engineering can be accompanied with approaches, like chloroplast editing, synthetic biology, nanoparticle-mediated rapid chloroplast transformation tools, and horizontal gene transfer to target the endogenous metabolic routes to biosynthesize novel nutritional compounds (Tanwar et al. 2022).

6.5 Biofortification Through Agronomic Practices

The cost-effective technique of biofortification may improve and enhance the level of micronutrients, i.e. minerals and vitamins in plants and crops by genetic engineering and/or breeding. The biofortification process is applied in agronomic practices to enhance micronutrients and/or minerals in edible sections of crop plants by mineral fertilizers. Biofortification via agronomic methods is a simple and cost-effective process as it emphasizes on effective nutrient accessibility to crops (Shahzad et al. 2021). A better soil constitution and enhanced soil biodiversity are generated through sustainable agronomic practices like organic agriculture, minimum or no tillage, and intercropping (Chojnacka et al. 2011). The biofortification strategies used to enhance the nutritional value of crop plants in agricultural processes include genetic biofortification, that rely on traditional plant breeding and genetic engineering for better nutrient content and optimal fertilizer concentrations (Cakmak and Kutman 2018). Biofortified crops can be used as an alternative strategy to address malnutrition problems in humans and have the potential to provide sustainable solutions to overcome the micronutrient deficiencies prevalent in various rural poor populations that have inadequate nutrient availability (Simkin 2019).

Mostly biofortification strategies focus on necessary micronutrients that are organic compounds (vitamins) or minerals required in amounts <1 mg/day (Zhu et al. 2013). According to Welch and Graham (2004), an effective biofortification plan should have ultimate harvested grains of the biofortified genotypes that should be sustainable for the acceptance of farmers, and the levels of micronutrients be enhanced in such a way that it has considerable beneficial effect on health of humans, and the levels of micronutrients should be stable throughout diverse climatic regions and locations. Saini et al. (2020) emphasized biofortification techniques for improving the accessibility of important components in consumable crops by genomic interpolations and agronomic procedures.

Rice was the initial transgenic crop biofortified with provitamin A, (rice endosperm does not have provitamin A inherently), by metabolic engineering to enhance the content of provitamin A. "Golden rice" was designated as such due to the yellow color of the transgenic rice kernels. The invention of golden rice has provided access for the advancement of other "golden" crops, like wheat, sorghum, corn, potato, and cassava (Blancquaert et al. 2017). The biosynthesis of astaxanthin in the rice endosperm offers another effective model (Zhu et al. 2018). Agronomic biofortification approach has been suggested essential by Cakmak (2008) for maintaining adequate quantity of zinc present in soil and providing enough zinc carried in the seeds throughout reproductive development stage of cereal grains. To overcome the zinc deficiency in humans, zinc fertilizers are considered as an effective agronomic means for biofortification of wheat. It is reported that foliar Zn application caused effective biofortification of wheat with Zn without having any yield loss irrespective of the soil and environmental conditions. Zou et al. (2012) suggested to adopt foliar Zn fertilizer approach for improving nutritive Zn intake and compensating the lack of Zn for humans in rural areas. The prompt increase of zinc levels in field pea grains has been reported by Poblaciones and Rengel (2016) when tested for biofortification of soil foliar Zn fertilizers. The agronomic biofortification with sodium selenate by foliar application in wheat (Lara et al. 2019). The combination of iron sulfate with poultry manure and biochar could also be a successful method to enhance the yield and grain Fe biofortification of wheat in pH-influenced saline soils (Ramzani et al. 2016). The enhanced nutritional value of edible portions of Solanaceae vegetables was achieved by improving the absorptions of K, Zn, Fe, and Ca through agronomic biofortification practices (Adu et al. 2018). Ali et al. (2021) reported that treatment of foliar Zn and Fe improved the growth and yield of wheat crop. Sordi et al. (2021) reported 7% increase in grain yield and 53% increase of zinc contents in wheat and rice achieved through agronomic biofortification. The application of foliar Zn and Se stimulated better deposition of these minerals in whole grains compared to polished grains. Further, mapping images and quantification revealed the deposition of these minerals in the outer part of grains irrespective of the genotypes ensuring agronomic biofortification (Mangueze et al. 2018). The supplementation of sulfur fertilizers to spring wheat is suggested by Klikocka et al. (2018) as a source of agronomic biofortification with micronutrients (Fe, Mn, Zn, and Cu) Barczak et al. (2019) showed the need of agronomic method of biofortification, wherein it was established that fertilization of spring barley by supplementation of selected micronutrient s (Mn, Fe, Zn, and Cu) along with sulfur led to a rise in the constitution of these micronutrients in the grain of crops fertilized with sulfur in contrast to crops not nourished with same. Mandi et al. (2022)

demonstrated successful approach to ensure biofortification by sulfur fertilization and green manure in wheat and rice crops.

The practice of agronomic biofortification is emphasized by Peramaiyan et al. (2022) for genetically incompetent cultivars and for biofortified plants that are produced by conventional and genetic techniques. Agronomic biofortification is a practical way to enhance yields and nutritional attributes for various crop-micronutrient constituents. The treatment of plants with improved micronutrient fertilizers does not have intense undesirable environmental impacts if applied at with suitable ratios and usually has agronomic advantages since it makes better soil potency and crop yield. Biofortification is considered as a sustainable approach (Jha and Warkentin 2020) to improve and enhance the nutritional profile of food crops. The success of agronomic biofortification mainly rely on the bioavailability of micronutrients during the complete route from soil to crop, food, and to the consumer. In general, concentration of micronutrient-supplemented fertilizers does not affect environment if applied at appropriate ratios in addition to being agronomically advantageous by enhancing soil fertility and crop conditions (De Valença et al. 2017). Bhardwaj et al. (2022) emphasized that investment for the improvement of agronomic biofortification practices could be more promising compared to genetic and transgenic biofortification due to ensure availability of micronutrients to the genetically biofortified hidden hunger crops. Shahane and Shivay (2022) compiled a comprehensive list of areas that required urgent consideration of agronomic biofortification, among them some are selection of current types for their micronutrient, enhancement of quality parameters for crops with micronutrient fertilization responsiveness, soil micronutrient status across cropping system, micronutrient fertilization by rocks and mineral flours for organic farming, crop and ecosystem scheme particular recognition of approaches, duration and ratio of micronutrient treatment, enhancement of low-cost modified fertilizers for micronutrient treatment, investigations on improving the consumption effectiveness of mineral micronutrient fertilizers, opportunities for rising the bioavailability of micronutrients in main edibles, and finding possibilities for controlling the toxicity of micronutrients. Agronomic biofortification method is considered as one of the future strategies by Szerement et al. (2022) due to effectively improving the micronutrient malnourishment by rising the mineral contents in the plants and instantaneously improvement of their bioavailability by reduction of non-nutritional components as well as improving the application of mineral inclusion agents.

6.6 Genome Editing Is a Powerful Tool

The agricultural field has been revolutionized through genome editing of crops in the recent past. The advancement in genetic manipulation of DNA or RNA created a unique improvement in the editing of the genome that has progressed rapidly during the last 10 years and has provided remarkable advantages in numerous fields of basic and applied biotechnology (Cornu et al. 2017). Gene editing can be accomplished both *in vivo* and *in vitro* by using an editing tool that specifically identifies, cuts, and corrects the genes as well as achieves additional specific genomic modifications (Ghosh et al. 2019a). According to Li et al. (2020), this technology has also made advancements in the fields of imaging cell development, functional gene screening, epigenetic modification, gene expression regulation, pharmaceutical drug development, and disease diagnosis.

The association of synthetic biology and genome editing permits the development of purposely devised plants with specific sequence modifications. Crops developed by means of GE technology have gained substantial consideration regarding their safety since GE is a specific alteration that has no or slight impacts on traditional crop breeding. In comparison to GM (gene modification), it is likely that GE improves numerous crop characteristics with a better success ratio along with the absence of external gene insertion. The superiority of gene transfer and mutagenesis has been boosted considerably during the last five years due to the introduction of designer nucleases that permit the targeted initiation of double-strand breaks (DSBs) at precise genomic sites (Najera et al. 2019). The target genes are recognized, incised, and modified accurately by these mechanisms. GE achieved by the enzymes that are collectively known as site-directed nucleases (SDNs) that are guided by RNA molecules or DNA sequences to bind to the genome's target site, hence editing specific endogenous sequences. (Abdallah et al. 2021). These mechanisms involve endonucleases guided by RNA–DNA or protein–DNA interactions and have been applied to various crop species like rice, maize, barley, wheat, and sorghum. These nucleases are designed to identify a genomic target sequence and accordingly split the DNA, resulting in DSBs (double strand breaks). The obtained DSBs are repaired by either the homologous recombination (HR) or nonhomologous end joining (NHEJ) causing targeted gene insertions, knockout mutations, or precise site-directed point mutations (Li et al. 2016). Genome editing tools are mostly utilized in many plants for the elucidation of gene characteristics and enhancement of agronomical characteristics. This encompasses the production of plants with compositional distinctiveness and properties that enhance the efficiency and quality of crops along with providing resistance to biotic and abiotic stresses (Iqbal et al. 2020). New genomic techniques have progressively led to the accomplishment of modifications in plant genomes that are not analogous to conventional breeding or transgenesis techniques (Hudig et al. 2022). The utilization of GE mechanisms has been extensively developed in crop plants with numerous examples of specific mutagenesis involving gene knockins and knockouts, alteration, activation, or repression of target genes (Ahmed et al. 2022). Meganucleases (MegNs), transcription activator-like effector nucleases (TALENs), zinc finger nucleases (ZFNs), and clustered regularly interspaced short palindromic repeats (CRISPR)/CRISPR-associated protein 9 (Cas9) (CRISPR/Cas-9) are the mainly used methods of site-specific genome editing, which have gained enormous interest for new agricultural advancements.

6.6.1 Meganucleases (MegNs)

Meganucleases have shown promise in a variety of applications, including repairing disease-related genetic mutations and modifying crops to have better features. It can be difficult to create meganucleases for a particular DNA sequence. Custom MegNs with desired DNA recognition and cutting properties can now be made because of improvements in protein engineering approaches like directed evolution and rational design.

Meganucleases apply protein-based identification to detect target DNA sequences having a distinct nuclease domain that initiates a staggered DSB. These MegNs identify longer sites based on actual specificity, thus certain variations are obtained by mutagenesis (Zhu et al. 2017). Baltes and Voytas (2015) considered that the MegNs (also known as homing endonucleases) were originally established to be encoded by mobile introns, afterward, they have been used for producing targeted DSBs within genomes. The comparatively small size (_165 aa) and large DNA recognition sequence (_18 bp) of MegNs makes them a desirable tool for genome engineering. MegNs are usually present in bacteria, archaea, and lower eukaryotes (Silva et al. 2011). SceI meganuclease was developed in tobacco cells to create DSBs in a reporter gene carrying SceI site (Pucta 1999). In cotton, efficient gene insertions have been achieved by engineered MegNs (Hulluin et al. 2013).

6.6.2 Zinc Finger Nucleases

The identification of target sequence and distinctiveness of ZFNs are governed mainly by the amino acid sequence of each finger, the nuclease domain interaction, and by the number of fingers. Due to the modular structure of ZFNs, the DNA-binding and catalytic domains can be independently improved, thus enabling new modular assemblies (Li et al. 2020). ZFNs are composed of elements responsible for DNA recognition joined with DNA-cleavage domain of the FokI endonuclease. ZFNs depend upon chimerically constructed nucleases that possess Cys2-His2 zinc finger domain (Gaj et al. 2013). ZFNs have been successfully employed for gene editing of *Arabidopsis thaliana*, soybean, nicotiana, petunia, rapeseed, rice, apple, maize, and fig (Martinez-Fortun et al. 2017; Ran et al. 2017). By ZFNs, insertional disruption of ZmIPK1 gene modified the inositol phosphate pathway in maize seeds and provided herbicide resistance (Shukla et al. 2009). ZFNs were also utilized in the rice genome for conducting gene insertion studies and combining desirable traits (Cantos et al. 2014). Although the fundamental challenge of targeting DSBs to the specific site in the genome was resolved with the introduction of ZFNs, still the mechanism showed various disadvantages. The engineering of the enzymes is expensive, their specificity is limited, and they cause DSBs at genomic sites that are analogous but not exactly the target site; therefore, undesirable secondary, "off-target" mutations are initiated frequently (Puchta 2017).

6.6.3 TALENs

TALENs, or transcription activator-like effector nucleases, became known as an alternative to ZFNs with vast capability for precise genome modification (Christian et al. 2010). TALENs are designed for a protein that can identify and cleave DNA sequences. TALENs are composed of a nuclease domain that splits the DNA at a specific site and a DNA-binding domain resulting from TAL effectors, which are bacterial proteins and could identify DNA sequences. The repair mechanisms of the cell then close that cleavage, making tiny indels, or a new DNA sequence may be initiated at that cleavage site. TALENs are operated by re-engineering a different protein for individual targets; however, the target procedure is complicated by constructing procedures of repeat combinations even though it is currently believed as one of the most specific and exact genome-editing systems in addition to less cytotoxicity from the off-targeting outcomes (Shah et al. 2018). TALENs were first time applied to *Oryza sativa* with the consideration of crop improvement. Gene *OsSWEET14* (bacterial blight susceptibility gene) was disrupted by TALEN. The result was the production of *Oryza sativa* mutant, which was resistant to blight disease (Li et al. 2012), TALEN-based interruption of BADH2 gene (betaine aldehyde dehydrogenase), resulted in the production of flavorful rice (Shan et al. 2015). Cell wall composition and saccharification efficiency were upgraded in sugarcane by TALENS (Jung and Altepeter 2016; Kannan et al. 2018). TALEN technology was also useful in wheat (on 3 TaMLO homoeologs) for inferring resistance to powdery mildew (Wang et al. 2014). Nicotiana, *Arabidopsis*, soybean, rapeseed, potato, barley, flax, sugarcane, tomato, rice, maize, and wheat have all been modified using TALENs (Martinez-Fortun et al. 2017; Ran et al. 2017).

6.6.4 CRISPR/Cas-9

CRISPR is clustered regularly interspaced short palindromic repeats, of DNA sequences, which were first described in *Escherichia coli*, afterward in other bacteria, including *Streptococcus pyogenes*, and was realized to be a part of the prokaryotic adaptive immune defense against plasmids and bacteriophages (Horvath and Barrangou 2010). Cas9 is additionally an effective means for altering genomic sequences included in regulation (Swinnen et al. 2016). The CRISPR/Cas9 system has two main constituents: Cas9 protein and guide RNA (gRNA). Cas9 protein is an RNA-dependent DNA endonuclease to develop a complex with gRNA. The gRNA is a tiny RNA comprised of 20 nucleotides corresponding to target sequences and is essential for recruiting Cas9 protein to the target site (Wada et al. 2020; Alok et al. 2021). The recognition sites in CRISPR/Cas are constituted by a nucleic acid that permits the easier formation of CRISPR plasmids unlike the DNA recognition sites in ZFNs and TALENs that are constituted by proteins. The main absolute function of CRISPR/cas9 system in plants encompasses gene knockouts. These mutants are attained by combining indels causing in-frame shift mutations to form premature stop codons (Iqbal et al. 2020; Sushmita et al. 2021). The extension of distinct Cas effectors has made improvements in increased efficacy, target range, and improved capability for exact DNA modifications with base editing and prime editing (Laforest and Nadakuduti 2020). According to Kantor et al. (2020), CRISPR/Cas technology is universally employed as the enhancement of genome editing mechanisms owing to its effortless manipulation and simplicity. Currently, utilization of RNA viruses to provide CRISPR/Cas9 communicating RNA in plants establishes a more proficient and easy approach to producing transgene-free genome-edited plants (Ma et al. 2020); however, at present RNA virus-mediated CRISPR/Cas9 transport is valid in tobacco plants only (Gu et al. 2021). This editing tool has been employed in numerous plants for treating biotic stress. To develop wheat plants resistant to powdery mildew, CRISPR/cas9 technology was used simultaneously to transform three homologs of *EDR1* gene. The Taedr1 wheat plants obtained were tolerant to powdery mildew and cell death (Zhang et al. 2017). Nekrasov et al. (2017) produced powdery mildew-resistant tomatoes by manipulating *SlMLO1* gene. Resistance to blast by mutating gene *OsERF922* (Wang et al. 2016) and to bacterial blight by mutating gene

OsSWEET13 (Zhou et al. 2015) in rice was obtained by CRISPR/Cas9 technology. The CRISPR/Cas9 system has been effectively applied in several crops and plant species like *Arabidopsis*, rice, tobacco, sorghum, wheat, maize, soybean, tomato, potato, apple, banana, flax, grapes, Camelina, rapeseed, lettuce, cucumber, grapefruit, oranges, and watermelon (Osakabe and Osakabe 2015; Zhang et al. 2016; Jaganathan et al. 2018; Bhalothia et al. 2020; Lin et al. 2020).

In conclusion, the area of gene engineering has been completely transformed by gene-editing techniques including CRISPR-Cas9, zinc finger nucleases (ZFNs), MegNs, and TALENs. With the help of these potent tools, scientists may alter DNA sequences with pinpoint accuracy and speed, enabling targeted gene alterations, gene insertions, and trait stacking. Their adaptability and diversity of uses have created new opportunities for biotechnology, medicinal, and agriculture breakthroughs. These gene-editing technologies have a lot of potential for treating various genetic diseases, enhancing agricultural attributes, and increasing our knowledge of the intricate workings of the genome with further study and development. These tools are probably going to be crucial in determining the direction of genetic research and applications in the future, resulting in game-changing discoveries in a variety of sectors as we continue to explore their possibilities.

6.7 Examples of Biofortification Using Genome Editing Technologies

6.7.1 Amino Acid Biofortification

CRISPR/Cas9-based genome editing has been used to perform grain biofortification to augment their nutritional values which play a vital role in the control of diseases caused by nutrient deficiencies (Nerkar et al. 2022). The genes DHPS (*dapA*) and AK (*lysC*), encoding key enzymes involved in lysine biosynthesis have been edited to effectively improve lysine content in rice by as much as 25-fold (Yang et al. 2016). Additionally, these lines with high lysine content demonstrated enhanced physico-chemical characteristics while not impacting starch concentration. Also, these plants exhibited normal growth patterns in field trials with minor difference in grain color and plant height (Yang et al. 2016). In maize plants, the genes *ZmALS1* and *ZmALS2* involved in acetolactate synthase expression were edited, which resulted in accelerated biosynthesis of the branched chain amino acids such as isoleucine and leucine (Svitashev et al. 2015).

6.7.2 GABA Biofortification

Biofortification of tomatoes has been performed by editing the glutamate decarboxylase gene, which deleted the C-terminal autoinhibitory domain of this key enzyme involved in γ-aminobutyric acid (GABA) biosynthesis. This resulted in mutant tomatoes in which the accumulation of GABA was augmented sevenfold, the GABA being an important neurotransmitter associated with controlling blood pressure and anxiety (Nonaka et al. 2017a). Table 6.1 shows a few examples of the biofortification of tomato using genome editing and RNAi technologies.

Table 6.1 Biofortification of tomato using genome editing and RNAi technologies.

Gene	Pathway	Function	Source	Approach	Outcome
Stay-green 1 (SGR1), lycopene e-cyclase (LCY-E), lycopene b-cyclase (LCY-B1 and B2), and b-lycopene cyclase (blc)	Biosynthesis of carotenoids	SGR1 negatively regulates SlPSY gene; LCY-E catalyzes cyclization of lycopene to a-carotene; LCY-B converts lycopene into b-carotene	Endogenous	CRISPR/Cas9	Lycopene and b-carotene content remarkably increased, with maximum lycopene in SGR1 knockout lines (Li et al. 2018b)
Glutamate decarboxylase (GAD) genes, (SlGAD2 and SlGAD3)	Biosynthesis of GABA	Catalyzes GABA synthesis from glutamate	Endogenous	CRISPR/Cas9	1.5 to 10-fold higher GABA accumulation in red ripe fruits Nonaka et al. (2017b)
Lycopene b-cyclase (CYC-B)	Biosynthesis of carotenoids	Converts lycopene into b-carotene	Endogenous	CRISPR/Cas9	110–500% higher lycopene than cherry and the wild S. pimpinellifolium tomatoes Zsogon et al. (2018)
Pyruvate-dependent GABA-T 1 (GABA-TP1), GABA-TP2, GABA-TP3, CAT9 and Succinate semialdehyde dehydrogenase (SSADH)	GABA shunt	GABA-T converts GABA to succinic semialdehyde (SSA), which is in turn converted to succinate by SSADH; CAT9 transports GABA from vacuoles to mitochondria for GABA to catabolize	Endogenous	CRISPR/Cas9	Significantly greater GABA accumulation. Further, downregulated glutamate and upregulated alanine in some lines (Li et al. 2018a)
Anthocyanin mutant 1 (SlANT1)	Biosynthesis of anthocyanin	Encodes a Myb transcription factor that is involved in anthocyanin accumulation	Endogenous	CRISPR/Cas9 and TALEN	Dark purple plants yielded due to anthocyanin accretion Cermak et al. (2015)
DE-ETIOLATED 1 (DET1)	Regulation of photomorphogenesis	A component of light signal transduction pathway (photomorphogenesis regulatory gene)	Endogenous	Fruit specific RNAi	Augmented flavonoid and carotenoid content Davuluri et al. (2005)

(*Continued*)

Table 6.1 (Continued)

Gene	Pathway	Function	Source	Approach	Outcome
Stay-green 1 (SlSGR1)	Biosynthesis of carotenoids	Negatively regulate SlPSY	Endogenous	RNAi	Enhanced b-carotene and lycopene contents; delayed onset of fruit ripening; diminished chlorophyll breakdown Luo et al. (2013)
9-cis-epoxycarotenoid dioxygenase 1 (SlNCED1)	Degradation of carotenoids/ biosynthesis of ABA	Cleaves 9-cis-violaxanthin and 9-cis-neoxanthin into xanthoxin (C15) and C25-epoxy-apo-aldehyde/ C25-allenic-apo-aldehyde	Endogenous	Fruit specific RNAi	Increased accumulation of lycopene and b-carotene; also pectin due to diminished accretion of ABA in fruits Sun et al. (2012)

Source: Adapted from Vats et al. (2022).

6.7.3 Improvement of Oil Content and Quality

In rapeseed, mutants with yellow seed have been generated through CRISPR-Cas9 gene editing of the BnTT8 homologs that augmented the oil content by 9.47% in the T2 generation. The BnTT8 double mutants contained elevated oil yield potential as well as modified composition of fatty acids in addition to improved nutritional quality and could therefore serve beneficially in breeding of rapeseed (Zhai et al. 2020).

6.7.4 Improvement of Resistant Starch Content

In tetraploid potatoes, starch-branching enzymes have been mutated using CRISPR-Cas9 mutagenesis, and this has led to the development of mutant potato lines free of transgenes and having augmented levels of resistant starch that can enable the improvement of insulin regulation of blood sugar levels. These findings show that CRISPR-Cas9 mutagenesis holds great potential for generating commercially efficient crops.

6.7.5 Improvement of Micronutrient Bioavailability

The CRISPR-Cas9-mediated knockout of the gene OsITPK1-6 encoding the nitrogen transporter associated with accumulation of protein in rice led to augmented nitrogen uptake in addition to decreased phytic acid accretion in rice grain, thereby resulting in enhanced availability of micronutrients (Lu and Zhu 2017; Jiang et al. 2019). Karunarathne et al. (2022) generated the barley abnormal cytokinin response 1 repressor (HvARE1) mutants having elevated nitrogen content in the shoots through Agrobacterium-enabled genetic transformation of immature embryos (cv. Golden Promise) using sgRNAs that targeted HvARE1 wherein the plants were grown on nitrogen-deficient soil. Such types of crops having nitrogen use efficiency can help reduce the input of fertilizers in the soil, making them a cost-efficient alternative that can preclude environmental pollution caused by excessive applications of fertilizers. Plants with diminished cytokinin levels generated by gene editing have been shown to contain enhanced levels of elements such as Ca, Cu, Fe, Mn, Zn, and P within their biomass (Ramireddy et al. 2018a).

There is increasing realization of the potential for the use of genome editing technology for zinc and iron fortification of pulses through editing ferric-chelate reductase oxidase, iron-regulated transporter, zinc-regulated transporters, YELLOW STRIPE 1-like (YSL), natural resistance-associated macrophage protein (NRAMP), as well as ZIP proteins (Tan et al. 2017).

6.7.6 Crops Enriched in Iron

Iron is a vital element involved in metabolic pathways such as electron transport chain and oxygen transport. Disorders in iron metabolism result in diseases that are most frequent in humans such as neurodegenerative diseases and anemia (Abbaspour et al. 2014). Crop biofortification and dietary practices can help fulfill the requirement of iron (Liberal et al. 2020). In wheat plants, CRISPR-Cas9 technique has been used to engineer iron biofortification by disruption of the inositol pentakisphosphate 2-kinase 1 (IPK1) gene

(Ibrahim et al. 2021). Fe bioavailability was found to be increased by the knockdown of the rice OsVIT2 gene using gene editing (Ludwig and Slamet-Loedin 2019; Shahzad et al. 2021).

In rice seeds, iron content can be increased by overexpressing the nicotianamine synthase (NAS) gene, catalyzing the trimerization of S-adenosyl methionine to synthesize nicotianamine (NA) and nicotianamine aminotransferase (Bashir et al. 2006; Kim et al. 2006; Haydon and Cobbett 2007). NAS overexpression augments the phytosiderophore secretion and iron uptake. NA is responsible for chelating Fe(II) and Fe(III) cations and plays a vital role in its translocation and stability (Takahashi et al. 2001; Koike et al. 2004). A permease in the chloroplasts transports iron from the cytoplasm into the plastid (Duy et al. 2007), whereupon it associates with the iron-storage protein, ferritin localized in the plastid (Briat et al. 1999; Petit et al. 2001). A sixfold enhancement of iron content was reported in the endosperm upon combined expression of AtNAS1 and Pvferritin. Phytase does not preclude this accumulation of iron, but contrarily helps diminish the iron anti-nutrient phytate. Therefore, it is conclusive that ferritin and NAS overexpression in transgenic plants can improve metal translocation to seeds.

6.7.7 Zn-enriched Crops

Zinc is an important mineral associated with several metabolic pathways and is involved in the catalytic activities of nearly 100 enzymes (Sandstead 1994). It is required for DNA synthesis, cell division (Prasad 1995), protein synthesis (Prasad 1995; Heyneman 1996), immune functions (Prasad 1995; Solomons 1998), and wound healing (Heyneman 1996). Transgenic strategy was used to break down plant cytokinin to generate Zn-fortified barley plants (Ramireddy et al. 2018b). Knockout of the CYP735A gene associated with the synthesis of trans-zeatin cytokinins using CRISPR-Cas9 resulted in increased Zn concentrations (Gao et al. 2019). The concentration of zinc was augmented in wheat through editing of the TaVIT2 gene (Mourad et al. 2019).

6.7.8 Crops Enriched in Vitamin A

Among the widely disseminated isoprenoid pigments vital for photosynthetic organisms are carotenoids. Humans do not biosynthesize carotenoids de novo; however, they need them in the food material they consume, mainly as vitamin A precursors and β-carotene (Maoka 2020; Kumar et al. 2022). Vitamin A is essential for several biological activities in humans including light transduction associated with vision, immunological functions, embryonic development, and health (Timoneda et al. 2018). VAD or vitamin A deficiency is among the most concerning health issues worldwide and can lead to pediatric blindness, night blindness, xerophthalmia, as well as augmented likelihood of mortality and morbidity, especially in young children (Sommer 2008; Reddy et al. 2022). To confront deficiency of vitamin A, golden crops enriched with carotenoids were developed using conventional breeding as well as genetic engineering (Zheng et al. 2020a). In particular, the CRISPR-Cas9 technique has provided the most efficacious and widely used method for generating these golden crops (Zheng et al. 2020a). For the Kitaake rice line, the 5.2 kb of the cassette for cartenogenesis containing the maize *Ctrl* and *PSY* genes under the control of the endosperm-specific promoter was inserted using the CRISPR-Cas9 technology (Dong et al. 2020), wherein it was reported that these mutants accrued βcarotene in the endosperm at levels as high as 7.9 µg/g dry weights (DWs), that is equivalent to the levels demonstrated in Golden Rice2, the transgenically developed Kaybonnet rice variety (Zheng et al. 2020b). In another study, the maize ZmPSY1 gene encoding phytoene synthase associated with the carotenoid biosynthesis pathway was mutated using the type II CRISPR-Cas9 technology resulting in stable transformation in the progeny (Zhu et al. 2016).

6.7.9 Crops Enriched in Vitamin E

Among the essential constituents of the human diet is vitamin E (tocopherol), which is a major lipid-soluble antioxidant and deficiency of vitamin E leads to the development of cardiovascular complications and certain types of cancers (Rizvi et al. 2014). Overexpression of the enzymes homogentisate phytyltransferase (HvHPT) and homogentisate geranylgeranyltransferase (HvHGGT) in *Hordeum vulgare* using the CRISPR-Cas9 technique resulted in a significant enhancement of tocotrienol and tocopherol contents (Zeng et al. 2020). This strategy can be used for improving the vitamin E content of other crops.

6.7.10 Engineering Crops Adapted to Growing in Toxic Environments

Cadmium has high carcinogenic properties which upon long-term intake can cause renal failure. Novel cultivars of Indica rice capable of accumulating diminished grain Cd levels were generated by CRISPR-Cas9 mutagenesis of the OsNramp5 gene that regulates the uptake of Cd through the roots (Tang et al. 2017). Evaluation of these mutants showed that under conditions of high Cd content, the grain yield and agronomic traits were not impacted (Tang et al. 2017; Liu et al. 2021).

6.7.11 CRISPR-Cas9-enabled Decrease in Anti-nutrients

An alternate approach for biofortification using the genome editing technology lies in targeting the anti-nutrient genes that are responsible for the diminished bioavailability of micronutrients. In rapeseeds, 35% reduction in the anti-nutrient phytic acid content was achieved by the CRISPR-Cas9-mediated knockout of the ITPK gene that codes for an enzyme catalyzing the penultimate step of phytate biosynthesis and this mutation did not impact plant performance (Sashidhar et al. 2020). In *Triticum aestivum*, the Inositol Pentakisphosphate 2-kinase 1 (TaIPK1) gene was disrupted using the CRISPR-Cas9 technology, which resulted in the reduction of phytic acid leading to the enhancement of zinc accretion in wheat grains (Ibrahim et al. 2021). Knockout of genes associated with the phytate biosynthesis pathway in crops, which could result in the increase of bioavailable zinc and iron. Such an approach has been promising in wheat and rice, where varieties with low phytate were generated by the knockdown of Inositol 1, 3, 4, 5, 6-pentakisphosphate 2-kinase (IPK1) gene using RNAi or CRISPR-Cas technologies (Ali et al. 2013; Aggarwal et al. 2018; Ibrahim et al. 2021; Husaini 2022).

The wheat gluten protein is another major anti-nutritional substance that can result in coeliac disease in individuals who are gluten intolerant. Since over 100 loci within the wheat genome are responsible for the biosynthesis of gluten, traditional breeding strategies present great challenges. Sanchez-Leon et al. (2018), used CRISPR-Cas9 mutagenesis of the conserved sequences in the α-gliadin genes that enabled the generation of transgene-free, low-gluten wheat lines.

Saponins are good examples of anti-nutrient compounds when consumed at greater amounts (Kumar and Pandey 2020) despite being beneficial at lower amounts. It was reported that 246 P450 genes, 13 OSC genes, and 112 uridine diphosphate glycosyltransferases (UGTs) are associated with saponin biosynthesis in *Arabidopsis thaliana*. In this context, there is a necessity to determine the key regulatory genes involved with the saponin biosynthesis pathway toward eliminating them for decreased production of saponins (Guo et al. 2021). Likewise, genes responsible for biosynthesis of other anti-nutrients such as α-amylase, protease inhibitor and lathyrogens need to be recognized and further modified for reducing the production of anti-nutrient compounds (Robinson et al. 2019).

Aside from the generation of nutrient-enriched crops, several crops have been developed to augment crop production, resistance to biotic and abiotic stresses, enhance nutritional value and crop quality. Over many decades, innovations in agricultural biotechnology have significantly improved crop productivity. Consumers are more conscious of crop quality as it is directly associated with human health by providing nutrients such as vitamins, proteins, minerals, fiber and bioactive substances (Slavin and Lloyd 2012). CRISPR-mediated systems have improved the quality of horticultural, staple and oilseed crops with tremendous efficiency and accuracy (Ku and Ha 2020).

6.7.12 Benefits of Genome Editing over Other Technologies for Biofortification

Genome editing as a technology has many benefits for the generation of biofortified crops. Traditional biofortification requires the use of plant breeding approaches or else the generation of transgenic plants. Traditional plant breeding has limitations as it can be a lengthy process and in certain cases, the improvement of one trait such as iron content could lead to the loss of another trait, such as disease resistance. Transgenic technologies such as genetic modification can mitigate this hurdle but can be cumbersome due to the expensive and lengthy regulatory requirements that must be undertaken in some countries. Other countries continue to ban the cultivation of genetically modified crops.

On the other hand, genome editing is emerging as a technology that is facile and rapid, and thus more amenable as a biofortification strategy. Genome editing is easier to regulate, as often the genetic change is just through the inclusion or removal of a single nucleotide. As a result, barriers such as length of time and cost of regulations are not an issue that will delay this important technology from making worthwhile changes in people's lives.

6.8 Regulation of Genome Editing

While genetically modified crops (GMOs) produced so far have largely focused on cereals and oilseeds, genome edited crops have been initiated with far more plant diversity. This is facilitated by the speed, low cost and reduced regulatory burden associated with genome editing in general. Approximately 23% of the products being developed using genome editing are targeting the vegetable segment; 7% are for fruits; and 3% each are targeting ornamentals, legumes, and forage and grasses (Krishat Jagat 2023).

In terms of vegetables, GABA tomatoes mentioned above were commercialized in Japan by Sanatech Seed in 2021. This year, the company Pairwise produced a gene-edited green leafy vegetable range, known as Conscious™ Foods, that has been edited to enhance flavor and colors. Due to the current conflict in the Ukraine and resulting food shortages, governments are looking at food security, and thus genome editing as a technology, through a new lens. Unfortunately, some countries have not budged in their rejection of any type of genome modification. Currently, countries which accept genome editing technologies as equivalent to traditional breeding include the USA, Brazil, Argentina, Paraguay, Ecuador, Israel, Columbia, and Chile. Other countries, such as Canada, Nigeria, Russian Federation, Japan, Australia, India, Pakistan, the Philippines, and Indonesia have rules that are clear cut and use a case-by-case evaluation of gene-edited products. On the other hand, the European Union, China (mainland), and the UK are developing a new regulatory framework for plants produced using gene-editing techniques.

6.9 Conclusions and Future Prospects

Gene-editing (GE) comprises a series of new technologies that can be used to manipulate the genome of an organism at highly specific locations. The most well-known of these technologies is CRISPR-Cas9, but others include TALEN, meganuclease and xinc-finger technology. Genome editing is being utilized extensively by both private and public sectors to develop novel traits, such as biofortification, in a diversity of crops. At least 500 crops are currently under development worldwide using genome editing technologies, with the private sector contributing to 43% of product development. Examples of companies, which are currently in this space, include Pairwise and Sanatech Seed, Corteva Agriscience, Yield10 Bioscience, Benson Hill, Arcadia Biosciences, Calyxt, and Inari Agriculture.

Biofortification of crops will be critical in helping the world reach its Sustainable Development Goals. By altering one or a few nucleotides, improved crops with higher levels of minerals and vitamins can quickly be realized. This chapter has provided examples of biofortification of crops using genome editing and offers some insights into the global regulatory landscape.

Biofortification using genome editing has great potential for improving nutritional status in low- and middle-income countries. The examples provided in this chapter offer a view of the benefits of genome editing applications. It remains to be seen, however, how some countries who are reluctant to use new plant breeding technologies will address the needs of their populaces, as this technology continues to take hold.

References

Abbaspour, N., Hurrell, R., and Kelishadi, R. (2014). Review on iron and its importance for human health. *Journal of Research in Medical Sciences: The Official Journal of Isfahan University of Medical Sciences* 19 (2): 164–174.

Abdallah, N.A., Hamwieh, A., Radwan, K. et al. (2021). Genome editing techniques in plants: a comprehensive review and future prospects toward zero hunger. *GM Crops & Food* 12 (2): 601–615. https://doi.org/10.1080/21645698.2021.2021724.

Adu, M.O., Asare, P.A., Yawson, D.O. et al. (2018). Agronomic biofortification of selected underutilised solanaceae vegetables for improved dietary intake of potassium (K) in Ghana. *Heliyon* 4 (8): e00750.

Aggarwal, S., Kumar, A., Bhati, K.K. et al. (2018). RNAi-mediated downregulation of inositol pentakisphosphate kinase (IPK1) in wheat grains decreases phytic acid levels and increases Fe and Zn accumulation. *Frontiers in Plant Science* 9: 259.

Ahmad, Y., Haider, S., Iqbal, J. et al. (2022). The mechanisms of genome editing technologies in crop plants. In: *Principles and Practices of OMICS and Genome Editing for Crop Improvement* (ed. C.S. Prakash, S. Fiaz, and S. Fahad). Cham: Springer https://doi.org/10.1007/978-3-030-96925-7_13.

Ahmed, F., Prendiville, N., and Narayan, A. (2016). Micronutrient deficiencies among children and women in Bangladesh: progress and challenges. *Journal of Nutritional Science* 5: e46. https://doi.org/10.1017/jns.2016.39.

Akombi, B.J., Agho, K.E., Merom, D. et al. (2017). Child malnutrition in sub-Saharan Africa: a meta-analysis of demographic and health surveys (2006-2016). *PLoS One* 12 (5): e0177338. https://doi.org/10.1371/journal.pone.0177338.

Ali, N., Paul, S., Gayen, D. et al. (2013). Development of low phytate rice by RNAi mediated seed-specific silencing of inositol 1, 3, 4, 5, 6-pentakisphosphate 2-kinase gene (IPK1). *PLoS One* 8 (7): e68161.

Ali, S., Shah, S., and Arif, M. (2021). Agronomic biofortification with zinc and iron for the improvement of wheat phenology and yield. *Sarhad Journal of Agriculture* 37 (3): 901–914.

Alok, A., Chauhan, H., Upadhyay, S.K. et al. (2021). Compendium of plant-specific CRISPR vectors and their technical advantages. *Life* 11: 1021. https://doi.org/10.3390/life11101021.

Andersson, M.S., Saltzman, A., Virk, P.S., and Pfeiffer, W.H. (2017). Progress update: crop development of biofortified staple food crops under harvestplus. *African Journal of Food, Agriculture, Nutrition and Development* 17 (2): 11905–11935.

Apel, W. and Bock, R. (2009). Enhancement of carotenoid biosynthesis in transplastomic tomatoes by induced lycopene-to-provitamin A conversion. *Plant Physiology* 151 (1): 59–66.

Baltes, N.J. and Voytas, D.F. (2015). Enabling plant synthetic biology through genome engineering. *Trends in Biotechnology* 33 (2): 120–131.

Bansal, K.C. and Saha, D. (2012). Chloroplast genomics and genetic engineering for crop improvement. *Agricultural Research* 1: 53–66.

Barczak, B., Jastrzębska, M., and Kostrzewska, M.K. (2019). Biofortification of spring barley grain with microelements through sulfur fertilization. *Journal of Chemistry* https://doi.org/10.1155/2019/8214298.

Bashir, K., Inoue, H., Nagasaka, S. et al. (2006). Cloning and characterization of deoxymugineic acid synthase genes from graminaceous plants. *The Journal of Biological Chemistry* 281 (43): 32395.

Bhardwaj, A.K., Chejara, S., Malik, K. et al. (2022). Agronomic biofortification of food crops: an emerging opportunity for global food and nutritional security. *Frontiers in Plant Science* 13: 1055278.

Bhalothia, P., Yajnik, K., Alok, A., and Upadhyay, S.K. (2020). The current progress of CRISPR/Cas9 development in plants. *Elsevier*. https://doi.org/10.1016/B978-0-12-818140-9.00011-8.

Black, R.E., Victora, C.G., Walker, S.P. et al. (2013). Maternal and child undernutrition and overweight in low-income and middle-income countries. *The Lancet* 382 (9890): 427–451.

Blancquaert, D., De Steur, H., Gellynck, X., and Van Der Straeten, D. (2017). Metabolic engineering of micronutrients in crop plants. *Annals of the New York Academy of Sciences* 1390 (1): 59–73.

Bock, R. (2014). Genetic engineering of the chloroplast: novel tools and new applications. *Current Opinion in Biotechnology* 26: 7–13.

Bouis, H.E. and Saltzman, A. (2017). Improving nutrition through biofortification: a review of evidence from HarvestPlus, 2003 through 2016. *Global Food Security* 12: 49–58.

Briat, J.-F., Lobreaux, S., Grignon, N., and Vansuyt, G. (1999). Regulation of plant ferritin synthesis: how and why. *Cellular and Molecular Life Sciences: CMLS* 56 (1): 155–166.

Burkhardt, P.K., Beyer, P., Wunn, J. et al. (1997). Transgenic rice endosperm expressing daffodil phytoene synthase accumulates phytoene, a key intermediate of provitamin A biosynthesis. *The Plant Journal* 11 (5): 1071–1078.

Cakmak, I. (2008). Enrichment of cereal grains with zinc: agronomic or genetic biofortification? *Plant and Soil* 302: 1–17.

Cakmak, I. and Kutman, U.Á. (2018). Agronomic biofortification of cereals with zinc: a review. *European Journal of Soil Science* 69 (1): 172–180.

Cantos, C., Francisco, P., Trijatmiko, K.R. et al. (2014). Identification of "safe harbor" loci in indica rice genome by harnessing the property of zinc-finger nucleases to induce DNA damage and repair. *Frontiers in Plant Science* 5 (302): https://doi.org/10.3389/fpls.2014.00302.

Cermak, T., Baltes, N.J., Cegan, R. et al. (2015). High-frequency, precise modification of the tomato genome. *Genome Biology* 16 (1): 232. https://doi.org/10.1186/s13059-015-0796-9.

Chojnacka, K., Mikulewicz, M., and Cieplik, J. (2011). Biofortification of food with microelements. *American Journal of Agricultural and Biological Sciences* 6: 544–548.

Christian, M., Cermak, T., Doyle, E.L. et al. (2010). Targeting DNA double-strand breaks with TAL effector nucleases. *Genetics* 186: 757–761.

Christou, P. and Twyman, R.M. (2004). The potential of genetically enhanced plants to address food insecurity. *Nutrition Research Reviews* 17 (1): 23–42.

Cornu, T.I., Mussolino, C., and Cathomen, T. (2017). Refining strategies to translate genome editing to the clinic. *Nature Medicine* 23: 415–423.

D'Andrea, L. and Rodriguez-Concepcion, M. (2019). Manipulation of plastidial protein quality control components as a new strategy to improve carotenoid contents in tomato fruit. *Frontiers in Plant Science* 10: 1071.

Daniell, H., Kumar, S., and Dufourmantel, N. (2005). Breakthrough in chloroplast genetic engineering of agronomically important crops. *Trends in Biotechnology* 23 (5): 238–245.

Datta, K., Rai, M., Parkhi, V. et al. (2006). Improved'golden'indica rice and post-transgeneration enhancement of metabolic target products of carotenoids (β-carotene) in transgenic elite cultivars (IR64 and BR29). *Current Science* 91: 935–939.

Davuluri, G.R., Van Tuinen, A., Fraser, P.D. et al. (2005). Fruit-specific RNAi-mediated suppression of DET1 enhances carotenoid and flavonoid content in tomatoes. *Nature Biotechnology* 23 (7): 890–895. https://doi.org/10.1038/nbt1108.

De Valença, A.W., Bake, A., Brouwer, I.D., and Giller, K.E. (2017). Agronomic biofortification of crops to fight hidden hunger in sub-Saharan Africa. *Global Food Security* 12: 8–14.
De-la-Peña, C., León, P., and Sharkey, T.D. (2022). Chloroplast biotechnology for crop improvement. *Frontiers in Plant Science* 13: 109.
Dixit, S., Shukla, A., Singh, V. and Upadhyay, S.K. (2021) Engineering of plant metabolic pathway for nutritional improvement. *Wiley*. https://doi.org/10.1002/9781119672425.ch20.
Dong, O.X., Yu, S., Jain, R. et al. (2020). Marker-free carotenoid enriched rice generated through targeted gene insertion using CRISPR-Cas9. *Nature Communications* 11: 1–10. https://doi.org/10.1038/s41467-020-14981-y.
Duy, D., Wanner, G., Meda, A.R. et al. (2007). PIC1, an ancient permease in *Arabidopsis* chloroplasts, mediates iron transport. *Plant Cell* 19 (3): 986–1006.
FAO, IFAD, UNICEF, WFP, & WHO (2020). In brief to the state of food security and nutrition in the world 2021. In: *Transforming Food Systems for Food Security, Improved Nutrition and Affordable Healthy Diets for all* (ed. FAO), 5–31. FAO.
Foley, J.K., Michaux, K.D., Mudyahoto, B. et al. (2021). Scaling up delivery of biofortified staple food crops globally: paths to nourishing millions. *Food and Nutrition Bulletin* 42 (1): 116–132.
Gaj, T., Gersbach, C.A., and Barbas, C.F. (2013). ZFN, TALEN, and CRISPR/Cas-based methods for genome engineering. *Trends in Biotechnology* 31 (7): 397–405.
Gao, C. (2021). Genome engineering for crop improvement and future agriculture. *Cell* 184 (6): 1621–1635. https://doi.org/10.1016/j.cell.2021.01.005.
Gao, S., Xiao, Y., Xu, F. et al. (2019). Cytokinindependent regulatory module underlies the maintenance of zinc nutrition in rice. *The New Phytologist* 224 (1): 202–215. https://doi.org/10.1111/nph.15962.
Ghosh, D., Venkataramani, P., Nandi, S., and Bhattacharjee, S. (2019a). CRISPR-Cas9 a boon or bane: the bumpy road ahead to cancer therapeutics. *Cancer Cell International* 19: 12.
Ghosh, S., Datta, K., and Datta, S.K. (2019b). Rice vitamins. In: *Rice* (ed. J. Bao), 195–220. Elsevier.
Godfray, H.C.J. (2014). The challenge of feeding 9–10 billion people equitably and sustainably. *The Journal of Agricultural Science* 152 (S1): 2–8.
Gu, X., Liu, L., and Zhang, H. (2021). Transgene-free genome editing in plants. *Frontiers in Genome Editing* 3: https://doi.org/10.3389/fgeed.2021.805317.
Guo, C., Lyu, W., and Geng, Y. (2021). Bioinformatics-based screening of key genes for saponin metabolism in quinoa. *Research Square* https://doi.org/10.21203/rs.3.rs-139481/v1.
Gupta, O.P., Singh, A.K., Singh, A. et al. (2022). Wheat biofortification: utilizing natural genetic diversity, genome-wide association mapping, genomic selection, and genome editing technologies. *Frontiers in Nutrition* 9: 826131. https://doi.org/10.3389/fnut.2022.826131.
Haas, J.D., Luna, S.V., Lung'aho, M.G. et al. (2016). Consuming iron biofortified beans increases iron status in Rwandan women after 128 days in a randomized controlled feeding trial. *The Journal of Nutrition* 146 (8): 1586–1592.
Halluin, K.D., Vanderstraeten, C., Stals, E. et al. (2008). Homologous recombination: a basis for targeted genome optimization in crop species such as maize. *Plant Biotechnology Journal* 6 (1): 93–102.
Halluin, K.D., Vanderstraeten, C., Hulle, J.V. et al. (2013). Targeted molecular trait stacking in cotton through targeted double-strand break induction. *Plant Biotechnology Journal* 11 (8): 933–941.

Harada, H., Maoka, T., Osawa, A. et al. (2014). Construction of transplastomic lettuce (*Lactuca sativa*) dominantly producing astaxanthin fatty acid esters and detailed chemical analysis of generated carotenoids. *Transgenic Research* 23: 303–315.

Haydon, M.J. and Cobbett, C.S. (2007). Transporters of ligands for essential metal ions in plants. *New Phytologist* 174 (3): 499–506.

Heyneman, C.A. (1996). Zinc deficiency and taste disorders. *The Annals of Pharmacotherapy* 30 (2): 186–187. https://doi.org/10.1177/106002809603000215.

Hirschi, K.D. (2009). Nutrient biofortification of food crops. *Annual Review of Nutrition* 29: 401–421. https://doi.org/10.1146/annurev-nutr-080508-141143.

Horvath, P. and Barrangou, R. (2010). CRISPR/Cas, the immune system of bacteria and archaea. *Science* 327 (5962): 167–170. https://doi.org/10.1186/s12870-020-02385-5.

Hudig, M., Laibach, N., and Hein, A.C. (2022). Genome editing in crop plant research – alignment of expectations and current developments. *Plants* 11 (2): 212. https://doi.org/10.3390/plants11020212.

Husaini, A.M. (2022). High-value pleiotropic genes for developing multiple stress-tolerant biofortified crops for 21st-century challenges. *Heredity* 128 (6): 460–472. https://doi.org/10.1038/s41437-022-00500-w.

Hussein, H.S., Ruiz, O.N., Terry, N., and Daniell, H. (2007). Phytoremediation of mercury and organomercurials in chloroplast transgenic plants: enhanced root uptake, translocation to shoots, and volatilization. *Environmental Science & Technology* 41 (24): 8439–8446.

Ibrahim, S., Saleem, B., Rehman, N. et al. (2021). CRISPR/Cas9 mediated disruption of inositol pentakisphosphate 2-Kinase 1 (TaIPK1) reduces phytic acid and improves iron and zinc accumulation in wheat grains. *Journal of Advanced Research* https://doi.org/10.1016/j.jare.2021.07.006.

Iqbal, Z., Iqbal, M.S., Ahmad, A. et al. (2020). New prospects on the horizon: Genome editing to engineer plants for desirable traits. *Current Plant Biology* 24 (1–12): 100171.

Jaganathan, D., Ramasamy, K., Sellamuthu, G. et al. (2018). CRISPR for crop improvement: an update review. *Frontiers in Plant Science* 9: 985.

Jagat, K. (2023). Gene-edited crops market growth spurred by regulatory progress and approvals. January 18. Gene-edited crops market growth spurred by regulatory progress and approvals|Krishak Jagat.

Jha, A.B. and Warkentin, T.D. (2020). Biofortification of pulse crops: status and future perspectives. *Plants* 9 (1): 73.

Jiang, M., Liu, Y., Liu, Y. et al. (2019). Mutation of inositol 1, 3, 4- trisphosphate 5/6-kinase6 impairs plant growth and phytic acid synthesis in rice. *Plants* 8: 114. https://doi.org/10.3390/plants8050114.

Jung, J.H. and Altpeter, F. (2016). TALEN mediated targeted mutagenesis of the caffeic acid O-methyltransferase in highly polyploid sugarcane improves cell wall composition for production of bioethanol. *Plant Molecular Biology* 92 (1–2, 142): 131.

Kannan, B., Jung, J.H., Moxley, G.W. et al. (2018). TALEN-mediated targeted mutagenesis of more than 100 COMT copies/alleles in highly polyploid sugarcane improves saccharification efficiency without compromising biomass yield. *Plant Biotechnology Journal* 16 (4): 856–866.

Kantor, A., Mcclements, M.E., and Maclaren, R.E. (2020). CRISPR-Cas9 DNA base-editing and prime-editing. *International Journal of Molecular Sciences* 21: https://doi.org/10.3390/ijms21176240.

Karunarathne, S.D., Han, Y., Zhang, X.Q., and Li, C. (2022). CRISPR/Cas9 gene editing and natural variation analysis demonstrate the potential for HvARE1 in improvement of nitrogen use efficiency in barley. *Journal of Integrative Plant Biology* 64 (3): 756–770. https://doi.org/10.1111/jipb.13214.

Khan, M.S., Mustafa, G., and Joyia, F.A. (2019). Technical advances in chloroplast biotechnology. *Transgenic Crops-Emerging Trends and Future Perspectives*. https://doi.org/10.5772/intechopen.81240.

Kim, S.A., Punshon, T., Lanzirotti, A. et al. (2006). Localization of iron in *Arabidopsis* seed requires the vacuolar membrane transporter VIT1. *Science* 314 (5803): 1295–1298.

Kiran, A., Wakeel, A., Mahmood, K. et al. (2022). Biofortification of staple crops to alleviate human malnutrition: contributions and potential in developing countries. *Agronomy* 12 (2): 452.

Klikocka, H. and Marx, M. (2018). Sulphur and nitrogen fertilization as a potential means of agronomic biofortification to improve the content and uptake of microelements in spring wheat grain DM. *Journal of Chemistry* 2018: 1–12.

Koike, S., Inoue, H., Mizuno, D. et al. (2004). OsYSL2 is a rice metal-nicotianamine transporter that is regulated by iron and expressed in the phloem. *The Plant Journal* 39 (3): 415–424.

Ku, H.-K. and Ha, S.-H. (2020). Improving nutritional and functional quality by genome editing of crops: status and perspectives. *Frontiers in Plant Science* 11: 577313. https://doi.org/10.3389/fpls.2020.577313.

Kumar, S. and Pandey, G. (2020). Biofortification of pulses and legumes to enhance nutrition. *Heliyon* 6: e03682. https://doi.org/10.1016/j.heliyon.2020.e03682.

Kumar, S., Palve, A., Joshi, C. et al. (2019). Crop biofortification for iron (Fe), zinc (Zn) and vitamin A with transgenic approaches. *Heliyon* 5 (6): e01914. https://doi.org/10.1016/j.heliyon.2019.e01914.

Kumar, D., Yadav, A., Ahmad, R. et al. (2022). CRISPR-based genome editing for nutrient enrichment in crops: a promising approach toward global food security. *Frontiers in Genetics* 13: 932859. https://doi.org/10.3389/fgene.2022.932859.

Laforest, L.C. and Nadakuduti, S.S. (2020). Advances in delivery mechanisms of CRISPR gene-editing reagents in plants. *Frontiers in Genome Editing* 4: 830178. https://doi.org/10.3389/fgeed.2022.830178.

Lara, T.S., de Lima Lessa, J.H., de Souza, K.R.D. et al. (2019). Selenium biofortification of wheat grain via foliar application and its effect on plant metabolism. *Journal of Food Composition and Analysis* 81: 10–18.

Lau, W., Fischbach, M.A., Osbourn, A., and Sattely, E.S. (2014). Key applications of plant metabolic engineering. *PLoS Biology* 12 (6): e1001879.

Leung, A.M., Braverman, L.E., and Pearce, E.N. (2012). History of U.S. iodine fortification and supplementation. *Nutrients* 4 (11): 1740–1746. https://doi.org/10.3390/nu4111740.

Li, T., Liu, B., Spalding, M.H. et al. (2012). High-efficiency TALEN-based gene editing produces disease-resistant rice. *Nature Biotechnology* 30 (5): 390.

Li, J., Stoddard, T.J., Demorest, Z.L. et al. (2016). Multiplexed, targeted gene editing in Nicotianabenthamiana for glyco-engineering and monoclonal antibody production. *Plant Biotechnology Journal* 14: 533–542.

Li, X., Wang, Y., Chen, S. et al. (2018a). Lycopene is enriched in tomato fruit by CRISPR/Cas9-mediated multiplex genome editing. *Frontiers in Plant Science* 9: 559. https://doi.org/10.3389/fpls.2018.00559.

Li, Y., Wang, H., Zhang, Y., and Martin, C. (2018b). Can the world's favorite fruit, tomato, provide an effective biosynthetic chassis for highvalue metabolites? *Plant Cell Reports* 37 (10): 1443–1450. https://doi.org/10.1007/s00299-018-2283-8.

Li, H., Yang, Y., Hong, W. et al. (2020). Applications of genome editing technology in the targeted therapy of human diseases: mechanisms, advances and Prospects. *Signal Transduction and Targeted Therapy* 5 (1): 1.

Liberal, Â., Pinela, J., Vívar-Quintana, A.M. et al. (2020). Fighting iron-deficiency anemia: innovations in food fortificants and biofortification strategies. *Foods* 9 (12): 1871. https://doi.org/10.3390/foods9121871.

Lin, Q., Zong, Y., Xue, C. et al. (2020). Prime genome editing in rice and wheat. *Nature Biotechnology* 38 (5): 582–585.

Listman, G.M., Guzmán, C., Palacios-Rojas, N. et al. (2019). Improving nutrition through biofortification: preharvest and postharvest technologies. *Cereal Foods World* 64 (3): 1–7.

Liu, P., Bhatia, R., and Pachón, H. (2014). Food fortification in India: a literature review. *Indian Journal of Community Health* 26 (Supp 1): 59–74.

Liu, Q., Yang, F., Zhang, J. et al. (2021). Application of CRISPR/Cas9 in crop quality improvement. *International Journal of Molecular Sciences* 22 (8): 4206. https://doi.org/10.3390/ijms22084206.

Lockyer, S., White, A., and Buttriss, J.L. (2018). Biofortified crops for tackling micronutrient deficiencies–what impact are these having in developing countries and could they be of relevance within Europe? *Nutrition Bulletin* 43 (4): 319–357.

Lowe, N.M. (2021). The global challenge of hidden hunger: perspectives from the field. *The Proceedings of the Nutrition Society* 80 (3): 283–289. https://doi.org/10.1017/s0029665121000902.

Lu, Y. and Zhu, J.-K. (2017). Precise editing of a target base in the rice genome using a modified CRISPR/Cas9 system. *Molecular Plant* 10: 523–525. https://doi.org/10.1016/j.molp.2016.11.013.

Ludwig, Y. and Slamet-Loedin, I.H. (2019). Genetic biofortification to enrich rice and wheat grain iron: from genes to product. *Frontiers in Plant Science* 10: 833. https://doi.org/10.3389/fpls.2019.00833.

Luo, Z., Zhang, J., Li, J. et al. (2013). A STAY-GREEN protein SlSGR1 regulates lycopene and b-carotene accumulation by interacting directly with SlPSY1 during ripening processes in tomato. *The New Phytologist* 198 (2): 442–452. https://doi.org/10.1111/nph.12175.

Ma, X., Zhang, X., Liu, H., and Li, Z. (2020). Highly Efficient DNA-free plant genome editing using virally delivered CRISPR-Cas9. *Nature Plants* 6: 773–779. https://doi.org/10.1038/s41477-020-0704-5.

Mandi, S., Shivay, Y.S., Prasanna, R. et al. (2022). Improving micronutrient density in basmati rice and durum wheat through summer green manuring and elemental sulfur fertilisation. *Crop & Pasture Science* https://doi.org/10.1071/CP21313.

Mangueze, A.V.D.J., Pessoa, M.F.G., Silva, M.J. et al. (2018). Simultaneous zinc and selenium biofortification in rice. Accumulation, localization and implications on the overall mineral content of the flour. *Journal of Cereal Science* 82: 34–41.

Maoka, T. (2020). Carotenoids as natural functional pigments. *Journal of Natural Medicines* 74 (1): 1–16. https://doi.org/10.1007/s11418-019-01364-x.

Martínez-Fortún, J., Phillips, D.W., and Jones, H.D. (2017). Potential impact of genome editing in world agriculture. *Emerging Topics in Life Sciences* 1 (2): 117–133.

Masunaga, T. and Fong, J.D.M. (2018). Strategies for increasing micronutrient availability in soil for plant uptake. In: *Plant Micronutrient Use Efficiency* (ed. M.A. Hossain, T. Kamiya, D.J. Burritt, et al.), 195–208. Elsevier.

Mishra, G.P., Dikshit, H.K., Priti Kukreja, B. et al. (2022). Historical overview of biofortification in crop plants and its implications. In: *Biofortification of Staple Crops* (ed. S. Kumar, H.K. Dikshit, G.P. Mishra, and A. Singh), 31–61. Singapore: Springer https://doi.org/10.1007/978-981-16-3280-8_2.

Mourad, A.M., Alomari, D.Z., Alqudah, A.M. et al. (2019). Recent advances in wheat (Triticum spp.) breeding. In: *Advances in Plant Breeding Strategies: Cereals* (ed. J.M. Al-Khayri, J.S. Mohan, and D.V. Johnson), 559–593. New York: https://doi.org/10.1007/978-3-030-23108-8_15.

Mozafar, A. (1993). Nitrogen fertilizers and the amount of vitamins in plants: a review. *Journal of Plant Nutrition* 16 (12): 2479–2506.

Najera, V.A., Twyman, R.M., Christou, P., and Zhu, C. (2019). Applications of multiplex genome editing in higher plants. *Current Opinion in Biotechnology* 59: 93–102.

Nekrasov, V., Wang, C., Win, J. et al. (2017). Rapid generation of a transgene-free powdery mildew resistant tomato by genome deletion. *Scientific Reports* 7 (1): 1–6.

Nerkar, G., Devarumath, S., Purankar, M. et al. (2022). Advances in crop breeding through precision genome editing. *Frontiers in Genetics* 13: 880195. https://doi.org/10.3389/fgene.2022.88019.

Nonaka, S., Arai, C., Takayama, M. et al. (2017a). Efficient increase of C -aminobutyric acid (GABA) content in tomato fruits by targeted mutagenesis. *Scientific Reports* 7: 705. https://doi.org/10.1038/s41598-017-06400-y.

Nonaka, S., Arai, C., Takayama, M. et al. (2017b). Efficient increase of Ç-aminobutyric acid (GABA) content in tomato fruits by targeted mutagenesis. *Scientific Reports* 7 (1): 7057. https://doi.org/10.1038/s41598-017-06400-y.

Occhialini, A., Pfotenhauer, A.C., Frazier, T.P. et al. (2020). Generation, analysis, and transformation of macro-chloroplast potato (*Solanum tuberosum*) lines for chloroplast biotechnology. *Scientific Reports* 10 (1): 21144.

Okwuonu, I.C., Narayanan, N.N., Egesi, C.N., and Taylor, N.J. (2021). Opportunities and challenges for biofortification of cassava to address iron and zinc deficiency in Nigeria. *Global Food Security* 28: 100478. https://doi.org/10.1016/j.gfs.2020.100478.

Omari, R., Zotor, F., Tagwireyi, J., and Lokosang, L. (2019). Advocacy for scaling up biofortified crops for improved micronutrient status in Africa: approaches, achievements, challenges and lessons. *Proceedings of the Nutrition Society* 78 (4): 567–575. https://doi.org/10.1017/S0029665119000521.

Osakabe, Y. and Osakabe, K. (2015). Genome editing with engineered nucleases in plants. *Plant & Cell Physiology* 56 (3): 389–400.

de Pee, S. (2014). Proposing nutrients and nutrient levels for rice fortification. *Annals of the New York Academy of Sciences* 1324 (1): 55–66.

Peramaiyan, P., Craufurd, P., Kumar, V. et al. (2022). Agronomic biofortification of zinc in rice for diminishing malnutrition in South Asia. *Sustainability* 14 (13): 7747.

Petit, J.-M., Briat, J.-F., and Lobréaux, S. (2001). Structure and differential expression of the four members of the *Arabidopsis thaliana* ferritin gene family. *Biochemical Journal* 359 (3): 575–582.

Pixley, K., Rojas, N.P., Babu, R. et al. (2013). Biofortification of maize with provitamin A carotenoids. *Carotenoids and Human Health* 271–292.

Poblaciones, M.J. and Rengel, Z. (2016). Soil and foliar zinc biofortification in field pea (*Pisum sativum* L.): Grain accumulation and bioavailability in raw and cooked grains. *Food Chemistry* 212: 427–433.

Prasad, A.S. (1995). Zinc: an Overview. *Nutrition* 11 (1 Suppl. l): 93–99.

Prasanna, B.M., Palacios-Rojas, N., Hossain, F. et al. (2020). Molecular breeding for nutritionally enriched maize: status and prospects. *Frontiers in Genetics 10*.

Puchta, H. (1999). Use of I-Sce I to induce DNA double-strand breaks in Nicotiana. *Methods in Molecular Biology* https://doi.org/10.1385/1-59259-675-4:447.

Puchta, H. (2017). Applying CRISPR/Cas genome engineering in plants: the best yet to come. *Current Opinion in Plant Biology* 36: 1–8.

Ramireddy, E., Galuszka, P., and Schmülling, T. (2018a). Zn-fortified cereal grains in field-grown barley by enhanced root cytokinin breakdown. *Plant Signaling & Behavior* 13 (11): e1530023. https://doi.org/10.1080/15592324.2018.1530023.

Ramireddy, E., Hosseini, S.A., Eggert, K. et al. (2018b). Root engineering in barley: increasing cytokinin degradation produces a larger root system, mineral enrichment in the shoot and improved drought tolerance. *Plant Physiology* 177 (3): 1078–1095. https://doi.org/10.1104/pp.18.00199.

Ramzani, P.M.A., Khalid, M., Naveed, M. et al. (2016). Iron biofortification of wheat grains through integrated use of organic and chemical fertilizers in pH affected calcareous soil. *Plant Physiology and Biochemistry* 104: 284–293.

Ran, Y., Liang, Z., and Gao, C. (2017). Current and future editing reagent delivery systems for plant genome editing. *Science China. Life Sciences* 60 (5): 490–505.

Read, S.A., Obeid, S., Ahlenstiel, C., and Ahlenstiel, G. (2019). The role of zinc in antiviral immunity. *Advances in Nutrition* 10 (4): 696–710.

Reddy, G.B., Shalini, T., Ghosh, S. et al. (2022). Prevalence of vitamin A deficiency and dietary inadequacy in indian school-age children and adolescents. *European Journal of Nutrition* 61 (1): 197–209. https://doi.org/10.1007/s00394-021-02636-7.

Reed, K.M. and Bargmann, B.O.R. (2021). Protoplast regeneration and its use in new plant breeding technologies. *Frontiers in Genome Editing* 20: https://doi.org/10.3389/fgeed.2021.734951.

Řepková, J. (2010). Potential of chloroplast genome in plant breeding. *Czech Journal of Genetics and Plant Breeding* 46 (3): 103–113.

Rhowell, N.T. Jr., Fernie, A.R., and Sreenivasulu, N. (2021). Meeting human dietary vitamin requirements in the staple rice via strategies of biofortification and post-harvest fortification. *Trends in Food Science & Technology* 109: 65–82.

Rizvi, S., Raza, S.T., Ahmed, F. et al. (2014). The role of vitamin e in human health and some diseases. *Sultan Qaboos University Medical Journal* 14 (2): e157–e165.

Robinson, G., Balk, J., and Domoney, C. (2019). Improving pulse crops as a source of protein, starch and micronutrients. *Nutrition Bulletin* 44: 202–215. https://doi.org/10.1111/nbu.12399.

Roser, M. and Ritchie, H. (2019). *Hunger and Undernourishment*. Our World in Data.

Ruf, S., Kroop, X., and Bock, R. (2021). Chloroplast transformation in *Arabidopsis*. *Current Protocols* 1 (4): e103.

Saini, D.K., Devi, P., and Kaushik, P. (2020). Advances in genomic interventions for wheat biofortification: a review. *Agronomy* 10 (1): 62.

Sanchez-Leon, S., Gil-Humanes, J., Ozuna, C.V. et al. (2018). Low-gluten, nontransgenic wheat engineered with CRISPR/Cas9. *Plant Biotechnology Journal* 16: 902–910. https://doi.org/10.1111/pbi.12837.

Sandstead, H.H. (1994). Understanding zinc: recent observations and interpretations. *The Journal of Laboratory and Clinical Medicine* 124 (3): 322–327.

Sashidhar, N., Harloff, H.J., Potgieter, L., and Jung, C. (2020). Gene editing of three BnITPK genes in tetraploid oilseed rape leads to significant reduction of phytic acid in seeds. *Plant Biotechnology Journal* 18: 2241–2250. https://doi.org/10.1111/pbi.13380.

Scott, S.P., Murray-Kolb, L.E., Wenger, M.J. et al. (2018). Cognitive performance in Indian school-going adolescents is positively affected by consumption of iron-biofortified pearl millet: a 6-month randomized controlled efficacy trial. *The Journal of Nutrition* 148 (9): 1462–1471.

Shah, T., Andleeb, T., Latief, S., and Noor, M.A. (2018). Genome editing in plants: advancing crop transformation and overview of tools. *Plant Physiology and Biochemistry* 131: 12–21.

Shahane, A.A. and Shivay, Y.S. (2022). Agronomic biofortification of crops: current research status and future needs. *Indian Journal of Fertilisers* 18 (2): 164–179.

Shahzad, R., Jamil, S., Ahmad, S. et al. (2021). Biofortification of cereals and pulses using new breeding techniques: current and future perspectives. *Frontiers in Nutrition* 8: 721728. https://doi.org/10.3389/fnut.2021.721728.

Shan, Q., Zhang, Y., Chen, K. et al. (2015). Creation of fragrant rice by targeted knockout of the Os BADH 2 gene using TALEN technology. *Plant Biotechnology Journal* 13 (6): 791–800.

Sharwood, R.E. (2017). Engineering chloroplasts to improve Rubisco catalysis: prospects for translating improvements into food and fiber crops. *New Phytologist* 213 (2): 494–510.

Shukla, V.N., Doyon, Y., Miller, J.C. et al. (2009). Precise genome modification in the crop species Zea mays using zinc-finger nucleases. *Nature* 459 (7245): 437–441.

Silva, G., Poirot, L., Galetto, R. et al. (2011). Meganucleases and other tools for targeted genome engineering: perspectives and challenges for gene therapy. *Current Gene Therapy* 11 (1): 11–27.

Simkin, A.J. (2019). Genetic engineering for global food security: photosynthesis and biofortification. *Plants* 8 (12): 586.

Singh, R., Govindan, V., and Andersson, M.S. (2017). *Zinc-Biofortified Wheat: Harnessing Genetic Diversity for Improved Nutritional Quality*. Science Brief: Biofortification No. 1. Global Crop Diversity Trust > Briefs.

Slavin, J.L. and Lloyd, B. (2012). Health benefits of fruits and vegetables. *Advances in Nutrition* 3 (4): 506–516. https://doi.org/10.3945/an.112.002154.

Solomons, N.W. (1998). Mild human zinc deficiency produces an imbalance between cell-mediated and humoral immunity. *Nutrition Reviews* 56 (1 Pt 1): 27–28. https://doi.org/10.1111/j.1753-4887.1998.tb01656.x.

Sommer, A. (2008). Vitamin A deficiency and clinical disease: an historical overview. *The Journal of Nutrition* 138 (10): 1835–1839. https://doi.org/10.1093/jn/138.10.1835.

Sordi, E., Novakowiski, J.H., Rebesquini, R. et al. (2021). Agronomic biofortification of rice and wheat with zinc: a metanalytical study. *Research, Society and Development* 10 (6): e39210615133–e39210615133.

Ssentongo, P., Ssentongo, A.E., Ba, D.M. et al. (2021). Global, regional and national epidemiology and prevalence of child stunting, wasting and underweight in low- and middle-income countries, 2006-2018. *Scientific Reports* 11 (1): 5204. https://doi.org/10.1038/s41598-021-84302-w.

Stevens, G.A., Beal, T., Mbuya, M.N.N. et al. (2022). Micronutrient deficiencies among preschool-aged children and women of reproductive age worldwide: a pooled analysis of individual-level data from population-representative surveys. *The Lancet Global Health* 10 (11): e1590–e1599. https://doi.org/10.1016/S2214-109X(22)00367-9.

Sun, L., Yuan, B., Zhang, M. et al. (2012). Fruit-specific RNAi-mediated suppression of SlNCED1 increases both lycopene and b-carotene contents in tomato fruit. *Journal of Experimental Botany* 63 (8): 3097–3108. https://doi.org/10.1093/jxb/ers026.

Sushmita, K.G., Upadhyay, S.K., and Verma, P.C. (2021). An overview of genome-engineering methods. In: *Genome Engineering for Crop Improvement* (ed. S.K. Upadhyay). Wiley. https://doi.org/10.1002/9781119672425.ch1.

Svitashev, S., Young, J.K., Schwartz, C. et al. (2015). Targeted mutagenesis, precise gene editing, and site-specific gene insertion in maize using Cas9 and guide RNA. *Plant Physiology* 169: 931–945. https://doi.org/10.1104/pp.15.00793.

Swinnen, G., Goossens, A., and Pauwels, A. (2016). Lessons from domestication: targeting Cis-regulatory elements for crop improvement. *Trends in Plant Science* 21: 506–515. https://doi.org/10.1016/j.tplants.2016.01.014.

Szerement, J., Szatanik-Kloc, A., Mokrzycki, J., and Mierzwa-Hersztek, M. (2022). Agronomic biofortification with Se, Zn, and Fe: an effective strategy to enhance crop nutritional quality and stress defense – a Review. *Journal of Soil Science and Plant Nutrition* 22 (1): 1129–1159.

Takahashi, M., Nakanishi, H., Kawasaki, S. et al. (2001). Enhanced tolerance of rice to low iron availability in alkaline soils using barley nicotianamine aminotransferase genes. *Nature Biotechnology* 19 (5): 466–469.

Tan, G.Z., Das Bhowmik, S.S., Hoang, T.M. et al. (2017). Finger on the pulse: pumping iron into chickpea. *Frontiers in Plant Science* 8: 1755. https://doi.org/10.3389/fpls.2017.01755.

Tang, L., Mao, B., Li, Y. et al. (2017). Knockout of OsNramp5 using the CRISPR/Cas9 system produces low Cd-Accumulating indica rice without compromising yield. *Scientific Reports* 7: 14438. https://doi.org/10.1038/s41598-017-14832-9.

Tanumihardjo, S.A., McCulley, L., Roh, R. et al. (2020). Maize agro-food systems to ensure food and nutrition security in reference to the sustainable development goals. *Global Food Security* 25: 100327.

Tanwar, N., Arya, S.S., Rookes, J.E. et al. (2022). Prospects of chloroplast metabolic engineering for developing nutrient-dense food crops. *Critical Reviews in Biotechnology* 1–18. https://doi.org/10.1080/07388551.2022.2092717.

Timoneda, J., Rodríguez-Fernández, L., Zaragozá, R. et al. (2018). Vitamin A deficiency and the lung. *Nutrients* 10 (9): 1132. https://doi.org/10.3390/nu10091132.

Tiozon, R.N. Jr., Camacho, D.H., Bonto, A.P. et al. (2021). Efficient fortification of folic acid in rice through ultrasonic treatment and absorption. *Food Chemistry* 335: 127629.

Twumasi-Afriyie, S., Friesen, D., and Pixley, K. (2011). Quality protein maize: progress and prospects. *Plant Breed Rev* 34: 83.

UNICEF (2007). Vitamin A supplementation: a decade of progress. The United Nations Children's Fund (UNICEF). Global overview on progress and limitations of ongoing vitamin A supplementation programs.

Upadhyay, S.K. (2021). *Genome Engineering for Crop Improvement*. Wiley https://doi.org/10.1002/9781119672425.

Valkov, V.T., Gargano, D., Cardi, T., and Scotti, N. (2021). Plastid transformation in potato: an important source of nutrition and industrial materials. In: *Chloroplast Biotechnology: Methods and Protocols* (ed. P. Maliga), 247–256. Springer.

Vats, S., Bansal, R., Rana, N. et al. (2022). Unexplored nutritive potential of tomato to combat global malnutrition. *Critical Reviews in Food Science and Nutrition* 62 (4): 1003–1034. https://doi.org/10.1080/10408398.2020.1832954.

Wada, N., Ueta, R., Osakabe, Y., and Osakabe, K. (2020). Precision genome editing in plants: state of- the-art in CRISPR/Cas9-based genome engineering. *BMC Plant Biology* 20: 234.

Wang, Y., Cheng, X., Shan, Q. et al. (2014). Simultaneous editing of three homoeoalleles in hexaploid bread wheat confers heritable resistance to powdery mildew. *Nature Biotechnology* 32 (9): 947.

Wang, F., Wang, C., Liu, P. et al. (2016). Enhanced rice blast resistance by CRISPR/Cas9-targetd mutagenesis of the ERF transcription factor gene OsERF922. *PLoS One* 11 (4): e0154027.

Wang, Y., Wei, Z., and Xing, S. (2018). Stable plastid transformation of rice, a monocot cereal crop. *Biochemical and Biophysical Research Communications* 503 (4): 2376–2379.

Wang, T., Zhang, H., and Zhu, H. (2019). CRISPR technology is revolutionizing the improvement of tomato and other fruit crops. *Horticulture Research* 6 (1): 77. https://doi.org/10.1038/s41438-019-0159-x.

Welch, R.M. and Graham, R.D. (2004). Breeding for micronutrients in staple food crops from a human nutrition perspective. *Journal of Experimental Botany* 55 (396): 353–364.

Wessels, I., Maywald, M., and Rink, L. (2017). Zinc as a gatekeeper of immune function. *Nutrients* 9 (12): 1286.

Wurbs, D., Ruf, S., and Bock, R. (2007). Contained metabolic engineering in tomatoes by expression of carotenoid biosynthesis genes from the plastid genome. *The Plant Journal* 49 (2): 276–288.

Yang, Q.Q., Zhang, C.Q., Chan, M.L. et al. (2016). Biofortification of rice with the essential amino acid lysine: molecular characterization, nutritional evaluation, and field performance. *Journal of Experimental Botany* 67: 4285–4296. https://doi.org/10.1093/jxb/erw209.

Ye, X., Al-Babili, S., Klöti, A. et al. (2000). Engineering the provitamin A (beta-carotene) biosynthetic pathway into (carotenoid-free) rice endosperm. *Science* 287 (5451): 303–305. https://doi.org/10.1126/science.287.5451.303.

Yu, Y., Yu, P.-C., Chang, W.-J. et al. (2020). Plastid transformation: how does it work? Can it be applied to crops? What can it offer? *International Journal of Molecular Sciences* 21 (14): 4854.

Zaidi, S.S.E.A., Mahas, A., Vanderschuren, H., and Mahfouz, M.M. (2020). Engineering crops of the future: CRISPR approaches to develop climate-resilient and disease-resistant plants. *Genome Biology* 21 (1): 289. https://doi.org/10.1186/s13059-020-02204-y.

Zeng, Z., Han, N., Liu, C. et al. (2020). Functional dissection of HGGT and HPT in barley Vitamin E biosynthesis via CRISPR/Cas9- enabled genome editing. *Annals of Botany* 126 (5): 929–942. https://doi.org/10.1093/aob/mcaa115.

Zhai, Y., Yu, K., Cai, S. et al. (2020). Targeted mutagenesis of BnTT8 homologs controls yellow seed coat development for effective oil production in *Brassica Napus* L. *Plant Biotechnology Journal* 18 (5): 1153–1168. https://doi.org/10.1111/pbi.13281.

Zhang, D., Li, Z., and Li, J.F. (2016). Targeted gene manipulation in plants using the CRISPR/Cas technology. *Journal of Genetics and Genomics* 43 (5): 251–262.

Zhang, Y., Bai, Y., Wu, G. et al. (2017). Simultaneous modification of three homoeologs of Ta EDR 1 by genome editing enhances powdery mildew resistance in wheat. *The Plant Journal* 91 (4): 714–724.

Zhau, J., Peng, Z., Long, J. et al. (2015). Gene targeting by the TAL effector PthXo2 reveals cryptic resistance gene for bacterial blight of rice. *The Plant Journal* 82 (4): 632–643.

Zheng, X., Kuijer, H.N., and Al-Babili, S. (2020a). Carotenoid biofortification of crops in the CRISPR era. *Trends in Biotechnology* 39: 857–860. https://doi.org/10.1016/j.tibtech.2020.12.003.

Zheng, X., Giuliano, G., and Al-Babili, S. (2020b). Carotenoid biofortification in crop plants: citius, altius, fortius. biochimica biophysica acta (BBA). *Molecular and Cell Biology of Lipids* 1865 (11): 158664. https://doi.org/10.1016/j.bbalip.2020.158664.

Zhu, C., Sanahuja, G., Yuan, D. et al. (2013). Biofortification of plants with altered antioxidant content and composition: genetic engineering strategies. *Plant Biotechnology Journal* 11 (2): 129–141.

Zhu, J., Song, N., Sun, S. et al. (2016). Efficiency and inheritance of targeted mutagenesis in maize using CRISPR-Cas9. *Journal of Genetics and Genomics* 43: 25–36. https://doi.org/10.1016/j.jgg.2015.10.006.

Zhu, C., Bortesi, L., Baysal, C. et al. (2017). Characteristics of genome editing mutations in cereal crops. *Trends in Plant Science* 22 (1): https://doi.org/10.1016/j.tplants.2016.08.009.

Zhu, Q., Zeng, D., Yu, S. et al. (2018). From golden rice to aSTARice: bioengineering astaxanthin biosynthesis in rice endosperm. *Molecular Plant* 11 (12): 1440–1448.

Zou, C.Q., Zhang, Y.Q., Rashid, A. et al. (2012). Biofortification of wheat with zinc through zinc fertilization in seven countries. *Plant and Soil* 361: 119–130.

Zsogon, A., Cermak, T., Naves, E.R. et al. (2018). De novo domestication of wild tomato using genome editing. *Nature Biotechnology* 36 (12): 1211–1216. https://doi.org/10.1038/nbt.4272.

7

Genome Editing for Nutritional Improvement of Crops

Pooja Kanwar Shekhawat[1], *Hasthi Ram*[2], *and Praveen Soni*[1]

[1] *Department of Botany, University of Rajasthan, Jaipur, Rajasthan, India*
[2] *Laboratory of Functional Rice Genomics, National Institute of Plant Genome Research, New Delhi, Delhi, India*

Abbreviations

ANT	Anthocyanin
BADH	Betaine aldehyde dehydrogenase
BBL	Berberine bridge enzyme-like
BE	Base editing
BLC	Beta lycopene cyclase
bZIPs	Basic leucine zipper proteins
CAT	Cationic amino acid transporter
CLV	CLAVATA
CRISPR/Cas9	Clustered regularly interspaced short palindromic repeats (CRISPR)/ CRISPR-associated protein system 9
CKX	Cytokinin oxidase
CycB	Lycopene β-cyclase
DSB	Double strand breaks
DEP	Dense and erect panicle
DGAT	Diacylglycerol acyltransferase
DOX	Dioxygenase
EPSPS	5-Enolpyruvylshikimate-3-phosphate synthase
FAD	Fatty acid desaturase
FAE	Fatty acid elongase
FAS	FASCIATED
FW	Fruit weight
GAD	Glutamate decarboxylase
GBSS	Granule-bound starch synthase

Applications of Genome Engineering in Plants, First Edition. Edited by Santosh Kumar Upadhyay.
© 2024 John Wiley & Sons Ltd. Published 2024 by John Wiley & Sons Ltd.

GETs	Genome editing techniques
Gn1a	Grain number 1a
GR	Golden rice
GS	Grain size
GW	Grain weight
HAK	High-affinity K^+ transporter
HDR	Homology-directed recombination
HR	Homologous recombination
IdnDH	Idonate dehydrogenase
IPK	Inositol phosphate kinase
ISA	Isoamylase
ITPK	Inositol 1,3,4-trisphosphate 5/6-kinase
lncRNA	Long non-coding RNA
LCY	Lycopene ε-cyclase
MADS	MADS-box protein
MegaN	Meganuclease
MRP	Multidrug resistance-associated protein
MULT	MULTIFLORA
MYB12	MYB transcription factor 12
NHEJ	Non-homologous end joining
Nor	Non-ripening
NPBTs	Novel plant breeding techniques
OGD	oxoglutarate-dependent dioxygenase
PDAT	Phospholipid diacylglycerol acyltransferase
PE	Prime editing
PRX	Peroxidase
PPOs	Polyphenol oxidases
PPKL	Putative protein phosphatase Kelch-like
PSY	Phytoene synthase gene
PTST	Protein targeting to starch
RC	Red coloration
RIN	Ripening inhibitor
SBE	Starch branching enzyme
SGR	Stay-green
SP	Self-pruning
SP5G	Self-pruning 5G
SSADH	Succinate semialdehyde dehydrogenase
SSR	Sterol side chain reductase
TALENs	Transcription activator-like effector nucleases
TGW	Thousand-grain weight
VInv	Vacuolar invertase
WUS	WUSCHEL
Wx	Waxy protein
ZFNs	Zinc finger nucleases

7.1 Introduction

Malnutrition and hunger are becoming big problems as world population is rising. According to an estimation, around 800 million people are facing an inadequate supply of food and approximately 2 billion are suffering from undernourishment (FAO 2019). To make sure everyone has enough food and good nutrition, agriculture sector needs to produce more nutrient-rich food. To reach the second goal of the United Nations for sustainable development, which is "zero hunger and better nutrition," we need to change from traditional agricultural practices to the most recent ones (Meemken and Qaim 2018). In the middle of the twentieth century, agrochemicals and better agronomic practices helped farmers get the most out of their crops. Meanwhile, semi-dwarf crops providing higher yields were also developed by traditional breeding techniques, which led to "green revolution" (Bhargava and Srivastava 2019). However, constant use of agrochemicals has resulted into adverse effects on environment. Now in the current period, different genome editing techniques (GETs) have sped up the plant breeding process and have provided useful, time-saving, cost-effective, and flexible ways to improve available germplasm (Fiaz et al. 2021). As a consequence of use of new GETs, it is now possible to precisely edit agriculturally important genes. GETs are novel plant breeding techniques (NPBTs) of modern age, which include ZFNs (zinc finger nucleases), MegaN (meganucleases), TALENs (transcription activator-like effector nucleases), CRISPR/Cas9 (clustered regularly interspaced short palindromic repeats CRISPR/CRISPR associated protein system 9), the CRISPR from *Prevotella* and *Francisella*1 (Cpf1), and base editing (BE) and prime editing (PE) (Fiaz et al. 2021; Upadhyay 2021).

7.2 Evolution of Techniques for Improvement of Crops' Genomes

Conventional plant breeding and random mutagenesis (making use of physical or chemical mutagens) methods that rely on naturally occurring or artificially made genetic differences, spend a lot of time, money, and efforts figuring out superior offspring. The randomness of notable mutations that lead to the appearance of desirable traits makes it hard to create novel commercial varieties (Pacher and Puchta 2017). In 1996, era of transgenic crops started. GM (genetically modified) crops with beneficial traits were made by genetic engineering. Still, there are a lot of unproven worries about the safety of GM crops for health and the environment. Adding genes from other species also takes time and money (Prado et al. 2014). Poor science communication and existing mistrust about GM crops have fueled hot debates about them.

In the last two decades, four categories of nucleases have been engineered for genome editing: MegaN (Rouet et al. 1994), ZFNs (Kim et al. 1996), TALENs (Christian et al. 2010), and CRISPR/Cas9 (Jinek et al. 2012; Voytas and Gao 2014; Cermak et al. 2015; Sushmita et al. 2021). These endonucleases edit genomic sequences by introducing site-specific double-strand breaks (DSBs) in DNA. MegaNs are not used so much because it is hard to find the target locus. ZFN and TALEN are specially created restriction enzymes that are widely recognized for efficiently, effectively, and accurately altering the genome (Saurabh et al. 2014). Importantly, the DNA binding domain (DBD) proteins used in these two approaches are different, but the mechanisms for both GETs are the same. The discovery of CRISPR/Cas9 and the progress in biology have made it possible for biotechnologists to cause mutations in any genome of interest in a specific and efficient way (Bhalothia et al. 2020; Fiaz et al. 2021). CRISPR/Cas9 and its variations are third-generation GETs (Wolabu et al. 2020). It can change genomic features of bacteria, animal cells, plant cells, and among other living things (Alok et al. 2018, 2021; Kumar et al. 2022; Singh et al. 2022). This system needs a short guide RNA (sgRNA) to lead Cas9 to the specific site (Li et al. 2018g). Cas9 nuclease cut the double-stranded DNA (dsDNA) at that sequence, which matches with the sgRNA. It causes blunt-end DSBs. DSBs can be fixed by either precise homologous recombination (HR), which results in gene insertions and replacements, or imprecise nonhomologous end joining (NHEJ), which leads to gene knockouts by inserting or removing nucleotides (Symington and Gautier 2011; Upadhyay et al. 2013).

Cpf1 stands for CRISPR from *Prevotella* and *Francisella* 1, a class 2 CRISPR effector (previously known as Cas12a), has also helped agricultural research scientists in editing genomes (Zetsche et al. 2015). This endonuclease is an effector nuclease protein that cuts DNA at the proximal end of the protospacer adjacent motif (PAM) and is directed by a single RNA that binds at the upstream region of the PAM (Zetsche et al. 2015). As Cpf1's nuclease is smaller than that of Cas9, it needs shorter CRISPR RNA (crRNA) that works better and creates DSBs (Liu et al. 2017). CRISPR/Cpf1 enzymes have a key property called target-activated nonspecific ssDNase activity. This is carried out by the same active site that cuts dsDNA at a specific site (Liu et al. 2017). The Cpf1 mechanism avoids the trans-activating crRNA (tracrRNA) needed for crRNA maturation (Zhang et al. 2017). The Cpf1 also works well with the target site because it uses a T-rich PAM region, while Cas9 needs a G-rich PAM (Kim et al. 2021).

The CRISPR/Cas9 base editing (BE) method is significantly more exact and straightforward since it enables nucleotide conversion without leading to DSBs in the target DNA (Yin et al. 2017). Cytosine BE (CBE), which changes cytosine to thymine, shows great efficiency (Komor et al. 2016; Marx 2018). This system is made up of four components: a single sgRNA, dCas9, cytosine deaminase, and an inhibitor of uracil DNA glycosylase (UGI). Another method is adenine BE (ABE), which converts adenine to guanine (Gaudelli et al. 2017; Khanday et al. 2019). Without needing DSBs for DNA alteration, the BEs prevent indel formation at both target and off-target locations (Eid et al. 2018). Prime editing (PE) is the most advanced version of CRISPR. It has a high rate of target recognition, very low off-target effects, and a high rate of mutation (Fiaz et al. 2021).

7.3 Genome Editing for Nutritional Improvement

The ability of GETs to increase crop nutrient content has been proved in a variety of crop species (Wang et al. 2019d). GETs are better than RNAi technology as knocking out the corresponding gene doesn't cause pleiotropic effects (Ku and Ha 2020). Figure 7.1 shows the application of GETs approaches for crop improvement. The details of crops, which have been improved via GETs, have been provided in Table 7.1.

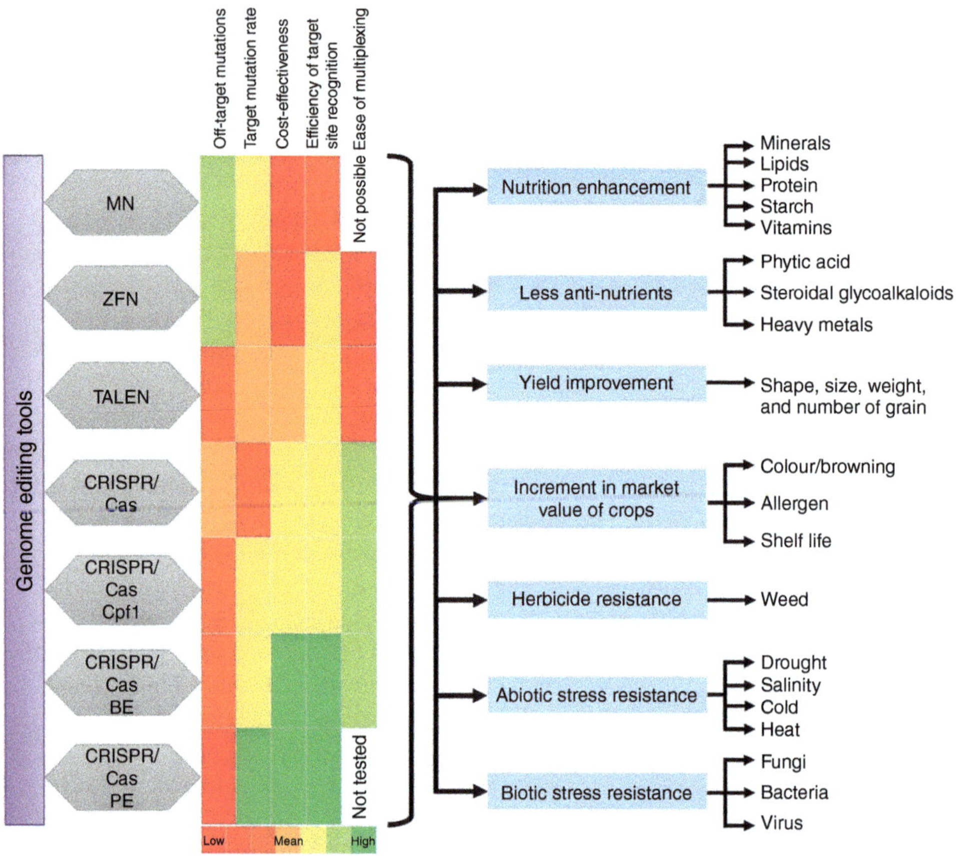

Figure 7.1 Representation of different genome editing tools and their applications for trait improvement.

7.3.1 Improvement in Cereal Crops

Several important cereal crops have been targeted for genome editing to increase food and nutrition security for the fast-growing population. Rice is one of them, which is consumed by millions of people all over the world on a daily basis and it accounts for 20% of the energy supply for the world's dietary needs (Birla et al. 2017). Rice grain fragrance is an important qualitative feature that influences its acceptance and market value (Ashokkumar et al. 2020). The *BADH2* allele was mutated through TALEN resulting in the enhancement of 2-acetyl-1-pyrroline (2AP) content up to 0.35–0.75 mg/kg in rice grains (Shan et al. 2015). Deletions in *BADH2* alleles through CRISPR/Cas9 produced aroma in the non-aromatic rice cultivar ASD16 (Ashokkumar et al. 2020). Similarly, by targeting another gene *EPSPS,* nipponbare cultivar was improved for aroma (Li et al. 2016). Apart from the fragrance, another parameter that greatly affects the quality of rice cooking and eating is starch content, which is related to glutinosity (Juliano 1998). It is made up of two polymers: amylose and amylopectin. Knockout of the *SBEIIb* gene, encoding a starch branching enzyme, produced rice with low viscosity

Table 7.1 List of genes edited for nutritional improvement of crops.

S.N.	Crop name	Name of editing tool	Targeted gene(s)	Trait improvement category	Trait group	Improved trait	Transformation method	Reference
1.	Rice (*Oryza sativa*)	TALEN	*BADH2*	Nutrition	Market value	Increased fragrance	*Agrobacterium*-mediated	Shan et al. (2015)
		CRISPR/Cas9	*GR-1, GR-2*	Nutrition	Antioxidant	Increased carotenoids content	Particle bombardment	Dong et al. (2020)
		CRISPR/Cas9	*Nramp5*	Nutrition	Anti-nutrient	Reduced cadmium uptake	*Agrobacterium*-mediated	Tang et al. (2017)
		CRISPR/Cas9	*SBEIIb*	Nutrition	Starch	Enhanced resistant starch and amylose content	*Agrobacterium*-mediated	Sun et al. (2017)
		CRISPR/Cas9 BEs	*Waxy*	Nutrition	Starch	Improved amylose/amylopectin ratio	*Agrobacterium*-mediated	Zhang et al. (2018a)
		CRISPR/Cas9	*rc*	Nutrition	Antioxidant	Improved levels of proanthocyanidins and anthocyanidins	*Agrobacterium*-mediated	Zhu et al. (2019)
		CRISPR/Cas9	*PRX2*	Nutrition	Minerals	Improved tolerance against potassium deficiency	*Agrobacterium*-mediated	Mao et al. (2018)
		CRISPR/Cas9	*HAK1*	Nutrition	Anti-nutrient	Reduced uptake of cesium	*Agrobacterium*-mediated	Nieves-Cordones et al. (2017)
		CRISPR/Cas9	*EPSPS*	Nutrition	Market value	Improved aroma	Biolistic transformation	Li et al. (2016)
		CRISPR/Cas9	*ISA1*	Nutrition	Starch	Reduced starch content	*Agrobacterium*-mediated	Chao et al. (2019)
		CRISPR/Cas9	*NRAMP2*	Nutrition	Minerals	Remobilized Fe and decreased distribution of Cd	*Agrobacterium*-mediated	Chang et al. (2022)

(Continued)

Table 7.1 (Continued)

S. N.	Crop name	Name of editing tool	Targeted gene(s)	Trait improvement category	Trait group	Improved trait	Transformation method	Reference
		CRISPR/Cas9	*FAD2-1*	Nutrition	Lipid	Increased oleic/linoleic ratio in bran oil	*Agrobacterium*-mediated	Abe et al. (2018)
		CRISPR/Cas9	*PPKL1*	Yield	Agronomic trait	Increased yield	*Agrobacterium*-mediated	Zhang et al. (2012)
		CRISPR/Cas9	*ITPK6*	Nutrition	Anti-nutrient	Low phytic acid	*Agrobacterium*-mediated	Jiang et al. (2019)
		CRISPR/Cas9	*GL2/G*, *RF4*, *GRF3*	Yield	Agronomic trait	Increased grain size	*Agrobacterium*-mediated	Hao et al. (2019)
		CRISPR/Cas9 Cpf1	*GS3*	Yield	Agronomic trait	Increased grain length	*Agrobacterium*-mediated	Xu et al. (2019)
		CRISPR/Cas9	*GS9*	Yield	Agronomic trait	Changed slender grains into round grains	*Agrobacterium*-mediated	Zhao et al. (2018)
		CRISPR/Cas9	*GS3, GL3.1*	Yield	Agronomic trait	Increase in grain size and length	*Agrobacterium*-mediated	Yuyu et al. (2020)
		CRISPR/Cas9 BEs	*SNB*, *SLR1*, *SPL17*, *Tms9-1*, *SPL14*	Yield	Agronomic trait	Improved spikelet, floral organ, male fertility, grain shape, size, and weight	*Agrobacterium*-mediated	Hua et al. (2020)
		CRISPR/Cas9	*TMS5*	Yield	Agronomic trait	Developed thermo-sensitive genic male sterile rice	*Agrobacterium*-mediated	Zhou et al. (2016)
		CRISPR/Cas9	*GW2*, *TGW6*, *GW5*	Yield	Agronomic trait	Increased weight of grain	*Agrobacterium*-mediated	Xu et al. (2016)
		CRISPR/Cas9	*GS3*, *Gn1a*	Yield	Agronomic trait	Improved grain size and length	*Agrobacterium*-mediated	Shen et al. (2018)

		CRISPR/Cas9	*AAP6*, *AAP10*	Nutrition	Protein	Reduced grain protein content	*Agrobacterium*-mediated	Wang et al. (2020)
		CRISPR/Cas9	*GBSSI or Wx*	Nutrition	Starch	Decreased grain amylose content	*Agrobacterium*-mediated	Huang et al. (2020)
		CRISPR/Cas9	*BADH2*	Nutrition	Market value	Produced aroma in a non-aromatic rice cultivar	*Agrobacterium*-mediated	Ashokkumar et al. (2020)
		CRISPR/Cas9 CPf1	*NAL*	Yield	Agronomic trait	Increased grain yield	*Agrobacterium*-mediated	Xu et al. (2019)
		CRISPR/Cas9	*Gs3*, *GW2*, *Gn1A*	Yield	Agronomic traits	Increased grain number, weight, width, and size	*Agrobacterium*-mediated	Zhou et al. (2019)
		CRISPR/Cas9	*pyl1/4/6*	Growth and Yield	Agronomic trait	Improved growth and grain yield	*Agrobacterium*-mediated	Miao et al. (2018)
		CRISPR/Cas9	*Gn1a*, *DEP1*	Yield	Agronomic trait	Increased number of grains and panicle	*Agrobacterium*-mediated	Huang et al. (2018)
2.	Wheat (*Triticum aestivum*)	CRISPR/Cas9	α-*Gliadin*	Nutrition	Anti-nutrient	Decreased gluten content	Biolistic transformation	Sanchez- Leon et al. (2018)
		CRISPR/Cas9	*GW2*	Nutrition and yield	Agronomic trait	Increased grain weight	Biolistic transformation	Zhang et al. (2018b)
		CRISPR/Cas9	*GASR7*	Yield	Agronomic trait	Increased grain length and weight	Biolistic bombardment	Zhang et al. (2016)
		CRISPR/Cas9	*GW2*, *LPX-1*, *MLO*	Yield	Agronomic trait	Increased seed size and grain weight	Protoplast transformation	Wang et al. (2018)
		CRISPR/Cas9	*GW7*	Yield	Agronomic trait	Increased grain width and weight	*Agrobacterium*-mediated	Wang et al. (2019c)
		CRISPR/Cas9	*VIT2*	Nutrition	Minerals	Enhanced Fe content	Biolistic transformation	Connorton et al. (2017)

(*Continued*)

Table 7.1 (Continued)

S. N.	Crop name	Name of editing tool	Targeted gene(s)	Trait improvement category	Trait group	Improved trait	Transformation method	Reference
		CRISPR/Cas9 BEs	*GW2*	Yield	Agronomic trait	Increased grain size and weight	*Agrobacterium*-mediated	Li et al. (2018e)
		CRISPR/Cas9 BEs	*DEP1*	Yield	Agronomic trait	Improved inflorescence architecture, panicle growth, and grain yield	*Agrobacterium*-mediated	Li et al. (2018e)
		CRISPR/Cas9 PEs	*GW2*, *GASR7*	Yield	Agronomic trait	Improved grain length and weight	*Agrobacterium*-mediated	Lin et al. (2020)
3.	Maize (*Zea mays*)	ZFNs	*IPK1*	Nutrition	Anti-nutrients	Reduced phytic acid content	*Agrobacterium*-mediated	Shukla et al. (2009)
		CRISPR/Cas9, TALEN	*PK1A*, *IPK*, *MRP4*	Nutrition	Anti-nutrients	Low phytic acid content	*Agrobacterium*-mediated	Liang et al. (2014)
		CRISPR/Cas9	*MADS47*	Nutrition	Anti-nutrients	Reduced zein protein	*Agrobacterium*-mediated	Qi et al. (2016)
		CRISPR/Cas9	*Wx1*	Nutrition	Starch	Increased amylopectin content	Biolistic transformation	Waltz, (2016a)
		CRISPR/Cas9	*Waxy*	Yield	Agronomic trait	Increased yield	*Agrobacterium*-mediated	Gao et al. (2020)
		CRISPR/Cas9	*ARGOS8*	Yield	Agronomic trait	Increased grain yield	*Agrobacterium*-mediated	Shi et al. (2017)
		CRISPR/Cas9	*MS45*	Yield	Agronomic trait	Improved yield	Biolistic method	Svitashev et al. (2015)
		CRISPR/Cas9 CPf1	*glossy2*	Yield	Agronomic trait	Improved yield	*Agrobacterium*-mediated	Lee et al. (2019)
4.	Peanut (*Arachis hypogaea*)	TALENs	*FAD2*	Nutrition	Lipid	Enhanced oleic acid content	*Agrobacterium*-mediated	Wen et al. (2018)

5.	Sorghum (*Sorghum bicolor*)	CRISPR/Cas9	*K1C*	Nutrition	Protein	Reduced kafirins	*Agrobacterium*-mediated	Li et al. (2018d)
6.	Barley (*Hordeum vulgar*)	CRISPR/Cas9	*GBSSI, PTST1*	Nutrition	Starch	Produced starch-free endosperm	*Agrobacterium*-mediated	Zhong et al. (2019)
7.	Rapeseed (*Brassica napus*)	CRISPR/Cas9	*FAD2_Aa*	Nutrition	Lipid	Enhanced oleic acid content	*Agrobacterium*-mediated	Okuzaki et al. (2018)
8.	Soybean (*Glycine max*)	TALEN	*FAD2-1B, FAD2-1A*	Nutrition	Lipid	Increased oleic acid and decreased linoleic acid contents	*Agrobacterium*-mediated	Haun et al. (2014)
		TALEN	*FAD2-1A, FAD2-1B, FAD3A*	Nutrition	Lipid	Reduced linolenic acid content	*Agrobacterium*-mediated, Biolistic transformation	Demorest et al. (2016)
		CRISPR/Cas9	*Bd 28 K, Bd30 K*	Nutrition	Market value	Produced hypoallergenic plants	*Agrobacterium*-mediated	Sugano et al. (2020)
9.	Camelina (*Camelina sativa*)	CRISPR/Cas9	*DGAT1, PDAT1*	Nutrition	Lipid	Altered composition of fatty acids	*Agrobacterium*-mediated	Aznar-Moreno and Durrett, (2017)
		CRISPR/Cas9	*FAD2*	Nutrition	Lipid	Decreased linoleic acid and linolenic acid while increased oleic acid content in oil	*Agrobacterium*-mediated	Morineau et al. (2017)
		CRISPR/Cas9	*FAD2*	Nutrition	Lipid	Decreased polyunsaturated fatty acids and increased oleic acid	*Agrobacterium*-mediated	Jiang et al. (2017)
		CRISPR/Cas9	*FAE1*	Nutrition	Lipid	Improved fatty acids composition	*Agrobacterium*-mediated	Ozseyhan et al. (2018)

(Continued)

Table 7.1 (Continued)

S. N.	Crop name	Name of editing tool	Targeted gene(s)	Trait improvement category	Trait group	Improved trait	Transformation method	Reference
10.	Potato (*Solanum tuberosum*)	CRISPR/Cas9	*GBSS*	Nutrition	Starch	Increased amylopectin content	PEG-mediated transfection	Andersson et al. (2017)
		CRISPR/Cas9	*GBSSI*	Nutrition	Starch	Decreased amylose content	Ribonucleoprotein mediated	Andersson et al. (2018)
		CRISPR/Cas9	*GBSSI*	Nutrition	Starch	Decreased amylose content	*Agrobacterium*-mediated	Kusano et al. (2018)
		CRISPR/Cas9	*PPO2*	Nutrition	Market value	Reduction in browning	Ribonucleoprotein mediated	Gonzalez et al. (2020)
		CRISPR/Cas9	*16DOX*	Nutrition	Anti-nutrient	Reduced steroidal glycoalkaloids	*Agrobacterium*-mediated	Nakayasu et al. (2018)
		TALEN	*SSR2*	Nutrition	Anti-nutrient	Reduced steroidal glycoalkaloids	*Agrobacterium*-mediated	Yasumoto et al. (2019)
		TALEN	*VInv*	Nutrition	Starch	Decreased reducing sugars content	PEG-mediated transfection	Clasen et al. (2016)
		CRISPR/Cas9 and CBE	*GBSSI*	Nutrition	Starch	Increased amylose content	*Agrobacterium*-mediated	Veillet et al. (2019)
11.	Tomato (*Solanum lycopersicum*)	CRISPR/Cas9	*ALC*	Nutrition	Market value	Improved shelf life	*Agrobacterium*-mediated	Yu et al. (2017)
		TALEN, CRISPR/Cas9	*ANT1*	Nutrition	Antioxidant	Increased anthocyanin content	*Agrobacterium*-mediated	Cermak et al. (2015)

CRISPR/Cas9	*GAD2*, *GAD3*	Nutrition	Amino acid	Increased γ -aminobutyric acid content	*Agrobacterium*-mediated	Nonaka et al. (2017)
CRISPR/Cas9	*CAT9*, *GABA-TP1*, *SSADH*, *GABA-TP3*	Nutrition	Amino acid	Higher γ -aminobutyric acid content	*Agrobacterium*-mediated	Li et al. (2018a)
CRISPR/Cas9	*lncRNA1459*	Nutrition	Market value and antioxidant	Reduced ethylene production, lycopene accumulation and delayed fruit ripening	*Agrobacterium*-mediated	Li et al. (2018b)
ZFN	*L1L4/NF- YB6*	Nutrition	Anti-nutrients	Reduced content of oxalic acid	Electroporation	Gago et al. (2017)
CRISPR/Cas9	*fas*, *lc*	Nutrition	Agronomic traits	Improved plant architecture, inflorescence branching, and fruit size	*Agrobacterium*-mediated	Rodriguez-Leal et al. (2017)
CRISPR/Cas9	*MYB12*	Nutrition	Metabolites	Altered flavonoid levels	*Agrobacterium*-mediated	Deng et al. (2018)
CRISPR/Cas9	*RIN*	Nutrition	Market value	Delay in fruit ripening	*Agrobacterium*-mediated	Ito et al. (2015)
CRISPR/Cas9	*RIN/LeMADS*	Nutrition	Market value	Delay in fruit ripening	*Agrobacterium*-mediated	Jung et al. (2018)
CRISPR/Cas9	*LCY-B2*, *SGR1*, *Blc LCY-E*	Nutrition	Antioxidant	Enhanced levels of beta-carotenoids and lycopene	*Agrobacterium*-mediated	Li et al. (2018c)

(*Continued*)

Table 7.1 (Continued)

S. N.	Crop name	Name of editing tool	Targeted gene(s)	Trait improvement category	Trait group	Improved trait	Transformation method	Reference
		CRISPR/Cas9	*FAS*, *CycB*, *FW2.2*, *SP*, *MULT*	Nutrition	Nutrition and agronomic traits	Increased lycopene content, size and number of the fruits	*Agrobacterium*-mediated	Zsogon et al. (2018)
		CRISPR/Cas9	*SP*, *SP5*, *WUS*, *CLV3*, *SlGGP1*	Nutrition	Vitamins	Enhanced level of vitamin C	*Agrobacterium*-mediated	Li et al. (2018f)
		CRISPR/Cas9	*PSY1*	Nutrition	Antioxidant	Increased anthocyanin contents	*Agrobacterium*-mediated	Filler Hayut et al. (2017)
		CRISPR/Cas9	*PL*, *PG2a*, *TBG4*	Nutrition	Market value	Increased shelf life	*Agrobacterium*-mediated	Wang et al. (2019a)
		CRISPR/Cas9	*HYS*	Nutrition	Antioxidant	Reduction in anthocyanin level	*Agrobacterium*-mediated	Qiu et al. (2019)
		CRISPR/Cas9 CBE	*SlDDB1*, *SlDET1*, *SlCYC-B*	Nutrition	Antioxidant	Enhanced lycopene, β carotene, and carotenoid	*Agrobacterium*-mediated	Hunziker et al. (2020)
		CRISPR/Cas9	*SlANT2*, *SlAN2-like*	Nutrition	Antioxidant	Decreased anthocyanin content	*Agrobacterium*-mediated	Yan et al. (2020), Zhi et al. (2020)
		CRISPR/Cas9	*FLORAL4*	Nutrition	Metabolites	Increased phenylalanine-derived volatile content	*Agrobacterium*-mediated	Tikunov et al. (2020)
		CRISPR/Cas9	*ENO*	Yield	Agronomic traits	Increased size of the floral meristem	*Agrobacterium*-mediated	Yuste-Lisbona et al. (2020)
		CRISPR/Cas9	*SlINVINH1*, *SlVPE5*	Nutrition	Starch	Increased glucose, fructose, and total soluble solids	*Agrobacterium*-mediated	Wang et al. (2021)

12.	Eggplant (*Solanum melongena*)	CRISPR/Cas9	*PPO4*, *PPO5*, *PPO6*	Nutrition	Market value	Decreased browning	*Agrobacterium*-mediated	Maioli et al. (2020)
13.	*Mustard (Brassica rapa)*	CRISPR/Cas9	*OG1A*, *OG1B*	Nutrition	Sugar	Decreased fructose, glucose and increased sucrose contents	*Agrobacterium*-mediated	Jiang et al. (2020)
14.	Carrot (*Daucus carota*)	CRISPR/Cas9	*F3H*	Nutrition	Antioxidant	Improved anthocyanin contents	*Agrobacterium*-mediated	Klimek-Chodacka et al. (2018)
15.	Sweet Potato (*Ipomoea batatas*)	CRISPR/Cas9	*GBSSI*	Nutrition	Starch	Improved starch quality	*Agrobacterium*-mediated	Wang et al. (2019b)
16.	Grape (*Vitis vinifera*)	CRISPR/Cas9	*IdnDH*	Nutrition	Organic acid	Reduced tartaric acid content	*Agrobacterium*-mediated	Ren et al. (2016)
17.	Banana (*Musa acuminata*)	CRISPR/Cas9	*LCYε*	Nutrition	Antioxidant	Increased β-carotene	*Agrobacterium*-mediated	Kaur et al. (2020)
		CRISPR/Cas9	*MaACO1*	Nutrition	Market value	Increased shelf life	*Agrobacterium*-mediated	Hu et al. (2021)
18.	Apple (*Malus pumila*)	CRISPR/Cas9	*IdnDH*	Nutrition	Organic acid	Increase tartaric acid content	Ribonucleoproteins mediated	Osakabe et al. (2018)
19.	Lettuce (*Lactuca sativa*)	CRISPR/Cas9	*GGP2*	Nutrition	Vitamin	Increased ascorbate content	*Agrobacterium*-mediated	Zhang et al. (2018c)
20.	Tobacco (*Nicotiana tabacum*)	Meganuclease	*BBL*	Nutrition	Anti-nutrients	Reduced level of nicotine	*Agrobacterium*-mediated	USDA-APHIS, (2020)
21.	Mushroom (*Agaricus bisporus*)	CRISPR/Cas9	*PPO*	Nutrition	Market value	Reduced browning	PEG-mediated transfection	Waltz (2016b)

(*Continued*)

Table 7.1 (Continued)

S. N.	Crop name	Name of editing tool	Targeted gene(s)	Trait improvement category	Trait group	Improved trait	Transformation method	Reference
22.	Ground Cherry (*Physalis pruinosa*)	CRISPR/Cas9	*AGO7*, *SP*, *SP5g*	Yield and nutrition	Agronomic traits	Improved plant architecture, number of flowers, and fruit size	*Agrobacterium*-mediated	Lemmon et al. (2018)
23.	Cassava (*Manihot esculenta*)	CRISPR/Cas9	*GBSSI*	Nutrition	Starch	Reduced amylose content in root	*Agrobacterium*-mediated	Bull et al. (2018)
24.	Strawberry (*Fragaria vesca)*	CRISPR/Cas9 BE	*ebZIPs1.1*	Nutrition	Starch	Increased sugar content	*Agrobacterium*-mediated	Xing et al. (2020)
25.	Melon (*Cucumis melo*)	CRISPR/Cas9	*CmOr*	Nutrition	Antioxidant	Improved β–carotene accumulation	*Agrobacterium*-mediated	Tzuri et al. (2015)

and high amylose (Sun et al. 2017). Zhang et al. (2018a) found that knockout of the *Waxy* gene in rice improves the ratio of amylose to amylopectin. Reduction in amylose content in rice grains has also been achieved via genome editing (Chao et al. 2019; Huang et al. 2020). Starch content has also been modified in other crops apart from rice. CRISPR/Cas9-mediated repression of the *GBSS1* and *PTST1* produced endosperm with very low amount of starch in barley (Zhong et al. 2019). Similarly, elite maize cultivars were modified using CRISPR/Cas9 to generate high-amylopectin variants by knocking out the *Waxy* gene (Waltz 2016a).

Phenolic chemicals in dietary grains increase antioxidant activity and give health advantages such as a lower risk of chronic illnesses (Gunaratne et al. 2013; Gonzalez et al. 2020). It has been found that pigmented rice has more phenolics and works better as an antioxidant than white rice. Peonidin-3-glucoside and cyaniding-3-glucoside are the key anthocyanins in colored rice that have powerful antioxidant and anti-inflammatory effects (Gonzalez et al. 2020). CRISPR/Cas9-mediated restoration of the frame of *rc* gene improved proanthocyanidin and anthocyanidin contents in rice grains (Zhu et al. 2019). Biofortification of rice has also been achieved via genome editing. For example, β-carotene in rice grains was increased and golden rice (Kitaake cultivar) was produced by targeted insertion of about 5.2 kb sequence having a carotenoid biosynthesis pathway-related cassette (Dong et al. 2020).

Through gene editing, wheat and soybean have been made less allergenic (Sanchez-Leon et al. 2018; Sugano et al. 2020). Additionally, editing of genes has been done in a variety of crops in order to avoid the buildup of anti-nutritional components inside plants, for instance, phytic acid content was decreased in rice and maize (Shukla et al. 2009; Liang et al. 2014; Jiang et al. 2019). Phytic acid chelates metal ions of micronutrients such as Fe^{2+} and Zn^{2+} and reduces their bioavailability (Aggarwal et al. 2018). Moreover, humans lack phytase enzyme to digest it (Ali et al. 2013). Iron is required for plant development as well as a variety of cellular functions involving redox reactions or electron transfer catalysis (Connorton et al. 2017). In humans, Fe deficiency causes anemia especially in females. Similarly, Zn is also essential for proper growth and development (Mayer et al. 2008). To promote iron and zinc accumulation in wheat grains, the gene editing was done to disrupt *inositol pentakisphosphate 2-kinase 1* (*TaIPK1*), which decreased phytic acid (Ibrahim et al. 2021). Similarly, CRISPR/Cas9 and ZFNs were employed in maize to mutate genes related to the phytic acid pathway to lessen the negative effects of phytic acid on Fe, Zn, and calcium (Ca) absorption (Liang et al. 2014). Zn uptake is negatively regulated by cytokinins (Gao et al. 2019). Therefore, using CRISPR/Cas9, Zn concentration was increased in rice by knocking down *CYP735A*, which encodes an enzyme involved in the biosynthesis of trans-zeatins (Gao et al. 2019). Heavy metals are also anti-nutrients which are quite abundant in soil and are easily absorbed by food crops, posing a threat to human health. Accumulation of these anti-nutrients has also been addressed by the genome editing (Qi et al. 2016; Nieves-Cordones et al. 2017; Chang et al. 2022). Researchers have minimized cadmium contamination in rice grains through targeting Os*Nramp5* gene (Tang et al. 2017). The main storage protein in sorghum grains is known as kafirin, which forms protein bodies that have low digestibility and are devoid of essential amino acids (Li et al. 2018d). Researchers have effectively decreased levels of kafirin and enhanced the digestibility of sorghum grains by targeting the *k1c* gene that produces kafirins (Li et al. 2018d).

Initiatives are being made to boost the oleic acid content of peanuts using TALEN-mediated knockdown of the *FAD2* gene (Wen et al. 2018). For the same reason, this gene has also been knocked out in rice (Abe et al. 2018). The amount of tocopherols was increased in barley by editing *HGGT* and *HPT* genes (Zeng et al. 2020).

With the help of genome editing technology, it is possible to engineer the architecture of many staple crops leading to enhancements related to number of spikelets, floral organs, shape, and size of grains (Zhang et al. 2012; Endo et al. 2016; Rodrguez-Leal et al. 2017; Lemmon et al. 2018; Miao et al. 2018; Zhao et al. 2018; Hao et al. 2019; Xu et al. 2019; Hua et al. 2020). Grain protein, grain length, and many other yield-related agronomic traits were also improved through the application of CRISPR/Cas9 (Svitashev et al. 2015; Shi et al. 2017; Li et al. 2018e; Zhang et al. 2018b; Lee et al. 2019; Lin et al. 2020; Wang et al. 2020). Negative loci that affect rice's grain yield, such as *GW2*, *GW5*, *GS3*, *GL3.1*, *TMS5*, *TGW6*, *GS5*, *Gnla*, and *DEP1*, have been knocked out, and mutant plants with much higher grain yields have been developed (Xu et al. 2016; Zhou et al. 2016; Liu et al. 2017; Huang et al. 2018; Shen et al. 2018; Xu et al. 2019; Zhou et al. 2019; Hua et al. 2020; Yuyu et al. 2020).

7.3.2 Improvement in Oilseed Crops

Oilseed crops like soybeans, rapeseed, sunflowers, and nuts are source of oils for cooking and industry. Biotechnologists and crop breeders have tried to change the profiles of fatty acids of oil of their seeds to make them better for their intended use. Earlier transgenic and RNAi techniques were used to change the level of fatty acids in soybean and rapeseed seed oils (Ku and Ha 2020). Now different GETs are being used to turn down or turn off the desaturase genes (Ku and Ha 2020).

One of the most prominent genome editing objectives is to enhance the level of monounsaturated fatty acids (MUFAs) for instance oleic acid that possess a longer shelf life compared to polyunsaturated fatty acids (PUFAs) like linoleic acid and linolenic acid (Clemente and Cahoon 2009). Soybean is one of the most prominent edible oil crops around the globe. Presence of high PUFA in it reduces the quality of cooking oil due to its poor oxidative and frying stability (Haun et al. 2014). To change the fatty acid profile in soybean, rapeseed, and camelina via improving the oleic-to-linoleic acid ratio, *FAD2*, which codes for delta-12 desaturase, was targeted and disrupted (Haun et al. 2014; Jiang et al. 2017; Morineau et al. 2017; Okuzaki et al. 2018). Genome editing via TALENs was carried out to mutate *FAD2-1A* and *FAD2-1B* genes simultaneously. As an outcome, the level of monounsaturated oleic acid improved, while the level of linoleic and linolenic acids decreased in soybean oil (Haun et al. 2014). TALENs were again used to mutate *FAD3A* gene (that encodes the delta-15 desaturase) to make triple mutant *fad2-1a fad2-1b fad3a* in soybean (Demorest et al. 2016). This further reduced the amount of linoleic acid and increased oleic acid level (Demorest et al. 2016). Many people in the US and Europe are allergic to the soybean storage proteins 7S, 11S globulins, and 2S albumin (Sugano et al. 2020). Through CRISPR/Cas9, the knockdown of *GlymBd 28 K* and *GlymBd 30 K* genes generated less allergenic soybean (Sugano et al. 2020).

A plant known as camelina (*Camelina sativa*) has been getting a lot of attention over the last few years. It has unique oil characteristics as it contains a high level of omega-3 fats and tocopherols (Faure and Tepfer 2016). Two research groups in 2017 individually knocked out the *FAD2* gene in camelina for an improvement in seed oil content (Jiang et al. 2017; Morineau et al. 2017). Even though Camelina is a hexaploid plant, these papers showed that gRNA can be used to disrupt all three *FAD2* loci at the same time. As a result, oleic acid levels got enhanced and PUFA levels declined in camelina oil. Similarly, using a designed gRNA identical to all three homologue, *DGAT1* or *PDAT1* genes of camelina involved in triacylglycerol synthesis were edited through CRISPR/Cas9 to produce seeds with modified fatty acid composition (Aznar-Moreno and Durrett 2017). The seed oil of camelina contains a high concentration of very long-chain fatty acids (VLCFAs), which is unwanted. By knocking out the *FAE1* gene, researchers reduced VLCFAs and improved levels of oleic acid, which is suitable for food and industrial purposes (Ozseyhan et al. 2018; Lee et al. 2021).

The *FAD2* gene was also mutated in rapeseed (Okuzaki et al. 2018). It brought out a slight rise in oleic acid and a reduction in linoleic acid. Rapeseed has numerous homologs of the *FAD2* gene such as *FAD2Ca* and *FAD2Cb,* which can also be targeted to increase the amount of oleic acid in rapeseed (Okuzaki et al. 2018).

7.3.3 Improvements in Horticulture Crops

Horticultural crops include fruits and vegetables, which are rich sources of vitamins, carbohydrates, minerals, proteins, and organic acids. These crops are usually harvested quickly. After harvest, respiration causes the fruit to continue to ripe (Ku and Ha 2020). Breeders try to increase the market value of these crops by making them more nutritious, keeping them fresh for a longer period, and making them easier to process in terms of flavour, size, color, and texture (Ku and Ha 2020). Tomato is the most important crop in horticulture because of its nutritional and economic value (Cermak et al. 2015). Numerous genome editing tests have been done on tomato and many genes have been targeted to increase its agronomic traits (Rodriguez-Leal et al. 2017; Zsogon et al. 2018; Yuste-Lisbona et al. 2020). Homology-based repair mechanism after creating DSBs has been more successful in this crop than other horticultural crops so far (Ito et al. 2015). In this crop, CRISPR/Cas9 system was used to insert CaM35S promoter upstream of *ANT1* gene, which encodes for a MYB transcription factor. It brought out increase in anthocyanin accumulation and tomatoes turned purple (Cermak et al. 2015). Researchers have also tried to discover ways to slow down the ripening of tomato fruits so that they can be stored for longer time. MADS-box transcription factor is encoded by *RIN* gene. Mutants of this gene produce late ripening fruits. Ito et al. (2015) and Jung et al. (2018) employed CRISPR/Cas9 to knock down this gene in different cultivars like Golden Bell, Mamirio, and Ailsa Craig. Similarly, targeted mutation in the *Alcobaca* (*ALC*) gene resulted in an enhancement in the shelf life of tomato (Yu et al. 2017). In another study, *lncRNA1459* was targeted to delay ripening. Li et al. (2018d) targeted this gene and found that the mutants produced less ethylene, stored less lycopene, and showed extended shelf life of the tomato fruit (Li et al. 2018b).

Improving tomato quality has also been the goal of genome editing (Yu et al. 2017). For this, specific metabolites have been targeted. For example, tomatoes contain a fair amount of GABA, a metabolite that lowers blood pressure when eaten every day (Matsumoto et al. 1997). *SlGAD2* and *SlGAD3* encode glutamate decarboxylases (GADs), which are the key enzymes for the production of GABA (Nonaka et al. 2017). They contain a C-terminal auto-inhibitory domain, which controls GAD activity in a negative way. This domain was deleted using CRISPR/Cas9 to increase GABA content in tomatoes. Nonaka et al. (2017)

found that adding a stop codon right before the auto-inhibitory domain made 15 times more GABA content. In another study, GABA-rich tomato was made via multiplexed CRISPR/Cas9 simultaneously targeting four key genes: *GABA-TP1, GABA-TP3, CAT9, SSADH* (Li et al. 2018a). Tomatoes also make a lot of oxalic acid, which is a type of anti-nutrient as it stops calcium from being absorbed by making its insoluble salts (Gago et al. 2017). Its content was reduced by ZFNs-mediated disruption of *L1L4* gene that encodes a transcription factor (Gago et al. 2017). In general, the color of the tomato peel is determined by the amount of naringenin chalcone (NarCh), a yellowish flavonoid. In its absence, tomatoes turn pink. The gene responsible for this is *SlMYB12*. A R2R3-MYB transcription factor is encoded by this gene (Ballester et al. 2010). *SlMYB12* gene was mutated via CRISPR/Cas9 and pink tomatoes were made (Deng et al. 2018). Multiplexed genome editing was carried out for genes *LCY-B1*, *SGR1*, β-lycopene cyclase (*Blc*), *LCY-E*, and *LCY-B2,* which are related to carotenoid metabolic pathway (Li et al. 2018c). As a result, lycopene content in tomato fruit increased up to 5.1 folds.

There are few genes that are vital for desired agronomic features and nutritional value of tomato such as *CycB* for nutritional quality, *FAS* and *FW2.2* for fruit size, *MULT* for fruit number, *OVATE* for fruit shape, and *SP* for plant growth habit (Zsogon et al. 2018). Mutations in these genes altered fruit morphology and shape and increased fruit size. Besides fruit architecture, fruit's nutritional content increased by five fold in terms of lycopene accumulation. In another study, multiplex CRISPR/Cas9 system-mediated mutations in *SP*, *SP5G*, *CLV3*, *WUS*, and *SIGGP1* genes increased fruit size, synchronized the ripening process, increased ascorbic acid synthesis, and made tomato plant compact and light insensitive (Li et al. 2018f).

β-carotene is important for eye health and other immune system functions because it is a source of vitamin A. β-carotene accumulation was enhanced in melon by site-directed mutagenesis of the *CmOr* gene (Tzuri et al. 2015). Anthocyanins help in the reduction of inflammation and act as anti-oxidants (Tanaka and Ohmiya 2008). Genome editing has altered anthocyanin production and resulted in creation of tomatoes that are pink, yellow, and purple (Cermak et al. 2015; Filler Hayut et al. 2017; Deng et al. 2018; Qiu et al. 2019; Hunziker et al. 2020; Tikunov et al. 2020; Yan et al. 2020; Zhi et al. 2020). Likewise, editing of *F3H* gene in carrot genome increased the amount of anthocyanin (Klimek-Chodacka et al. 2018).

Knocking out of *LCYε* gene increased β-carotene levels in banana fruit (Kaur et al. 2020). In another study, the disruption of *aminocyclopropane-1- carboxylate oxidase* (*ACO1*) increased the shelf life of banana (Hu et al. 2021). The nutritional values and fruit size of ground cherry were enhanced via genome editing (Lemmon et al. 2018). Jiang et al. (2020) changed the content of sugar in *Brassica rapa* plants through CRISPR/Cas9. Zhang et al. (2018c) found that when *LsGGP2* was overexpressed in lettuce using GET, the amount of ascorbate in the lettuce increased significantly. To increase market value, the gene editing has been done to reduce the browning of many horticulture crops (Gonzalez et al. 2020; Maioli et al. 2020). Postharvest browning is caused by the quick oxidation of phenols to melanin with the help of PPO enzyme. It reduces the commercial value of mushrooms. Using CRISPR/Cas9, researchers edited one of the six *PPO* genes in the common white button mushroom to stop it from turning brown (Waltz 2016b).

Low-nicotine tobacco was made by using a custom-made MegaN to target *berberine bridge enzyme-like* genes. These genes code for enzymes related to biosynthesis of nicotine

(USDA-APHIS 2020). The CRISPR/ Cas9 system was used to generate very effective targeted mutations in grapevine and apple (Osakabe et al. 2018). Tartaric acid (TA) is a powerful antioxidant found in grape fruits that also contributes to their flavour. TA, malic acid, and citric acid are the three main acids present in wine grapes that are important for the development of wine flavour. *IdnDH* gene which controls TA production was knocked out in grapes by genome editing resulting in 36% reduction in its amount (Ren et al. 2016).

Refrigeration has become more important to the world's food supply because it extends the time that food can be stored after harvest and prevents it from sprouting while it's being stored (Ku and Ha 2020). But when the food is cooked at high temperatures, this cold treatment can cause formation of more reducing sugars like glucose and fructose. These sugars react with amino acids to make more of a possible carcinogen called acrylamide (Tareke et al. 2002). Potato stands at fourth number in the world as a staple food crop after maize, wheat, and rice (Ku and Ha 2020). Due to high concentrations of reducing sugars, heat-processed potatoes that undergo cold-induced sweetening contains high acrylamide (Clasen et al. 2016). Legal action has been attempted to restrict the amount of acrylamide in fried potato goods due to concerns about the safety of acrylamide in potato chips and French fries on a worldwide scale (Clasen et al. 2016). In order to solve this issue, TALEN-mediated knockout of *Vacuolar invertase* (*VInv*) gene was carried out, which decreased reducing sugars contents and resulted in low acrylamide level (Clasen et al. 2016). Potato produces large amounts of the toxic steroidal glycoalkaloids (SGAs)-solanine and chaconine in most plant tissues, particularly tuber sprouts (Nakayasu et al. 2018). Knockdown of *16DOX, SSR2,* and *VlnV* through CRISPR/Cas9 decreased these anti-nutrients and enhanced post-harvest storage (Clasen et al. 2016; Nakayasu et al. 2018; Yasumoto et al. 2019). CRISPR/Cas9 mediated deletion of the *GBSSI* gene in potatoes also changed the starch content (Andersson et al. 2017; Andersson et al. 2018; Kusano et al. 2018; Veillet et al. 2019).

CRISPR/Cas9 system-mediated knockout of *GBSSI* and *SBEII* genes increased amylose percentage in sweet potato (Wang et al. 2019b). GETs have also been utilized in strawberries to create a range of sugar contents (Xing et al. 2020). Editing of *PTST1* also decreased the amount of amylose in the root starch of cassava (Bull et al. 2018).

7.4 Regulation of Genome Edited Crops: Current Status

The governments of different countries need to frame clear and uniform policy about genome-edited crops. Different governments have different ideas about how GE varieties should be developed, made, used, and sold (El-Mounadi et al. 2020). Development of transgene-free plants through GETs can help in avoiding biosafety regulations that apply to conventional GM plants (Van et al. 2019). At the moment, there are two ways to regulate GE crops: 1) based on the process, and 2) based on the end product (Eckerstorfer et al. 2019; Van et al. 2019). Some countries consider GE crops to be genetically modified, while others do not (Eckerstorfer et al. 2019). For example, Brazil and the United States have decided to regulate GE crops in the same way that conventional breeding does (Waltz 2016b). Similarly, the Court of Justice of the European Union (ECJ) ruled that GE crops must follow the same rules that employ on GM crops (Spicer and Molnar 2018). In the same way, regulatory bodies in Japan and New Zealand classify the GE crops analogous to GM crops and have strict biosafety rules (Chimata and Bharti 2019; Zannoni 2019). China is yet to announce policy regarding GE crops though China is the world leader regarding publications on genome editing (Hundleby and Harwood 2022). Recently, GE crops have been exempted from GM biosafety assessment in India. To sum up, the regulatory framework and laws should not only protect plants, animals, and the environment but at the same time should not make it difficult to develop and timely release improved varieties.

7.5 Future Perspectives and Conclusion

Today, food shortage and poor nutrition are major problems whose severity is increasing due to climate change and the rising population. Therefore, they need to be fixed and genome editing has a lot of potential to bring agricultural innovation to make the world healthier and free of hunger. It can fulfill the necessity of bio-fortification of crops with healthy fatty acids, important vitamins, and minerals. With time, breakthroughs are being made in GETs, which are making them more advanced; for example, different versions of CRISPR/Cas9 are being replaced by new ways to edit genes. The global community should take advantage of this technique by making regulatory frameworks and support systems easy to use for scientists. GETs have several edges over traditional plant breeding practices. In this way, base editors (that can make precise changes to nucleotides), epigenome modifiers (that can change DNA confirmation and the expression levels that go with it) and prime editors (that can put short pieces of DNA in the right place) are all good candidates. However, economic feasibility, consumer acceptance, and policy frameworks will play a role in the ultimate success of the applications of GETs. New developments in genome editing technology have addressed several regulatory concerns. In prime editing, the number of transgene-free plants has grown a lot. It can make a big difference in a short amount of time. To completely realize the potential of GETs, a multifaceted policy encompassing the adoption of research outcomes, information dissemination, technology development, and social acceptance of the edited product is required.

Author Contribution

PKS wrote the first draft of the chapter and prepared the figure. PS conceptualized, worked out the outline of the chapter, and contributed to editing and proofreading. HR contributed to proofreading.

Acknowledgment

PS acknowledges the Ministry of human resource development, Department of higher education, Government of India for the RUSA 2.0 programme (Thematic Project III) for financial support. PKS acknowledges National Institute of Plant Genome Research (NIPGR), New Delhi for providing research facilities. HR acknowledges the support received from core grant of NIPGR.

References

Abe, K., Araki, E., Suzuki, Y. et al. (2018). Production of high oleic/low linoleic rice by genome editing. *Plant Physiology and Biochemistry* 131: 58–62. https://doi.org/10.1016/j.plaphy.2018.04.033.

Aggarwal, S., Kumar, A., Bhati, K.K. et al. (2018). RNAi-mediated downregulation of inositol pentakisphosphate kinase (IPK1) in wheat grains decreases phytic acid levels and increases Fe and Zn accumulation. *Frontiers in Plant Science* 9: https://doi.org/10.3389/fpls.2018.00259.

Ali, N., Paul, S., Gayen, D. et al. (2013). Development of low phytate rice by RNAi mediated seed-specific silencing of inositol 1,3,4,5,6-pentakisphosphate 2-kinase gene (IPK1). *PLoS One* 8 (7): e68161. https://doi.org/10.1371/journal.pone.0068161.

Alok, A., Kumar, J., and Upadhyay, S.K. (2018). Engineering in hairy roots using CRISPR/Cas9-mediated editing. In: *Hairy Roots* (ed. V. Srivastava, S. Mehrotra, and S. Mishra). Singapore: Springer. https://doi.org/10.1007/978-981-13-2562-5_15.

Alok, A., Chauhan, H., Upadhyay, S.K. et al. (2021). Compendium of plant-specific CRISPR vectors and their technical advantages. *Life* 11: 1021. https://doi.org/10.3390/life11101021.

Andersson, M., Turesson, H., Nicolia, A. et al. (2017). Efficient targeted multiallelic mutagenesis in tetraploid potato (*Solanum tuberosum*) by transient CRISPR-Cas9 expression in protoplasts. *Plant Cell Reports* 36: 117–128.

Andersson, M., Turesson, H., Olsson, N. et al. (2018). Genome editing in potato via CRISPR-Cas9 ribonucleoprotein delivery. *Physiologia Plantarum* 164: 378–384. https://doi.org/10.1111/ppl.12731.

Ashokkumar, S., Jaganathan, D., Ramanathan, V. et al. (2020). Creation of novel alleles of fragrance gene OsBADH2 in rice through CRISPR/Cas9 mediated gene editing. *PLoS One* 15 (8): e0237018.

Aznar-Moreno, J.A. and Durrett, T.P. (2017). Simultaneous targeting of multiple gene homeologs to alter seed oil production in *Camelina sativa*. *Plant and Cell Physiology* 58 (7): 1260–1267.

Ballester, A.R., Molthoff, J., de Vos, R. et al. (2010). Biochemical and molecular analysis of pink tomatoes: deregulated expression of the gene encoding transcription factor SlMYB12 leads to pink tomato fruit color. *Plant Physiology* 152: 71–84. https://doi.org/10.1104/pp.109.147322.

Bhalothia, P., Yajnik, K., Alok, A., and Upadhayay, S.K. (2020). The current progress of CRISPR/Cas9 development in plants. In: *Genome Engineering via CRISPR-Cas9 Systems* (ed. S. Dhar). Elsevier. https://doi.org/10.1016/B978-0-12-818140-9.00011-8.

Bhargava, A. and Srivastava, S. (2019). *Participatory Plant Breeding: Concept and Applications*, vol. 1. Singapore: *Springer*.

Birla, D.S., Malik, K., Sainger, M. et al. (2017). Progress and challenges in improving the nutritional quality of rice (*Oryza sativa* L.). *Critical Reviews in Food Science and Nutrition* 57: 2455–2481. https://doi.org/10.1080/10408398.2015.1084992.

Bull, S.E., Seung, D., Chanez, C. et al. (2018). Accelerated ex-situ breeding of *GBSS*-and *PTST1*-edited cassava for modified starch. *Science Advances* 4 (9): eaat6086.

Cermak, T., Baltes, N.J., Čegan, R. et al. (2015). High-frequency, precise modification of the tomato genome. *Genome Biology* 16: 1–15.

Chang, J.-D., Xie, Y., Zhang, H. et al. (2022). The vacuolar transporter *OsNRAMP2* mediates Fe remobilization during germination and affects Cd distribution to rice grain. *Plant and Soil* 1–17. https://doi.org/10.21203/rs.3.rs1206604/v1.

Chao, S., Cai, Y., Feng, B. et al. (2019). Editing of rice Isoamylase gene *ISA1* provides insights into its function in starch formation. *Rice Science* 26: 77–87.

Chimata, M.K. and Bharti, G. (2019). Regulation of genome edited technologies in India. *Transgenic Research* 28 (Suppl 2): 175–181. https://doi.org/10.1007/s11248-019-00148-z.

Christian, M., Cermak, T., Doyle, E.L. et al. (2010). Targeting DNA double-strand breaks with TAL effector nucleases. *Genetics* 186: 757–761. https://doi.org/10.1534/genetics.110.120717.

Clasen, B.M., Stoddard, T.J., Luo, S. et al. (2016). Improving cold storage and processing traits in potato through targeted gene knockout. *Plant Biotechnology Journal* 14 (1): 169–176.

Clemente, T.E. and Cahoon, E.B. (2009). Soybean oil: genetic approaches for modification of functionality and total content. *Plant Physiology* 151: 1030–1040. https://doi.org/10.1104/pp.109.146282.

Connorton, J.M., Jones, E.R., Rodríguez-Ramiro, I. et al. (2017). Wheat vacuolar iron transporter *TaVIT2* transports Fe and Mn and is effective for biofortification. *Plant Physiology* 174 (4): 2434–2444.

Demorest, Z.L., Coffman, A., Baltes, N.J. et al. (2016). Direct stacking of sequence-specific nuclease-induced mutations to produce high oleic and low linolenic soybean oil. *BMC Plant Biology* 16: 1–8.

Deng, L., Wang, H., Sun, C. et al. (2018). Efficient generation of pink-fruited tomatoes using CRISPR/Cas9 system. *Journal of Genetics and Genomics* 45 (1): 51–54.

Dong, O.X., Yu, S., Jain, R. et al. (2020). Marker-free carotenoid-enriched rice generated through targeted gene insertion using CRISPR-Cas9. *Nature Communications* 11 (1): 1178.

Eckerstorfer, M.F., Engelhard, M., Heissenberger, A. et al. (2019). Plants developed by new genetic modification techniques – comparison of existing regulatory frameworks in the EU and non-EU countries. *Frontiers in Bioengineering and Biotechnology* 7: 26.

Eid, A., Alshareef, S., and Mahfouz, M.M. (2018). CRISPR base editors: genome editing without double-stranded breaks. *Biochemical Journal* 475 (11): 1955–1964.

El-Mounadi, K., Morales-Floriano, M.L., and Garcia-Ruiz, H. (2020). Principles, applications, and biosafety of plant genome editing using CRISPR-Cas9. *Frontiers in Plant Science* 11: 56.

Endo, A., Masafumi, M., Kaya, H., and Toki, S. (2016). Efficient targeted mutagenesis of rice and tobacco genomes using Cpf1 from *Francisella novicida*. *Scientific Reports* 6 (1): 38169.

FAO (2019). *The State of Food Security and Nutrition in the World 2019: Safeguarding Against Economic Slowdowns and Downturns*, vol. 2019. Food & Agriculture Organization.

Faure, J. and Tepfer, M. (2016). *Camelina*, a swiss knife for plant lipid biotechnology. *OCL* 23: D503. https://doi.org/10.1051/ocl/2016023.

Fiaz, S., Ahmar, S., Saeed, S. et al. (2021). Evolution and application of genome editing techniques for achieving food and nutritional security. *International Journal of Molecular Sciences* 22 (11): 5585.

Filler Hayut, S., Melamed Bessudo, C., and Levy, A.A. (2017). Targeted recombination between homologous chromosomes for precise breeding in tomato. *Nature Communications* 8 (1): 15605.

Gago, C., Drosou, V., Paschalidis, K. et al. (2017). Targeted gene disruption coupled with metabolic screen approach to uncover the *LEAFY COTYLEDON1-LIKE4* (*L1L4*) function in tomato fruit metabolism. *Plant Cell Reports* 36: 1065–1082.

Gao, S., Xiao, Y., Xu, F. et al. (2019). Cytokinin-dependent regulatory module underlies the maintenance of zinc nutrition in rice. *The New Phytologist* 224 (1): 202–215. https://doi.org/10.1111/nph.15962.

Gao, H., Gadlage, M.J., Lafitte, H.R. et al. (2020). Superior field performance of waxy corn engineered using CRISPR – Cas9. *Nature Biotechnology* 38: 579–581. https://doi.org/10.1038/s41587-020-0444-0.

Gaudelli, N.M., Komor, A.C., Rees, H.A. et al. (2017). Programmable base editing of A• T to G• C in genomic DNA without DNA cleavage. *Nature* 551 (7681): 464–471.

González, M.N., Massa, G.A., Andersson, M. et al. (2020). Reduced enzymatic browning in potato tubers by specific editing of a polyphenol oxidase gene via ribonucleoprotein complexes delivery of the CRISPR/Cas9 system. *Frontiers in Plant Science* 10: 1649. https://doi.org/10.3389/fpls.2019.01649.

Gunaratne, A., Wu, K., Li, D. et al. (2013). Antioxidant activity and nutritional quality of traditional red-grained rice varieties containing proanthocyanidins. *Food Chemistry* 138 (2–3): 1153–1161.

Hao, L.I., Ruiying, Q.I.N., Xiaoshuang, L.I.U. et al. (2019). CRISPR/Cas9-mediated adenine base editing in rice genome. *Rice Science* 26 (2): 125–128. https://doi.org/10.1016/j.rsci.2018.07.002.

Haun, W., Coffman, A., Clasen, B.M. et al. (2014). Improved soybean oil quality by targeted mutagenesis of the *fatty acid desaturase 2* gene family. *Plant Biotechnology Journal* 12 (7): 934–940.

Hu, C., Sheng, O., Deng, G. et al. (2021). CRISPR/Cas9-mediated genome editing of *MaACO1* (*Aminocyclopropane-1- carboxylate Oxidase* 1) promotes the shelf life of banana fruit. *Plant Biotechnology Journal* 19 (4): 654–656. https://doi.org/10.1111/pbi.13534.

Hua, K., Tao, X., Liang, W. et al. (2020). Simplified adenine base editors improve adenine base editing efficiency in rice. *Plant Biotechnology Journal* 18 (3): 770–778.

Huang, L., Zhang, R., Huang, G. et al. (2018). Developing superior alleles of yield genes in rice by artificial mutagenesis using the CRISPR/Cas9 system. *The Crop Journal* 6 (5): 475–481.

Huang, L., Li, Q., Zhang, C. et al. (2020). Creating novel Wx alleles with fine-tuned amylose levels and improved grain quality in rice by promoter editing using CRISPR/Cas9 system. *Plant Biotechnology Journal* 18 (11): 2164–2166. https://doi.org/10.1111/pbi.13391.

Hundleby, P. and Harwood, W. (2022). Regulatory constraints and differences of genome-edited crops around the globe. In: *Genome Editing* (ed. S.H. Wani and G. Hensel), 319–341. Cham: Springer International Publishing.

Hunziker, J., Nishida, K., Kondo, A. et al. (2020). Multiple gene substitution by target-AID base-editing technology in tomato. *Scientific Reports* 10 (1): 20471. https://doi.org/10.1038/s41598-020-77379-2.

Ibrahim, S., Saleem, B., Rehman, N. et al. (2021). CRISPR/Cas9 mediated disruption of *Inositol Pentakisphosphate 2- Kinase 1* (*TaIPK1*) reduces phytic acid and improves iron and zinc accumulation in wheat grains. *Journal of Advanced Research* 37: 33–41. https://doi.org/10.1016/j.jare.2021.07.006.

Ito, Y., Nishizawa-Yokoi, A., Endo, M. et al. (2015). CRISPR/Cas9-mediated mutagenesis of the RIN locus that regulates tomato fruit ripening. *Biochemical and Biophysical Research Communications* 467 (1): 76–82.

Jiang, W.Z., Henry, I.M., Lynagh, P.G. et al. (2017). Significant enhancement of fatty acid composition in seeds of the allohexaploid, *Camelina sativa*, using CRISPR/Cas9 gene editing. *Plant Biotechnology Journal* 15 (5): 648–657.

Jiang, M., Liu, Y., Liu, Y. et al. (2019). Mutation of inositol 1, 3, 4-trisphosphate 5/6-kinase6 impairs plant growth and phytic acid synthesis in rice. *Plants* 8 (5): 114.

Jiang, M., Zhan, Z., Li, H. et al. (2020). *Brassica Rapa* orphan genes largely affect soluble sugar metabolism. *Horticulture Research* 7: 181. https://doi.org/10.1038/s41438-020-00403-z.

Jinek, M., Chylinski, K., Fonfara, I. et al. (2012). A programmable dual-RNA-guided DNA endonuclease in adaptive bacterial immunity. *Science* 337: 816–821. https://doi.org/10.1126/science.1225829.

Juliano, B.O. (1998). Varietal impact on rice quality. *Cereal Foods World* 43: 207–222.

Jung, Y.J., Lee, G.J., Bae, S., and Kang, K.K. (2018). Reduced ethylene production in tomato fruits upon CRSPR/Cas9-mediated LeMADS-RIN mutagenesis. *Horticultural Science and Technology* 36 (3): 396–405.

Kaur, N., Alok, A., Kumar, P. et al. (2020). CRISPR/Cas9 directed editing of lycopene epsilon-cyclase modulates metabolic flux for β-carotene biosynthesis in banana fruit. *Metabolic Engineering* 59: 76–86.

Khanday, I., Skinner, D., Yang, B. et al. (2019). A male-expressed rice embryogenic trigger redirected for asexual propagation through seeds. *Nature* 565 (7737): 91–95.

Kim, Y.G., Cha, J., and Chandrasegaran, S. (1996). Hybrid restriction enzymes: zinc finger fusions to Fok I cleavage domain. *Proceedings. National Academy of Sciences. United States of America* 93: 1156–1160. https://doi.org/10.1073/pnas.93.3.1156.

Kim, D., Hager, M., Brant, E., and Budak, H. (2021). Efficient genome editing in wheat using Cas9 and Cpf1 (AsCpf1 and LbCpf1) nucleases. *Functional & Integrative Genomics* 21: 355–366.

Klimek-Chodacka, M., Oleszkiewicz, T., Lowder, L.G. et al. (2018). Efficient CRISPR/Cas9-based genome editing in carrot cells. *Plant Cell Reports* 37 (4): 575–586. https://doi.org/10.1007/s00299-018-2252-2.

Komor, A.C., Kim, Y.B., Packer, M.S. et al. (2016). Programmable editing of a target base in genomic DNA without double-stranded DNA cleavage. *Nature* 533 (7603): 420–424.

Ku, H.-K. and Ha, S.-H. (2020). Improving nutritional and functional quality by genome editing of crops: status and perspectives. *Frontiers in Plant Science* 11: 577313. https://doi.org/10.3389/fpls.2020.577313.

Kumar, D., Yadav, A., Ahmad, R. et al. (2022). CRISPR-based genome editing for nutrient enrichment in crops: a promising approach toward global food security. *Frontiers in Genetics* 13: 932859. https://doi.org/10.3389/fgene.2022.932859.

Kusano, H., Ohnuma, M., Mutsuro-Aoki, H. et al. (2018). Establishment of a modified CRISPR/Cas9 system with increased mutagenesis frequency using the translational enhancer dMac3 and multiple guide RNAs in potato. *Scientific Reports* 8 (1): 13753.

Lee, K., Zhang, Y., Kleinstiver, B.P. et al. (2019). Activities and specificities of CRISPR/Cas9 and Cas12a nucleases for targeted mutagenesis in maize. *Plant Biotechnology Journal* *17* (2): 362–372.

Lee, K.R., Jeon, I., Yu, H. et al. (2021). Increasing monounsaturated fatty acid contents in hexaploid *Camelina Sativa* seed oil by FAD2 gene knockout using CRISPR-Cas9. *Frontiers in Plant Science* 12: 702930. https://doi.org/10.3389/fpls.2021.702930.

Lemmon, Z.H., Reem, N.T., Dalrymple, J. et al. (2018). Rapid improvement of domestication traits in an orphan crop by genome editing. *Nature Plants* 4: 766–770. https://doi.org/10.1038/s41477-018-0259-x.

Li, J., Meng, X., Zong, Y. et al. (2016). Gene replacements and insertions in rice by intron targeting using CRISPR–Cas9. *Nature Plants* 2 (10): 1–6.

Li, R., Li, R., Li, X. et al. (2018a). Multiplexed CRISPR/Cas9-mediated metabolic engineering of γ-aminobutyric acid levels in *Solanum lycopersicum*. *Plant Biotechnology Journal* 16 (2): 415–427.

Li, R., Fu, D., Zhu, B. et al. (2018b). CRISPR/Cas9-mediated mutagenesis of *lncRNA1459* alters tomato fruit ripening. *The Plant Journal* 94 (3): 513–524.

Li, X., Wang, Y., Chen, S. et al. (2018c). Lycopene is enriched in tomato fruit by CRISPR/Cas9-mediated multiplex genome editing. *Frontiers in Plant Science* 9: 559.

Li, A., Jia, S., Yobi, A. et al. (2018d). Editing of an alpha-kafirin gene family increases, digestibility and protein quality in sorghum. *Plant Physiology* 177 (4): 1425–1438.

Li, C., Zong, Y., Wang, Y. et al. (2018e). Expanded base editing in rice and wheat using a Cas9-adenosine deaminase fusion. *Genome Biology* 19: 1–9.

Li, T., Yang, X., Yu, Y. et al. (2018f). Domestication of wild tomato is accelerated by genome editing. *Nature Biotechnology* 36 (12): 1160–1163.

Li, L., Hu, S., and Chen, X. (2018g). Non-viral delivery systems for CRISPR/Cas9-based genome editing: Challenges and opportunities. *Biomaterials* 171: 207–218.

Liang, Z., Zhang, K., Chen, K., and Gao, C. (2014). Targeted mutagenesis in *Zea mays* using TALENs and the CRISPR/Cas system. *Journal of Genetics and Genomics* 41 (2): 63–68.

Lin, Q., Zong, Y., Xue, C. et al. (2020). Prime genome editing in rice and wheat. *Nature Biotechnology* 38: 582–585. https://doi.org/10.1038/s41587-020-0455-x.

Liu, Y., Han, J., Chen, Z. et al. (2017). Engineering cell signaling using tunable CRISPR–Cpf1-based transcription factors. *Nature Communications* 8 (1): 2095.

Maioli, A., Gianoglio, S., Moglia, A. et al. (2020). Simultaneous CRISPR/Cas9 editing of three *PPO* genes reduces fruit flesh browning in *Solanum Melongena* L. *Frontiers in Plant Science* 11: 607161. https://doi.org/10.3389/fpls.2020.607161.

Mao, X., Zheng, Y., Xiao, K. et al. (2018). OsPRX2 contributes to stomatal closure and improves potassium deficiency tolerance in rice. *Biochemical and Biophysical Research Communications* 495 (1): 461–467.

Marx, V. (2018). Base editing a CRISPR way. *Nature Methods* 15 (10): 767–770.

Matsumoto, Y., Ohno, K., and Hiraoka, Y. (1997). Studies on the utilization of functional food materials containing high levels of gamma-aminobutyric acid (part 1). *Ehime Kougi Kenkyu Houkoku* 35: 97–100.

Mayer, J.E., Pfeiffer, W.H., and Beyer, P. (2008). Biofortified crops to alleviate micronutrient malnutrition. *Current Opinion in Plant Biology* 11 (2): 166–170. https://doi.org/10.1016/j.pbi.2008.01.007.

Meemken, E.M. and Qaim, M. (2018). Organic agriculture, food security, and the environment. *Annual Review of Resource Economics* 10: 39–63.

Miao, C., Xiao, L., Hua, K. et al. (2018). Mutations in a subfamily of abscisic acid receptor genes promote rice growth and productivity. *Proceedings of the National Academy of Sciences* 115 (23): 6058–6063.

Morineau, C., Bellec, Y., Tellier, F. et al. (2017). Selective genc dosage by CRISPR-Cas9 genome editing in hexaploid *Camelina sativa*. *Plant Biotechnology Journal* 15 (6): 729–739.

Nakayasu, M., Akiyama, R., Lee, H.J. et al. (2018). Generation of α-solanine-free hairy roots of potato by CRISPR/Cas9 mediated genome editing of the *St16DOX* gene. *Plant Physiology and Biochemistry* 131: 70–77.

Nieves-Cordones, M., Mohamed, S., Tanoi, K. et al. (2017). Production of low-Cs+ rice plants by inactivation of the K+ transporter *Os HAK 1* with the CRISPR-Cas system. *The Plant Journal* 92 (1): 43–56.

Nonaka, S., Arai, C., Takayama, M. et al. (2017). Efficient increase of G-aminobutyric acid (GABA) content in tomato fruits by targeted mutagenesis. *Scientific Reports* 7: 7057. https://doi.org/10.1038/s41598-017-06400-y.

Okuzaki, A., Ogawa, T., Koizuka, C. et al. (2018). CRISPR/Cas9-mediated genome editing of the *fatty acid desaturase 2* gene in *Brassica napus*. *Plant Physiology and Biochemistry* 131: 63–69.

Osakabe, Y., Liang, Z., Ren, C. et al. (2018). CRISPR-Cas9-mediated genome editing in apple and grapevine. *Nature Protocols* 13 (12): 2844–2863. https://doi.org/10.1038/s41596-018-0067-9.

Ozseyhan, M.E., Kang, J., Mu, X., and Lu, C. (2018). Mutagenesis of the FAE1 genes significantly changes fatty acid composition in seeds of *Camelina sativa*. *Plant Physiology and Biochemistry* 123: 1–7.

Pacher, M. and Puchta, H. (2017). From classical mutagenesis to nuclease-based breeding – directing natural DNA repair for a natural end-product. *The Plant Journal* 90: 819–833. https://doi.org/10.1111/tpj.13469.

Prado, J.R., Segers, G., Voelker, T. et al. (2014). Genetically engineered crops: from idea to product. *Annual Review of Plant Biology* 65: 769–790. https://doi.org/10.1146/annurev-arplant-050213-040039.

Qi, W., Zhu, T., Tian, Z. et al. (2016). High-efficiency CRISPR/Cas9 multiplex gene editing using the glycine tRNA-processing system-based strategy in maize. *BMC Biotechnology* 16: 1–8.

Qiu, Z., Wang, H., Li, D. et al. (2019). Identification of candidate HY5-dependent and -independent regulators of anthocyanin biosynthesis in tomato. *Plant and Cell Physiology* 60 (3): 643–656. https://doi.org/10.1093/pcp/pcy236.

Ren, C., Liu, X., Zhang, Z. et al. (2016). CRISPR/Cas9-mediated efficient targeted mutagenesis in Chardonnay (*Vitis vinifera* L.). *Scientific Reports* 6 (1): 32289.

Rodriguez-Leal, D., Lemmon, Z.H., Man, J. et al. (2017). Engineering quantitative trait variation for crop improvement by genome editing. *Cell* 171 (2): 470–e8. https://doi.org/10.1016/j.cell.2017.08.030.

Rouet, P., Smih, F., and Jasin, M. (1994). Introduction of double-strand breaks into the genome of mouse cells by expression of a rare-cutting endonuclease. *Molecular and Cellular Biology* 14 (12): 8096–8106. https://doi.org/10.1128/mcb.14.12.8096-8106.1994.

Sánchez-León, S., Gil-Humanes, J., Ozuna, C.V. et al. (2018). Low-gluten, nontransgenic wheat engineered with CRISPR/Cas9. *Plant Biotechnology Journal* 16: 902–910. https://doi.org/10.1111/pbi.12837.

Saurabh, S., Vidyarthi, A.S., and Prasad, D. (2014). RNA interference: concept to reality in crop improvement. *Planta* 239: 543–564.

Shan, Q., Zhang, Y., Chen, K. et al. (2015). Creation of fragrant rice by targeted knockout of the *Os BADH 2* gene using TALEN technology. *Plant Biotechnology Journal* 13 (6): 791–800.

Shen, L., Wang, C., Fu, Y. et al. (2018). QTL editing confers opposing yield performance in different rice varieties. *Journal of Integrative Plant Biology* 60 (2): 89–93. https://doi.org/10.1111/jipb.12501.

Shi, J., Gao, H., Wang, H. et al. (2017). ARGOS 8 variants generated by CRISPR-Cas9 improve maize grain yield under field drought stress conditions. *Plant Biotechnology Journal* 15 (2): 207–216.

Shukla, V.K., Doyon, Y., Miller, J.C. et al. (2009). Precise genome modification in the crop species *Zea mays* using zinc-finger nucleases. *Nature* 459 (7245): 437–441.

Singh, S., Rahangdale, S., Pandita, S. et al. (2022). CRISPR/Cas9 for insect pests management: a comprehensive review of advances and applications. *Agriculture* 12: 1896. https://doi.org/10.3390/agriculture12111896.

Spicer, A. and Molnar, A. (2018). Gene editing of microalgae: scientific progress and regulatory challenges in Europe. *Biology* 7 (1): 21.

Sugano, S., Hirose, A., Kanazashi, Y. et al. (2020). Simultaneous induction of mutant alleles of two allergenic genes in soybean by using site-directed mutagenesis. *BMC Plant Biology* 20 (1): 513–515. https://doi.org/10.1186/s12870-020-02708-6.

Sun, Y., Jiao, G., Liu, Z. et al. (2017). Generation of high-amylose rice through CRISPR/Cas9-mediated targeted mutagenesis of starch branching enzymes. *Frontiers in Plant Science* 8: 298.

Sushmita, K.G., Upadhyay, S.K., and Verma, P.C. (2021). An overview of genome-engineering methods. In: *Genome Engineering for Crop Improvement* (ed. S.K. Upadhyay). Wiley. https://doi.org/10.1002/9781119672425.ch1.

Svitashev, S., Young, J.K., Schwartz, C. et al. (2015). Targeted mutagenesis, precise gene editing, and site-specific gene insertion in maize using Cas9 and guide RNA. *Plant Physiology* 169 (2): 931–945.

Symington, L.S. and Gautier, J. (2011). Double-strand break end resection and repair pathway choice. *Annual Review of Genetics* 45: 247–271. https://doi.org/10.1146/annurevgenet-110410-132435.

Tanaka, Y. and Ohmiya, A. (2008). Seeing is believing: engineering anthocyanin and carotenoid biosynthetic pathways. *Current Opinion in Biotechnology* 19: 190–197. https://doi.org/10.1016/j.copbio.2008.02.015.

Tang, L., Mao, B., Li, Y. et al. (2017). Knockout of *OsNramp5* using the CRISPR/Cas9 system produces low Cd-accumulating indica rice without compromising yield. *Scientific Reports* 7 (1): 14438.

Tareke, E., Rydberg, P., Karlsson, P. et al. (2002). Analysis of acrylamide, a carcinogen formed in heated foodstuffs. *Journal of Agricultural and Food Chemistry* 50: 4998–5006.

Tikunov, Y.M., Roohanitaziani, R., Meijer-Dekens, F. et al. (2020). The genetic and functional analysis of flavor in commercial tomato: the FLORAL4 gene underlies a QTL for floral aroma volatiles in tomato fruit. *The Plant Journal* 103 (3): 1189–1204. https://doi.org/10.1111/tpj.14795.

Tzuri, G., Zhou, X., Chayut, N. et al. (2015). A 'golden' SNP in CmOrgoverns the fruit flesh color of melon (*Cucumis Melo*). *The Plant Journal* 82 (2): 267–279. https://doi.org/10.1111/tpj.12814.

Upadhyay, S.K. (2021). *Genome Engineering for Crop Improvement*. Wiley. https://doi.org/10.1002/9781119672425.

Upadhyay, S.K., Kumar, J., Alok, A. et al. (2013). RNA guided genome editing for multiple target gene mutations in wheat. *G3: Genes| Genomes| Genetics* 3 (12): 2233–2238.

USDA-APHIS FAS (2020). Oilseeds: world markets and trade. United States Department of Agriculture Foreign Agricultural Service. https://apps.fas.usda.gov/psdonline/circulars/oilseeds.pdf (accessed September 2020).

Van Vu, T., Sung, Y.W., Kim, J. et al. (2019). Challenges and perspectives in homology-directed gene targeting in monocot plants. *Rice* 12 (1): 1–29.

Veillet, F., Chauvin, L., Kermarrec, M.P. et al. (2019). The *Solanum tuberosum GBSSI* gene: a target for assessing gene and base editing in tetraploid potato. *Plant Cell Reports* 38 (9): 1065–1080. https://doi.org/10.1007/s00299-019-02426-w.

Voytas, D.F. and Gao, C. (2014). Precision genome engineering and agriculture: opportunities and regulatory challenges. *PLoS Biology* 12 (6): e1001877.

Waltz, E. (2016a). CRISPR-edited crops free to enter market, skip regulation. *Nature Biotechnology* 34 (6): 582–583.

Waltz, E. (2016b). Gene-edited CRISPR mushroom escapes US regulation. *Nature* 532 (7599): 293.

Wang, W., Pan, Q., He, F. et al. (2018). Transgenerational CRISPR-Cas9 activity facilitates multiplex gene editing in allopolyploid wheat. *The CRISPR Journal* 1 (1): 65–74.

Wang, D., Samsulrizal, N.H., Yan, C. et al. (2019a). Characterisation of CRISPR mutants targeting genes modulating pectin degradation in ripening tomato. *Plant Physiology* 179 (2): 544–557. https://doi.org/10.1104/pp.18.01187.

Wang, H., Wu, Y., Zhang, Y. et al. (2019b). CRISPR/Cas9-based mutagenesis of starch biosynthetic genes in sweet potato (*Ipomoea Batatas*) for the improvement of starch quality. *International Journal of Molecular Sciences* 20 (19): 4702. https://doi.org/10.3390/ijms20194702.

Wang, W., Pan, Q., Tian, B. et al. (2019c). Gene editing of the wheat homologs of TONNEAU 1-recruiting motif encoding gene affects grain shape and weight in wheat. *The Plant Journal* 100 (2): 251–264. https://doi.org/10.1111/tpj.14440.

Wang, T., Zhang, H., and Zhu, H. (2019d). CRISPR technology is revolutionizing the improvement of tomato and other fruit crops. *Horticulture Research* 6: https://doi.org/10.1038/s41438-019-0159-x.

Wang, S., Yang, Y., Guo, M. et al. (2020). Targeted mutagenesis of amino acid transporter genes for rice quality improvement using the CRISPR/Cas9 system. *The Crop Journal* 8 (3): 457–464.

Wang, B., Li, N., Huang, S. et al. (2021). Enhanced soluble sugar content in tomato fruit using CRISPR/Cas9-mediated *SlINVINH1* and *SlVPE5* gene editing. *PeerJ* 9: e12478. https://doi.org/10.7717/peerj.12478.

Wen, S., Liu, H., Li, X. et al. (2018). TALEN-mediated targeted mutagenesis of *fatty acid desaturase 2* (*FAD2*) in peanut (*Arachis hypogaea* L.) promotes the accumulation of oleic acid. *Plant Molecular Biology* 97: 177–185.

Wolabu, T.W., Park, J.J., Chen, M. et al. (2020). Improving the genome editing efficiency of CRISPR/Cas9 in *Arabidopsis* and *Medicago truncatula*. *Planta* 252: 1–14.

Xing, S., Chen, K., Zhu, H. et al. (2020). Fine-tuning sugar content in strawberry. *Genome Biology* 21 (1): 1–14.

Xu, R., Yang, Y., Qin, R. et al. (2016). Rapid improvement of grain weight via highly efficient CRISPR/Cas9-mediated multiplex genome editing in rice. *Journal of Genetics and Genomics= Yi chuan xue bao* 43 (8): 529–532.

Xu, R., Qin, R., Li, H. et al. (2019). Enhanced genome editing in rice using single transcript unit CRISPR-LbCpf1 systems. *Plant Biotechnology Journal* 17 (3): 553.

Yan, S., Chen, N., Huang, Z. et al. (2020). Anthocyanin fruit encodes an R2R3-MYB transcription factor, *SlAN2-like*, activating the transcription of *SlMYBATV* to fine-tune anthocyanin content in tomato fruit. *The New Phytologist* 225 (5): 2048–2063. https://doi.org/10.1111/nph.16272.

Yasumoto, S., Umemoto, N., Lee, H.J. et al. (2019). Efficient genome engineering using platinum TALEN in potato. *Plant Biotechnology* 36 (3): 167–173.

Yin, K., Gao, C., and Qiu, J.L. (2017). Progress and prospects in plant genome editing. *Nature Plants* 3 (8): 1–6.

Yu, Q.H., Wang, B., Li, N. et al. (2017). CRISPR/Cas9-induced targeted mutagenesis and gene replacement to generate long-shelf life tomato lines. *Scientific Reports* 7 (1): 11874.

Yuste-Lisbona, F.J., Fernández-Lozano, A., Pineda, B. et al. (2020). *ENO* regulates tomato fruit size through the floral meristem development network. *Proceedings. National Academy of Sciences. United States of America* 117 (14): 8187–8195. https://doi.org/10.1073/pnas.1913688117.

Yuyu, C., Aike, Z., Pao, X. et al. (2020). Effects of GS3 and GL3.1 for grain size editing by CRISPR/Cas9 in rice. *Rice Science* 27 (5): 405–413. https://doi.org/10.1016/j.rsci.2019.12.010.

Zannoni, L. (2019). Evolving regulatory landscape for genome-edited plants. *The CRISPR Journal* 2 (1): 3–8.

Zeng, Z., Han, N., Liu, C. et al. (2020). Functional dissection of *HGGT* and *HPT* in barley vitamin E biosynthesis via CRISPR/Cas9-enabled genome editing. *Annals of Botany* 126 (5): 929–942.

Zetsche, B., Gootenberg, J.S., Abudayyeh, O.O. et al. (2015). Cpf1 is a single RNA-guided endonuclease of a class 2 CRISPR-Cas system. *Cell* 163 (3): 759–771.

Zhang, X., Wang, J., Huang, J. et al. (2012). Rare allele of *OsPPKL1* associated with grain length causes extra-large grain and a significant yield increase in rice. *Proceedings of the National Academy of Sciences* 109 (52): 21534–21539.

Zhang, Y., Liang, Z., Zong, Y. et al. (2016). Efficient and transgene-free genome editing in wheat through transient expression of CRISPR/Cas9 DNA or RNA. *Nature Communications* 7 (1): 12617.

Zhang, X., Wang, J., Cheng, Q. et al. (2017). Multiplex gene regulation by CRISPR-ddCpf1. *Cell Discovery* 3 (1): 1–9.

Zhang, J., Zhang, H., Botella, J.R., and Zhu, J.K. (2018a). Generation of new glutinous rice by CRISPR/Cas9-targeted mutagenesis of the *Waxy* gene in elite rice varieties. *Journal of Integrative Plant Biology* 60 (5): 369–375.

Zhang, Y., Li, D., Zhang, D. et al. (2018b). Analysis of the functions of *TaGW2* homoeologs in wheat grain weight and protein content traits. *The Plant Journal* 94 (5): 857–866.

Zhang, H., Si, X., Ji, X. et al. (2018c). Genome editing of upstream open reading frames enables translational control in plants. *Nature Biotechnology* 36 (9): 894–898. https://doi.org/10.1038/nbt.4202.

Zhao, D.S., Li, Q.F., Zhang, C.Q. et al. (2018). GS9 acts as a transcriptional activator to regulate rice grain shape and appearance quality. *Nature Communications* 9 (1): 1240. https://doi.org/10.1038/s41467-018-03616-y.

Zhi, J., Liu, X., Li, D. et al. (2020). CRISPR/Cas9- mediated SlAN2 mutants reveal various regulatory models of anthocyanin biosynthesis in tomato plant. *Plant Cell Reports* 39 (6): 799–809. https://doi.org/10.1007/s00299-020-02531-1.

Zhong, Y., Blennow, A., Kofoed-Enevoldsen, O. et al. (2019). Protein targeting to starch 1 Is essential for starchy endosperm development in barley. *Journal of Experimental Botany* 70 (2): 485–496. https://doi.org/10.1093/jxb/ery398.

Zhou, H., He, M., Li, J. et al. (2016). Development of commercial thermo-sensitive genic male sterile rice accelerates hybrid rice breeding using the CRISPR/Cas9-mediated TMS5 editing system. *Scientific Reports* 6 (1): 37395.

Zhou, J., Xin, X., He, Y. et al. (2019). Multiplex QTL editing of grain-related genes improves yield in elite rice varieties. *Plant Cell Reports* 38: 475–485.

Zhu, Y., Lin, Y., Chen, S. et al. (2019). CRISPR/Cas9-mediated functional recovery of the recessive rc allele to develop red rice. *Plant Biotechnology Journal* 17 (11): 2096–2105.

Zsögön, A., Čermák, T., Naves, E.R. et al. (2018). De novo domestication of wild tomato using genome editing. *Nature Biotechnology* 36 (12): 1211–1216.

8

Genome-Editing Tools for Engineering of MicroRNAs and Their Encoded Peptides, miPEPs, in Plants

Ravi Shankar Kumar[1], *Hiteshwari Sinha*[2,3], *Tapasya Datta*[1], *Ashish Sharma*[1], *and Prabodh Kumar Trivedi*[1,2,3]

[1] *CSIR-Central Institute of Medicinal and Aromatic Plants (CSIR-CIMAP), Lucknow, Uttar Pradesh, India*
[2] *CSIR-National Botanical Research Institute, Council of Scientific and Industrial Research (CSIR-NBRI), Rana Pratap Marg, Lucknow, Uttar Pradesh, India*
[3] *Academy of Scientific and Innovative Research (AcSIR), Ghaziabad, Uttar Pradesh, India*

8.1 Introduction

Genome editing is a robust technique in terms of its ability to bring not only revolution in biotechnological approaches in the field of crop development but also human pathology. Genome editing (also called gene editing) is the tool by which the genome sequence of any organism can be precisely modified by using several techniques. These techniques may enable the regulation of the pattern of genes expression in the pre-determined region and enrich the understanding of the functional genomics of specific organisms. The advancement in gene editing technologies has offered researchers the capability to the rapid and economic establishment of sequence-specific modifications into the genome of various organisms. Plant breeding is the conventional method that has been performed by crossing and selection over the years, but these methods are protracted and require high labor, whereas genome editing targets and modifies the specific DNA sequences with high efficiency (Osakabe and Osakabe 2015; Wang et al. 2016). As genome-editing technology is often used for various organisms and plants, this technique generally includes three steps. First is the identification of target DNA by the exogenous nuclease with its recognition domain and then its binding to the sequence targeted, which leads to the formation of double-stranded breaks (DSBs) at or nearby the target sequences. Thereafter the nonhomologous end-joining (NHEJ) process mediated repair of the DSBs, often introduces mutation such as insertion or deletion due to NHEJ, which is a repair process that is error-prone (Wang et al. 2016; Jaganathan et al. 2018; Upadhyay 2021). The establishment of genome-editing was done in the 1990s, prior to the initiation of current nuclease-based gene editing platforms (Woolf 1998); however, due to low efficiencies of editing its use was restricted. The genome-editing includes three technologies with different nucleases, which are programmable nucleases (Cox et al. 2015). The first is meganuclease zinc finger nuclease

Applications of Genome Engineering in Plants, First Edition. Edited by Santosh Kumar Upadhyay.
© 2024 John Wiley & Sons Ltd. Published 2024 by John Wiley & Sons Ltd.

(ZFN), the second is the transcription activator-like effector nuclease (TALEN), and the third is the CRISPR/Cas9, which is clustered regularly interspaced short palindromic repeat with associated Cas9 (Gaj et al. 2016).

8.1.1 ZINC Finger Nucleases

ZFNs are two fused proteins, including Cys_2-His_2 zinc-finger protein and FokI restriction endonuclease with the cleavage domain, the first nuclease used for a targeted mutation in the genome (Kim et al. 1996; Porteus and Carroll 2005). The ZFNs work as a dimer, in which each monomer recognizes the 9–18 base pairs of target DNA containing the ZINC-finger domain (Figure 8.1). The dimerization of ZFNs is mediated by the FokI domain and cleaves the DNA sequence within the five to seven spacer sequences, which are separated by two flanking zinc finger domains (Smith et al. 2000). The ZFNs have three or four zinc finger domains; each domain having ~30 amino acid residues, which are arranged into a ββα motif. However, the residues that are recognized by the DNA are found in the α-helical domain and these interact with the DNA triplets (Pavletich and Pabo 1991; Wolfe et al. 2000). The zinc finger domain generally gets fused with each other to form canonical linker peptides and produce polydactyl zinc finger proteins that can target a wide range of the possible DNA sequences (Liu et al. 1997; Beerli et al. 1998; Kim et al. 2009). One of the chief drawbacks of ZFNs is their off-targeting effects on genome editing, which enhanced the development to get more specificity in the editing (Pattanayak et al. 2011). To minimize

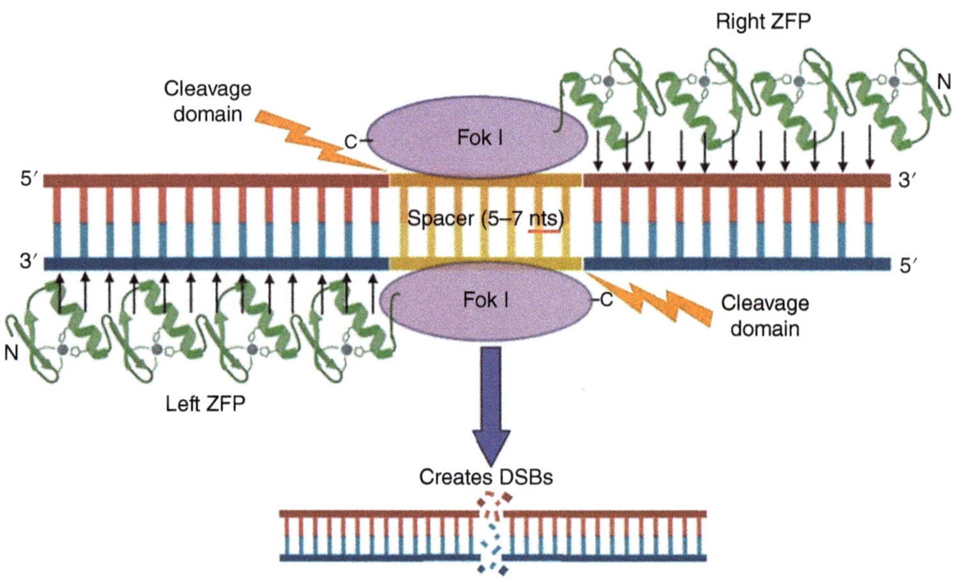

Figure 8.1 Mechanism of zinc finger nuclease (ZFN). ZFNs recognize the target with the two zinc-finger binding domains having the flanking region of 5–7 bp (base pair) spacer sequence recognized by the FokI cleavage domain. Then the paired FokI endonuclease introduces double-strand breaks (DSBs) into the target sequence. Figure created using BioRender (https://biorender.com/).

the limitation of ZFNs, several advances have been carried out. Among them, the generation of obligate heterodimeric ZFN architectures, which depends upon the repulsion of charges that restrict the surplus homodimerization of the FokI domain, thus reducing the off-target effect. Tools of protein engineering have also been employed to increase the potential of the cleavage domain of FokI (Miller et al. 2007; Guo et al. 2010; Doyon et al. 2011). ZFNs activity can be improved by injecting them as protein into the cell because of their cell permeable capability, which assists editing in the genome with lesser off-target effects. ZFNickases enzyme facilitates homology-directed repair by nicking at one strand of DNA. The enzyme composed of two monomers: one ZFN monomer is catalytically inactive, and another one is active, which directs the repair mechanism (Gaj et al. 2012; Kim et al. 2012; Ramirez et al. 2012; Wang et al. 2012). In the genome-editing platform, the generation of zinc-finger domains is a big challenge that can efficiently identify the triplet codons, particularly the 5′-TNN-3′ and 5′-CNN-3′ array, due to which ZFNs be deficient in the targeting flexibility.

8.1.2 TALE Nucleases

TALE protein is the bacterial effector present in the plant pathogenic bacteria belonging to the genus Xanthomonas, which was discovered in the year 2009. Soon after its discovery, this protein was customized that is capable to modify any genomic sequences (Boch et al. 2009), and successfully genome editing was done in 2011. Like the ZFNs, TALENs have a DNA binding domain with an amino acid terminal, which is fused to the FokI cleavage domain present at the carboxy terminal (Christian et al. 2010). The dimerization is also performed by the cleavage domain of FokI, which cuts genomic sequences within 12–19 bp spacer sequences (Figure 8.2) (Miller et al. 2011). The TALENs have higher genome-editing specificity, which recognizes 12–20 bps sequences (Guilinger et al. 2014). TALE's binding domain comprises of repeat domain series of ~34 amino acids each. Each repeat of the binding domain interacts with the DNA at the 12 and 13 amino acid residues called the repeat variable di-residues (RVDs). To arrange artificial TALE arrays, the commonly used RVDs are NI for adenine, NN or HN for guanine or adenine, HD for cytosine, and NG for thymine (Boch et al. 2009; Moscou and Bogdanove 2009; Streubel et al. 2012; Cong et al. 2013). The TALE repeat identifies only a single base pair, unlike the ZINC fingers, which recognize triplets with no site overlaps from adjoining domains (Mak et al. 2012). The construction of TALE DNA-binding domains can be done by various methods like Golden Gate assembly, iterative capped assembly, FLASH assembly, and ligation-independent cloning (Cermak et al. 2011; Briggs et al. 2012; Reyon et al. 2012; Schmid-Burgk et al. 2013). TALEN assembly focuses on method development that increases their function that comprises specificity profiling to reveal nonconventional RVDs, directed evolution, and generation of chimeric nucleases by combining TALE domains into homing endonuclease variants to enhance the specificity (Guilinger et al. 2014; Yang et al. 2014; Hubbard et al. 2015; Juillerat et al. 2015; Miller et al. 2015). TALENs possess two advantages in comparison to ZFNs for genome editing; first is an assembly of functional nuclease needs, lesser time, and experience. Second, compared to the ZFNs, TALENs show higher affinity to the target DNA, are more specific, and have reduced toxicity (Meckler et al. 2013; Mussolino et al. 2014). TALENs are considerably more than ZFNs with highly repetitive

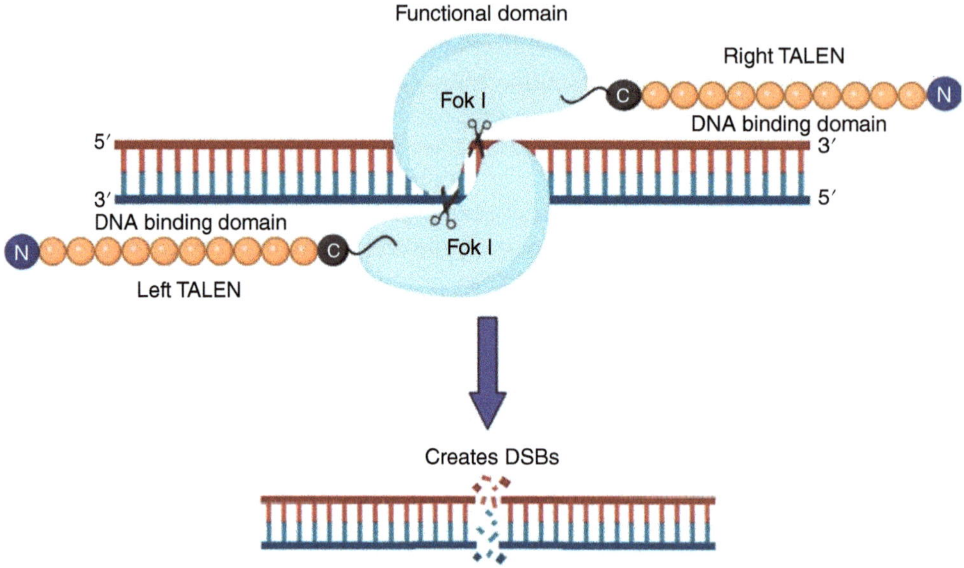

Figure 8.2 Mechanism of transcription activator-like effector nucleases (TALENs). TALENs recognize target sites that have two TALE DNA-binding sites with the flanking region of 12–20 bp spacer sequence recognized and cleaved by FokI domain. Then the paired FokI endonuclease introduces double strand breaks (DSBs) into the target sequence. Figure created using BioRender (https://biorender.com/).

structures, making them competent to deliver efficiently into the cell via lentivirus or AAV (single adeno-associated virus) (Holkers et al. 2013).

8.1.3 CRISPR/Cas 9

The clustered regularly interspaced short palindromic repeats (CRISPR)/CRISPR-associated protein 9 (Cas9) system is based on the sophisticated RNA-guided adaptive immune system of bacteria against attacking bacteriophages and the transfer of plasmids (Garneau et al. 2010; Wiedenheft et al. 2012; Sorek et al. 2013). This system, in nature, employs the dual RNA-guided DNA endonuclease Cas9 that introduces double-stranded breaks (DSBs) in the target DNA in a site-specific manner. Target specificity is acquired from the dual-RNA guide, which is transcribed from the CRISPR loci in bacteria that have acquired fragments of foreign DNA from the invading phages (Gasiunas et al. 2012). The dual RNA hybrid then binds to the complementary target sequence, flanked by a short protospacer adjacent motif (PAM) and guides Cas9 to introduce DSBs to cleave the foreign DNA (Jinek et al. 2012).

The CRISPR-Cas9 system has been devised as a powerful tool for genome-editing through RNA-guided DNA targeting. Through this technology, one can specifically introduce mutations in almost any genomic sequence as directed by a short stretch of guide RNA, leading to the effective characterization of the function of the gene being studied. Since the first evidence of its potential in the field of genetic engineering (Jinek et al. 2012), the CRISPR–Cas9 system has been devised immensely as a significant tool for introducing

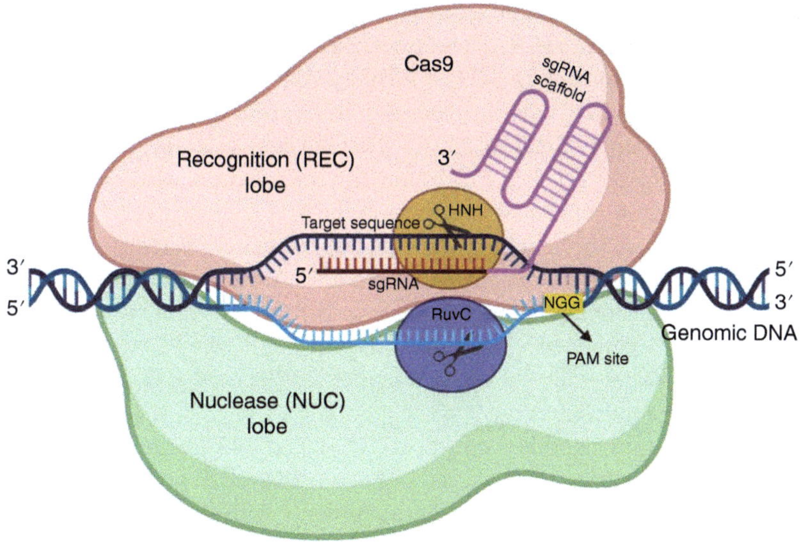

Figure 8.3 Clustered regularly interspaced short palindromic repeats-Cas9 system. CRISPR technology employs a guide RNA to direct the Cas9 enzyme to a target DNA sequence. Cas9 is guided by the sgRNA to the target sequence upstream of PAM (protospacer-adjacent motif) site in the DNA to introduce double strand breaks and cleave the target DNA. Figure created using BioRender (https://biorender.com/).

genome modifications in varied organisms due to its specificity, high efficiency, and ease of use (Figure 8.3) (Hsu et al. 2014; Bhalothia et al. 2020; Singh et al. 2022).

8.2 CRISPR–Cas9-Mediated DNA Interference in Bacterial Adaptive Immunity

CRISPR loci are broadly distributed and present in 85% of archaea, and 42% of bacteria are known to be present within bacteriophages as well as other mobile genetic elements (Makarova et al. 2019). A CRISPR locus typically present in a bacterial system (type II CRISPR-Cas system) comprises of an array of short, constant, partially palindromic, repetitive noncoding DNA sequences (repeats) by short stretches of non-repetitive variable sequences known as spacers, as well as a set of CRISPR-associated (cas) genes. The constant repeat regions are used to differentiate crRNA from other cellular RNAs and direct the specific recruitment of crRNA for its process of maturation. The trans-activating CRISPR RNA (tracrRNA) gene lies upstream of the cas operon and transcribes a unique noncoding RNA homologous to the repeat sequences. In the first phase of this acquired immune response, called immunization or spacer acquisition, after exposure to the invading phage or plasmids, short segments of foreign DNA are integrated as new spacers into the CRISPR repeat-spacer array within the host bacterial chromosome (Amitai et al. 2016; Nussenzweig et al. 2020). Once incorporated, the new spacer is transcribed along with all the other previously present spacers into a long precursor CRISPR RNA (pre-crRNA) comprising of both the repeats and spacer sequences. The tracrRNA is separately transcribed that associates with the pre-crRNA. The tracrRNA plays a fundamental role in recruiting Cas9 endonuclease. Further, the CRISPR transcripts undergo endonucleolytic processing (RNase III cleavage and 5′ trimming) to yield mature CRISPR RNAs (crRNAs) whose length gets limited to approximately 20 nts (Deltcheva et al. 2011). Finally, the mature crRNA contains the spacer at the 5′ end, a short RNA segment that is complementary to a portion from the foreign genetic element, and at the 3′ end comprises of a portion of the CRISPR repeat sequence. In the interference phase, the mature crRNA–tracrRNA hydrides to the foreign target sequence complementary to crRNA (protospacer) and then it associates with Cas9 endonuclease and guides it to incise the foreign DNA (protospacer) present 5′ upstream of the PAM sequence, upon a second infection. The PAM sequence is essential to discriminate between self and nonself sequences (Sternberg et al. 2014). In most of the cases of Cas9 endonucleases the PAM sequence is 5′-NGG-3′, where N can be any one of the four (A, T, C, G) nucleotide (Wang et al. 2015). The CRISPR loci hence function as molecular memory system for the bacterial immune system that retains records of its previous encounters with invading foreign genetic elements. This immunologic memory generated by CRISPRs is heritable and passed down to daughter cells, thereby rapidly transmitting immunity across the microbial community (Heller et al. 2014).

8.2.1 Types of CRISPR Systems

At present, CRISPRs are classified into two classes comprising six major types (I–VI) and 50 subtypes on the basis of their sequences. In each of the cases, a unique set of Cas proteins associated with their crRNA is utilized for CRISPR mediated interference. The Class 1 comprising of type I, type III, and IV systems employs a large multi-Cas protein complex for crRNA binding and cleavage of target sequence (Musharova et al. 2019), whereas Class 2 comprising of types II, V, and VI have a single crRNA:Cas ribonucleoprotein complex. In the type II CRISPR system, a single DNA endonuclease, Cas9, identifies target dsDNA substrates and degrade each of the strands with a distinct nuclease domain (HNH and RuvC).

8.2.2 The Cas9 Enzyme

In genome engineering, the Cas9 enzyme obtained from *Streptococcus pyogenes* is the most commonly studied and widely utilized Cas endonuclease enzyme. *Streptococcus pyogenes* Cas9 (SpyCas9) is a large (1368-amino-acid) DNA endonuclease that has multiple domains with varied functions. It utilizes its two different nuclease domains to cleave the dsDNA 3 bp upstream of the PAM sequence. The HNH-like nuclease domain introduces cuts in the DNA strand complementary to the guide RNA sequence (target strand), and the RuvC-like nuclease domain cleaves the DNA strand lying opposite to the complementary strand (non-target strand). Moreover, apart from participating in CRISPR-mediated interference, Cas9 also contributes to the maturation of crRNA and acquisition of the spacer sequences (Gasiunas et al. 2012; Chen et al. 2014; Jiang et al. 2015).

Cas9 in its apo state consists of two distinct lobes the alpha-helical recognition (REC) lobe and the nuclease (NUC) lobe comprising of the conserved HNH domain and the well-conserved RuvC domain having three split RuvC motifs, as well as variable C-terminal domain (CTD). The CTD has a distinct Cas9-specific fold and also PAM-interacting sites required for locating PAM sites. However, this PAM-identifying region remains disarranged in the Cas9 apoenzyme form and only upon associating with the sgRNA complex, it acquires its active configuration that recognizes target DNA and finally leads to its cleavage. Each of the HNH domain and RuvC domain cleaves one strand of the target dsDNA at a specific site that lies 3 bp upstream from the PAM sequence to generate a primarily blunt-ended DSB (Jinek et al. 2014).

8.3 CRISPR/Cas9 Effector Complex Assembly

The site-specific recognition of DNA and its cleavage requires the proper assembly of Cas9 with the guide RNA (a native crRNA–tracrRNA or sgRNA) to give rise to an active DNA surveillance complex. DNA target specificity is provided by the 20-nt spacer sequence of crRNA, and the tracrRNA has a fundamental role in the recruitment of Cas9 enzyme. Charpentier and Doudna, with their colleagues, in the year 2012 reported that in the process of target identification, Cas9 protein requires majorly a seed sequence of RNA nucleotides present within the spacer region of crRNAs, which essentially confers target sequence specificity (Semenova et al. 2011; Wiedenheft et al. 2011). In type II CRISPR systems, the seed region is mainly the sequence of 10–12 nucleotides near the PAM site lying at the 3′ end of the 20-nt spacer sequence. Mismatches within the seed sequence drastically modify or lead to complete elimination of binding and consequently the cleavage of the target DNA sequence is affected, whereas close homology in the seed region causes major off-target binding events even when several mismatches are located elsewhere (Jiang et al. 2015).

8.4 The Mechanism of CRISPR/Cas9-Mediated Genome Engineering

In recent years for genetic engineering, a simplified system of CRISPR/Cas9 has been developed, which only requires a chimeric sgRNA that unifies the crRNA and tracrRNA into a single RNA transcript and also retains the completely functional Cas9 nuclease associated with it (Doudna et al. 2014). Alterations in the sequence of the guide RNA (spacer) within the crRNA in the most basic two-component CRISPR–Cas9 system can be devised to target and cleave possibly any DNA sequence of interest in the genome (Cong et al. 2013; Mali et al. 2013).

The sgRNA guides the Cas9 endonuclease to investigate arbitrary DNA sequences in the genome utilizing the synthetically designed 20-nt guide RNA sequence. DSBs are generated within the targeted genomic DNA by the Cas9 as directed by the guide RNA. Host-mediated DNA repair mechanisms repair the DSB produced by the two distinct Cas9 nuclease domains. Without a repair template, the error-prone nonhomologous end joining (NHEJ) pathway is activated. It causes random insertions and deletions (indels) or even substitutions at the DSB site, generally causing a disruption in the function of the gene. But when a donor template having a sequence of interest flanked by homology arms is present, the error-free homology-directed repair (HDR) pathway is utilized to generate desired mutations by the process of homologous recombination, which leads to the introduction of precise editing within target sequence, like gene knockin, correction, deletion, or mutagenesis (Sashital et al. 2012; Sushmita et al. 2021).

8.4.1 Comparison with Other Technologies for Genome Editing

CRISPR/Cas9 system has a number of potential advantages over ZFNs and TALENs for genetic manipulation owing to the ease of designing, simplicity in use, and enhanced efficiency (Table 8.1) (Gaj et al. 2013; Gupta et al. 2014; Jiang et al. 2017; Malzahn et al. 2017; Alok et al. 2021).

- **Ease of customization**: Cas9 can be retargeted easily to new DNA sequences by only designing a pair of oligos encoding the 20-nt guide sequence. Whereas retargeting of TALEN for a new DNA sequence necessitates the development of two new TALEN genes.
- **Specificity and cleavage pattern towards target sequence**: Cas9 (SpCas9) introduces blunt cuts within the target sequence (3 bp 5′ of the PAM). Contrastingly, TALENs cleave in a nonspecific manner in the 12–24-bp linker between the pair of TALEN monomer-binding sites.
- **Simplicity in the basic mechanism**: Unlike the conventional DNA editing techniques mediated by nucleases – zinc finger nucleases (ZFNs) and transcription activator-like effector nucleases (TALENs) – DNA recognition by CRISPR/Cas9 is not specified by protein but rather by the 20-nt guide RNA sequence. This alleviates the requirement for tedious protein engineering of DNA-recognition domains for each DNA sequence to be targeted, thus can be applied for genomic manipulations and screening on a wide spectrum.
- **Editing efficiency**: CRISPR/Cas9 system can be devised to manipulate a number of genomic loci simultaneously, by delivering together a combination of sgRNAs to the cells targeted.

8.4.2 Limitations of the Cas9 System

- **Targeting site limited by PAM sequence**: The unique PAM sequence of SpCas9 requires a 5′-NGG PAM sequence that is present only approximately after every 8–12 bp thus becomes a limitation for the range that can be targeted.
- **Potential for off-target Mutagenesis**: The reduction or elimination of any unwanted off-target mutations at sites having a sequence homologous to the target sequence is a major pre-requisite for the successful use of genetic engineering using CRISPR technology sgRNA seldom identifies nontarget sequences that are nearly similar with few base mismatches leading to nondesirable, off-target mutations. Mutations at off-target sites can be reduced, if a mutant form of Cas9 possessing only one domain that retains activity (HNH or RuvC) is devised accompanied with two sgRNAs having close target sites in opposite strands. This causes a nick in only one strand at the target site by one of the sgRNA. But the use of the mutated Cas9 (nickase) having two sgRNAs with close target site makes close nicks in opposite strands, which causes a DSB that can be repaired through NHEJ, introducing mutation at the target site (Fu et al. 2013; Cho et al. 2014; Tsai et al. 2014).

Table 8.1 Summary of the mechanisms governing the genome-editing processes mediated by various nucleases along with their advantages and limitations.

Synthetic nuclease system	Key features	Advantages	Limitations
Zinc finger nucleases (ZFNs)	• ZFNs are fusion proteins comprising an array of site-specific DNA-binding zinc finger domains that are attached to the endonuclease domain of the bacterial FokI restriction enzyme • Each zinc finger domain binds to a 3- to 4-bp DNA sequence, and tandem domains can potentially bind to an extended nucleotide sequence (typically with a length that is a multiple of 3, usually 9 bp–18 bp • Target site identification is by protein–DNA interaction • Each zinc finger nuclease introduces a single-strand break	• High specificity to the desired genomic DNA sequences	• Laborious and time-consuming engineering • Off-target binding • Some nucleotide triplets lack Zing Finger binding domains • Pairing of ZFNs required to introduce double strand breaks (DSBs) in target sequence
Transcription activator-like effector nucleases (TALENs)	• TALE repeats comprise tandem arrays with 10–30 repeats that bind and recognize extended DNA sequences • Fusion of a domain of TALE repeats to the FokI endonuclease domain, forms TAL effector nucleases (TALENs), Each TALEN domain recognizes a DNA base pair • Target site identification is by protein–DNA interaction • Each TALEN introduces a single-strand break	• Higher specificity than ZFNs to the desired genomic DNA sequences	• Laborious and time-consuming engineering • Off-target binding • 5′ base of TALEN target sequence must be a thymine • TALEN binding is negatively impacted by DNA methylation • Pairing of TALENs required to introduce double strand breaks (DSBs) in target sequence
Clustered regularly interspaced short palindromic repeats (CRISPR)/Cas9	• Cas9 endonuclease derived from *Streptococcus pyogenes* aligns with a single chimeric guide RNA • a 20-bp sequence of the target DNA (protospacer) with an adjacent downstream NGG nucleotide sequence termed (PAM) site binds to the complementary sequence of the sgRNA • Target site identification is by RNA–DNA interaction • Results in DSBs at target sites	• CRISPR-Cas9 system can be devised to manipulate multiple genomic loci simultaneously, by co-delivering a combination of sgRNAs • Easy to design, that is it eliminates the need for tedious protein engineering of DNA-recognition domains for each DNA sequence to be targeted • Higher target specificity based on sgRNA and Cas9 binding to protospacer and PAM site, respectively	• Higher chances of off-target mutagenesis • Target range limited by PAM sites • CRISPR/Cas9 binding efficiency is impacted by chromatin accessibility

8.4.3 miRNAs

MicroRNAs (miRNAs) are known to be as small (21–24 nucleotides), noncoding RNAs that negatively regulate their target mRNA by repressing translation of or by degradation of the target mRNAs. These miRNAs regulate various biological processes critical for plant growth and development and also the responses to different biotic and abiotic stresses in plants (Budak et al. 2015; Shriram et al. 2016; Budak and Akpinar 2015). Small RNA is divided into four categories like miRNA, small interfering RNA (siRNA), transfer RNAs (tRNAs), and PIWI-interacting RNA (piRNA) (Zhu et al. 2018; Treiber et al. 2019). These small RNAs can induce silencing of genes either through silencing at transcriptional (TGS) level or at the post-transcriptional level (PTGS). However, miRNA regulates gene expression mostly posttranscription (Borges and Martienssen 2015).

8.4.4 Biogenesis of miRNA

On the basis of the genome location, the classification of miRNAs is done either into intronic or intergenic types. The miRNAs transcribed from the two genes are called intergenic miRNAs. In contrast, transcription of miRNA was performed from the noncoding region of the genome by the RNA Polymerase II (Pol II), called intergenic miRNAs (Millar and Waterhouse 2005). After transcription from the MIR gene, miRNA undergoes a series of posttranscriptional modifications, such as the removal of the 5′ cap as well as 3′ tail to form primary RNA (pri-miRNAs) (Kurihara and Watanabe 2004; Kim et al. 2011). The primary RNA has a self-complementary hairpin structure. The downstream processing of pri-miRNAs is performed by multiple protein complexes like dicer-like1(DCL1), HYPONASTIC LEAVES (HYL1), and SERRATE (SE) (Dong et al. 2008; Ren et al. 2012). The nascent duplex miRNA/miRNA* generated through the DCL-mediated processing of pri-miRNAs are further processed by HUA Enhancer 1 (HEN1), which methylates duplexes at the 3′ end (Yu et al. 2005). HASTY (plant homolog of EXPO5 protein) enables these duplexes with methylation move from the nucleus to the cytoplasm (Park et al. 2005). The miRNA/miRNA* duplex is degraded and forms a single-stranded miRNA known as guide RNA, which selectively gets incorporated with the AGO (Argonaute) protein. Finally, the AGO protein and a group of other proteins with assembled guide RNA for mRNA-induced silencing complex (RISC) (Zhang et al. 2015). This RISC binds with the target genes with the help of miRNA and silences the genes through translational inhibition and target cleavage (Figure 8.4).

There are 10 AGO proteins found in the Arabidopsis; however, AGO1 plays an important role in modulating silencing activity (Mi et al. 2008; Eamens et al. 2009). Four major domains are present in the AGO protein: the N-terminal domain, the middle domain, the PIWI domain, and the PAZ domain. The PAZ domain binds directly with the 3′ end of miRNAs whereas, at the 5′ phosphate end of miRNAs, the mid domain interacts, and the PIWI domain show endoribonuclease activity (Hutvagner and Simard 2008). The plant miRNAs undergo modifications at the 3′ terminal ribose with 2′-O-methylation by methyltransferase, Hua Enhancer1 (HEN1), an Mg^{2+} dependent methyltransferase that identifies RNAs (22 nt) with the two nucleotide overhangs at each end (Yang et al. 2006; Huang et al. 2009). The methyl group, attached to each strand of miRNAs, protects it from exonuclease activity (Ramachandran and Chen 2008).

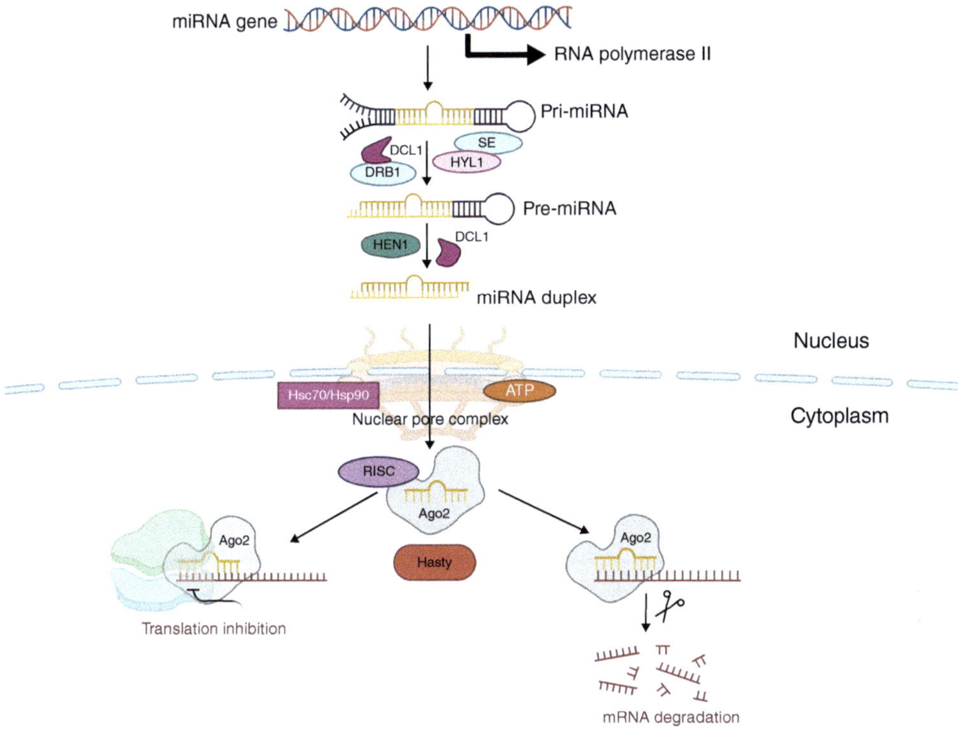

Figure 8.4 miRNA biogenesis and regulation in plants. miR gene transcribed to form pri-miRNA by the RNA polymerase II (Pol II). With the help of DCL-1 and other proteins like SEs, RBP, and HYL1 pri miRNAs cleaved into the pre miRNAs. Pre-miRNAs are processed further to form duplex miRNA of 21–24 nucleotides by the DCL1 and HEN1. miRNAs duplex is methylated by HEN1 and exported to the cytoplasm by plant HASTY (exportin 5 ortholog). The guide strand is incorporated into the AGO protein to form a RISC. Plant miRNA act by two methods: one is translational inhibition and the other is miRNA-guided target cleavage. Figure created using BioRender (https://biorender.com/).

There are 10 AGO proteins found in Arabidopsis; however, in plants, AGO1 plays an important role in modulating silencing activity (Mi et al. 2008; Eamens et al. 2009). There are four major domains in the AGO protein, the N terminal domain, the middle domain, the PIWI domain, and the PAZ domain. The PAZ domain binds directly with the 3′ end of miRNAs, whereas, at the 5′ phosphate end of miRNAs, the mid domain interacts, and the PIWI domain show endoribonuclease activity (Hutvagner and Simard 2008). The plant miRNAs undergo modifications at the 3′ terminal ribose with 2′-O-methylation by methyltransferase, Hua Enhancer1 (HEN1), an Mg^{2+} dependent methyltransferase which identifies RNAs (22 nt) with the two nucleotide overhangs at each end (Yang et al. 2006; Huang et al. 2009). The methyl group attached to each strand of miRNAs protects it from exonuclease activity (Ramachandran and Chen 2008).

8.4.5 miRNA and Gene Regulations

Many of the miRNAs play a fundamental role in plant life; they control growth and development of plants and their responses to different biotic and biotic stresses, which also includes how miRNAs adapt to nutrient availability. Some of the key miRNAs that play a central role in the development of Arabidopsis, such as miR172, translationally repress *APETALA2 2 (AP2)* and regulate floral meristem identity and flowering time (Aukerman and Sakai 2003; Zhao et al. 2007). However, miR156 targets SPL *(SQUAMOSA PROMOTER BINDING PROTEIN-LIKE)* genes like SPL13 and SPL9, which regulates vegetative growth development (He et al. 2018).

In various cases, the miRNAs core group plays developmental roles across many tissues, regions, and organs (Hashimoto et al. 2018). For example, miR165/miR166 in *Arabidopsis* regulates several developmental pathways like the development of leaf, embryo SAM (shoot apical meristem), and root (Reinhart et al. 2002; Du and Wang 2015). miR396 targets different GRF (growth-regulating factor), which regulates leaf primordia and cell proliferation (Kim et al. 2003).

Recently, various miRNAs have been identified under different environmental stresses and drawing the attention of researchers working in this field. Deep sequencing technologies made small RNA sequencing very easy, effectively using the profiling of miRNA expression to recognize several novel and stress-responsive miRNAs. Different analytical approaches showed the differential expression of several miRNAs under various abiotic stresses. The miR395 targets APS, which significantly modulates the sulfur assimilation pathway and maintains the sulfur pool in the plant (Liang et al. 2010; Jones-Rhoades and Bartel 2004). In Arabidopsis and Medicago, a higher accumulation of miR408 and miR156 was observed under conditions of drought stress, but in rice, it is downregulated (Liu et al. 2008; Shuai et al. 2013). Under salinity stress, miR169 is induced that targets NFYA5, whereas the miR398 expression is decreased that inducing the expression of the target gene encoding copper/superoxide dismutases (Cu/SODs) in *Arabidopsis* (Jia et al. 2009; Gao et al. 2016). These studies show that miRNAs are responsive to abiotic stresses and regulate tolerance of plants against different abiotic stresses.

8.5 Role of Genome-Editing in miRNA Expression

However, constitutive overexpression of miRNA changes the expression of their targets thay may be involved in plant growth and development and different stresses, which may show undesirable phenotypes. It may also show the cause pleiotropic effect undesirably and also regulate a number of miRNAs (Trumbo et al. 2015; Ferdous et al. 2017).

Functional characterization the miRNAs is done through overexpression or gain of function or loss of function through knockdown/knockout. Regulation of miRNAs could be possible by applying technique such as RNAi, amiRNA, TALEN, ZFN, and most recently added CRISPR/Cas9 (Schwab et al. 2006; Shan et al. 2013). Several gene-editing tools, like TALENs, zinc-finger nucleases and CRISPR/Cas systems, has been developed in recent years based on sequence-specific engineered endonucleases (SSNs). Among these gene editing techniques, some have been adopted in the functional characterization of miRNAs in different plants.

TALENs and ZFNs use protein motifs for target recognition and need to assemble specific proteins for each target, which hinders with widespread uses in contrast to the CRISPR/Cas9 technique, which recognizes and cleaves target site in sequence-specific manner using the single guide RNA (sgRNA), which greatly enhances its application (Gaj et al. 2013; Xie and Yang 2013). In animals, many miRNAs have been well characterized using TALENs, but only few studies are available, which show TALEN-mediated miRNA gene editing (Bi et al. 2020). However, the CRISPR/Cas9 system is successfully used in Arabidopsis, tobacco and rice for gene editing in the initial stages of its establishment (Li et al. 2013; Nekrasov et al. 2013; Shan et al. 2013). Recently, CRISPR/Cas9 nonhomologous end joining (NHEJ) system has been utilized in successful editing of many miRNAs by introducing indels at MIR gene in different plant species (Damodharan et al. 2018; Miao et al. 2019). Although using CRISPR/Cas9 system knockout and full-length deletion of MIR gene can be done efficiently by generating the null mutant plants, the one limitation with miRNA editing is the shorter sequence to design gRNA that targets MIR gene. Multiple miRNAs in one family in the genome may increase its limitation in knocking out the complete MIR gene family.

8.6 Applications of the CRISPR/Cas9 System in miRNA Editing

CRISPR/Cas system can be used to target miRNA genes, thus it can target pri-/pre-miRNA and mature miRNA and show miRNA silencing in plants (Aman et al. 2018; Sharma et al. 2020; Kumar et al. 2023). Genetically modified organisms (GMOs) are known to be a threat to the environment and human health. Therefore, it is included within the stringent regulations of the government (Prado et al. 2014). However, CRISPR/Cas9 edited crops receive regulatory waivers in many countries across the globe (Zhu et al. 2020). The progress regarding the transgene-free edited system of protein-coding genes is immensely increased. Recently, various studies showed that the editing of miRNA through CRISPR/Cas9 would accumulate the knowledge of transgene-free edited system in plants (Zhu et al. 2020).

Os-miR535 was edited using CRISPR/Cas9 genome editing technique. Editing of Os-miR535 showed significant drought tolerance in rice (Yue et al. 2020). At the same time, in *Glycine max*, gma-GmCCS and GmCSDs are targeted by gma-miR398c edited plants, which is negatively regulated in drought tolerance (Zhou et al. 2020). In soybean and Arabidopsis, gma-miR398c overexpressing plants exhibited decreased drought tolerance, while the gma-miR398c edited plants showed increased drought sensitivity. In tomato, *Phytophthora infestans* responsible for the late blight badly affects the production of the crop (Fry et al. 2015). miR482 targets NBS–LRR disease-resistance genes and provide sensitivity toward to *P. infestans* (Jiang et al. 2018; Canto-Pastor et al. 2019).

The edited lines of miR482b and miR482b/c showed tolerance against *P. infestans,* which clearly indicated that miR482 acts as a key factor in developing pathogen-resistant tomatoes using gene-editing tools, such as CRISPR/Cas9 (Hong et al. 2021). Recent studies showed successful editing of miR858 and coding region of miPEP858a via CRISPR/Cas9 system exhibited specific phenotypes, such as retarded plant growth and delayed flowering that also affects the anthocyanin, flavonoids, and lignin accumulation (Sharma et al. 2020; Badola et al. 2022). Among the well-studied families of miRNA, miR408 has been shown to be conserved and has a crucial role with its diverse responses to, cold, drought, osmotic stresses, deficiency of nutrients as well as oxidative stress (Ma et al. 2015; Song et al. 2018). Recently, it has been demonstrated that miR408-edited plants through CRISPR/Cas9 showed a higher degree of survival than miR408 overexpressing plants in nutrient deficiency and Arsenic (As) stress (Kumar et al. 2023).

8.6.1 microRNA-Encoded Peptide

MicroRNAs (miRNAs) are known to negatively affect the expression of target genes in plants and animals. Recently, it has been studied that the miRNA's primary transcript encodes small peptides known as microRNA-encoded peptides (miPEPs) that positively regulates the transcription of the associated miRNA (Lauressergues et al. 2015). Possibly the role of miPEPs is that it acts as a trans-acting factor or transcription factor that positively modulates the MIR gene transcription (Figure 8.5).

8.6.2 Biogenesis of miPEPs

The transcription of miRNA is occurred by the RNA polymerase to form pri-miRNA (Chapman and Carrington 2007). Further, pri-miRNA goes under enzymatic activity, such as RNase III enzyme DCL1 and HYL1, to form smaller miRNA segments forming a

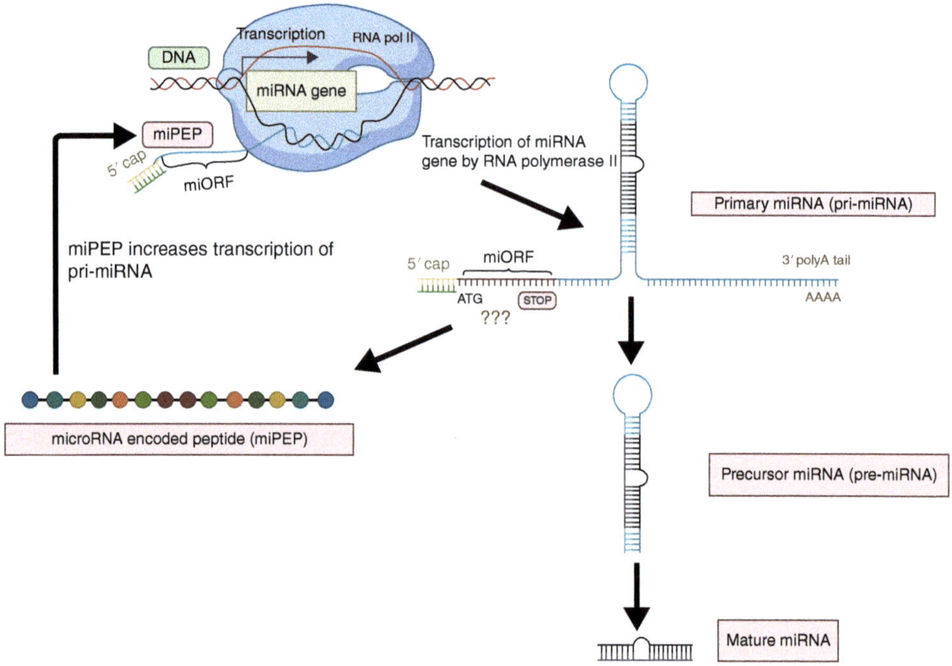

Figure 8.5 miPEP biogenesis and their transcriptional regulation. Schematic representation of miPEP biogenesis, RNA-pol-II transcribe miRNA into pri-miRNA. This pri-miRNA undergoes a series of cleavage and modification to form mature miRNA. A part of pri-miRNA contains small functional ORFs, which encode small peptides known as micro-encoded peptides (miPEPs), which regulate the transcription of the miRNA gene. Figure created using BioRender (https://biorender.com/).

hairpin loop structure called pre-miRNA. The pre-miRNA, by the downstream processing, is converted into mature miRNA, incorporated into the AGO protein to produce the RISC complex and guide mRNA cleavage or translational inhibition (Rogers and Chen 2013; O'Brien et al. 2018). Previously, it was thought that after biosynthesis of mature miRNA from the pri-miRNA, the sequence present in the upstream and downstream has no functional role. However, recent studies show that small ORFs (open reading frames) are present in the pri-miRNA, which encode small peptides known as microRNA-encoded peptides (miPEPs).

8.6.3 Role of miPEP

The processing of mature miRNA is mainly done in the nucleus and it gets exported to the cytoplasm (Wang et al. 2019). In the starting phases of miRNA biogenesis, pri-miRNA is processed by the DCL complexes into three parts, first being the upstream region of the pre-miRNA, second the pre-miRNA, and the third and last is the 3′ end having the poly-A tail region. The pre-miRNA region gets further converted into the mature miRNA and 3′ is possibly get degraded (Rogers and Chen 2013). However, the upstream region of pre-RNA contains small ORFs, which encode small peptides, for example, in grapes, the pri-miR171d gets accumulated in the nucleus, and in the cytoplasm, it is also detected in a small amount (Lauressergues et al. 2015; Chen et al. 2020; Sharma et al. 2020; Lauressergues et al. 2022). Recent studies suggested that miPEPs act as endogenous peptides that positively regulates their corresponding miRNA generation in the autoregulatory feedback mechanism (Ormancey et al. 2020). the miPEPs enhance the transcription of their corresponding pri-miRNAs and regulate the level of their accumulation. For functional characterization of miPEP858a and miPEP408, the CRISPR-Cas9 genome editing technique was used (Sharma et al. 2020; Kumar et al. 2023).

8.7 miPEPs Act as the Master Regulator in Plant Growth and Development

Recent studies suggest that miPEPs regulate miRNA biogenesis positively; likewise, in the Arabidopsis plant, miPEP165a positively regulates the corresponding pri-miR165a accumulation resulting in the downregulations of its target genes involved in root developments (Lauressergues et al. 2015). The promoter of miR858a enhances the expression of GUS gene in the different reporter lines (PromiR858a: ORF1:GUS and PromiR858a: ATG1:GUS), which are fused with an entire ORF region that encodes miPEP858a or only with the start codon (ATG). Furthermore, when the GUS assay was performed in the media supplemented with synthetic exogenous peptide, enhanced GUS activity was observed, indicating that miPEP858a positively regulates the transcription level of the gene (Sharma et al. 2020). Recently, it was suggested that a network of small molecules comprising of miPEP858a/miR858a and phytosulfokine (PSK4) controls the plant growth and development in Arabidopsis, showing the big impact of small peptides and miRNAs on the growth of plants (Badola et al. 2022). The study used the CRISPR/Cas9 approach to establish the role of this network in plant growth and development.

The application of synthetic peptides (miPEPs) exogenously can show the same autoregulatory effect without additional processing and modification as the pre-miRNAs-derived peptide. The exogenously supplemented miPEPs are internalized as per studies through endocytosis and passive diffusion, unlike the precursor-derived peptide (Okuda et al. 2009; Oh et al. 2018; Ormancey et al. 2020). Ormancey and her team showed that the internalization of miPEP165a within the differentiation zone and a mature zone involves endocytosis-associated internalization and passive diffusion through the root apex. Such results were also observed in miPEP858a regarding the internalization and transport of miPEPs (Sharma et al. 2020; Badola et al. 2022). Likewise, miPEP171 regulates the development and nodule formation in soybean is the key research for crop plants (Couzigou and Combier 2016).

The miPEP positively regulates its associated miRNA expression that may regulate the gene that governs the physiological, developmental, and agronomically important traits in crop plants. Though the application of miPEPs in the field is an expensive process, it can overcome complex procedures like cloning and genetic transformation. All these studies suggest that miPEPs are small regulatory molecules, which might contribute to the regulation of miRNAs and their associated traits.

8.8 Conclusions and Future Prospect

With the advancement in technologies for sequencing, functional genomics has undergone a massive revolution with new editing tools to develop novel allelic of mutants for the improvement of desired traits in plants. Out of several genome editing tools, CRISPR/Cas technology has acquired most of the attention as compared to the other existing techniques for genome editing because of its specific nuclease domains and precise editing. In recent years, different Cas proteins (endonucleases) were identified and functionally characterized, which makes CRISPR editing technique easier and more precise. CRISPR/Cas9-mediated genome editing has been applied to enhance the yield and nutritional quality of crops as well as provide tolerance toward different abiotic and biotic stresses.

However, various miRNAs have been validated for many agricultural traits, but only a few pieces of information have been identified for some of the important troubles. These include miRNAs associated with false smut and rice sheath blight disease, heat stress-related miRNAs, miRNAs associated in providing resistance to viruses and insects, miRNAs linked to biofortification, nutrients and water-use efficiency, etc. In addition, much of the studies have focused on a variety of model plants like tomato, rice, and Arabidopsis, but the importance of miRNAs in other agricultural crops (such as oilseed crops and legumes) needed to be done. To find natural allelic variations that provide trait multiplicity in germplasm, allele mining of miRNAs provides an indication for crop improvement by utilizing genome editing of miRNAs. Including this, to understand host–pathogen interaction mediated by miRNAs, regulation of plant miRNAs via pathogen genes and vice versa needed to be explored. In particular, previous and current findings suggest that miRNAs are powerful genetic elements that can be employed to improve the traits of crop plants. The challenges faced in using miRNAs in genome editing can be addressed by better characterization of the functions of miRNAs and their regulation in various crops. The function of miRNAs that are associated with crops of economic importance can be validated through genome editing.

Acknowledgments

This work was supported by the Council of Scientific and Industrial Research, New Delhi (Grant number: MLP07). PKT acknowledges Department of Science and Technology for JC Bose National Fellowship.

References

Alok, A., Chauhan, H., Upadhyay, S.K. et al. (2021). Compendium of plant-specific CRISPR vectors and their technical advantages. *Life* 11: 1021. https://doi.org/10.3390/life11101021.

Aman, R., Ali, Z., Butt, H. et al. (2018). RNA virus interference via CRISPR/Cas13a system in plants. *Genome Biology* 19 (1): 1–9.

Amitai, G. and Sorek, R. (2016). CRISPR-Cas adaptation: insights into the mechanism of action. *Nature Reviews. Microbiology* 14 (2): 67–76.

Aukerman, M.J. and Sakai, H. (2003). Regulation of flowering time and floral organ identity by a microRNA and its APETALA2-like target genes. *The Plant Cell* 15 (11): 2730–2741.

Badola, P.K., Sharma, A., Gautam, H., and Trivedi, P.K. (2022). MicroRNA858a, its encoded peptide, and phytosulfokine regulate *Arabidopsis* growth and development. *Plant Physiology* 138: 1397–1415.

Bi, H., Fei, Q., Li, R. et al. (2020). Disruption of miRNA sequences by TALENs and CRISPR/Cas9 induces varied lengths of miRNA production. *Plant Biotechnology Journal* 18 (7): 1526–1536.

Beerli, R.R., Segal, D.J., Dreier, B., and Barbas, C.F. (1998). Toward controlling gene expression at will: specific regulation of the erbB-2/HER-2 promoter by using polydactyl zinc finger proteins constructed from modular building blocks. *Proceedings of the National Academy of Sciences, USA* 95 (25): 14628–14633.

Bhalothia, P., Yajnik, K., Alok, A., and Upadhyay, S.K. (2020). The current progress of CRISPR/Cas9 development in plants. In: *Genome Engineering via CRISPR-Cas9 System*, 123–129. *Elsevier*. https://doi.org/10.1016/B978-0-12-818140-9.00011-8.

Boch, J., Scholze, H., Schornack, S. et al. (2009). Breaking the code of DNA binding specificity of TAL-type III effectors. *Science* 326 (5959): 1509–1512.

Borges, F. and Martienssen, R.A. (2015). The expanding world of small RNAs in plants. *Nature Reviews Molecular Cell Biology* 16 (12): 727–741.

Briggs, A.W., Rios, X., Chari, R. et al. (2012). Iterative capped assembly: rapid and scalable synthesis of repeat-module DNA such as TAL effectors from individual monomers. *Nucleic Acids Research* 40 (15): e117–e117.

Budak, H. and Akpinar, B.A. (2015). Plant miRNAs: biogenesis, organization and origins. *Functional & Integrative Genomics* 15 (5): 523–531.

Budak, H., Kantar, M., Bulut, R., and Akpinar, B.A. (2015). Stress responsive miRNAs and isomiRs in cereals. *Plant Science* 235: 1–13.

Canto-Pastor, A., Santos, B.A., Valli, A.A. et al. (2019). Enhanced resistance to bacterial and oomycete pathogens by short tandem target mimic RNAs in tomato. *Proceedings of the National Academy of Sciences* 116 (7): 2755–2760.

Cermak, T., Doyle, E.L., Christian, M. et al. (2011). Efficient design and assembly of custom TALEN and other TAL effector-based constructs for DNA targeting. *Nucleic Acids Research* 39 (12): e82–e82.

Chapman, E.J. and Carrington, J.C. (2007). Specialization and evolution of endogenous small RNA pathways. *Nature Reviews Genetics* 8 (11): 884–896.

Chen, H., Choi, J., and Bailey, S. (2014). Cut site selection by the two nuclease domains of the Cas9 RNA-guided endonuclease. *Journal of Biological Chemistry* 289 (19): 13284–13294.

Chen, Q.J., Deng, B.H., Gao, J. et al. (2020). A miRNA-encoded small peptide, vvi-miPEP171d1, regulates adventitious root formation. *Plant Physiology* 183 (2): 656–670.

Cho, S.W., Kim, S., Kim, Y. et al. (2014). Analysis of off-target effects of CRISPR/Cas-derived RNA-guided endonucleases and nickases. *Genome Research* 24 (1): 132–141.

Christian, M., Cermak, T., Doyle, E.L. et al. (2010). Targeting DNA double-strand breaks with TAL effector nucleases. *Genetics* 186 (2): 757–761.

Cong, L., Ran, F.A., Cox, D. et al. (2013). Multiplex genome engineering using CRISPR/Cas systems. *Science* 339 (6121): 819–823.

Couzigou, J.M. and Combier, J.P. (2016). Plant micro RNAs: key regulators of root architecture and biotic interactions. *New Phytologist* 212 (1): 22–35.

Cox, D.B.T., Platt, R.J., and Zhang, F. (2015). Therapeutic genome editing: prospects and challenges. *Nature Medicine* 21 (2): 121–131.

Damodharan, S., Corem, S., Gupta, S.K., and Arazi, T. (2018). Tuning of Sl ARF 10A dosage by sly-miR160a is critical for auxin-mediated compound leaf and flower development. *The Plant Journal* 96 (4): 855–868.

Deltcheva, E., Chylinski, K., Sharma, C.M. et al. (2011). CRISPR RNA maturation by trans-encoded small RNA and host factor RNase III. *Nature* 471 (7340): 602–607.

Dong, Z., Han, M.H., and Fedoroff, N. (2008). The RNA-binding proteins HYL1 and SE promote accurate in vitro processing of pri-miRNA by DCL1. *PNAS* 105 (29): 9970–9975.

Doudna, J.A. and Charpentier, E. (2014). The new frontier of genome engineering with CRISPR-Cas9. *Science* 346 (6213): 1258096.

Doyon, Y., Vo, T.D., Mendel, M.C. et al. (2011). Enhancing zinc-finger-nuclease activity with improved obligate heterodimeric architectures. *Nature Methods* 8 (1): 74–79.

Du, Q. and Wang, H. (2015). The role of HD-ZIP III transcription factors and miR165/166 in vascular development and secondary cell wall formation. *Plant Signaling & Behavior* 10 (10): e1078955.

Eamens, A.L., Smith, N.A., Curtin, S.J. et al. (2009). The *Arabidopsis thaliana* double-stranded RNA binding protein DRB1 directs guide strand selection from microRNA duplexes. *RNA* 15 (12): 2219–2235.

Ferdous, J., Sanchez-Ferrero, J.C., Langridge, P. et al. (2017). Differential expression of microRNAs and potential targets under drought stress in barley. *Plant, Cell & Environment* 40 (1): 11–24.

Fry, W.E., Birch, P.R.J., Judelson, H.S. et al. (2015). Five reasons to consider *Phytophthora infestans* a reemerging pathogen. *Phytopathology* 105 (7): 966–981.

Fu, Y., Foden, J.A., Khayter, C. et al. (2013). High-frequency off-target mutagenesis induced by CRISPR-Cas nucleases in human cells. *Nature Biotechnology* 31 (9): 822–826.

Gaj, T., Gersbach, C.A., and Barbas, C.F. (2013). ZFN, TALEN, and CRISPR/Cas-based methods for genome engineering. *Trends in Biotechnology* 31 (7): 397–405.

Gaj, T., Guo, J., Kato, Y. et al. (2012). Targeted gene knockout by direct delivery of zinc-finger nuclease proteins. *Nature Methods* 9 (8): 805–807.

Gaj, T., Sirk, S.J., Shui, S.L., and Liu, J. (2016). Genome-editing technologies: principles and applications. *Cold Spring Harbor Perspectives in Biology* 8 (12): a023754.

Gao, F., Wang, N., Li, H. et al. (2016). Identification of drought-responsive microRNAs and their targets in *Ammopiptanthus mongolicus* by using high-throughput sequencing. *Scientific Reports* 6 (1): 1–16.

Garneau, J.E., Dupuis, M.È., Villion, M. et al. (2010). The CRISPR/Cas bacterial immune system cleaves bacteriophage and plasmid DNA. *Nature* 468: 67–71.

Gasiunas, G., Barrangou, R., Horvath, P., and Siksnys, V. (2012). Cas9–crRNA ribonucleoprotein complex mediates specific DNA cleavage for adaptive immunity in bacteria. *Proceedings of the National Academy of Sciences* 109 (39): E2579–E2586.

Guilinger, J.P., Pattanayak, V., Reyon, D. et al. (2014). Broad specificity profiling of TALENs results in engineered nucleases with improved DNA-cleavage specificity. *Nature Methods* 11 (4): 429–435.

Guo, J., Gaj, T., and Barbas, C.F. III (2010). Directed evolution of an enhanced and highly efficient FokI cleavage domain for zinc finger nucleases. *Journal of Molecular Biology* 400 (1): 96–107.

Gupta, R.M. and Musunuru, K. (2014). Expanding the genetic editing tool kit: ZFNs, TALENs, and CRISPR-Cas9. *The Journal of Clinical Investigation* 124 (10): 4154–4161.

Hashimoto, K., Miyashima, S., Sato-Nara, K. et al. (2018). Functionally diversified members of the MIR165/6 gene family regulate ovule morphogenesis in *Arabidopsis thaliana*. *Plant and Cell Physiology* 59 (5): 1017–1026.

He, J., Xu, M., Willmann, M.R. et al. (2018). Threshold-dependent repression of SPL gene expression by miR156/miR157 controls vegetative phase change in *Arabidopsis thaliana*. *PLoS Genetics* 14 (4): e1007337.

Heler, R., Marraffini, L.A., and Bikard, D. (2014). Adapting to new threats: the generation of memory by CRISPR-Cas immune systems. *Molecular Microbiology* 93 (1): 1–9.

Holkers, M., Maggio, I., Liu, J. et al. (2013). Differential integrity of TALE nuclease genes following adenoviral and lentiviral vector gene transfer into human cells. *Nucleic Acids Research* 41 (5): e63–e63.

Hong, Y., Meng, J., He, X. et al. (2021). Editing miR482b and miR482c simultaneously by CRISPR/Cas9 enhanced tomato resistance to *Phytophthora infestans*. *Phytopathology* 111 (6): 1008–1016.

Hsu, P.D., Lander, E.S., and Zhang, F. (2014). Development and applications of CRISPR-Cas9 for genome engineering. *Cell* 157 (6): 1262–1278.

Huang, S.Q., Peng, J., Qiu, C.X., and Yang, Z.M. (2009). Heavy metal-regulated new microRNAs from rice. *Journal of Inorganic Biochemistry* 103 (2): 282–287.

Hubbard, B.P., Badran, A.H., Zuris, J.A. et al. (2015). Continuous directed evolution of DNA-binding proteins to improve TALEN specificity. *Nature Methods* 12 (10): 939–942.

Hutvagner, G. and Simard, M.J. (2008). Argonaute proteins: key players in RNA silencing. *Nature Reviews Molecular Cell Biology* 9 (1): 22–32.

Jaganathan, D., Ramasamy, K., Sellamuthu, G. et al. (2018). CRISPR for crop improvement: an update review. *Frontiers in Plant Science* 9: 985.

Jones-Rhoades, M.W. and Bartel, D.P. (2004). Computational identification of plant microRNAs and their targets, including a stress-induced miRNA. *Molecular Cell* 14 (6): 787–799.

Jia, X., Wang, W.X., Ren, L. et al. (2009). Differential and dynamic regulation of miR398 in response to ABA and salt stress in *Populus tremula* and *Arabidopsis thaliana*. *Plant Molecular Biology* 71 (1): 51–59.

Jiang, F. and Doudna, J.A. (2015). The structural biology of CRISPR-Cas systems. *Current Opinion in Structural Biology* 30: 100–111.

Jiang, F. and Doudna, J.A. (2017). CRISPR–Cas9 structures and mechanisms. *Annual Review of Biophysics* 46: 505–529.

Jiang, N., Meng, J., Cui, J. et al. (2018). Function identification of miR482b, a negative regulator during tomato resistance to *Phytophthora infestans*. *Horticulture Research* 5: 9.

Jinek, M., Chylinski, K., Fonfara, I. et al. (2012). A programmable dual-RNA–guided DNA endonuclease in adaptive bacterial immunity. *Science* 337 (6096): 816–821.

Jinek, M., Jiang, F., Taylor, D.W. et al. (2014). Structures of Cas9 endonucleases reveal RNA-mediated conformational activation. *Science* 343 (6176): 1247997.

Juillerat, A., Pessereau, C., Dubois, G. et al. (2015). Optimized tuning of TALEN specificity using non-conventional RVDs. *Scientific Reports* 5 (1): 1–7.

Kim, E., Kim, S., Kim, D.H. et al. (2012). Precision genome engineering with programmable DNA-nicking enzymes. *Genome Research* 22 (7): 1327–1333.

Kim, H.J., Lee, H.J., Kim, H. et al. (2009). Targeted genome editing in human cells with zinc finger nucleases constructed via modular assembly. *Genome Research* 19 (7): 1279–1288.

Kim, J.H., Choi, D., and Kende, H. (2003). The AtGRF family of putative transcription factors is involved in leaf and cotyledon growth in *Arabidopsis*. *The Plant Journal* 36 (1): 94–104.

Kim, Y.G., Cha, J., and Chandrasegaran, S. (1996). Hybrid restriction enzymes: zinc finger fusions to Fok I cleavage domain. *Proceedings of the National Academy of Sciences, USA* 93: 1156–1160.

Kim, Y.J., Zheng, B., Yu, Y. et al. (2011). The role of mediator in small and long noncoding RNA production in *Arabidopsis thaliana*. *The EMBO Journal* 30: 814–822.

Kumar, R.S., Sinha, H., Datta, T. et al. (2023). microRNA408 and its encoded peptide regulate sulfur assimilation and arsenic stress response in *Arabidopsis*. *Plant Physiology* 192: 837–856.

Kurihara, Y. and Watanabe, Y. (2004). *Arabidopsis* micro-RNA biogenesis through Dicer-like 1 protein functions. *Proceedings of the National Academy of Sciences, USA* 101 (34): 12753–12758.

Lauressergues, D., Couzigou, J.M., Clemente, H.S. et al. (2015). Primary transcripts of microRNAs encode regulatory peptides. *Nature* 520 (7545): 90–93.

Lauressergues, D., Ormancey, M., Guillotin, B. et al. (2022). Characterization of plant microRNA-encoded peptides (miPEPs) reveals molecular mechanisms from the translation to activity and specificity. *Cell Reports* 38 (6): 110339.

Li, J.F., Norville, J.E., Aach, J. et al. (2013). Multiplex and homologous recombination–mediated genome editing in *Arabidopsis* and *Nicotiana benthamiana* using guide RNA and Cas9. *Nature Biotechnology* 31 (8): 688–691.

Liang, G., Yang, F., and Yu, D. (2010). MicroRNA395 mediates regulation of sulfate accumulation and allocation in *Arabidopsis thaliana*. *The Plant Journal* 62 (6): 1046–1057.

Liu, H.H., Tian, X., Li, Y.J. et al. (2008). Microarray-based analysis of stress-regulated microRNAs in *Arabidopsis thaliana*. *RNA* 14 (5): 836–843.

Liu, Q., Segal, D.J., Ghiara, J.B., and Barbas, C.F. (1997). Design of polydactyl zinc-finger proteins for unique addressing within complex genomes. *Proceedings of the National Academy of Sciences, USA* 94 (11): 5525–5530.

Ma, C., Burd, S., and Lers, A. (2015). miR408 is involved in abiotic stress responses in *Arabidopsis*. *The Plant Journal* 84 (1): 169–187.

Mak, A.N.S., Bradley, P., Cernadas, R.A. et al. (2012). The crystal structure of TAL effector PthXo1 bound to its DNA target. *Science* 335 (6069): 716–719.

Makarova, K.S., Wolf, Y.I., Iranzo, J. et al. (2019). Evolutionary classification of CRISPR-Cas systems: a burst of class 2 and derived variants. *Nature Reviews. Microbiology* 18: 67–83.

Mali, P., Aach, J., Stranges, P.B. et al. (2013). CAS9 transcriptional activators for target specificity screening and paired nickases for cooperative genome engineering. *Nature Biotechnology* 31: 833–838.

Malzahn, A., Lowder, L., and Qi, Y. (2017). Plant genome editing with TALEN and CRISPR. *Cell & Bioscience* 7 (1): 1–18.

Meckler, J.F., Bhakta, M.S., Kim, M.S. et al. (2013). Quantitative analysis of TALE–DNA interactions suggests polarity effects. *Nucleic Acids Research* 41 (7): 4118–4128.

Mi, S., Cai, T., Hu, Y. et al. (2008). Sorting of small RNAs into *Arabidopsis* argonaute complexes is directed by the 5′ terminal nucleotide. *Cell* 133 (1): 116–127.

Miao, L., Yin, R.X., Zhang, Q.H. et al. (2019). A novel circRNA-miRNA-mRNA network identifies circ-YOD1 as a biomarker for coronary artery disease. *Scientific Reports* 9 (1): 18314.

Millar, A.A. and Waterhouse, P.M. (2005). Plant and animal microRNAs: similarities and differences. *Functional & Integrative Genomics* 5 (3): 129–135.

Miller, J.C., Tan, S., Qiao, G. et al. (2011). A TALE nuclease architecture for efficient genome editing. *Nature Biotechnology* 29 (2): 143–148.

Miller, J.C., Holmes, M.C., Wang, J. et al. (2007). An improved zinc-finger nuclease architecture for highly specific genome editing. *Nature Biotechnology* 25 (7): 778–785.

Miller, J.C., Zhang, L., Xia, D.F. et al. (2015). Improved specificity of TALE-based genome editing using an expanded RVD repertoire. *Nature Methods* 12 (5): 465–471.

Moscou, M.J. and Bogdanove, A.J. (2009). A simple cipher governs DNA recognition by TAL effectors. *Science* 326 (5959): 1501–1501.

Musharova, O., Sitnik, V., Vlot, M. et al. (2019). Systematic analysis of Type I-E *Escherichia coli* CRISPR-Cas PAM sequences ability to promote interference and primed adaptation. *Molecular Microbiology* 111: 1558–1570.

Mussolino, C., Alzubi, J., Fine, E.J. et al. (2014). TALENs facilitate targeted genome editing in human cells with high specificity and low cytotoxicity. *Nucleic Acids Research* 42 (10): 6762–6773.

Nekrasov, V., Staskawicz, B., Weigel, D. et al. (2013). Targeted mutagenesis in the model plant *Nicotiana benthamiana* using Cas9 RNA-guided endonuclease. *Nature Biotechnology* 31 (8): 691–693.

Nussenzweig, P.M. and Marraffini, L.A. (2020). Molecular mechanisms of CRISPR-Cas immunity in bacteria. *Annual Review of Genetics* 54: 93–120.

O'Brien, J., Hayder, H., Zayed, Y., and Peng, C. (2018). Overview of microRNA biogenesis, mechanisms of actions, and circulation. *Frontiers in Endocrinology* 9: 402.

Oh, E., Seo, P.J., and Kim, J. (2018). Signaling peptides and receptors coordinating plant root development. *Trends in Plant Science* 23 (4): 337–351.

Okuda, K., Chateigner-Boutin, A.L., Nakamura, T. et al. (2009). Pentatricopeptide repeat proteins with the DYW motif have distinct molecular functions in RNA editing and RNA cleavage in *Arabidopsis* chloroplasts. *The Plant Cell* 21 (1): 146–156.

Ormancey, M., Le Ru, A., Duboé, C. et al. (2020). Internalization of miPEP165a into *Arabidopsis* roots depends on both passive diffusion and endocytosis-associated processes. *International Journal of Molecular Sciences* 21 (7): 2266.

Osakabe, Y. and Osakabe, K. (2015). Genome editing with engineered nucleases in plants. *Plant and Cell Physiology* 56 (3): 389–400.

Park, M.Y., Wu, G., Gonzalez-Sulser, A. et al. (2005). Nuclear processing and export of microRNAs in *Arabidopsis. Proceedings of the National Academy of Sciences* 102 (10): 3691–3696.

Pattanayak, V., Ramirez, C.L., Joung, J.K., and Liu, D.R. (2011). Revealing off-target cleavage specificities of zinc-finger nucleases by in vitro selection. *Nature Methods* 8 (9): 765–770.

Pavletich, N.P. and Pabo, C.O. (1991). Zinc finger-DNA recognition: crystal structure of a Zif268-DNA complex at 2.1 Å. *Science* 252 (5007): 809–817.

Porteus, M.H. and Carroll, D. (2005). Gene targeting using zinc finger nucleases. *Nature Biotechnology* 23 (8): 967–973.

Prado, J.R., Segers, G., Voelker, T. et al. (2014). Genetically engineered crops: from idea to product. *Annual Review of Plant Biology* 65: 769–790.

Ramachandran, V. and Chen, X. (2008). Degradation of microRNAs by a family of exoribonucleases in *Arabidopsis. Science* 321 (5895): 1490–1492.

Ramirez, C.L., Certo, M.T., Mussolino, C. et al. (2012). Engineered zinc finger nickases induce homology-directed repair with reduced mutagenic effects. *Nucleic Acids Research* 40 (12): 5560–5568.

Reinhart, B.J., Weinstein, E.G., Rhoades, M.W. et al. (2002). MicroRNAs in plants. *Genes & Development* 16 (13): 1616–1626.

Ren, G., Xie, M., Dou, Y. et al. (2012). Regulation of miRNA abundance by RNA bindingprotein TOUGH in *Arabidopsis. PNAS* 109 (31): 12817–12821.

Reyon, D., Tsai, S.Q., Khayter, C. et al. (2012). FLASH assembly of TALENs for high-throughput genome editing. *Nature Biotechnology* 30 (5): 460–465.

Rogers, K. and Chen, X. (2013). Biogenesis, turnover, and mode of action of plant microRNAs. *The Plant Cell* 25 (7): 2383–2399.

Sashital, D.G., Wiedenheft, B., and Doudna, J.A. (2012). Mechanism of foreign DNA selection in a bacterial adaptive immune system. *Molecular Cell* 46: 606–615.

Schmid-Burgk, J.L., Schmidt, T., Kaiser, V. et al. (2013). A ligation-independent cloning technique for high-throughput assembly of transcription activator–like effector genes. *Nature Biotechnology* 31 (1): 76–81.

Schwab, R., Ossowski, S., Riester, M. et al. (2006). Highly specific gene silencing by artificial microRNAs in *Arabidopsis. The Plant Cell* 18 (5): 1121–1133.

Semenova, E., Jore, M.M., Datsenko, K.A. et al. (2011). Interference by clustered regularly interspaced short palindromic repeat (CRISPR) RNA is governed by a seed sequence. *PNAS* 108 (25): 10098–10103.

Shan, Q., Wang, Y., Li, J. et al. (2013). Targeted genome modification of crop plants using a CRISPR-Cas system. *Nature Biotechnology* 31 (8): 686–688.

Sharma, A., Badola, P.K., Bhatia, C. et al. (2020). Primary transcript of miR858 encodes regulatory peptide and controls flavonoid biosynthesis and development in *Arabidopsis. Nature Plants* 6 (10): 1262–1274.

Shriram, V., Kumar, V., Devarumath, R.M. et al. (2016). MicroRNAs as potential targets for abiotic stress tolerance in plants. *Frontiers in Plant Science* 7: 817.

Shuai, P., Liang, D., Zhang, Z. et al. (2013). Identification of drought-responsive and novel *Populus trichocarpa* microRNAs by high-throughput sequencing and their targets using degradome analysis. *BMC Genomics* 14 (1): 1–14.

Singh, S., Rahangdale, S., Pandita, S. et al. (2022). CRISPR/Cas9 for insect pests management: a comprehensive review of advances and applications. *Agriculture* 12: 1896. https://doi.org/10.3390/agriculture12111896.

Smith, J., Bibikova, M., Whitby, F.G. et al. (2000). Requirements for double-strand cleavage by chimeric restriction enzymes with zinc finger DNA-recognition domains. *Nucleic Acids Research* 28: 3361–3369.

Song, Z., Zhang, L., Wang, Y. et al. (2018). Constitutive expression of miR408 improves biomass and seed yield in *Arabidopsis*. *Frontiers in Plant Science* 8: 2114.

Sorek, R., Lawrence, C.M., and Wiedenheft, B. (2013). CRISPR-mediated adaptive immune systems in bacteria and archaea. *Annual Review of Biochemistry* 82: 237–266.

Sternberg, S., Redding, S., Jinek, M. et al. (2014). DNA interrogation by the CRISPR RNA-guided endonuclease Cas9. *Nature* 507: 62–67.

Streubel, J., Blücher, C., Landgraf, A., and Boch, J. (2012). TAL effector RVD specificities and efficiencies. *Nature Biotechnology* 30 (7): 593–595.

Sushmita, K.G., Upadhyay, S.K., and Verma, P.C. (2021). An overview of genome-engineering methods. In: *Genome Engineering for Crop Improvement*, 1–21. Wiley Online Library.

Treiber, T., Treiber, N., and Meister, G. (2019). Regulation of microRNA biogenesis and its crosstalk with other cellular pathways. *Nature Reviews Molecular Cell Biology* 20 (1): 5–20.

Trumbo, J.L., Zhang, B., and Stewart, C.N. Jr. (2015). Manipulating micro RNA s for improved biomass and biofuels from plant feedstocks. *Plant Biotechnology Journal* 13 (3): 337–354.

Tsai, S.Q., Wyvekens, N., Khayter, C. et al. (2014). Dimeric CRISPR RNA-guided FokI nucleases for highly specific genome editing. *Nature Biotechnology* 32 (6): 569–576.

Upadhyay, S.K. (2021). *Genome Engineering for Crop Improvement*. Wiley. https://doi.org/10.1002/9781119672425.

Wang, J., Li, J., Zhao, H. et al. (2015). Structural and mechanistic basis of PAM-dependent spacer acquisition in CRISPR-Cas systems. *Cell* 163: 840–853.

Wang, H., La Russa, M., and Qi, L.S. (2016). CRISPR/Cas9 in genome editing and beyond. *Annual Review of Biochemistry* 85: 227–264.

Wang, J., Friedman, G., Doyon, Y. et al. (2012). Targeted gene addition to a predetermined site in the human genome using a ZFN-based nicking enzyme. *Genome Research* 22 (7): 1316–1326.

Wang, J., Mei, J., and Ren, G. (2019). Plant microRNAs: biogenesis, homeostasis, and degradation. *Frontiers in Plant Science 10*: 360.

Wiedenheft, B., van Duijn, E., Bultema, J.B. et al. (2011). RNA-guided complex from a bacterial immune system enhances target recognition through seed sequence interactions. *PNAS* 108 (25): 10092–10097.

Wiedenheft, B., Sternberg, S.H., and Doudna, J.A. (2012). RNA-guided genetic silencing systems in bacteria and archaea. *Nature* 482: 331–338.

Wolfe, S.A., Nekludova, L., and Pabo, C.O. (2000). DNA recognition by Cys2His2 zinc finger proteins. *Annual Review of Biophysics and Biomolecular Structure* 29 (1): 183–212.

Woolf, T. (1998). Therapeutic repair of mutated nucleic acid sequences. *Nature Biotechnology* 16: 341–344.

Xie, K. and Yang, Y. (2013). RNA-guided genome editing in plants using a CRISPR–Cas system. *Molecular Plant* 6 (6): 1975–1983.

Yang, H., Wang, H., and Jaenisch, R. (2014). Generating genetically modified mice using CRISPR/Cas-mediated genome engineering. *Nature Protocols* 9 (8): 1956–1968.

Yang, W., Chendrimada, T.P., Wang, Q. et al. (2006). Modulation of microRNA processing and expression through RNA editing by ADAR deaminases. *Nature Structural & Molecular Biology* 13 (1): 13–21.

Yu, B., Yang, Z., Li, J. et al. (2005). Methylation as a crucial step in plant microRNA biogenesis. *Science* 307 (5711): 932–935.

Yue, E., Cao, H., and Liu, B. (2020). OsmiR535, a potential genetic editing target for drought and salinity stress tolerance in *Oryza sativa*. *Plants* 9 (10): 1337.

Zhang, J., Li, S., Li, L. et al. (2015). Exosome and exosomal microRNA: trafficking, sorting, and function. *Genomics, Proteomics & Bioinformatics* 13 (1): 17–24.

Zhao, L., Kim, Y., Dinh, T.T., and Chen, X. (2007). miR172 regulates stem cell fate and defines the inner boundary of APETALA3 and PISTILLATA expression domain in *Arabidopsis* floral meristems. *The Plant Journal* 51 (5): 840–849.

Zhou, Y., Liu, W., Li, X. et al. (2020). Integration of sRNA, degradome, transcriptome analysis and functional investigation reveals gma-miR398c negatively regulates drought tolerance via GmCSDs and GmCCS in transgenic *Arabidopsis* and soybean. *BMC Plant Biology* 20: 1–19.

Zhu, H., Li, C., and Gao, C. (2020). Applications of CRISPR–Cas in agriculture and plant biotechnology. *Nature Reviews Molecular Cell Biology* 21 (11): 661–677.

Zhu, L., Ow, D.W., and Dong, Z. (2018). Transfer RNA-derived small RNAs in plants. *Science China Life Sciences* 61 (2): 155–161.

9

Genome Editing for Trait Improvement in Ornamental Plants

Yang Zhou, Yuxin Li, and Wen Liu

School of Horticulture, Hainan University/Key Laboratory for Quality Regulation of Tropical Horticultural Crops of Hainan Province, Haikou, China

9.1 Introduction

With the improvement of people's living standard, flowers, as one of the main factors of the decoration environment, have been more and more popular. Increasing the number of flowers, regulating the flowering time, extending the flowering period and creating novelty in flower structure, flower color, flower shape and floral scent are the main goals of ornamental plant breeding. Among all the ornamental characters, flower color is the main factor affecting its commercial value. Color is a complex character that is mainly determined by four categories of pigments including chlorophyll, flavonoids, carotenoids, and betalains (Yu et al. 2019). Chlorophyll is present in all photosynthesizing plants and can absorb most red and violet light while reflecting green light, thus rendering plant leaves green (Albuquerque et al. 2021). Carotenoids can be divided into beta-carotenoid, violaxanthin, lutein, and neoxanthin (Grassino et al. 2023). Different types and contents of carotenoids affect the red or yellow color of plant flowers and leaves (Xia et al. 2021). Flavonoids are a very important class of substances in plants and are the main source of pigments in most plant flowers (Liu et al. 2020). Betalains only exist in plants belonging to the Caryophyllales, except for Kewaceae and Macarthuriaceae (Brockington et al. 2015), and are a kind of pigments that have been less studied so far (Montes-Lora et al. 2018).

Although classical breeding strategies (such as intraspecific and interspecific hybridization, mutation breeding, polyploid breeding) have been widely used to develop new plant varieties. However, these breeding methods have some limitations such as long breeding time, difficult realization of polyploid homozygous, low accuracy, and low frequency (Azadi et al. 2016; Sharma and Messar 2017; Kushi-Kaboshi et al. 2018). Many valuable flower crops (such as rose, chrysanthemum, and carnation) have high heterozygosity, large genome (chrysanthemum, lily) and large chromosome numbers (*Hibiscus mutabilis*), long life cycles

Applications of Genome Engineering in Plants, First Edition. Edited by Santosh Kumar Upadhyay.
© 2024 John Wiley & Sons Ltd. Published 2024 by John Wiley & Sons Ltd.

(*Anthurium andraeanum*), and sterility in most varieties of some plants (carnation, chrysanthemum, and rose), which greatly limited the applicability of traditional breeding methods. Therefore, finding new ways to improve the traits of ornamental plants is an urgent need for current breeding work (Giovannini et al. 2021).

Gene editing is an emerging biotechnology with the advantage of being able to precisely edit one or more specific genes to control gene function, gene expression, and epigenetic state. The principle of gene editing is to cut the DNA double strand of the target gene through specific guiding sequences to form double-strand breakage (DSB). When the body makes inaccurate repair of DSB through its own repair mechanism, the deletion or insertion of the repair site base will be caused, thus resulting in the loss of gene function (Gaudelli et al. 2017). Currently, gene editing technology mainly includes zinc finger nuclease (ZFN) (Carroll 2011), transcription-activator-like effector nuclease (TALEN) (Joung and Sander 2013), and CRISPR/Cas9 system (Richter et al. 2012; Upadhyay 2021). CRISPR/Cas9 system has been extensively studied in the improvement of ornamental plant characters. The CRISPR/Cas 9 system has three main components: the Cas9 protein, CRISPR RNA (crRNA), and trans-activated crRNA (tracrRNA) (Richter et al. 2012; Sushmita et al. 2021). Cas9 is an endonuclease responsible for cutting DNA sequences. In the engineered CRISPR/Cas9 system, tracrRNA is joined to crRNA to become a single stranded guide RNA (sgRNA), which guides Cas9 to achieve DNA sequence-specific cleavage (Alok et al. 2021). The CRISPR/Cas9 system has been used in horticultural plants such as *Ipomoea nil* (Shibuya et al. 2018; Watanabe et al. 2018), *Petunia hybrida* (Subburaj et al. 2016; Sun and Kao 2018; Xu et al. 2020), *Lilium pumilum* (Yan et al. 2019), *Torenia fournieri* (Nishihara et al. 2018), and *Phalaenopsis* sp. (Tong et al. 2020). In addition, gene editing in ornamental plants is not as controversial as it might be in food crops, and it is expected that public acceptance may be higher (Nishihara and Nakatsuka 2011).

9.2 Application of Gene Editing Technology in Color Regulation of Ornamental Plants

Flower color is one of the important indicators to evaluate flower commercial values, which is mainly controlled by plant pigments such as chlorophyll, betaine, carotenoid, and flavonoid (Tanaka et al. 2010; Boutigny et al. 2020). It has been shown that inhibition of anthocyanin deposition can be achieved by insertion of antisense the chalcone synthase gene (CHS) or dihydroflavonol 4-reductase (DFR) gene in transgenic plants. Thus, the color of flowers and veins of 89% transgenic lines were changed (Aida et al. 2000). Watanabe et al. (2017) altered the expression of the dihydroflavonol-4-reductase-B (*DFR-B*) gene of anthocyanin biosynthetase in Japanese morning glory Ipomoea by gene editing technology and observed changes in stem color throughout the initial stage of plant tissue culture. The carotenoid cleavage dioxygenases (CCDs) genes were modified with CRISPR/Cas9-mediated mutagenesis technology, and the plants with pale yellow petals were obtained (Watanabe et al. 2018). Anthocyanin is the most important flavonoid pigment that can render red, pink, purple, and blue color (Peng et al. 2021). Anthocyanin mainly includes pelargonidin, cyanidin, and delphinidin, which was regulated by flavanone 3-hydroxylase (F3H), flavonoid 3′-hydroxylase (F3′H), and flavonoid-3′,

5′-hydroxylase (F3′5H), respectively (Wang et al. 2021). Modification of *Torenia fournieri F3′H* gene produced lilac to white flowers in gene-edited lines (Nishihara et al. 2018). F3′H is necessary for the synthesis of anthocyanins, which can form red bract color of *Euphorbia pulcherrima*. Nitarska et al. (2021) knocked out the *F3′H* gene from red *E. pulcherrima* using the CRISPR/Cas9 system, and the bract turned orange. Although F3′H was not completely inactivated, the bracts changed color from vivid red to bright red-orange and anthocyanin levels were significantly reduced in the transgenic plants compared to the wild type. Phytoene desaturase (PDS) controls the generation of key enzymes in carotenoid synthesis and is also essential for chlorophyll biosynthesis. Albino plants can be obtained by modifying *PDS* gene using CRISPR/Cas9 technology (Zhang et al. 2016; Yan et al. 2019; Bhalothia et al. 2020). Zhang et al. (2021) used CRISPR/Cas9 to edit the *DEEP PURPLE* (*DPL*) gene in Petunia hybrida and found that anthocyanins disappeared from the dorsal veins of flower buds in mutant lines.

9.3 Application of Gene Editing Technology in Ornamental Plants Preservation

The postharvest preservation time of ornamental plants is one of the important indexes to evaluate their commercial property. Ethylene plays an important role in plant growth, development, and senescence (Yang and Hoffman 1984). In plants, the enzymes required for catalytic processes in ethylene biosynthesis pathway include S-adenosine methionine synthase (SAM), 1-aminocyclopropane-1-carboxylate (ACC) synthase, and ACC oxidase (ACO). Methionine was a precursor of ethylene synthesis and formed into S-amido-methionine in the catalytic of SAM, and then formed into ACC under the action of ACC synthetase. Finally, ACC formed ethylene under the catalytic of ACO (Lieberman and Mapson 1964). ACO is a key enzyme in ethylene biosynthesis. Most ornamental plants have a shorter flowering period during bottling due to the increased amount of ethylene released (Pandey et al. 2000). Flower life can be increased by reducing ethylene production, which can be reduced by knocking the expression of *ACO* (John 1997; Boatright 2000). Xu et al. (2020) knocked out *ACO1* gene in *Petunia* cv. "mirage" rose by gene editing technology and obtained a series of knockout mutants of different genotypes. The wild-type mutant was found to wither and fade naturally after six days of flowering, while the phenotype of the mutant showed an extended lifespan. The *Petunia hybrida* lines with editing *PhACO3* and *PhACO4* genes generated by CRISPR/Cas9 displayed reduced ethylene amounts and prolonged flower life (Xu et al. 2021). Xu et al. (2020) edited *ACO1* gene of *Petunia* by CRISPR/Cas9 system and found that 31.5% of target point mutation frequency was detected in T0 generation lines, and ethylene production was reduced and flowering period was significantly prolonged in gene-edited lines. In *Dianthus* "Carnation," vase life was approximately doubled from five to nine days by silencing *ACO* compared to wild type plants (Chandler 2007; Savin et al. 2019). EPHEMERAL1 (EPH1), a NAC transcription factor, is believed to be a key regulator of petal senescence in Japanese morning glory (*Ipomoea nil*). The petal senescence was delayed after mutation of EPH1 gene (Shibuya et al. 2018).

9.4 Application of Gene Editing Technology in Shape and Organ Regulation of Ornamental Plants

As one of the most important horticultural traits of ornamental plants, flower shape is an important breeding target. Su et al. (2017) analyzed the function of *Torenia fournieri* TfRAD1 by CRISPR/Cas9-mediated genome editing to create a function loss allele (Tfrad1-Cas9). The results showed that the lateral petals of the editing lines became ventrally lateralized in shape (Su et al. 2017). It has been reported that the MADS gene family encoding DNA binding proteins are highly expressed in the floral organs and are involved in the formation and development of floral organs (Lin et al. 2016). The *MADS* gene of *Phalaenopsis* was edited, and target *MADS* mutations (insertion/deletion) were detected in all but a few lines, 60% of which were triple-mutated homozygous or biallelic (Tong et al. 2020). The MADS box gene *AGAMOUS* (*AG*) played an important role in the development of stamens and carpels in *Arabidopsis thaliana*. Mutations of the *AG* genes cause stamens and carpels to change into petal-like organs. The *Torenia fournieri* contains two *AG* genes: *TfPLE* and *TfFAR*. The single knockout of *TfPLE* only caused morphological changes in the carpel, while the mutation of *TfFAR* did not. When the two genes were knocked out at the same time, *T. fournieri* showed the transformation of stamen and carpel to petal-like structure, indicating that the two genes played a cooperative and synergistic role in the development of reproductive organs. The multi-petal phenotype is expected to be a highly valuable commercial flower trait in horticultural flowers (Sasaki and Ohtsubo 2020).

9.5 Application of Gene Editing Technology in Other Traits of Ornamental Plants

In addition to color, flower preservation, flower shape, and flower organ development, gene editing technology has also been applied to other ornamental plant traits. In the molecular breeding study of *Dendrobium catenatum*, the vector synthesized by CaMV35s promoter, GUS, and green fluorescent protein as reporter genes was transferred into *D. catenatum* mediated by *Agrobacterium*-transformation method. Mutant plants with a mutation rate of 10–100% were obtained by gene editing technology (Yan et al. 2015). The *C3H*, *C4H*, *4CL*, *CCR*, and *IRX* genes in the lignin biosynthesis pathway were edited using the CRISPR/Cas9 system in *D. catenatum*, and the mutation rates of these genes reached 10–100% (mutation types included base insertion, deletion, or replacement) (Kui et al. 2017). The results established an effective model for genetic research and molecular breeding of *D. catenatum* and other orchids.

Chrysanthemum morifolium is an important ornamental plant with high ornamental value and broad market prospect. However, chrysanthemum is a hexaploid plant with a huge genome, which leads to the lack of whole genome information, which brings difficulties to the genetic breeding of chrysanthemum. In order to optimize the CRISPR/Cas9 gene editing system in chrysanthemum, Kushi-Kaboshi et al. (2017) created transgenic chrysanthemum plants overexpressing *YGFP* gene and then edited the *YGFP* gene in transgenic plants using CRISPR/Cas9 system. After comparing the activity of CaMV35S promoter and

PcUbi promoter in chrysanthemum callus, it was found that PcUbi promoter was more effective for gene editing in chrysanthemum callus. PcUbi promoter was further used to express Cas9 protein, and different sgRNAs were designed for two sites of YGFP, and YGFP mutant chrysanthemum plants were obtained, with gene editing efficiency of 0–33%. This is the first report of gene editing in chrysanthemum using CRISPR/Cas9 system. The results provided a reference for the genetic improvement of chrysanthemum. Shinoyama et al. (2020) knocked out *DMC1* gene in chrysanthemum using gene editing method. The result showed that stable male sterile lines appeared under the condition of 10–30 °C. Self-incompatibility in wild *Petunia inflata* is regulated by the polymorphic S locus, which contained multiple pollen specific S-site F box (SLF) genes and a pistil-specific S-RNase gene. Each SLF protein in pollen was assembled into an SCF (Skp1-Cullin1-F-box) E3 ubiquitin ligase complex. SUN and Kao (2018) used CRISPR/Cas9 system to edit the Skp1 subunit (SSK1) gene in *Petunia* SCF^{SLF} complex to study the mechanism of self-incompatibility in *Petunia*, and the target gene of 1 of the 3 regenerated plants was successfully edited.

9.6 Conclusions and Perspectives

Different from traditional genetic engineering breeding methods, gene editing breeding can change plant genes in a targeted and efficient way, so as to obtain new germplasm with improved characters. Through genetic isolation after stable transformation or some other transformation techniques, the gene editing vector can be separated or unintegrated into the plant genome, thus people can obtain the material without any exogenous transgenic elements, which is fundamentally different from the transgenic material. However, most ornamental plants are unable to carry out genetic transformation at present, which greatly restricts the application of gene editing system. With the continuous improvement of gene editing technology, the application and optimization of more genetic transformation methods and the deepening of functional genome research of ornamental plants, gene editing technology will become an important means and play an increasingly important role in ornamental plant breeding in the future.

Acknowledgments

This work was supported by Hainan Provincial Natural Science Foundation of China (318QN189, 319MS009), and the Startup funding from Hainan University (KYQD(ZR)1845).

References

Aida, R., Yoshida, K., Kondo, T. et al. (2000). Copigmentation gives bluer flowers on transgenic torenia plants with the antisense dihydroflavonol-4-reductase gene. *Plant Science* 160: 49–56.

Albuquerque, B.R., Oliveira, M.B.P.P., Barros, L., and Ferreira, I.C.F.R. (2021). Could fruits be a reliable source of food colorants? Pros and cons of these natural additives. *Critical Reviews in Food Science and Nutrition* 61 (5): 805–835.

Alok, A., Chauhan, H., Upadhyay, S.K. et al. (2021). Compendium of plant-specific CRISPR vectors and their technical advantages. *Life* 11: 1021. https://doi.org/10.3390/life11101021.

Azadi, P., Bagheri, H., Nalousi, A.M. et al. (2016). Current status and biotechnological advances in genetic engineering of ornamental plants. *Biotechnology Advances* 34: 1073–1090.

Bhalothia, P., Yajnik, K., Alok, A., and Upadhyay, S.K. (2020). The current progress of CRISPR/Cas9 development in plants. *Elsevier*. https://doi.org/10.1016/B978-0-12-818140-9.00011-8.

Boatright, J.L. (2000). Transcriptional regulation of *PhACO1* in *Petunia hybrida*. Doctoral dissertation. Purdue University.

Boutigny, A.L., Dohin, N., Pornin, D., and Rolland, M. (2020). Overview and detectability of the genetic modifications in orna-mental plants. *Horticulture Research* 7: 11.

Brockington, S.F., Yang, Y., Gandia-Herrero, F. et al. (2015). Lineage-specific gene radiations underlie the evolution of novel betalain pigmentation in *Caryophyllales*. *The New Phytologist* 207: 1170–1180.

Carroll, D. (2011). Genome engineering with zinc-finger nucleases. *Genetics* 188 (4): 773–782.

Chandler, S.F. (2007). Practical lessons in the commercialization of genetically modified plants-long vase-life carnation. *Acta Horticulturae* 764: 71–82.

Gaudelli, N.M., Komor, A.C., Rees, H.A. et al. (2017). Programmable base editing of A•T to G•C in genomic DNA without DNA cleavage. *Nature* 551 (7681): 464–471.

Giovannini, A., Laura, M., Nesi, B. et al. (2021). Genes and genome editing tools for breeding desirable phenotypes in ornamentals. *Plant Cell Reports* 40 (3): 461–478.

Grassino, A.N., Brncic, S.R., Sabolovic, M.B. et al. (2023). Carotenoid content and profiles of pumpkin products and by-products. *Molecules* 28 (2): 858.

John, P. (1997). Ethylene biosynthesis: the role of 1-aminocy-clopropane-1-carboxylate (ACC) oxidase, and its possible evolutionary origin. *Physiologia Plantarum* 100: 583–592.

Joung, J.K. and Sander, J.D. (2013). TALENs: a widely applicable technology for targeted genome editing. *Nature Reviews. Molecular Cell Biology* 14 (1): 49–55.

Kishi-Kaboshi, M., Aida, R., and Sasaki, K. (2017). Generation of gene-edited *Chrysanthemum morifolium* using multicopy transgenes as targets and markers. *Plant & Cell Physiology* 58 (2): 216–226.

Kishi-Kaboshi, M., Aida, R., and Sasaki, K. (2018). Genome engineering in ornamental plants: current status and future prospects. *Plant Physiology and Biochemistry* 131: 47–52.

Kui, L., Chen, H., Zhang, W. et al. (2017). Building a genetic manipulation tool box for orchid biology: identification of constitutive promoters and application of CRISPR/Cas9 in the orchid, *Dendrobium officinale*. *Frontiers in Plant Science* 7: 2036.

Lieberman, M. and Mapson, L.W. (1964). Genesis and biogenesis of ethylene. *Nature* 204 (4956): 343–345.

Lin, C.S., Hsu, C.T., Liao, D.C. et al. (2016). Transcriptome-wide analysis of the MADS-box gene family in the orchid *Erycina pusilla*. *Plant Biotechnology Journal* 14: 284–298.

Liu, C., Yu, Q., Li, Z. et al. (2020). Metabolic and transcriptomic analysis related to flavonoid biosynthesis during the color formation of *Michelia crassipes* tepal. *Plant Physiology and Biochemistry* 155: 938–951.

Montes-Lora, S., Rodríguez-Pulido, F.J., Cejudo-Bastante, M.J., and Heredia, F.J. (2018). Implications of the red beet ripening on the colour and betalain composition relationships. *Plant Foods for Human Nutrition* 73 (3): 216–221.

Nishihara, M. and Nakatsuka, T. (2011). Genetic engineering of flavonoid pigments to modify flower color in floricultural plants. *Biotechnology Letters* 33 (3): 433–441.

Nishihara, M., Higuchi, A., Watanabe, A., and Tasaki, K. (2018). Application of the CRISPR/Cas9 system for modification of flower color in *Torenia fournieri*. *BMC Plant Biology* 18 (1): 331.

Nitarska, D., Boehm, R., Debener, T. et al. (2021). First genome edited poinsettias: targeted mutagenesis of flavonoid 3'-hydroxylase using CRISPR/Cas9 results in a colour shift. *Plant Cell, Tissue and Organ Culture* 147: 49–60.

Pandey, S., Ranade, S.A., Nagar, P.K., and Kumar, N. (2000). Role of polyamines and ethylene as modulators of plant senes-cence. *Journal of Biosciences* 25: 291–299.

Peng, J., Dong, X., Xue, C. et al. (2021). Exploring the molecular mechanism of blue flower color formation in *Hydrangea macrophylla* cv. "Forever Summer". *Frontiers in Plant Science* 12: 585665.

Richter, C., Chang, J.T., and Fineran, P.C. (2012). Function and regulation of clustered regularly interspaced short palindromic repeats (CRISPR)/CRISPR associated (Cas) systems. *Viruses* 4 (10): 2291–2311.

Sasaki, K. and Ohtsubo, N. (2020). Production of multi-petaled *Torenia fournieri* flowers by functional disruption of two class-C MADS-box genes. *Planta* 251: 1–16.

Savin, K.W., Baudinette, S.C., Graham, M.W. et al. (2019). Antisense ACC oxidase RNA delays carnation petal senescence. *HortScience* 30: 970–972.

Sharma, R. and Messar, Y. (2017). Transgenics in ornamental crops: creating novelties in economically important cut flowers. *Current Science* 113: 43–52.

Shibuya, K., Watanabe, K., and Ono, M. (2018). CRISPR/Cas9-mediated mutagenesis of the EPHEMERAL1 locus that regulates petal senescence in Japanese morning glory. *Plant Physiology and Biochemistry* 131: 53–57.

Shinoyama, H., Ichikawa, H., Nishizawa-Yokoi, A. et al. (2020). Simultaneous TALEN-mediated knockout of chrysanthemum DMC1 genes confers male and female sterility. *Scientific Reports* 10 (1): 16165.

Su, S., Xiao, W., Guo, W. et al. (2017). The CYCLOIDEA-RADIALIS module regulates petal shape and pigmentation, leading to bilateral corolla symmetry in *Torenia fournieri* (*Linderniaceae*). *The New Phytologist* 215 (4): 1582–1593.

Subburaj, S., Chung, S.J., Lee, C. et al. (2016). Site-directed mutagenesis in *Petunia* × *hybrida* protoplast system using direct delivery of purified recombinant Cas9 ribonucleoproteins. *Plant Cell Reports* 35 (7): 1535–1544.

Sushmita, Kaur, G., Upadhyay, S.K., and Verma, P.C. (2021). An overview of genome-engineering methods. *Wiley*. https://doi.org/10.1002/9781119672425.ch1.

Sun, L. and Kao, T.H. (2018). CRISPR/Cas9-mediated knockout of PiSSK1 reveals essential role of S-locus F-box protein-containing SCF complexes in recognition of non-self S-RNases during cross-compatible pollination in self-incompatible *Petunia inflata*. *Plant Reproduction* 31 (2): 129–143.

Tanaka, Y., Brugliera, F., Kalc, G. et al. (2010). Flower color modification by engineering of the flavonoid biosynthetic pathway: practical perspectives. *Bioscience, Biotechnology, and Biochemistry* 74: 1760.

Tong, C.G., Wu, F.H., Yuan, Y.H. et al. (2020). High-efficiency CRISPR/Cas-based editing of *Phalaenopsis* orchid *MADS* genes. *Plant Biotechnology Journal* 18 (4): 889–891.

Upadhyay, S.K. (2021). *Genome Engineering for Crop Improvement*. Wiely. https://doi.org/10.1002/9781119672425.

Wang, R., Ren, C., Dong, S. et al. (2021). Integrated metabolomics and transcriptome analysis of flavonoid biosynthesis in safflower (*Carthamus tinctorius* L.) with different colors. *Frontiers in Plant Science* 12: 712038.

Watanabe, K., Kobayashi, A., Endo, M. et al. (2017). CRISPR/Cas9-mediated mutagenesis of the dihydroflavonol-4-reductase-B (DFR-B) locus in the Japanese morning glory *Ipomoea (Pharbitis) nil*. *Scientific Reports* 7: 1–9.

Watanabe, K., Oda-Yamamizo, C., Sage-Ono, K. et al. (2018). Alteration of flower colour in Ipomoea nil through CRISPR/Cas9-mediated mutagenesis of carotenoid cleavage dioxygenase 4. *Transgenic Research* 27 (1): 25–38.

Xia, Y., Chen, W., Xiang, W. et al. (2021). Integrated metabolic profiling and transcriptome analysis of pigment accumulation in *Lonicera japonica* flower petals during colour-transition. *BMC Plant Biology* 21 (1): 98.

Xu, J., Kang, B.C., Naing, A.H. et al. (2020). CRISPR/Cas 9-mediated editing of 1-aminocyclopropane-1-carboxylate oxidase1enhances *Petunia* flower longevity. *Plant Biotechnology Journal* 18 (1): 287–297.

Xu, J., Naing, A.H., Bunch, H. et al. (2021). Enhancement of the flower longevity of petunia by CRISPR/Cas9-mediated targeted editing of ethylene biosynthesis genes. *Postharvest Biology and Technology* 174: 111460.

Yan, L., Wang, X., Liu, H. et al. (2015). The genome of *Dendrobium officinale* illuminates the biology of the important traditional Chinese orchid herb. *Molecular Plant* 8: 922–934.

Yan, R., Wang, Z., Ren, Y. et al. (2019). Establishment of efficient genetic gransformation systems and application of CRISPR/Cas9 genome editing technology in *Lilium pumilum* DC. Fisch. and *Lilium longiflorum* white heaven. *International Journal of Molecular Sciences* 20 (12): 2920.

Yang, S.F. and Hoffman, N.E. (1984). Ethylene biosynthesis and its regulation in higher plants. *Annual Review of Plant Physiology* 35: 155–189.

Yu, X., Hu, S., He, C. et al. (2019). Chlorophyll metabolism in postharvest tea (*Camellia sinensis* L.) leaves: variations in color values, chlorophyll derivatives, and gene expression levels under different withering treatments. *Journal of Agricultural and Food Chemistry* 67 (38): 10624–10636.

Zhang, G.Q., Xu, Q., Bian, C. et al. (2016). The *Dendrobium catenatum* Lindl. genome sequence provides insights into polysaccharide synthase, floral development and adaptive evolution. *Scientific Reports* 6: 1–10.

Zhang, B., Xu, X., Huang, R. et al. (2021). CRISPR/Cas9-mediated targeted mutation reveals a role for *AN4* rather than *DPL* in regulating venation formation in the corolla tube of *Petunia hybrida*. *Horticulture Research* 8: 116.

10

Abiotic Stress Tolerance in Plants by Genome Editing Applications

Elif Karlik Urhan

Institution of Science, Department of Biotechnology, Istanbul University, Istanbul, Turkey

10.1 Introduction

The global environmental changes threat the plant productivity by suppressing growth and development and is directly affecting the agricultural food production (Liang et al. 2017a; Debbarma et al. 2019). Increasing world population is estimated to reach approximately 10 billion, which will enhance food demand up to 70% by 2050 (Alexandratos and Bruinsm 2012; McGuire 2015). Abiotic stress, including drought, flooding, nutrient deficiencies, high temperature, low temperature, salinity, and heavy metals, already become grounds for up to 50% yield loss in many major crops (Yadav et al. 2020). Many abiotic stress factors disturb plant growth and yield by affecting the synthesis, metabolism, transport, and storage of nutrients (Cramer et al. 2011). The effects of several abiotic stresses on plants have been shown in the Figure 10.1. Besides, the occurrence of abiotic stress affects plant–pathogen interaction by increasing the host plant susceptibility to various biotic stress factors (Peters et al. 2014; Pandey et al. 2017).

Plants normally adapt to changing environmental conditions by evolving acquired adaptive traits that are morphological and physiological adaptations, as well as signaling pathways (Zhu 2016; Shumayla et al. 2019; Upadhyay and Shumayla 2023). However, extreme environmental changes require long-term solutions to cope with abiotic stress, especially drought, salinity, and heat stress. Traditional crop breeding techniques such as hybridization and marker-assisted selection (MAS) are time consuming, laborious, cost intensive, and not sufficient (Collard and Mackill 2008; Breseghello and Coelho 2013). Traditional techniques mainly depend on exploring the knowledge of genes and quantitative trait loci (QTLs) controlling different traits that are natural variations existing in the primary gene pool (Khush 2013). Therefore, there is a limit how far agriculture can adapt to the changing environment (Anderson et al. 2020). Although traditional crop improvement approaches made major contributions against abiotic stress (Fita et al. 2015; Driedonks et al. 2016; Hussain et al. 2017; Shumayla et al. 2023), advanced biotechnological approaches are the only possible effective tool that they are efficient, fast, and specific (Saikia et al. 2020; Razzaq et al. 2022).

Applications of Genome Engineering in Plants, First Edition. Edited by Santosh Kumar Upadhyay.
© 2024 John Wiley & Sons Ltd. Published 2024 by John Wiley & Sons Ltd.

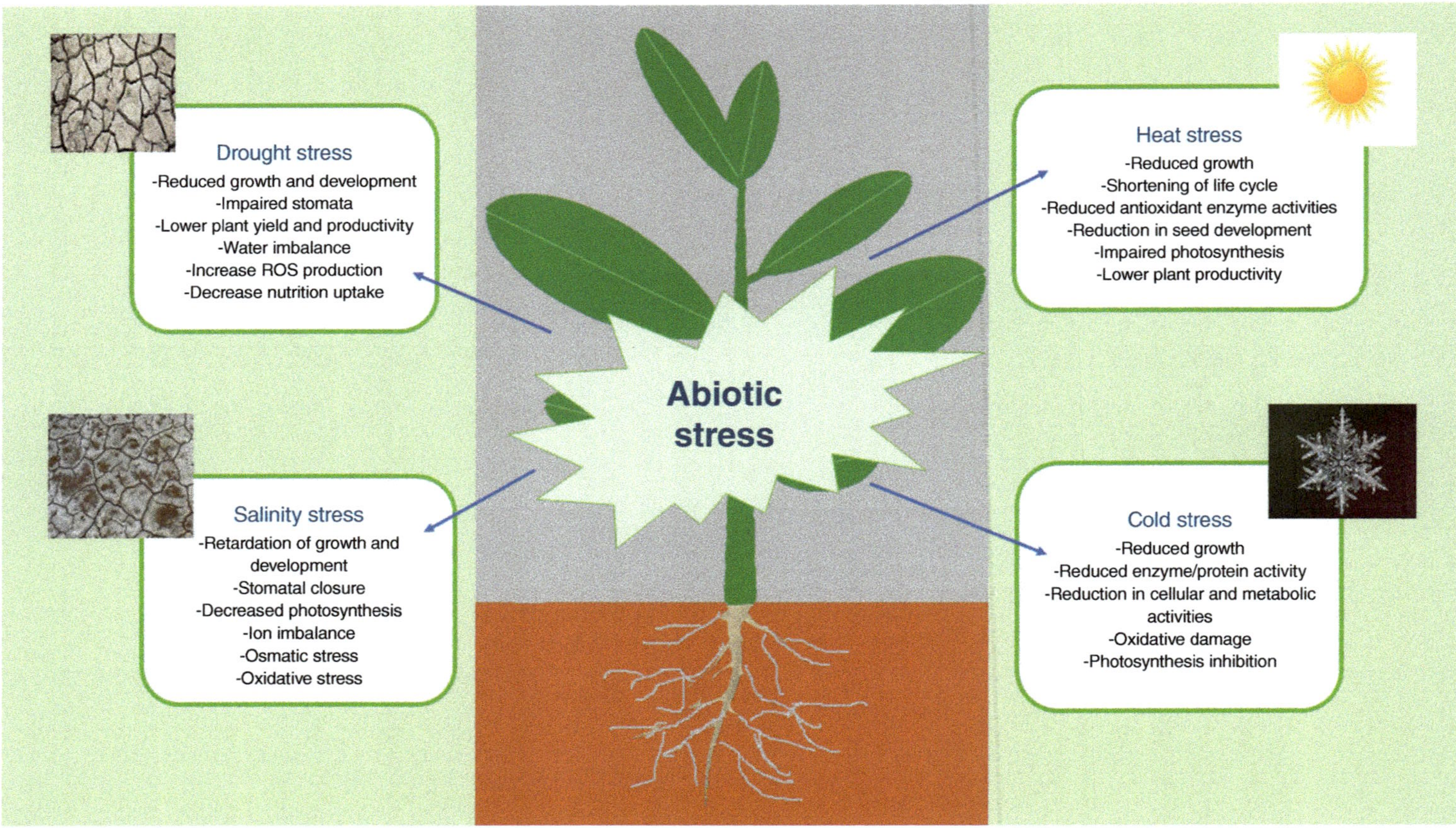

Figure 10.1 The effects of different abiotic stresses on plants.

Nowadays, recent genome editing tools have removed the barriers of traditional crop breeding strategies and started in a new era of crop improvement. Especially, site specific nucleases such as zinc finger nucleases (ZFNs), transcriptional activator-effector nucleases (TALENs), and CRISPR-related endonuclease Cas9 (CRISPR/Cas9) are used to target specific locations of genes in the genome. They create double stranded break (DSB) in the targeted DNA. The repair of DSB is carried out through homologous recombination (HR) or nonhomologous end joining (NHEJ) (Miglani 2017; Ansari et al. 2020; Shushmita et al. 2021; Upadhyay 2021). The repair pathway mediated through NHEJ is an error creating arbitrary changes causing mutations either insertions or deletions (InDels). This kind of frames shift alterations results in knockouts of genes and models to investigate gene function. However, HR is significantly more accurate in the process of homologous sequence exchange which is considerably useful for gene knockin or gene replacement studies (Ansari et al. 2020). CRISPR/Cas9 is the most effective and easiest genome editing tool that is naturally a prokaryotic adaptive immune system, including a guide RNA (gRNA) associated with a DNA endonuclease known as Cas9 (Barrangou et al. 2007; Weeks et al. 2016). The gRNA is composed of two components: a fusion of a CRISPR RNA (crRNA) and a trans-activating crRNA (tracrRNA) (Deltcheva et al. 2011; Jinek et al. 2012; Alok et al. 2018, 2021). DNA cutting site is targeted by the crRNA that serves as a guiding sequence that has a homology region base pair with the tracrRNA. Both crRNA and tracrRNA are required to integrate along with Cas9 that cleaves DNA with DSBs at specific target site. In genome editing method, Jinek and colleagues created a single chimeric RNA formed by the fusion of crRNA and tracrRNA, also known as single-guided RNA (sgRNA) coordinating the sequence dependent Cas9 DSBs (Jinek et al. 2012, 2014). To target the specific site in genome, sgRNA recognizes a protospacer-adjacent motif (PAM), which flanks the 3′ end of the crRNA-targeted sequence (Gasiunas et al. 2012; Cong et al. 2013; Upadhyay et al. 2013).

The traditional techniques were performed successfully. Besides such breeding applications, genome editing took breeding studies much further by improving the frequency of targeted integration. For instance, *ALS2* gene (*acetolactate synthase*) edited for conferring resistance to herbicides by CRISPR/Cas9 system in maize (Svitashev et al. 2015). Genome editing techniques, especially CRISPR/Cas9 has been widely used to improve higher yield, quality and nutritional value, and biotic and abiotic stress tolerance of plants (Table 10.1) (Shan et al. 2014; Liang et al. 2017b; Zhang et al. 2019a; Ansari et al. 2020; Bhalothia et al. 2020; Sushmita et al. 2021; Singh et al. 2022). In this chapter, we surveyed the catalog of latest published studies related to abiotic stress tolerance.

10.2 Drought Tolerance

Drought is one of the main restrictions on agricultural productivity and is expected to further increase worldwide. Drought tolerance is a complex quantitative polygenic trait, including physiological and molecular mechanisms (Fang and Xiong 2015). To cope with drought stress, several adaptations and mitigation strategies are required. One of these strategies includes genome editing technologies to reveal natural drought tolerance mechanism. In this concept, *AREB1* promoter were activated by the CRISPRa dCas9^{HAT}

Table 10.1 Examples of plants improved using the CRISPR-Cas system.

Plant	Target gene	Mode of action	Improved trait	Reference
Rice	*SAPK2*	Knockout	Salinity tolerance	Lou et al. (2017)
	OsOTS1			Zhang et al. (2019c)
	OsNAC041			Bo et al. (2019)
	OsRR22			Zhang et al. (2019b)
	OsAnn3		Cold tolerance	Shen et al. (2017)
	OsPIN5b, GS3 and OsMYB30			Liao et al. (2019)
	OsPYL1/4/6		Heat tolerance	Miao et al. (2018)
	OsERF922		Blast fungus resistance	Wang et al. (2016)
	eIF4G		Rice tungro spherical virus resistance	Macovei et al. (2018)
	MADS78 and MADS79	Knockin	Early seed development	Paul et al. (2020)
	OsGRF4		Grain size	Li et al. (2016)
Tomato	*ALC*	Gene replacement	Shelf life	Yu et al. (2017)
	SINPR1	Knockout	Drought tolerance	Li et al. (2019)
	SIMAPK3			Wang et al. (2017)
	SICBF1		Cold tolerance	Li et al. (2018a,b)
	SIMAPK3		Heat tolerance	Nekrasov et al. (2017)
	SlMlo1		Powdery mildew resistance	Yu et al. (2019)
Arabidopsis	*OST2*	Knockout	Drought tolerance	Osakabe et al. (2016)
	AVP1	Knockin		Park et al. (2017)
	AREB1			Roca Paixão et al. (2019)
Wheat	*GASR7*	Knockout	Kernel weight	Zhang et al. (2016)
	α-gliadin		Low gluten	Sánchez-León et al. (2018)
Cucumber	*eIF4E*	Knockout	Broad virus resistance	Chandrasekaran et al. (2016)
Soybean	*FAD2-1A, FAD2-1B*	Knockout	Improved oil quality	Kim et al. (2017b)
Maize	*ARGOS8*	Knockin	Drought tolerance	Shi et al. (2017)
Cotton	*Gh14-3-3d*	Knockin	Cotton verticillium wilt resistance	Zhang et al. (2018b)
Barley	*HvITPK1*	Knockout	Salinity tolerance	Vlčko and Ohnoutková (2020)

system, which is a fusion of Cas9 and *Arabidopsis histone acetyltransferase 1* (*AtHAT1*) regulating gene expression activation by switching chromatin to a relaxed state. The plants generated by CRISPRa dCas9^{HAT} enhanced chlorophyll content and stomatal aperture under drought stress (Roca Paixão et al. 2019). Studies demonstrated that roles of the homeodomain-leucine zipper transcription factors are crucial for plant development, secondary metabolism, and abiotic stress responses (Perotti et al. 2017; Gong et al. 2019). In soybean, Zhong et al. (2022) indicated that CRISPR/Cas9-mediated gene editing of *GmHdz4* transcription factor enhanced drought tolerance. Gene-edited *gmhdz4* exhibited enhanced growth of total root length, the root superficial area, and the number of root tips than overexpressed *GmHdz4* and control plants. Additionally, higher activity of antioxidant enzymes in *gmhdz4* conferred improved drought tolerance compared with both overexpressed *GmHdz4* and control plants (Zhong et al. 2022).

As a key stress hormone, abscisic acid (ABA) regulates abiotic stress responses by activating ABA response genes (Chen et al. 2021). A novel transcription factor AITRs was identified as ABA-induced transcription repressors functioning as negative regulators in regulating ABA signaling and response to abiotic stress conditions (Tian et al. 2017). Knockout of the entire family of *AITR* genes via CRISPR/Cas9 reduced the sensitivity to ABA and enhanced both drought and salinity stress tolerance in *Arabidopsis*. Besides, growth and development in mutants was not affected. Additionally, one of these mutant types *aitr123456 sextuple* exhibited a wild-type response to the pathogen infection (Chen et al. 2021). In another study, tobacco *NtAITRs* genes were analyzed by generating CRIPR/Cas9 mutants. Interestingly, *ntaitr12356* quintuple mutants are hypersensitive to ABA on contrary to Chen et al. (2021), although drought stress tolerance was similarly enhanced (Li et al. 2022). In plants, NAC (NAM, ATAF1/2, CUC2) family is one of the largest transcription factors regulating a variety of biological process, including abiotic stress response (Liu et al. 2022a). In wheat, overexpressed *TaNAC071-A* significantly improved drought stress tolerance by increasing stress-responsive genes, whereas knockdown of this gene resulted with reduced drought tolerance (Mao et al. 2022). In 2020, Wang and colleagues performed knockout studies of *OsNAC006* via CRISPR/Cas9 in rice. Edited *OsNAC006* enhanced rice sensitivity to drought and heat stress. Also, RNA sequencing (RNA-seq) of mutants demonstrated that *OsNAC006* regulated the stress response by changing many stress-responsive genes involved in response to stimuli, oxidoreductase activity, cofactor binding, and membrane-related pathways (Wang et al. 2020a, b). Additionally, Wang et al. (2022a) also characterized another NAC family gene *OsNAC092* by generating knockout lines via CRISPR/Cas9. According to RNA-seq data, knock-out of *OsNAC092* resulted in alterations of a global expression, and differential gene expression was mainly associated with response to light stimulus, MAPK signaling pathway, plant hormone signal transduction, response to oxidative stress, photosynthesis, and water deprivation. Besides, antioxidants enzyme activity was enhanced under drought stress conditions (Wang et al. 2022a). Similarly, *CINNAMOYL-CoA REDUCTASE 10* (*OsCCR10*) was reported to be activated by *OsNAC5* modulating drought tolerance through regulating lignin accumulation in rice. Silencing of OsCCR10 via CRISPR/Cas9 resulted in reduced accumulation in roots causing the reduction of drought tolerance (Bang et al. 2022). In another study, Yang et al. (2022) established NAC transcription factor (TF) *GmNAC12* overexpression and knockout lines to reveal NAC TF function in soybean. Gene *GmNAC12* knockout lines exhibited reduced drought stress tolerance with lower survival rate than overexpressed *GmNAC12* and wild type. Functional analysis demonstrated that *GmNAC12* interacted with proteins were associated with chitin, chlorophyll, ubiquitin–protein transferase, and peroxidase activity, indicating that *GmNAC12* plays crucial roles in drought stress tolerance (Yang et al. 2022).

Ubiquitin-mediated proteolysis plays crucial roles via regulating the abundance of numerous enzymes, structural and regulatory proteins ensuring proper cellular function in plant growth, and adaptation to environmental stress (Stone 2014, 2019). However, the mechanism of E3 ubiquitin ligases regulating abiotic stress response via phytohormone signaling, especially ABA and jasmonic acid (JA) is still limited. In rice, Lv et al. (2022) identified and characterized U-box E3 ubiquitin ligase PLANT U-BOX PROTEIN 16 (OsPUB16) negatively modulating drought stress tolerance. Gene-editing of *OsPUB16* created by CRISPR/Cas9 system showed significantly enhanced drought stress tolerance. Moreover, ABA and JA responses were both regulated in *ospub16* mutants that ABA and JA levels were higher than wild type under drought stress. The study exhibited *OsPUB16* modulated the ubiquitination and degradation of OsMADS23, which is the substrate of OSMOTIC STRESS/ABA-ACTIVATED PROTEIN KINASE 9 (SAPK9) by inducing ABA biosynthesis. Additionally, the ChIP-qPCR analysis exhibited that the expression of JA-biosynthetic gene *OsAOC* was activated by binding of OsMADS23 to its promoter. According to mechanistic studies, the phosphorylation of OsMADS23 by SAPK9 resulted in decreased ubiquitination level by interfering with the OsPUB16-OsMADS23 interaction, which thus increased the stability of OsMADS23 and promoted *OsAOC* expression. Therefore, *OsPUB16* regulates both ABA and JA signaling pathways that are integrated by the "SAPK9-OsMADS23-*OsAOC*" module during drought stress tolerance in rice (Lv et al. 2022). Several adaptation mechanisms are associated with alterations in endogenous hormone levels are affected by drought stress (Hauser et al. 2011). *ZmLBD5, a class-II lateral organ boundaries domain* gene, was reported to be involved in the regulation of growth and the drought response by affecting gibberellin (GA) and ABA synthesis in maize. Interestingly, silencing of *lbd5* via CRISPR/Cas9 were exhibited dwarf phenotype; however, *lbd5* was drought tolerant with higher yield compared to wild type (Feng et al. 2022).

In cereals, timing of flowering is critical physiological process affected by endogenous and environmental signals (Izaw 2007; Sun et al. 2014). The main signals encoding via *Flowering Locus T-like* (*FT-like*) genes have been widely studied and revealed to be critical for flowering (Taoka et al. 2013). A rice *FT-like* gene, *OsFTL4,* was functionally characterized by generating *osftl4* mutants via CRISPR/Cas9. Knockout of *OsFTL4* caused early flowering compared to wild type, whereas it enhanced drought stress tolerance by reducing stomatal conductance and water loss (Gu et al. 2022). Additionally, creating of CRISPR/Cas9 knockout mutant lines of *Regulator of Flowering and Stress* (*RFS*) showed also early flowering phenotypes; however, *RFS* mutants reduced drought tolerance in *Arabidopsis* and *Brachypodium distachyon* (Ying et al. 2023). When plants are exposed to drought stress, plant responses involving large reconfiguration of hormonal pathways that orchestrate physiological changes toward tolerance including cellular senescence process (Huang et al. 2019). The regulation of *senescence-associated protein* genes (*SAP*) is also affected by drought stress causing substantial yield loss (Uji et al. 2017; Jiang et al. 2018). Silencing of drought-induced gene rice *Senescence-associated protein* (*OsSAP*) via CRIPR/Cas9 resulted in more susceptible in response to drought stress by reducing survival rate, shoot and root lengths, chlorophyll content, tiller number, and 1000-grain weight. These findings implied that rice *OsSAP* possibly acts as a positive regulator during drought stress tolerance (Park et al. 2022).

Stomata are tiny openings or pores in leaves that allow for gas exchange and water transport, which thus impact crucial functions such as photosynthesis and drought tolerance (Berry et al. 2010; Lin et al. 2022). Stomatal movements are regulated by Ca^{2+-}dependent signaling in response to drought stress (Yoo et al. 2019). DORN1, also known as P2K1, is a legume-type lectin receptor kinase identified as the first purinoceptor for the plant response to extracellular ATP through its role in increasing $[Ca^{2+}]_{cyt}$ (Choi et al. 2014). The knowledge about roles of eATP and its receptor DORN1 is limited, and therefore *DORN1* knockout mutants were generated by the CRISPR/Cas 9 system. Loss of function study exhibited the effective quantum yield of PSII [Y(II)], the photochemical quenching coefficient (qP), and the rate of photosynthetic electron transport through PSII (ETR) were decreased, whereas nonphotochemical fluorescence quenching $Y_{(NO)}$ values demonstrated obvious enhancement under drought stress conditions. These results exhibited that drought tolerance in *dorn1* mutants were impaired; however, application of exogenous calcium ions improved the drought tolerance, suggesting that drought tolerance may depend on calcium and calcium-related signaling pathways by modulating via DORN1 (Wang et al. 2022b).

Developing stress-tolerant cultivars requires a deep understanding of plant genetics. Traditional breeding has been affectively used until today; however, it is time-consuming, labor-intensive, and costly. To solve this problem, novel CRISPR/Cas9 system is a substantial editing tool to generate plant varieties with a wide range of traits in response to drought stress (Fiaz et al. 2021; Park et al. 2022; Han et al. 2023). Further studies hold the key to solving these problems.

10.3 Salinity Tolerance

Salinity in growing environment is a widespread factor limiting the productivity through negatively affecting seed germination and disrupts several physiological and metabolic processes such as alterations in the enzymatic activity (Tarchoun et al. 2022). To cope with salinity stress, plants adapted complex mechanisms, including modulation of reactive oxygen species (ROS) levels, ion homeostasis, and accumulation of compatible solutes (Sun et al. 2009; Wani et al. 2013; Chan et al. 2016). In plants, ion homeostasis is sustained by modulating the acquisition and distribution of sodium (Na^+) or potassium (K^+) under salt stress conditions (Wu et al. 2018; Isayenkov and Maathuis 2019; Rubio et al. 2020). Na^+ and K^+ influx and efflux processes are regulated by the membrane transporters, especially the monovalent cation/proton antiporter (CPA) family (Chanroj et al. 2012; Ye et al. 2013). The plasma membrane Na^+/H^+ antiporter SOS1 is crucial for salt tolerance (Quintero et al. 2011). The regulation of SOS1 is particularly modulated via signaling molecules such as cyclic nucleotides in response to osmatic and salt stress (Świeżawska et al. 2018). In addition to *Arabidopsis*, physiological roles of SOS1 have been widely studied in several plant species, including tomato, rice, and cotton (Shi et al. 2002; Razzaque et al. 2014; Chen et al. 2017b; Wang et al. 2020b). To investigate function of SOS1, Zhang et al. (2022a) created soybean *GmSOS1* mutant lines by CRISPR/Cas9. Silencing of *GmSOS1* resulted in hypersensitivity in response to salinity with excessive Na^+ intake and accumulation in roots and leaves. Additionally, *gmsos1* mutant lines exhibited dramatic decrease in K^+ contents in roots. Furthermore, the ability of Na^+ efflux of *gmsos1* reduced in the root meristem zone. These results indicated that soybean GmSOS1 as a Na^+/H^+ exchanger had critical roles for maintaining of ion homeostasis in response to salt stress (Zhang et al. 2022a).

K^+ transporters are also crucial to maintain Na^+/K^+ homeostasis, modulating root K^+ uptake (Wu et al. 2018). Especially, Na^+ could replace K^+ according to their physicochemical similarities, affecting the enzymatic reactions (Zhao et al. 2020). Disruption of K^+ homeostasis by Na^+ stress prevents various metabolic processes, including photosynthesis and protein synthesis, and triggers ROS accumulation and programmed cell death (Shabala 2009; Demidchik et al. 2010). However, the knowledge about the effects of K^+ long distance translocation on salinity tolerance is still limited. The Shaker K^+ channel (AKT2) is a voltage sensing ion channel, leading to K^+ to pass through the membrane (Zhao and Blunck 2016). In *Arabidopsis*, AtAKT2 regulates K^+ distribution via phloem and promotes the sucrose loading to phloem (Gajdanowicz et al. 2011; Dreyer et al. 2017). Moreover, studies implied that phloem-mediated K^+ transport via AtAKT2 possibly play important roles in salinity tolerance (Wu et al. 2018; Isayenkov and Maathuis 2019; Rubio et al. 2020). In rice, loss of function of *osakt2* mutant lines created by CRISPR/Cas9 exhibited reduced K^+ concentration in phloem and altered the K^+ shoot-to-root redistribution, causing increased sensitivity to salt stress. As a result, accumulation of Na^+ increased under salt stress conditions, indicating that OsAKT2 function in maintenance of Na^+/K^+ homeostasis by inhibiting the Na^+ uptake in roots (Tian et al. 2021). As a secondary ion transporters, Na^+/H^+ antiporters (NHXs) play critical roles in plant development and stress responses, especially salt stress, and K^+ homeostasis (Hernández et al. 2009; Leidi et al. 2010; Bassil et al. 2011). NHXs are categorized into three main classes due to their subcellular localization: plasma membrane (PM)-class, endosomal (Endo)-class, and vacuole (Vac)-class (Yokoi et al. 2002). Furthermore, NHX5 and NHX6 are members of the Endo-class NHXs localized in the *trans*-Golgi network, playing important roles in vesicle trafficking, especially to the vacuole (Bassil et al. 2011). Under salinity conditions, Na^+ is highly accumulated that Na^+ is actively extruded out of the cell by NHX7/SOS1 (PM-class NHX) system, and the sequestration of Na^+ is mediated by the Vac-class NHXs (Shi et al. 2002; Pardo et al. 2006; Apse and Blumwald 2007). The *nhx1 nhx2* double mutant in *Arabidopsis* exhibited higher sensitivity to external K^+ with curled root, poorly expanded and yellow leaves (Bassil et al. 2011; Barragán et al. 2012). In soybean, GmNHX5 is the member of Endo-class NHXs located in Golgi apparatus (Sun et al. 2019). Overexpression of *GmNHX5* in hairy roots enhanced the salt tolerance, whereas its CRISPR/Cas9 knockout mutants were more sensitive to salinity than wild type. Moreover, the expression of *GmSOS1*, *GmSKOR*, and *GmHKT1* genes increased in overexpressed *GmNHX5* seedlings showed higher K^+/Na^+ ratio and osmotic adjustment substances accumulation under salt stress (Sun et al. 2021b). Furthermore, Na^+ pumping activity of Vac-class NHXs is classified as two tonoplast protein pumps: H^+-pumping adenosine triphosphatase (H^+-ATPase) and H^+-pumping pyrophosphatase (H^+-PPase; Pasapula et al. 2011; Baisakh et al. 2012). In barley, *HVP10* encoding *vacuolar* H^+*-PPase* gene is mainly expressed in roots and involved in salt stress responses. Knockdown (RNAi) and CRISPR/Cas9 knockout barley plants of *HVP10* were created to explore the function under salt stress. Both knockdown and knockout of *HVP10* transgenic lines demonstrated inhibited growth with increased shoot Na^+ concentration, Na^+ transportation rate, and xylem Na^+ loading compared to wild type. The study indicated that HVP10 enhanced salt tolerance via promoting Na^+ sequestration into the root vacuole by acting synergistically with HvNHX1 and HvNHX4 (Fu et al. 2022). In 2022, two QTLs were observed to differentiate for total K^+ levels in Horkuch (salt tolerant donor) and IR29 (salt sensitive) at the seedling stage. Both the QTL intervals consist of a variety of transmembrane transporters, including vacuolar two pore potassium channel (*OsTPKa*) and HAK type transporter (*OsHAK_like*) (Haque et al. 2022). For further insight, precise mutations were introduced to *OsTPKa* and *OsHAK_like* in Horkuch via CRISPR/Cas9. Loss of function of *OsTPKa* and *OsHAK_like* resulted in reduced salinity tolerance, suggesting that they might have important roles in defending the rice landrace Horkuch against salt stress (Haque et al. 2023).

Polyploid plants such as wheat (*Triticum aestivum* L.) may have better adaptations to various environmental conditions that they usually exhibit enhanced tolerance to biotic and abiotic stresses compared to their diploid progenitors (Yang et al. 2014; Van de Peer et al. 2017; Wu et al. 2020). According to previous studies, *TaHKT1;5-D* from chromosome 4DL encoding high-affinity K^+ transporter limited Na^+ accumulation in hexaploidy bread wheat showing more salt tolerance than its tetraploid progenitor *Triticum turgidum* (Dvorak et al. 1994; Munns et al. 2012; Byrt et al. 2014). Nevertheless, salt tolerance mechanism is a complex trait that epigenetic modifications, including DNA/RNA methylation, histone modification, and chromatin remodeling, are additionally important in response to abiotic stress, such as drought and salinity (Chang et al. 2020). Histone acetylation, especially acetylation of lysine residues on histone tail, is one of these modifications, acting as a new key regulator factor to cope with abiotic stress conditions (Ueda et al. 2017; Li et al. 2019). Zheng et al. (2021) created CRISPR/Cas9 *histone acetyltransferase TaHAG1* mutant line to reveal the function of histone acetylation in salt tolerance regulation of polyploidy wheat. CRISPR/Cas9 *TaHAG1* mutant line exhibited excessive salt sensitivity, whereas overexpression of *TaHAG1* enhanced salinity tolerance by regulating ROS production and signal specificity. The study showed that overexpression of TaHAG1 triggered enhanced H3 acetylation and upregulation of a sunset of genes involved in hydrogen peroxide production under salinity (Zheng et al. 2021). A novel gene cluster containing ACQUIRED OSMOTOLERANCE (*ACQOS*; *AT5G46520*) and neighboring nucleotide-binding leucine-rich repeats (*NLRs*; *AT5G46510*) are paralogs that likely resulted from a recent duplication event that represents around 27% of genes in *Arabidopsis* (Kim et al. 2013). Plants harboring wild-type *ACQOS* alleles tend to be sensitive to salt stress, whereas plants containing full deletions exhibit considerable tolerance to salt stress, indicating that ACQOS may suppress osmotic tolerance (Ariga et al. 2017). To reveal the functions of *ACQOS* and *AT5G46510* in osmotic tolerance, Kim et al. (2021) introduced small insertion/deletion mutations into two target genes via CRISPR/Cas9. The study confirmed that *ACQOS* was directly related to salt tolerance, while the neighboring *AT5G46510* was not.

Protein phosphatase 2C (PP2C) is a protein phosphatase that is responsible for dephosphorylation of phosphorylated protein molecules involved in various signaling pathway, including growth, development, and ABA-mediated stress acclimation (Xue et al. 2008; Xiu-Yun and Tian 2017). In *Betula platyphylla, BpPP2C1* gene was silenced by CRISPR/Cas9 that higher salt sensitivity was observed in Knock-out *BpPP2C1* mutant lines. Additionally, transcriptome analysis exhibited that differentially expressed genes *in BpPP2C1* mutant lines were associated with abscisic acid signaling pathway, oxidative stress, flavonoid biosynthetic pathway, and anion transport (Xing et al. 2021). In another study, *SlHyPRP1* encoding hybrid proline-rich proteins (HyPRPs, putative plant cell wall glycoproteins) were edited in tomato by CRISPR/Cas9 (Tran et al. 2021). HyPRPs, containing a repetitive proline-rich N-terminal domain (PRD) and a conserved eight-cysteine motif (8CM) C-terminal domain, act as a positive regulator for cell death, but as negative regulator for biotic stress in pepper and tobacco (Yeom et al. 2012). When Li et al. (2016) suppressed *SlHyPRP1* via RNAi, they observed multi-stress (drought and salinity) tolerance without obvious developmental phenotype changes. Furthermore, Tran et al. (2021) conducted precise elimination of both domains (PRD and 8CM) via CRIPR/Cas9 system, and furthermore precise deletion of each domain resulted in salt stress-tolerant tomato lines. However, PRD and/or 8CM domains were contributed to salinity stress during the different stages of growth.

Plants have evolved sophisticated network of transcriptional and hormonal components against to changing environments. TFs such as WRKY regulate simultaneously multiple responses against to several abiotic stresses such as drought, salinity, and temperature fluctuations (Banerjee and Roychoudhury 2015; Ullah et al. 2018; Kiranmai et al. 2018). Previous studies indicated that WRKY TFs are induced by various signaling substances, including ABA, JA, and salicylic acid (SA) (Qiu et al. 2007; Tang et al. 2013; Ullah et al. 2018). In *Arabidopsis*, *AtWRKY3* and *AtWRKY4* were characterized in response to salt and methyl jasmonate (MeJa) by generating CRISPR/Cas9 knockout mutant lines. Silencing of *AtWRKY3* and *AtWRKY4* exhibited an attenuated tolerance of salt and MeJa, especially double mutant lines showed excessive salt and MeJa sensitivity with reduced activity of antioxidant enzymes and enhanced relative electrolyte leakage (Li et al. 2021). Furthermore, *AtWRKY66*, salt- and ABA-inducible transcription activator, was chosen for loss of function analysis using the CRISPR/Cas9. Knockout *AtWRKY66* mutant lines showed significantly lower activities of antioxidant enzymes than wild type plants under salt stress and ABA treatments, indicating serious oxidative damage may reduce the tolerance to salt stress. Consequently, *AtWRKY66* acted as a positive regulator in modulating salt tolerance through ABA signaling pathway (Zhang et al. 2023). Regulation of TFs is also one of the main mechanisms ensuring the stress tolerance by contributing the absorption, transportation, or compartmentation of Na^+ and K^+ (Chen and Zhu 2004). The plant basic-helix–loop–helix (bHLH) is one of the largest TFs involving salt stress tolerance (Guo et al. 2021). Previous studies showed that *OsbHLH035*, *OsbHLH062*, and *OsbHLH068* genes among bHLH family were found to be associated with salt stress response in rice (Wu et al. 2015; Chen et al. 2017a, 2018). However, the function of subfamily members of bHLH has not been uncovered yet. In 2022, Alam and colleagues investigated the function of rice *OsbHLH024* in response to salt stress, and therefore they created loss of function mutant *osbhlh024* by using CRISPR/Cas9 editing system. The mutant *osbhlh024* demonstrated enhanced salinity tolerance with less accumulation of ROS, MDA, and Na^+ but higher K^+. Additionally, ion transporter genes of *OsHKT1;3, OsHAK7*, and *OsSOS1* were highly expressed under salt stress conditions, indicating that *OsbHLH024* gene was presumably a negative regulator in response to salt stress (Alam et al. 2022a). However, another study by Alam et al. (2022b) demonstrated that other gene family member of bHLH *OsbHLH044* was specifically a positive regulator in response to salt stress. Similarly, CRISPR-Cas9-mediated gene-editing transgenic lines *osbhlh044* was generated to address the function of it under salt stress. The *osbhlh044* mutants exhibited lower antioxidant activities but enhanced lipid peroxidation and hydrogen peroxide (H_2O_2) accumulation causing the salinity sensitivity due to increased ROS accumulation. Moreover, higher Na^+ in both shoots and roots were observed to related to differentiated gene expression levels of ion homeostasis-related genes (*OsHKTs*, *OsHAK*, *OsSOSs*, and *OsNHX*) and ABA-responsive gene (*OsLEA3*) in mutant lines after exposing the salt stress. Moreover, genes coding for starch and seed storage proteins were significantly downregulated causing the grain chalkiness in the *osbhlh044* mutants. Salt- and starch-related genes were possibly activated by OsbHLH044 and was supported via Yeast one Hybrid approach (Alam et al. 2022b).

In plants, physiological responses are basically regulated by ABA, and therefore salinity stress increases the ABA levels by activating ABA biosynthetic genes, which in turn regulates the ABA-responsive genes (Verma et al. 2016; Vishwakarma et al. 2017). In chloroplasts, two key enzymes, violaxanthin de-epoxidase (VDE) and zeaxanthin epoxidase (ZEP), directly controlled the xanthophyll cycle and ABA biosynthesis play critical roles in growth and response to stress responses (Buch et al. 1995; Grzyb et al. 2006; Ye et al. 2012). In rice, *OsVDE* gene were overexpressed and gene edited via CRISPR/Cas9 by Wang et al. (2021b). Silencing of *OsVDE* resulted in the dwarfism, shorter panicle, and lower seed-setting rate than the wild type. However, overexpression of *OsVDE* did not show any difference from the wild type. Moreover, *OsVDE* mutants had enhanced ABA levels and survival rate in response to salinity. According to the RT–PCR analysis, *OsVDE* negatively regulated ABA biosynthesis by affecting the *OsNECD* TFs members, leading to increase in stomatal closure percentage in response to salinity (Wang et al. 2021b). In *Arabidopsis*, knockout of the entire family of *AITR* genes resulted in increased drought and salt stress tolerance (Tian et al. 2017; Chen et al. 2021). Furthermore, CRISPR-Cas9-mediated gene-editing *GmAITRs* enhanced salinity tolerance by changing the expression of ABA signaling key regulator genes, leading to ABA sensitivity in soybean (Wang et al. 2021a). As we mentioned previously in drought tolerance section, NAC TFs involve stress tolerance mechanisms, especially drought and salinity (Nakashima et al. 2012; Liu et al. 2022a). Wen et al. (2022) created CRISPR/Cas9 mutant lines of *NtNAC028* to reveal the function in stress tolerance mechanisms in tobacco. Silencing of *NtNAC028* caused delayed leaf senescence and enhanced drought and salt stress tolerance by leading to higher antioxidant enzyme activities, upregulation of expression of ROS-scavenging and abiotic stress-related genes. However, overexpression of *NtNAC028* in *Arabidopsis* showed early senescence and hypersensitivity to both drought and salinity, suggesting that *NtNAC028* acts as a negative regulator of senescence and abiotic stress tolerance (Wen et al. 2022).

Methylglyoxal (MG) is a reactive α-ketoaldehyde generated as a byproduct of glycolysis that MG causes oxidative damage by interacting with DNA, RNA, and proteins (Rai et al. 2021). MG is detoxified via glyoxalase pathway, including glyoxalase I (GLYI) and glyoxalase II (GLYII) enzymes, is responsible for reactive carbonyl species (RCS) scavenging (Ghosh et al. 2014; Manoj et al. 2019). In rice, a *GLY OsGLYI3* gene, were identified as a seed protein enhanced under natura storage conditions (Gao et al. 2016). Moreover, CRISPR/Cas9 knockout *OsGLYI3* mutant line exhibited enhanced MG levels, and greater sensitivity to salt stress, suggesting that *OsGLYI3* contributed to seed longevity and salinity tolerance (Liu et al. 2022b). In rice, *betaine aldehyde dehydrogenase 2, OsBadh2,* is a key gene associated with the synthesis of 2-acetyl-1-pyrroline (2AP) determining the rice grain is whether aromatic or non-aromatic (Baicharoen et al. 2018). Previous studies speculated that aroma genes, *OsBadh1* and *OsBadh2,* might play roles in abiotic stress tolerance (Fitzgerald et al. 2008; Singh et al. 2010). To reveal the relationship between aroma genes and salt tolerance, Prodhan et al. (2022) introduced precise mutations to *OsBadh2* via CRISPR/Cas9 in non-aromatic rice varieties. Consequently, knockout *OsBadh2* mutant lines became aromatic, showed enhanced salt tolerance, suggesting that more stress-tolerant aromatic rice lines might be generated from nonaromatic lines at further.

CRISPR/Cas9 as a newly emerging technique has been widely used to understand salt stress tolerance mechanisms to develop salt-tolerant plants (Di et al. 2019; Prodhan et al. 2022; Zhang et al. 2023). Targeted editing of salt tolerance-related genes via CRISPR/Cas9 has expanded insight in molecular breeding field. Integrating CRISPR/Cas9 technique with modern breeding approaches will contribute to exploring the multi-level nature of plant adaptations to salt stress.

10.4 Temperature Stress Tolerance

The optimal growth requires different temperatures during day and night, and over the whole growing season. Every plant species has its own optimal temperature conditions (Bratzel and Turck 2015). A fluctuation in optimal temperature causes significant reduction in the growth rate and grain yield by decreasing seed setting ratio and limiting various physiological processes (Waqas et al. 2021). Temperature stress is induced by a high- or low-temperature due to the duration of the exposure, the rate of temperature fluctuations, and the plant growth stage (Kai and Iba 2014). Especially, the reproductive stage of a plant is most sensitive to changing temperatures. Temperature stresses are categorized into two stages: 1) heat stress (short-term exposure to $> 37\,^{\circ}\mathrm{C}$ at vegetative stage, long-term exposure to $> 27\,^{\circ}\mathrm{C}$ at reproductive or vegetative stage), 2) cold stress (chilling [0–15 °C] and freezing [<0 °C]). However, plants have adapted several molecular mechanisms, including metabolites, antioxidants, proteins (such as heat shock proteins), membrane lipids and protectants, regulatory factors to cope with temperature stress (Kai and Iba 2014). Understanding the adaptation mechanisms for temperature fluctuations will give new insights to develop temperature stress-tolerant plants for agriculture.

10.4.1 Heat Stress Tolerance

All biological functions are dependent on optimum temperature that rising in temperature above threshold value is considered as heat stress causing irreversible damages to plants (Lippmann et al. 2019; Sarwar et al. 2019). In recent decades, global warming has accelerated the increase in air temperature potentiating the challenges for plant species (Lippman et al. 2019). Plants exposed to heat stress are negatively affected for several physiological processes, such as growth, development, photosynthesis, water availability, respiration, transpiration, and membrane stability (Zhao et al. 2021). Furthermore, heat stress causes disturbance on molecular and biochemical level promoting protein denaturation and accumulation of ROS (Hasanuzzaman et al. 2013). Heat stress is sensed via fluidity changes in the plasma membrane and heat-responsive cyclic nucleotide gated-ion channels, which use Ca^{2+} and ROS as messengers (Guihur et al. 2022). Three major families of Ca^{2+} sensors are calmodulins (CaMs), calcineurin B-like (CBL) proteins, and calcium-dependent protein kinases (CPKs). Among these Ca^{2+} sensors, CPKs function as Ca^{2+} responders that catalyze the direct phosphorylation of downstream target substrates (Boudsocq and Sheen 2013). To address the function of CPKs in plant thermotolerance, tomato *CPK28* knockout mutant lines were created via CRISPR/Cas9. The *cpk28* mutants exhibited reduced thermotolerance with increased ROS levels and protein oxidation. Interaction study demonstrated that CPK28 directly interacted with cytosolic ascorbate peroxidase 2 (APX2) that phosphorylated the APX2 at Thr-59 and Thr-164 may be promoted by Ca^{2+} stimuli (Hu et al. 2021).

Stress conditions induce the high level of ROS, which is highly toxic and involved in signal transduction pathway mediating stress response mechanisms (Torres 2010). To survive under stress conditions, maintaining of ROS homeostasis is critical to develop a variety of responses, including enzymatic and nonenzymatic antioxidants. Therefore, plant glutaredoxins (GRXs), belong to thioredoxins family, are components of the ROS scavenging network, functioning as oxidoreductases that use glutathione and its associated NADPH-dependent glutathione reductase (Mittler et al. 2004; Foyer and Noctor 2005; Dos Santos and Rey 2006; Li 2014). Kakeshpour et al. (2021) generated four GRXs (*SlGRXS14*, *SlGRXS15*, *SlGRXS16*, and *SlGRXS17*) mutant lines by using multiplex CRISPR/Cas9 system in tomato plants. All *Slgrxs* mutant lines exhibited more sensitive phenotype to several abiotic stresses than the wild-type tomatoes. Interestingly, *Slgrxs15* mutants were observed to be embryonic lethal. However, *Slgrxs14* and *Slgrx17* mutants demonstrated hypersensitivity to almost all stress, including heat stress and drought, whereas *Slgrxs16* mutant were mostly affected by chilling stresses (Kakeshpour et al. 2021). Another thioredoxins family member is nuclear-localized nucleoredoxin (NRX), which is a key component for the integration of the antioxidant system with the stress response (Marchal et al. 2014). In tomato, targeted mutations were introduced into *SlNRX1* via CRISPR/Cas9 system to investigate the functions in response to heat stress. Silencing of tomato *SlNRX1* resulted in reduced heat stress tolerance with higher electrolyte leakage and ROS accumulation than wild-type. Additionally, heat shock proteins (HSPs) and antioxidant enzymes were decreased in *slnrx1* mutant lines either in absence or presence or heat stress, indicating that SlNRX1 may act as a positive regulator in heat stress tolerance mechanism by increasing antioxidant capacity and inducing HSPs to protect cells against the protein denaturation (Cha et al. 2022).

NAC TFs are involved in various stress responses such as drought, salinity, and heat stress (Liu et al. 2020; Mao et al. 2022; Wen et al. 2022). Zhang et al. (2019c) indicated that two genes encoding NAC TFs specifically expressed in maize seeds, *ZmNAC128* and *ZmNAC130,* play important roles in starch and protein accumulation during grain filling in rice. Furthermore, Ren et al. (2021) selected NAC TFs *ZmNAC127* and *ZmNAC129* for overexpression and CRISPR/Cas9-mediated knockout studies to reveal their function in seed development and response to heat stress. Both knockout and overexpression of *zmnac127* and *zmnac129* mutant lines exhibited incomplete grain filling and shrunken grains that were deteriorated under heat stress. According to the transcriptome and proteomic analysis, ONAC127 and ONAC129 TFs regulated grain filling by mediating sugar transportation and stress responsive genes, including monosaccharide transporter gene *OsMST6*, sugar transporter gene *OsSWEET4*, calmodulin-like protein gene *OsMSR2,* and AP2/ERF factor gene *OsEATB* (Ren et al. 2021).

HSPs are a large family of molecular chaperones assisting the folding of newly synthesized proteins, protecting proteins from aggregation, and assisting re-folding of proteins to native state when they were denatured by heat stress (Heldt and Piechulla 2021). Moreover, HSPs also enhance the stability of membrane proteins such as the photosystems (Wu et al. 2022). Among HSPs, HSP70 is a superfamily assisted by a large chaperone system (Genevaux et al. 2007; Mayer 2010). Under stress conditions, binding of HSP70 to polypeptide chains requires ATP hydrolysis, which is mediated by J-proteins, components of HSP70 chaperone system (Wang et al. 2015). Bcl-2-associated athanogene (BAG) is a nucleotide exchange factors interacting with HSP70 on its ATPase domain to assist ATP binding and releasing ADP (Brive et al. 2001). Previous studies indicated that BAG family play critical roles in response to a variety of stresses, including heat, drought, and salinity (Pan et al. 2016; Arif et al. 2021; Zhou et al. 2021). However, function of BAG in thermotolerance of tomato is still unknown. Herein, Huang et al. (2022) generated overexpression and CRISPR/Ca9-mediated knockout *BAG9* mutant lines in tomato. Thermotolerance was reduced in *bag9* knockout mutant lines, whereas overexpressed *BAG9* plants showed enhanced thermotolerance. Interestingly, HSP20s were identified as BAG9-interacting proteins in the cytoplasm. Additionally, the accumulation of heat-induced HSPs and activities of antioxidant enzymes were decreased in *bag9* knockout mutant lines, indicating that BAG9 stimulated HSPs and cellular redox homeostasis for enhancing thermotolerance (Huang et al. 2022).

To alleviate heat stress effects, plants activate adaptive strategies leading to repression of gene expression, and only upregulating the expression of selected genes, including HSPs as molecular chaperones, heat shock transcription factors, and upstream transcription factors (Akerfelt et al. 2010). Most of the HSPs and heat shock TFs are alternately spliced and translated into truncated proteins localized to different subcellular compartments to act as positive or negative regulators in response to heat stress (Kim et al. 2017a; Wu et al. 2019; Zhang et al. 2020). In eukaryotes, splicing catalyzed by the spliceosome complex requires the precise splice sites. However, regulatory signals provide flexibility to spliceosome selecting for splice site (Shi 2017). The spliceosome complex consists of dynamic ribonucleoproteins (RNPs) and additional proteins (Wahl et al. 2009). Cyclophilin (CYP) family is one of the additional proteins associated with the spliceosome complex (Zhou et al. 2002). In *Arabidopsis*, *CYCLOPHILIN 18-2* (*AtCYP18-2*) was upregulated and interacted with spliceosome complex components, primarily associated with the small nuclear RNAs *U5* and *U6* under heat stress conditions. Loss of function mutants of *AtCYP18-2* was created via CRISPR/Cas9 by Lee et al. (2023). Consequently, silencing of *AtCYP18-2* resulted in higher sensitivity to heat stress via effecting the alternative splicing of heat-stress responsive genes, particularly causing intron retention (Lee et al. 2023).

Heat stress tolerance is a polygenic trait mediated by various complex signaling pathways. The exploration of genes involved in heat stress response requires advanced breeding tools such as CRISPR /Cas9 technology. Future studies will provide insight on how plants sense temperature and turn it into a signal to regulate transcription, posttranscription, or translational events for enhancing thermotolerance.

10.4.2 Cold Stress Tolerance

Temperature fluctuations affect the growth, fitness, and geographical distribution of plants that low as well as elevated temperatures reduce the yield and limit the natural occurrence (Powell et al. 2012; Jeon et al. 2013; Kreyling et al. 2015). Cold stress is basically categorized as either chilling stress (0–15 °C) or freezing stress (<0 °C) (Sanghera et al. 2011; Jin et al. 2017). Plants have evolved adaptation mechanisms that cold acclimation in plants is the increase of freezing tolerance during exposure to low but non-freezing temperatures (Hincha and Zuther 2014). Furthermore, injury mechanisms are completely different. Damage of chilling (or low temperature) is a direct effect of temperature leading to a series of biochemical and physiological changes in the cell. However, freezing damage is the result of osmotic dehydration generated by extracellular ice crystallization (Guy 1990; Hincha and Zuther 2014). Cold acclimation can be triggered by C-repeat/dehydration responsive element (DRE) binding factor (CBF/DREB)-dependent and -independent pathways that mediate the expression of *cold-responsive* (*COR*) genes also known as the CBF regulon (Thomashow 2010; Jia et al. 2016; Shi et al. 2018). Interestingly, *CBF* genes are almost always physically organized in tandem repeats present in many species such as wheat, barley, and pea (Tondelli et al. 2011; Beji et al. 2020). Three *CBF* (*CBF1*, *CBF2*, and *CBF3*) genes involved in cold acclimation are located on chromosome 4 and arranged in tandem repeats. In 2022, Géry and Téoulé created CRISPR/Cas9-mediated single knockout lines for *CBF1*, *CBF2*, and *CBF3*, double knockout lines for *CBF1* and *CBF3*, *CBF2* and *CBF3*, and triple knockout lines for *CBF1*, *CBF2*, and *CBF3* in various genetic backgrounds. Consequently, single knockout lines were slightly affected due to the expression of remaining *CBF* copies, whereas double or triple knockout lines exhibited excessive sensitivity in response to cold stress, indicating that *CBF* pathway play crucial roles in cold tolerance after acclimation (Géry and Téoulé 2022). In another study, *FvICE1* gene isolated from *Fragaria vesca "Hawaii 4"* that INDUCER OF CBF EXPRESSION (ICE) transcription factors regulate molecular mechanism of cold and drought stress tolerance in plants (Han et al. 2023). According to Chinnusamy et al. (2003), *ICE1* activate stress-relate genes by associating with MYC element in *CBF3/DREB1A* promoter. To reveal the function of *FvICE1*, its overexpression and CRISPR/cas9 vectors were constructed. The findings demonstrated that overexpressed *FvICE1* exhibited high tolerance to cold and drought in strawberries, although gene-editing *fvce1* had lower tolerance (Han et al. 2023).

The DNA binding with one finger (DOF) family is a plant-specific TFs playing critical roles in growth, development, as well as abiotic stress responses such as cold stress (da Silva et al. 2016; Molina-Hidalgo et al. 2017; Zhang et al. 2018c). In grapevine, *VaDof* gene family were identified and *Dof17d* gene was observed to be strongly induced by cold stress in grapes. The overexpression of *VaDof17d* resulted in increased cold tolerance, whereas CRISPR/Cas9 Knock-out lines exhibited reduced cold tolerance, indicating that *VaDof17d* acted as a positive regulator in response to cold stress (Wang et al. 2021c). Another TF family, GRAS TFs are called after the first three functionally characterized genes, *gibberellin insensitive* (*GAI*), *repressor* of *GA1* (*RGA*), and *scarecrow* (*SCR*) (Di Laurenzio et al. 1996; Peng et al. 1997; Silverstone et al. 1998). In plants, the GRAS TFs are involved in GA regulation and a variety of biological processes, including stress tolerance mechanisms (Richards et al. 2000; Sun et al. 2012; Liu et al. 2017). Among the GRAS TFs, PAT1 is significantly induced by cold stress in grapevines (Xin et al. 2013). The overexpression of grapevine *VaPAT1* increased cold tolerance in *Arabidopsis* (Yuan et al. 2016). However, how *VaPAT1* function in response to cold stress remained unknown. Therefore, Wang et al. (2021b) created both overexpression and CRISPR/Cas9-mediated knockout *VaPAT1* mutant lines. Cold tolerance was enhanced in overexpressed *VaPAT1* grape calli. Additionally, interaction study exhibited that VaPAT1 interacted with INDETERMINATE-DOMAIN 3 (VaIDD3). Both VaPAT1 and VaIDD3 coordinately regulated the stress-related genes expression, including *lipoxygenase 3* (*LOX3*) involved in jasmonate biosynthesis, which is possibly a downstream target gene of *VaPAT1*. On the contrary, silencing of *VaPAT1* resulted in the loss of the C-terminus dramatically inhibited the accumulation of *VaLOX3* and JA levels during cold stress (Wang et al. 2021d).

When plants are exposed to a variety of stresses, the expression of WRKY TFs are induced rapidly as a tissue-specific manner mediating the transcription of physiological processes (Banerjee and Roychoudhury 2015). WRKY TFs have been functionally studied in response to abiotic stress such as salinity, drought, and cold stress in rice that is highly sensitive to low temperatures (Nan et al. 2020; Huang et al. 2021; Viana et al. 2021; Tang et al. 2022). Zhang et al. (2022b) reported that OsWRKY63 negatively regulates chilling tolerance in rice. According to the overexpression and CRISPR/Cas9 knockout analysis, overexpressed *OsWRKY63* lines exhibited more sensitive phenotype in response to chilling, while *OsWRKY63* knockout mutants exhibited increased chilling tolerance. In addition to cold stress, the expression of *OsWRKY63* was also induced by dehydration stress, high salinity stress, and ABA treatment. Molecular and biochemical approaches indicated that OsWRKY63 located in the nucleus, acting as a transcription repressor by directly targeting the OsWRKY76 to downregulate many cold stress-related genes and reactive oxygen species scavenging-related genes. Furthermore, CRISPR/Cas9-mediating *OsWRKY76*-knockout lines showed excessive sensitivity to chilling stress. Interaction analysis revealed that OsbHLH148 was a target of OsWRKY76. In rice, OsbHLH148 induces the expression of OsDREB1B to increase chilling tolerance. The study suggested that OsWRKY63 negatively mediates chilling tolerance via the OsWRKY63-OsWRKY76-OsDREB1B transcriptional regulatory cascade in rice (Zhang et al. 2022b). Mitogen-activated protein kinase (MAPK) cascades play critical roles in signaling pathway modulating signals transferring through a relay of phosphorylation events (Zhang et al. 2021). A typical MAPK cascade consists of three types of protein kinases: MAPK, MAPK kinase (MKK), and MKK kinase (MKKK). Previous studies implied that MAPKs either positively or negatively regulate the cold tolerance by phosphorylating and stabilizing downstream proteins (Li et al. 2017c; Zhang et al. 2017; Zhang et al. 2018a). On further analysis, *osmkkk55*, *osmkkk62*, *osmkkk70*, *osmkkk62/70*, and *osmkkk55/62/70* mutant lines were created by CRISPR/Cas9 system. Double *osmkkk62/70* and triple *osmkkk55/62/70* mutants exhibited enhanced cold tolerance with a higher seed setting and pollen fertility. Under cold stress conditions, GA levels in anther were slightly increased in *osmkkk62/70* mutants, indicating that OsMKKK70 negatively mediated cold stress by fine-tuning GA levels in rice anthers (Mei et al. 2022).

Melatonin is considered as master regulator, which is an indole substance derived from serotonin mediating various physiological responses, including circadian rhythm, flowering organogenesis, photosynthesis, fruit ripening, reproduction, and plant acclimatization to changing environments (Akula and Mukherjee 2020; Sun et al. 2020). Previous study indicated that exogenous melatonin with Ca^{+2} alleviated chilling stress by enhancing photosynthetic and antioxidant activities (Pu et al. 2021). In watermelon, the positive effects of exogenous melatonin were also reported in response to various abiotic stress, including cold, salinity, and drought stress (Li et al. 2017a,b, 2020). Watermelon *ClCOMT1* gene involved in melatonin biosynthesis was characterized by creating overexpression and CRISPR/Cas9 knockout lines in both watermelon calli and *Arabidopsis*. *ClCOMT1* knockout lines exhibited reduced melatonin content, whereas melatonin levels increased in overexpressed *ClCOMT1* lines. Moreover, overexpression of *ClCOMT1* was increased in transgenic *Arabidopsis* lines in response to cold, drought, and salt stress, indicating that ClCOMT1 acted as a positive regulator in stress tolerance (Chang et al. 2021). In plants, ROS accumulation is a common phenomenon under stress conditions. To cope with ROS, plants synthesize ROS neutralizing substances that are mainly nonenzymatic and enzymatic antioxidants. Ascorbic acid (AsA) is one of the universal nonenzymatic antioxidants playing important roles as ROS scavenger under environmental stresses (Qian et al. 2014; Akram et al. 2017). A basic leucine zipper domain (bZIP) TF, AcePosF21 was observed to be involved in the regulation of AsA biosynthesis and cold stress-tolerance mechanisms. To reveal how cold stress elicits AsA biosynthesis, Liu et al. (2023) performed molecular and biochemical approaches, including silencing of *AcePosF21* via CRISPR/Cas9. Silencing of *AcePosF21* reduced AsA levels leading to increase the generation of ROS in kiwi fruit under cold stress. Moreover, AcePosF21 enhanced a key regulator *GDP-L-galactose phosphorylase 3* (*AceGGP3*) gene by interacting with MYB TF AceMYB102 to produce more AsA under cold stress (Liu et al. 2023).

Cold stress induces the compatible solutes called as osmolytes (such as proline and soluble sugars) protecting plant cells against cold stress (Huang et al. 2012). Soluble sugars, including sucrose, hexose, raffinose, glucose, fructose, and trehalose are known as osmoprotectants contributing to osmatic adjustment caused by cold stress (Liu et al. 2018). Many studies exhibited that overexpression of genes associated with soluble sugar accumulation enhances cold tolerance by providing carbon skeletons and energy for the synthesis of other compounds (Tarkowski et al. 2015; Wang et al. 2018; Ouyang et al. 2021). In plants, the starch is hydrolyzed by β-amylase (BAM) to generate maltose under cold stress conditions (Kaplan and Guy 2004; Sun et al. 2021a). BAM is a member of the glycosyl hydrolase family that hydrolyzes α-1,4-linked glucan chains, promoting starch degradation to enhance cold tolerance (Hou et al. 2017; Zhao et al. 2019). In grapevine, *VaBAM1* was characterized by overexpressing and editing using CRISPR/Cas9 approach in both *Arabidopsis* and grape calli. Under cold stress conditions, soluble sugar content and VaBAM1 activity were reduced in grape calli, whereas they were enhanced with high cold tolerance in both *VaBAM1*-overexpressed *Arabidopsis* and grape calli. As a result of interaction analysis, VaBAM1 interacted a serine/arginine-rich splicing factor 1 (VaSR1). Interestingly, the overexpression of *VaSR1* in grape calli decreased BAM activity with cold response genes, indicating that VaSR1 negatively regulated BAM activity leading to reduced cold tolerance in grapevine (Liang et al. 2023).

In plants, cold stress responses are regulated by a transcriptional cascade. The RING finger-containing E3 ubiquitin ligase high expression of osmotically responsive genes 1 (HOS1) functions as a negative regulator of the cold signaling mediating the ubiquitination and degradation of ICE1. HOS1 physically interacts with ICE1. The previous study indicated that cold stress stimulated the degradation of ICE1. However, disruption of ICE1 was inhibited in loss-of-function *hos1* mutants exhibited enhanced cold tolerance with upregulation of cold-responsive genes (Dong et al. 2006). Shkryl et al. (2021) introduced precise mutations into the first exon of *HOS1* via CRISPR/Cas9. CRISPR/Cas9-mediating *hos1* mutant lines were compared with the SALK T-DNA insertion mutant, line *hos1-3*, in terms of their tolerance to abiotic stresses. Both *hos1-3* and *hos1* mutant lines showed enhanced cold tolerance with increase of cold-responsive gene expressions. Additionally, phytoalexins synthesis altered in *hos1* mutant line that flavonol glycosides were upregulated, whereas glucosinolates content was downregulated, suggesting that HOS1 is a negative regulator of flavonoid biosynthesis (Shkryl et al. 2021). In *Brassica napus* L. (*B. napus*), silencing of *BnHOS1* via CRISPR/Cas9 also exhibited increased cold tolerance (Song et al. 2022). Recently, Gorpenchenko et al. (2023) generated CRISPR/Cas9- mediating *HOS1* knockout mutant lines in *Arabidopsis*. Therefore, *HOS1* knockout mutants demonstrated enhanced ROS accumulation under both light and high light conditions, whereas ROS levels were decreased in *HOS1* knockout mutants when plants exposed to cold stress. Under high light conditions, high basal level of AsA was observed in *HOS1* knockout mutants, generating a strong antioxidant response. Moreover, cold stress followed by high light exposure reduced ROS levels by increasing the expression of antioxidant enzyme genes accompanied by an increase in expression of NADPH oxidase genes *RbohD* and *RbohF*. ROS homeostasis maintained either cold-dependent and cold-independent in *HOS1* knockout mutants, suggesting that HOS1 inhibits the over-induction of defense mechanism to balance growth (Gorpenchenko et al. 2023).

Cold stress leads to reduction in plant growth, development, and productivity. It is essential to generate new cold-tolerant plants with greater adaptability to growth under cold stress conditions. Therefore, uncovering the underlying physiological and molecular mechanisms is crucial to improve cold stress tolerance that CRISPR/Cas9 can accelerate plant breeding.

10.5 Conclusions

The improvement of abiotic stress-tolerant plants is pivotal to sustain sufficient food supplies in a climate-changing world. Traditional plant breeding approaches are laborious and time consuming, requiring long-term effort to develop new tolerant plants. In contrast, CRISPR/Cas9 is easy and fast to design modifications of the genome in a precise and predictable manner. Recent years, CRISPR/Cas9 is mostly used to generate loss-of function and knockdown mutant lines to uncover abiotic stress tolerance mechanisms. However, abiotic stress tolerance is a complex trait due to multi-gene families, making it difficult to generate tolerant phenotype by knocking-out only a single gene that further improvement is needed.

References

Åkerfelt, M., Morimoto, R.I., and Sistonen, L. (2010). Heat shock factors: integrators of cell stress, development and lifespan. *Nature Reviews Molecular Cell Biology* 11 (8): 545–555. https://doi.org/10.1038/nrm2938.

Akram, N.A., Shafiq, F., and Ashraf, M. (2017). Ascorbic acid-a potential oxidant scavenger and its role in plant development and abiotic stress tolerance. *Frontiers in Plant Science* 8. https://doi.org/10.3389/fpls.2017.00613.

Akula, R. and Mukherjee, S. (2020). New insights on neurotransmitters signaling mechanisms in plants. *Plant Signaling & Behavior* 15 (6): 1737450. https://doi.org/10.1080/15592324.2020.1737450.

Alam, M.S., Kong, J., Tao, R. et al. (2022a). CRISPR/Cas9 mediated knock-out of the OsbHLH024 transcription factor improves salt stress resistance in rice (*Oryza sativa* L.). *Plants* 11 (9): 1184. https://doi.org/10.3390/plants11091184.

Alam, M.S., Yang, Z.-K., Li, C. et al. (2022b). Loss-of-function mutations of OsbHLH044 transcription factor lead to salinity sensitivity and a greater chalkiness in rice (*Oryza sativa* L.). *Plant Physiology and Biochemistry* 193: 110–123. https://doi.org/10.1016/j.plaphy.2022.10.033.

Alexandratos, N. and Bruinsma, J. (2012). *World Agriculture Towards 2030/2050: The 2012 Revision*, 12–03. Rome: FAO: ESA Working Paper No.

Alok, A., Kumar, J., and Upadhyay, S.K. (2018). Engineering in hairy roots using CRISPR/Cas9-mediated editing. In: *Hairy Roots* (ed. V. Srivastava, S. Mehrotra, and S. Mishra). Singapore: Springer. https://doi.org/10.1007/978-981-13-2562-5_15.

Alok, A., Chauhan, H., Upadhyay, S.K. et al. (2021). Compendium of plant-specific CRISPR vectors and their technical advantages. *Life* 11: 1021. https://doi.org/10.3390/life11101021.

Anderson, R., Bayer, P.E., and Edwards, D. (2020). Climate change and the need for agricultural adaptation. *Current Opinion in Plant Biology* 56: https://doi.org/10.1016/j.pbi.2019.12.006.

Ansari, W.A., Chandanshive, S.U., Bhatt, V. et al. (2020). Genome editing in cereals: approaches, applications and challenges. *International Journal of Molecular Sciences* 21 (11): 4040. https://doi.org/10.3390/ijms21114040.

Apse, M.P. and Blumwald, E. (2007). Na^+ transport in plants. *FEBS Letters* 581 (12): 2247–2254. https://doi.org/10.1016/j.febslet.2007.04.014.

Arif, M., Li, Z., Luo, Q. et al. (2021). The BAG2 and BAG6 genes are involved in multiple abiotic stress tolerances in *Arabidopsis thaliana*. *International Journal of Molecular Sciences* 22 (11): 5856. https://doi.org/10.3390/ijms22115856.

Ariga, H., Katori, T., Tsuchimatsu, T. et al. (2017). NLR locus-mediated trade-off between abiotic and biotic stress adaptation in *Arabidopsis*. *Nature Plants* 3 (6): https://doi.org/10.1038/nplants.2017.72.

Baicharoen, A., Vijayan, R., and Pongprayoon, P. (2018). Structural insights into betaine aldehyde dehydrogenase (BADH2) from *Oryza sativa* explored by modeling and simulations. *Scientific Reports* 8 (1): https://doi.org/10.1038/s41598-018-31204-z.

Baisakh, N., RamanaRao, M.V., Rajasekaran, K. et al. (2012). Enhanced salt stress tolerance of rice plants expressing a *vacuolar H+-ATPase subunit c1* (*SaVHAc1*) gene from the halophyte grass *Spartina alterniflora* Löisel. *Plant Biotechnology Journal* 10 (4): 453–464. https://doi.org/10.1111/j.1467-7652.2012.00678.x.

Banerjee, A. and Roychoudhury, A. (2015). WRKY proteins: signaling and regulation of expression during abiotic stress responses. *The Scientific World Journal* 2015 (807560): 1–17. https://doi.org/10.1155/2015/807560.

Bang, S.W., Choi, S., Jin, X. et al. (2022). Transcriptional activation of rice *CINNAMOYL-CoA REDUCTASE 10* by OsNAC5, contributes to drought tolerance by modulating lignin accumulation in roots. *Plant Biotechnology Journal* 20 (4): 736–747. https://doi.org/10.1111/pbi.13752.

Barragán, V., Leidi, E.O., Andrés, Z. et al. (2012). Ion exchangers NHX1 and NHX2 mediate active potassium uptake into vacuoles to regulate cell turgor and stomatal function in *Arabidopsis*. *The Plant Cell* 24 (3): 1127–1142. https://doi.org/10.1105/tpc.111.095273.

Barrangou, R., Fremaux, C., Deveau, H. et al. (2007). CRISPR provides acquired resistance against viruses in prokaryotes. *Science* 315 (5819): 1709–1712. https://doi.org/10.1126/science.1138140.

Bassil, E., Ohto, M., Esumi, T. et al. (2011). The *Arabidopsis* intracellular Na^+/H^+ antiporters NHX5 and NHX6 are endosome associated and necessary for plant growth and development. *The Plant Cell* 23 (1): 224–239. https://doi.org/10.1105/tpc.110.079426.

Beji, S., Fontaine, V., Devaux, R. et al. (2020). Genome-wide association study identifies favorable SNP alleles and candidate genes for frost tolerance in pea. *BMC Genomics* 21 (1): https://doi.org/10.1186/s12864-020-06928-w.

Berry, J.A., Beerling, D.J., and Franks, P.J. (2010). Stomata: key players in the earth system, past and present. *Current Opinion in Plant Biology* 13 (3): 232–239. https://doi.org/10.1016/j.pbi.2010.04.013.

Bhalothia, P., Yajnik, K., Alok, A., and Upadhayay, S.K. (2020). The current progress of CRISPR/Cas9 development in plants. In: *Genome Engineering via CRISPR-Cas9 Systems* (ed. S. Dhar). Elsevier. https://doi.org/10.1016/B978-0-12-818140-9.00011-8.

Bo, W., Zhaohui, Z., Huanhuan, Z. et al. (2019). Targeted mutagenesis of NAC transcription factor gene, *OsNAC041*, leading to salt sensitivity in rice. *Rice Science* 26 (2): 98–108. https://doi.org/10.1016/j.rsci.2018.12.005.

Boudsocq, M. and Sheen, J. (2013). CDPKs in immune and stress signaling. *Trends in Plant Science* 18 (1): 30–40. https://doi.org/10.1016/j.tplants.2012.08.008.

Bratzel, F. and Turck, F. (2015). Molecular memories in the regulation of seasonal flowering: from competence to cessation. *Genome Biology* 16 (1): https://doi.org/10.1186/s13059-015-0770-6.

Breseghello, F. and Coelho, A.S.G. (2013). Traditional and modern plant breeding methods with examples in rice (*Oryza sativa* L.). *Journal of Agricultural and Food Chemistry* 61 (35): 8277–8286. https://doi.org/10.1021/jf305531j.

Brive, L., Takayama, S., Briknarová, K. et al. (2001). The carboxyl-terminal lobe of Hsc70 ATPase domain is sufficient for binding to BAG1. *Biochemical and Biophysical Research Communications* 289 (5): 1099–1105. https://doi.org/10.1006/bbrc.2001.6087.

Büch, K., Stransky, H., and Hager, A. (1995). FAD is a further essential cofactor of the NAD(P)H and O2-dependent zeaxanthin-epoxidase. *FEBS Letters* 376 (1-2): 45–48. https://doi.org/10.1016/0014-5793(95)01243-9.

Byrt, C.S., Xu, B., Krishnan, M. et al. (2014). The Na^+ transporter, TaHKT1;5-D, limits shoot Na^+ accumulation in bread wheat. *The Plant Journal* 80 (3): 516–526. https://doi.org/10.1111/tpj.12651.

Cha, J.-Y., Ahn, G., Jeong, S.Y. et al. (2022). Nucleoredoxin 1 positively regulates heat stress tolerance by enhancing the transcription of antioxidants and heat-shock proteins in tomato. *Biochemical and Biophysical Research Communications* 635: 12–18. https://doi.org/10.1016/j.bbrc.2022.10.033.

Chan, Z., Yokawa, K., Kim, W.-Y., and Song, C.-P. (2016). Editorial: ROS regulation during plant abiotic stress responses. *Frontiers in Plant Science* 7: https://doi.org/10.3389/fpls.2016.01536.

Chandrasekaran, J., Brumin, M., Wolf, D. et al. (2016). Development of broad virus resistance in non-transgenic cucumber using CRISPR/Cas9 technology. *Molecular Plant Pathology* 17 (7): 1140–1153. https://doi.org/10.1111/mpp.12375.

Chang, Y., Zhu, C., Jiang, J. et al. (2020). Epigenetic regulation in plant abiotic stress responses. *Journal of Integrative Plant Biology* 62 (5): 563–580. https://doi.org/10.1111/jipb.12901.

Chang, J., Guo, Y., Yan, J. et al. (2021). The role of watermelon caffeic acid O-methyltransferase (ClCOMT1) in melatonin biosynthesis and abiotic stress tolerance. *Horticulture Research* 8 (1): https://doi.org/10.1038/s41438-021-00645-5.

Chanroj, S., Wang, G., Venema, K. et al. (2012). Conserved and diversified gene families of monovalent Cation/H^+ Antiporters from algae to flowering plants. *Frontiers in Plant Science* 3: https://doi.org/10.3389/fpls.2012.00025.

Chen, W.J. and Zhu, T. (2004). Networks of transcription factors with roles in environmental stress response. *Trends in Plant Science* 9 (12): 591–596. https://doi.org/10.1016/j.tplants.2004.10.007.

Chen, H.-C., Hsieh-Feng, V., Liao, P.-C. et al. (2017a). The function of OsbHLH068 is partially redundant with its homolog, AtbHLH112, in the regulation of the salt stress response but has opposite functions to control flowering in *Arabidopsis*. *Plant Molecular Biology* 94 (4–5): 531–548. https://doi.org/10.1007/s11103-017-0624-6.

Chen, X., Lu, X., Shu, N. et al. (2017b). GhSOS1, a plasma membrane Na^+/H^+ antiporter gene from upland cotton, enhances salt tolerance in transgenic *Arabidopsis thaliana*. *PLoS One* 12 (7): e0181450. https://doi.org/10.1371/journal.pone.0181450.

Chen, H.-C., Cheng, W.-H., Hong, C.-Y. et al. (2018). The transcription factor OsbHLH035 mediates seed germination and enables seedling recovery from salt stress through ABA-dependent and ABA-independent pathways, respectively. *Rice* 11 (1): https://doi.org/10.1186/s12284-018-0244-z.

Chen, S., Zhang, N., Zhou, G. et al. (2021). Knock-out of the entire family of AITR genes in *Arabidopsis* leads to enhanced drought and salinity tolerance without fitness costs. *BMC Plant Biology* 21 (1): https://doi.org/10.1186/s12870-021-02907-9.

Chinnusamy, V. (2003). ICE1: a regulator of cold-induced transcriptome and freezing tolerance in *Arabidopsis*. *Genes & Development* 17 (8): 1043–1054. https://doi.org/10.1101/gad.1077503.

Choi, J., Tanaka, K., Cao, Y. et al. (2014). Identification of a plant receptor for extracellular ATP. *Science* 343 (6168): 290–294. https://doi.org/10.1126/science.343.6168.290.

Collard, B.C.Y. and Mackill, D.J. (2008). Marker-assisted selection: an approach for precision plant breeding in the twenty-first century. *Philosophical Transactions of the Royal Society, B: Biological Sciences* 363 (1491): 557–572. https://doi.org/10.1098/rstb.2007.2170.

Cong, L., Ran, F.A., Cox, D. et al. (2013). Multiplex genome engineering using CRISPR/Cas systems. *Science* 339 (6121): 819–823. https://doi.org/10.1126/science.1231143.

Cramer, G.R., Urano, K., Delrot, S. et al. (2011). Effects of abiotic stress on plants: a systems biology perspective. *BMC Plant Biology* 11 (1): 163. https://doi.org/10.1186/1471-2229-11-163.

Debbarma, J., Sarki, Y.N., Saikia, B. et al. (2019). Ethylene response factor (ERF) family proteins in abiotic stresses and CRISPR-Cas9 genome editing of ERFs for multiple abiotic stress tolerance in crop plants: a review. *Molecular Biotechnology* 61 (2): 153–172. https://doi.org/10.1007/s12033-018-0144-x.

Deltcheva, E., Chylinski, K., Sharma, C.M. et al. (2011). CRISPR RNA maturation by trans-encoded small RNA and host factor RNase III. *Nature* 471 (7340): 602–607. https://doi.org/10.1038/nature09886.

Demidchik, V., Cuin, T.A., Svistunenko, D.A. et al. (2010). *Arabidopsis* root K^+-efflux conductance activated by hydroxyl radicals: single-channel properties, genetic basis and involvement in stress-induced cell death. *Journal of Cell Science* 123: 1468–1479. https://doi.org/10.1242/jcs.064352.

Di, Y.-H., Sun, X.-J., Hu, Z. et al. (2019). Enhancing the CRISPR/Cas9 system based on multiple GmU6 promoters in soybean. *Biochemical and Biophysical Research Communications* 519 (4): 819–823. https://doi.org/10.1016/j.bbrc.2019.09.074.

Dong, C.-H., Agarwal, M., Zhang, Y. et al. (2006). The negative regulator of plant cold responses, HOS1, is a RING E3 ligase that mediates the ubiquitination and degradation of ICE1. *Proceedings of the National Academy of Sciences of the United States of America* 103 (21): 8281–8286. https://doi.org/10.1073/pnas.0602874103.

Dreyer, I., Gomez-Porras, J.L., and Riedelsberger, J. (2017). The potassium battery: a mobile energy source for transport processes in plant vascular tissues. *New Phytologist* 216: 1049–1053. https://doi.org/10.1111/nph.14667.

Driedonks, N., Rieu, I., and Vriezen, W.H. (2016). Breeding for plant heat tolerance at vegetative and reproductive stages. *Plant Reproduction* 29 (1–2): 67–79. https://doi.org/10.1007/s00497-016-0275-9.

Dvořak, J., Noaman, M.M., Goyal, S., and Gorham, J. (1994). Enhancement of the salt tolerance of *Triticum turgidum* L. by the Kna1 locus transferred from the *Triticum aestivum* L. chromosome 4D by homoeologous recombination. *Theoretical and Applied Genetics* 87 (7): 872–877. https://doi.org/10.1007/bf00221141.

Fang, Y. and Xiong, L. (2015). General mechanisms of drought response and their application in drought resistance improvement in plants. *Cellular and Molecular Life Sciences* 72 (4): 673–689. https://doi.org/10.1007/s00018-014-1767-0.

Feng, X., Xiong, J., Zhang, W. et al. (2022). *ZmLBD5*, a class-II *LBD* gene, negatively regulates drought tolerance by impairing abscisic acid synthesis. *The Plant Journal* 112 (6): 1364–1376. https://doi.org/10.1111/tpj.16015.

Fiaz, S., Wang, X., Khan, S.A. et al. (2021). Novel plant breeding techniques to advance nitrogen use efficiency in rice: a review. *GM Crops & Food* 12 (2): 627–646. https://doi.org/10.1080/21645698.2021.1921545.

Fita, A., Rodríguez-Burruezo, A., Boscaiu, M. et al. (2015). Breeding and domesticating crops adapted to drought and salinity: a new paradigm for increasing food production. *Frontiers in Plant Science* 6: https://doi.org/10.3389/fpls.2015.00978.

Fitzgerald, T.L., Waters, D.L.E., and Henry, R.J. (2008). The effect of salt on betaine aldehyde dehydrogenase transcript levels and 2-acetyl-1-pyrroline concentration in fragrant and non-fragrant rice (*Oryza sativa*). *Plant Science* 175 (4): 539–546. https://doi.org/10.1016/j.plantsci.2008.06.005.

Foyer, C.H. and Noctor, G. (2005). Redox homeostasis and antioxidant signaling: a metabolic interface between stress perception and physiological responses. *The Plant Cell* 17 (7): 1866–1875. https://doi.org/10.1105/tpc.105.033589.

Fu, L., Wu, D., Zhang, X. et al. (2022). Vacuolar H^+-pyrophosphatase HVP10 enhances salt tolerance via promoting Na^+ translocation into root vacuoles. *Plant Physiology* 188 (2): 1248–1263. https://doi.org/10.1093/plphys/kiab538.

Gajdanowicz, P., Michard, E., Sandmann, M. et al. (2011). Potassium (K^+) gradients serve as a mobile energy source in plant vascular tissues. *Proceedings of the National Academy of Sciences of the United States of America* 108 (2): 864–869. https://doi.org/10.1073/pnas.1009777108.

Gao, J., Fu, H., Zhou, X. et al. (2016). Comparative proteomic analysis of seed embryo proteins associated with seed storability in rice (*Oryza sativa* L) during natural aging. *Plant Physiology and Biochemistry* 103: 31–44. https://doi.org/10.1016/j.plaphy.2016.02.026.

Gasiunas, G., Barrangou, R., Horvath, P., and Siksnys, V. (2012). Cas9-crRNA ribonucleoprotein complex mediates specific DNA cleavage for adaptive immunity in bacteria. *Proceedings of the National Academy of Sciences of the United States of America* 109 (39): E2579–E2586. https://doi.org/10.1073/pnas.1208507109.

Genevaux, P., Georgopoulos, C., and Kelley, W.L. (2007). The Hsp70 chaperone machines of *Escherichia coli:* a paradigm for the repartition of chaperone functions. *Molecular Microbiology* 66 (4): 840–857. https://doi.org/10.1111/j.1365-2958.2007.05961.x.

Géry, C. and Téoulé, E. (2022). Cold acclimation diversity in *Arabidopsis thaliana*: CRISPR/Cas9 as a tool to fine analysis of tandem gene arrays, application to CBF genes. *Development Genes and Evolution* 232 (5–6): 147–154. https://doi.org/10.1007/s00427-022-00693-4.

Ghosh, A., Pareek, A., Sopory, S.K., and Singla-Pareek, S.L. (2014). A glutathione responsive rice glyoxalase II, OsGLYII-2, functions in salinity adaptation by maintaining better photosynthesis efficiency and anti-oxidant pool. *The Plant Journal* 80 (1): 93–105. https://doi.org/10.1111/tpj.12621.

Gong, S., Ding, Y., Hu, S. et al. (2019). The role of HD-Zip class I transcription factors in plant response to abiotic stresses. *Physiologia Plantarum* 167 (4): 516–525. https://doi.org/10.1111/ppl.12965.

Gorpenchenko, T.Y., Veremeichik, G.N., Shkryl, Y.N. et al. (2023). Suppression of the HOS1 gene affects the level of ROS depending on light and cold. *Life* 13 (2): 524. https://doi.org/10.3390/life13020524.

Grzyb, J., Latowski, D., and Strzałka, K. (2006). Lipocalins – a family portrait. *Journal of Plant Physiology* 163 (9): 895–915. https://doi.org/10.1016/j.jplph.2005.12.007.

Gu, H., Zhang, K., Chen, J. et al. (2022). *OsFTL4*, an *FT-like* Gene, regulates flowering time and drought tolerance in rice (*Oryza sativa* L.). *Rice* 15 (1): https://doi.org/10.1186/s12284-022-00593-1.

Guihur, A., Rebeaud, M.E., and Goloubinoff, P. (2022). How do plants feel the heat and survive? *Trends in Biochemical Sciences* 47 (10): 824–838. https://doi.org/10.1016/j.tibs.2022.05.004.

Guo, J., Sun, B., He, H. et al. (2021). Current understanding of bHLH transcription factors in plant abiotic stress tolerance. *International Journal of Molecular Sciences* 22 (9): 4921. https://doi.org/10.3390/ijms22094921.

Guy, C.L. (1990). Cold acclimation and freezing stress tolerance: role of protein metabolism. *Annual Review of Plant Physiology and Plant Molecular Biology* 41 (1): 187–223. https://doi.org/10.1146/annurev.pp.41.060190.001155.

Han, J., Li, X., Li, W. et al. (2023). Isolation and preliminary functional analysis of *FvICE1*, involved in cold and drought tolerance in *Fragaria vesca* through overexpression and CRISPR/Cas9 technologies. *Plant Physiology and Biochemistry* 196: 270–280. https://doi.org/10.1016/j.plaphy.2023.01.048.

Haque, T., Elias, S.M., Razzaque, S. et al. (2022). Salt tolerance QTLs of an endemic rice landrace, Horkuch at seedling and reproductive stages. *Scientific Reports* 12 (1): https://doi.org/10.1038/s41598-022-21737-9.

Haque, U.S., Elias, S.M., Jahan, I., and Seraj, Z.I. (2023). Functional genomic analysis of K^+ related salt-responsive transporters in tolerant and sensitive genotypes of rice. *Frontiers in Plant Science* 13: https://doi.org/10.3389/fpls.2022.1089109.

Hasanuzzaman, M., Nahar, K., Alam, M. et al. (2013). Physiological, biochemical, and molecular mechanisms of heat stress tolerance in plants. *International Journal of Molecular Sciences* 14 (5): 9643–9684. https://doi.org/10.3390/ijms14059643.

Hauser, F., Waadt, R., and Schroeder, J.I. (2011). Evolution of abscisic acid synthesis and signaling mechanisms. *Current Biology* 21 (9): R346–R355. https://doi.org/10.1016/j.cub.2011.03.015.

Heldt, H.-W. and Piechulla, B. (2021). Biosynthesis, processing, and degradation of plant proteins. In: *Plant Biochemistry* (ed. H.-W. Heldt and B. Piechulla), 503–531. Academic Press. https://doi.org/10.1016/b978-0-12-818631-2.00021-0.

Hernández, A., Jiang, X., Cubero, B. et al. (2009). Mutants of the *Arabidopsis thaliana* Cation/H^+ antiporter AtNHX1 conferring increased salt tolerance in yeast. *Journal of Biological Chemistry* 284 (21): 14276–14285. https://doi.org/10.1074/jbc.m806203200.

Hincha, D.K. and Zuther, E. (2014). Introduction: plant cold acclimation and freezing tolerance. In: *Methods in Molecular Biology* (ed. D.K. Hincha and E. Zuther), 1–6. New York: Humana Press. https://doi.org/10.1007/978-1-4939-0844-8_1.

Hou, J., Zhang, H., Liu, J. et al. (2017). Amylases StAmy23, StBAM1 and StBAM9 regulate cold-induced sweetening of potato tubers in distinct ways. *Journal of Experimental Botany* 68 (9): 2317–2331. https://doi.org/10.1093/jxb/erx076.

Hu, Z., Li, J., Ding, S. et al. (2021). The protein kinase CPK28 phosphorylates ascorbate peroxidase and enhances thermotolerance in tomato. *Plant Physiology* 186 (2): 1302–1317. https://doi.org/10.1093/plphys/kiab120.

Huang, J., Sun, S., Xu, D. et al. (2012). A TFIIIA-type zinc finger protein confers multiple abiotic stress tolerances in transgenic rice (*Oryza sativa* L.). *Plant Molecular Biology* 80 (3): 337–350. https://doi.org/10.1007/s11103-012-9955-5.

Huang, H., Ullah, F., Zhou, D.-X. et al. (2019). Mechanisms of ROS regulation of plant development and stress responses. *Frontiers in Plant Science* 10. https://doi.org/10.3389/fpls.2019.00800.

Huang, K., Wu, T., Ma, Z. et al. (2021). Rice transcription factor OsWRKY55 is involved in the drought response and regulation of plant growth. *International Journal of Molecular Sciences* 22 (9): 4337. https://doi.org/10.3390/ijms22094337.

Huang, H., Liu, C., Yang, C. et al. (2022). BAG9 confers thermotolerance by regulating cellular redox homeostasis and the stability of heat shock proteins in *Solanum lycopersicum*. *Antioxidants* 11 (8): 1467. https://doi.org/10.3390/antiox11081467.

Hussain, F., Rafiq, M., Ghias, M. et al. (2017). Genetic diversity for seed yield and its components using principal component and cluster analysis in sunflower (*Helianthus annuus* L.). *Life Science Journal* 14 (5): 71–78. https://doi.org/10.7537/marslsj140517.10.

Isayenkov, S.V. and Maathuis, F.J.M. (2019). Plant salinity stress: many unanswered questions remain. *Frontiers in Plant Science* 10: https://doi.org/10.3389/fpls.2019.00080.

Izawa, T. (2007). Adaptation of flowering-time by natural and artificial selection in *Arabidopsis* and rice. *Journal of Experimental Botany* 58 (12): 3091–3097. https://doi.org/10.1093/jxb/erm159.

Jeon, J. and Kim, J. (2013). Cold stress signaling networks in *Arabidopsis*. *Journal of Plant Biology* 56 (2): 69–76. https://doi.org/10.1007/s12374-013-0903-y.

Jia, Y., Ding, Y., Shi, Y. et al. (2016). The cbfs triple mutants reveal the essential functions of CBFs in cold acclimation and allow the definition of CBF regulons in *Arabidopsis*. *New Phytologist* 212 (2): 345–353. https://doi.org/10.1111/nph.14088.

Jiang, Y., Ling, L., Zhang, L. et al. (2018). Comparison of transgenic Bt rice and their non-Bt counterpart in yield and physiological response to drought stress. *Field Crops Research* 217: 45–52. https://doi.org/10.1016/j.fcr.2017.12.007.

Jin, J., Zhang, H., Zhang, J. et al. (2017). Integrated transcriptomics and metabolomics analysis to characterize cold stress responses in *Nicotiana tabacum*. *BMC Genomics* 18 (1): https://doi.org/10.1186/s12864-017-3871-7.

Jinek, M., Chylinski, K., Fonfara, I. et al. (2012). A programmable dual-RNA-guided DNA endonuclease in adaptive bacterial immunity. *Science* 337 (6096): 816–821. https://doi.org/10.1126/science.1225829.

Jinek, M., Jiang, F., Taylor, D.W. et al. (2014). Structures of Cas9 endonucleases reveal RNA-mediated conformational activation. *Science* 343 (6176): 1247997–1247997. https://doi.org/10.1126/science.1247997.

Kai, H. and Iba, K. (2014). *Temperature Stress in Plants. eLS*. Chichester: Wiley. https://doi.org/10.1002/9780470015902.a0001320.pub2.

Kakeshpour, T., Tamang, T.M., Motolai, G. et al. (2021). CGFS-type glutaredoxin mutations reduce tolerance to multiple abiotic stresses in tomato. *Physiologia Plantarum* 173 (3): 1263–1279. https://doi.org/10.1111/ppl.13522.

Kaplan, F. (2004). Amylase induction and the protective role of maltose during temperature shock. *Plant Physiology* 135 (3): 1674–1684. https://doi.org/10.1104/pp.104.040808.

Khush, G.S. (2013). Strategies for increasing the yield potential of cereals: case of rice as an example. *Plant Breeding* 132: 433–436. https://doi.org/10.1111/pbr.1991.

Kim, G.-D., Cho, Y.-H., Lee, B.-H., and Yoo, S.-D. (2017a). STABILIZED1 modulates pre-mRNA splicing for thermotolerance. *Plant Physiology* 173 (4): 2370–2382. https://doi.org/10.1104/pp.16.01928.

Kim, H., Kim, S.-T., Ryu, J. et al. (2017b). CRISPR/Cpf1-mediated DNA-free plant genome editing. *Nature Communications* 8 (1): https://doi.org/10.1038/ncomms14406.

Kim, S.-T., Choi, M., Bae, S.-J., and Kim, J.-S. (2021). The functional association of ACQOS/VICTR with salt stress resistance in *Arabidopsis thaliana* was confirmed by CRISPR-mediated mutagenesis. *International Journal of Molecular Sciences* 22 (21): 11389. https://doi.org/10.3390/ijms222111389.

Kiranmai, K., Lokanadha Rao, G., Pandurangaiah, M. et al. (2018). A novel WRKY transcription factor, MuWRKY3 (*Macrotyloma uniflorum* Lam. Verdc.) enhances drought stress tolerance in transgenic groundnut (*Arachis hypogaea* L.) plants. *Frontiers in Plant Science* 9: https://doi.org/10.3389/fpls.2018.00346.

Kreyling, J., Schmid, S., and Aas, G. (2015). Cold tolerance of tree species is related to the climate of their native ranges. *Journal of Biogeography* 42 (1): 156–166. https://doi.org/10.1111/jbi.12411.

Laurenzio, L.D., Wysocka-Diller, J., Malamy, J.E. et al. (1996). The SCARECROW gene regulates an asymmetric cell division that is essential for generating the radial organization of the *Arabidopsis* root. *Cell* 86 (3): 423–433. https://doi.org/10.1016/S0092-8674(00)80115-4.

Lee, A., Park, H.J., Jo, S.H. et al. (2023). The spliceophilin CYP18-2 is mainly involved in the splicing of retained introns under heat stress *Arabidopsis*. *Journal of Integrative Plant Biology* https://doi.org/10.1111/jipb.13450.

Leidi, E.O., Barragán, V., Rubio, L. et al. (2010). The AtNHX1 exchanger mediates potassium compartmentation in vacuoles of transgenic tomato. *The Plant Journal* 61 (3): 495–506. https://doi.org/10.1111/j.1365-313x.2009.04073.x.

Li, S. (2014). Redox modulation matters: emerging functions for glutaredoxins in plant development and stress responses. *Plants* 3 (4): 559–582. https://doi.org/10.3390/plants3040559.

Li, S., Gao, F., Xie, K. et al. (2016). The OsmiR396c-OsGRF4-OsGIF1 regulatory module determines grain size and yield in rice. *Plant Biotechnology Journal* 14 (11): 2134–2146. https://doi.org/10.1111/pbi.12569.

Li, H., Chang, J., Chen, H. et al. (2017a). Exogenous melatonin confers salt stress tolerance to watermelon by improving photosynthesis and redox homeostasis. *Frontiers in Plant Science* 8: 295. https://doi.org/10.3389/fpls.2017.00295.

Li, H., Chang, J., Zheng, J. et al. (2017b). Local melatonin application induces cold tolerance in distant organs of *Citrullus lanatus* L. via long distance transport. *Scientific Reports* 7 (1): 40858. https://doi.org/10.1038/srep40858.

Li, H., Ding, Y., Shi, Y. et al. (2017c). MPK3- and MPK6-mediated ICE1 phosphorylation negatively regulates ICE1 stability and freezing tolerance in *Arabidopsis*. *Developmental Cell* 43 (5): 630–642. e4https://doi.org/10.1016/j.devcel.2017.09.025.

Li, R., Zhang, L., Wang, L. et al. (2018a). Reduction of tomato-plant chilling tolerance by CRISPR–Cas9-mediated SlCBF1 mutagenesis. *Journal of Agricultural and Food Chemistry* 66 (34): 9042–9051. https://doi.org/10.1021/acs.jafc.8b02177.

Li, S., Lin, Y.-C.J., Wang, P. et al. (2018b). The AREB1 transcription factor influences histone acetylation to regulate drought responses and tolerance in *Populus trichocarpa*. *The Plant Cell* 31 (3): 663–686. https://doi.org/10.1105/tpc.18.00437.

Li, R., Liu, C., Zhao, R. et al. (2019). CRISPR/Cas9-mediated SlNPR1 mutagenesis reduces tomato plant drought tolerance. *BMC Plant Biology* 19 (1): https://doi.org/10.1186/s12870-018-1627-4.

Li, H., Guo, Y., Cui, Q. et al. (2020). Alkanes (C_{29} and C_{31})-mediated intracuticular wax accumulation contributes to melatonin- and ABA-induced drought tolerance in watermelon. *Journal of Plant Growth Regulation* 39: 1441–1450. https://doi.org/10.1007/s00344-020-10099-z.

Li, P., Li, X., and Jiang, M. (2021). CRISPR/Cas9-mediated mutagenesis of WRKY3 and WRKY4 function decreases salt and Me-JA stress tolerance in *Arabidopsis thaliana*. *Molecular Biology Reports* 48 (8): 5821–5832. https://doi.org/10.1007/s11033-021-06541-4.

Li, G., Ma, Y., Wang, X. et al. (2022). CRISPR/Cas9 GENE EDITING OF NtAITRs, a family of transcription repressor genes, leads to enhanced drought tolerance in tobacco. *International Journal of Molecular Sciences* 23 (23): 15268. https://doi.org/10.3390/ijms232315268.

Liang, X.-Z., Wu, Y., Chambers, R.G. et al. (2017a). Determining climate effects on US total agricultural productivity. *Proceedings of the National Academy of Sciences* 114 (12): E2285–E2292. https://doi.org/10.1073/pnas.1615922114.

Liang, Z., Chen, K., Li, T. et al. (2017b). Efficient DNA-free genome editing of bread wheat using CRISPR/Cas9 ribonucleoprotein complexes. *Nature Communications* 8 (1): https://doi.org/10.1038/ncomms14261.

Liang, G., Hou, Y., Wang, H. et al. (2023). VaBAM1 weakens cold tolerance by interacting with the negative regulator VaSR1 to suppress β-amylase expression. *International Journal of Biological Macromolecules* 225: 1394–1404. https://doi.org/10.1016/j.ijbiomac.2022.11.197.

Liao, S., Qin, X., Luo, L. et al. (2019). CRISPR/Cas9-induced mutagenesis of semi-rolled Leaf1,2 confers curled leaf phenotype and drought tolerance by influencing protein expression patterns and ROS scavenging in rice (*Oryza sativa* L.). *Agronomy* 9 (11): 728. https://doi.org/10.3390/agronomy9110728.

Lin, P.-A., Chen, Y., Ponce, G. et al. (2022). Stomata-mediated interactions between plants, herbivores, and the environment. *Trends in Plant Science* 27 (3): 287–300. https://doi.org/10.1016/j.tplants.2021.08.017.

Lippmann, R., Babben, S., Menger, A. et al. (2019). Development of wild and cultivated plants under global warming conditions. *Current Biology* 29 (24): R1326–R1338. https://doi.org/10.1016/j.cub.2019.10.016.

Liu, Y., Huang, W., Xian, Z. et al. (2017). Overexpression of SlGRAS40 in tomato enhances tolerance to abiotic stresses and influences auxin and gibberellin signaling. *Frontiers in Plant Science* 8: 1659. https://doi.org/10.3389/fpls.2017.01659.

Liu, C.-T., Wang, W., Mao, B.-G., and Chu, C.-C. (2018). Cold stress tolerance in rice: physiological changes, molecular mechanism, and future prospects. *Yi Chuan = Hereditas* 40 (3): 171–185. https://doi.org/10.16288/j.yczz.18-007.

Liu, X., Lyu, Y., Yang, W. et al. (2020). A membrane-associated NAC transcription factor OsNTL3 is involved in thermotolerance in rice. *Plant Biotechnology Journal* 18 (5): 1317–1329. https://doi.org/10.1111/pbi.13297.

Liu, G.-S., Li, H.-L., Grierson, D., and Fu, D.-Q. (2022a). NAC transcription factor family regulation of fruit ripening and quality: a Review. *Cells* 11 (3): 525. https://doi.org/10.3390/cells11030525.

Liu, S., Liu, W., Lai, J. et al. (2022b). OsGLYI3, a glyoxalase gene expressed in rice seed, contributes to seed longevity and salt stress tolerance. *Plant Physiology and Biochemistry* 183: 85–95. https://doi.org/10.1016/j.plaphy.2022.04.028.

Liu, X., Bulley, S.M., Varkonyi-Gasic, E. et al. (2023). Kiwifruit bZIP transcription factor*AcePosF21*elicits ascorbic acid biosynthesis during cold stress. *Plant Physiology, Kiad* 121: https://doi.org/10.1093/plphys/kiad121.

Lou, D., Wang, H., Liang, G., and Yu, D. (2017). OsSAPK2 confers abscisic acid sensitivity and tolerance to drought stress in rice. *Frontiers in Plant Science* 8: 993. https://doi.org/10.3389/fpls.2017.00993.

Lv, Q., Li, X., Jin, X. et al. (2022). Rice OsPUB16 modulates the 'SAPK9-OsMADS23-OsAOC' pathway to reduce plant water-deficit tolerance by repressing ABA and JA biosynthesis. *PLoS Genetics* 18 (11): e1010520. https://doi.org/10.1371/journal.pgen.1010520.

Macovei, A., Sevilla, N.R., Cantos, C. et al. (2018). Novel alleles of rice eIF4G generated by CRISPR/Cas9-targeted mutagenesis confer resistance to *Rice tungro spherical virus*. *Plant Biotechnology Journal* 16 (11): 1918–1927. https://doi.org/10.1111/pbi.12927.

Manoj, V.M., Anunanthini, P., Swathik, P.C. et al. (2019). Comparative analysis of glyoxalase pathway genes in *Erianthus arundinaceus* and commercial sugarcane hybrid under salinity and drought conditions. *BMC Genomics* 19 (S9): https://doi.org/10.1186/s12864-018-5349-7.

Mao, H., Li, S., Chen, B. et al. (2022). Variation in cis-regulation of a NAC transcription factor contributes to drought tolerance in wheat. *Molecular Plant* 15 (2): 276–292. https://doi.org/10.1016/j.molp.2021.11.007.

Marchal, C., Delorme-Hinoux, V., Bariat, L. et al. (2014). NTR/NRX Define a new thioredoxin system in the nucleus of *Arabidopsis thaliana* cells. *Molecular Plant* 7 (1): 30–44. https://doi.org/10.1093/mp/sst162.

Mayer, M.P. (2010). Gymnastics of molecular chaperones. *Molecular Cell* 39 (3): 321–331. https://doi.org/10.1016/j.molcel.2010.07.012.

McGuire, S. (2015). FAO, IFAD, and WFP. The state of food insecurity in the world 2015: meeting the 2015 international hunger targets: taking stock of uneven progress. Rome: FAO, 2015. *Advances in Nutrition* 6 (5): 623–624. https://doi.org/10.3945/an.115.009936.

Mei, E., Tang, J., He, M. et al. (2022). OsMKKK70 negatively regulates cold tolerance at booting stage in rice. *International Journal of Molecular Sciences* 23 (22): 14472. https://doi.org/10.3390/ijms232214472.

Miao, C., Xiao, L., Hua, K. et al. (2018). Mutations in a subfamily of abscisic acid receptor genes promote rice growth and productivity. *Proceedings of the National Academy of Sciences* 115 (23): 6058–6063. https://doi.org/10.1073/pnas.1804774115.

Miglani, G.S. (2017). Genome editing in crop improvement: present scenario and future prospects. *Journal of Crop Improvement* 31 (4): 453–559. https://doi.org/10.1080/15427528.2017.1333192.

Mittler, R., Vanderauwera, S., Gollery, M., and Van Breusegem, F. (2004). Reactive oxygen gene network of plants. *Trends in Plant Science* 9 (10): 490–498. https://doi.org/10.1016/j.tplants.2004.08.009.

Molina-Hidalgo, F.J., Medina-Puche, L., Cañete-Gómez, C. et al. (2017). The fruit-specific transcription factor FaDOF2 regulates the production of eugenol in ripe fruit receptacles. *Journal of Experimental Botany* 68 (16): 4529–4543. https://doi.org/10.1093/jxb/erx257.

Munns, R., James, R.A., Xu, B. et al. (2012). Wheat grain yield on saline soils is improved by an ancestral Na^+ transporter gene. *Nature Biotechnology* 30 (4): 360–364. https://doi.org/10.1038/nbt.2120.

Nakashima, K., Takasaki, H., Mizoi, J. et al. (2012). NAC transcription factors in plant abiotic stress responses. *Biochimica et Biophysica Acta (BBA)-Gene Regulatory Mechanisms* 1819 (2): 97–103. https://doi.org/10.1016/j.bbagrm.2011.10.005.

Nan, H., Li, W., Lin, Y., and Gao, L. (2020). Genome-wide analysis of WRKY genes and their response to salt stress in the wild progenitor of asian cultivated rice, *Oryza rufipogon*. *Frontiers in Genetics* 11: 359. https://doi.org/10.3389/fgene.2020.00359.

Nekrasov, V., Wang, C., Win, J. et al. (2017). Rapid generation of a transgene-free powdery mildew resistant tomato by genome deletion. *Scientific Reports* 7 (1): https://doi.org/10.1038/s41598-017-00578-x.

Osakabe, Y., Watanabe, T., Sugano, S.S. et al. (2016). Optimization of CRISPR/Cas9 genome editing to modify abiotic stress responses in plants. *Scientific Reports* 6 (1): https://doi.org/10.1038/srep26685.

Ouyang, L., Leus, L., De Keyser, E., and Van Labeke, M.-C. (2021). Dehydrins and soluble sugars involved in cold acclimation of *Rosa wichurana* and *Rose Cultivar* 'Yesterday'. *Horticulturae* 7 (10): 379. https://doi.org/10.3390/horticulturae7100379.

Pan, Y.-J., Liu, L., Lin, Y.-C. et al. (2016). ethylene antagonizes salt-induced growth retardation and cell death process via transcriptional controlling of Ethylene-, BAG- and Senescence-associated genes in *Arabidopsis*. *Frontiers in Plant Science* 7: 696. https://doi.org/10.3389/fpls.2016.00696.

Pandey, P., Irulappan, V., Bagavathiannan, M.V., and Senthil-Kumar, M. (2017). Impact of combined abiotic and biotic stresses on plant growth and avenues for crop improvement by exploiting physio-morphological traits. *Frontiers in Plant Science* 8. https://doi.org/10.3389/fpls.2017.00537.

Pardo, J.M., Cubero, B., Leidi, E.O., and Quintero, F.J. (2006). Alkali cation exchangers: roles in cellular homeostasis and stress tolerance. *Journal of Experimental Botany* 57 (5): 1181–1199. https://doi.org/10.1093/jxb/erj114.

Park, J.-J., Dempewolf, E., Zhang, W., and Wang, Z.-Y. (2017). RNA-guided transcriptional activation via CRISPR/dCas9 mimics overexpression phenotypes in *Arabidopsis*. *PLoS One* 12 (6): e0179410. https://doi.org/10.1371/journal.pone.0179410.

Park, J.-R., Kim, E.-G., Jang, Y.-H. et al. (2022). Applications of CRISPR/Cas9 as new strategies for short breeding to drought gene in rice. *Frontiers in Plant Science* 13: https://doi.org/10.3389/fpls.2022.850441.

Pasapula, V., Shen, G., Kuppu, S. et al. (2011). Expression of an *Arabidopsis Vacuolar H+-pyrophosphatase gene* (*AVP1*) in cotton improves drought- and salt tolerance and increases fibre yield in the field conditions. *Plant Biotechnology Journal* 9 (1): 88–99. https://doi.org/10.1111/j.1467-7652.2010.00535.x.

Paul, P., Dhatt, B.K., Miller, M. et al. (2020). *MADS78* and *MADS79* are essential regulators of early seed development in rice. *Plant Physiology* 182 (2): 933–948. https://doi.org/10.1104/pp.19.00917.

Peng, J., Carol, P., Richards, D.E. et al. (1997). The *Arabidopsis GAI* gene defines a signaling pathway that negatively regulates gibberellin responses. *Genes & Development* 11 (23): 3194–3205. https://doi.org/10.1101/gad.11.23.3194.

Perotti, M.F., Ribone, P.A., and Chan, R.L. (2017). Plant transcription factors from the homeodomain-leucine zipper family I. Role in development and stress responses. *IUBMB Life* 69 (5): 280–289. https://doi.org/10.1002/iub.1619.

Peters, K., Breitsameter, L., and Gerowitt, B. (2014). Impact of climate change on weeds in agriculture: a review. *Agronomy for Sustainable Development* 34 (4): 707–721. https://doi.org/10.1007/s13593-014-0245-2.

Powell, N., Ji, X., Ravash, R. et al. (2012). Yield stability for cereals in a changing climate. *Functional Plant Biology* 39 (7): 539. https://doi.org/10.1071/fp12078.

Prodhan, Z.H., Islam, S.A., Alam, M.S. et al. (2022). Impact of OsBadh2 mutations on salt stress response in rice. *Plants* 11 (21): 2829. https://doi.org/10.3390/plants11212829.

Pu, Y.-J., Cisse, E.H.M., Zhang, L.-J. et al. (2021). Coupling exogenous melatonin with Ca^{2+} alleviated chilling stress in *Dalbergia odorifera* T. Chen. *Trees* 35 (5): 1541–1554. https://doi.org/10.1007/s00468-021-02134-7.

Qian, H.F., Peng, X.F., Han, X. et al. (2014). The stress factor, exogenous ascorbic acid, affects plant growth and the antioxidant system in *Arabidopsis thaliana*. *Russian Journal of Plant Physiology* 61 (4): 467–475. https://doi.org/10.1134/s1021443714040141.

Qiu, D., Xiao, J., Ding, X. et al. (2007). OsWRKY13 mediates rice disease resistance by regulating defense-related genes in salicylate- and jasmonate-dependent signaling. *Molecular Plant-Microbe Interactions* 20 (5): 492–499. https://doi.org/10.1094/mpmi-20-5-0492.

Quintero, F.J., Martinez-Atienza, J., Villalta, I. et al. (2011). Activation of the plasma membrane Na/H antiporter salt-overly-sensitive 1 (SOS1) by phosphorylation of an auto-inhibitory C-terminal domain. *Proceedings of the National Academy of Sciences* 108 (6): 2611–2616. https://doi.org/10.1073/pnas.1018921108.

Rai, R., Singh, S., Rai, K.K. et al. (2021). Regulation of antioxidant defense and glyoxalase systems in cyanobacteria. *Plant Physiology and Biochemistry* 168: 353–372. https://doi.org/10.1016/j.plaphy.2021.09.037.

Razzaq, M.K., Akhter, M., Ahmad, R.M. et al. (2022). CRISPR-Cas9 based stress tolerance: new hope for abiotic stress tolerance in chickpea (*Cicer arietinum*). *Molecular Biology Reports* 49: 8977–8985. https://doi.org/10.1007/s11033-022-07391-4.

Razzaque, S., Elias, S.M., Biswas, S. et al. (2014). Cloning of the plasma membrane sodium/hydrogen antiporter SOS1 for its overexpression in rice. *Plant Tissue Culture and Biotechnology* 23 (2): 263–273. https://doi.org/10.3329/ptcb.v23i2.17527.

Ren, Y., Huang, Z., Jiang, H. et al. (2021). A heat stress responsive NAC transcription factor heterodimer plays key roles in rice grain filling. *Journal of Experimental Botany* 72 (8): 2947–2964. https://doi.org/10.1093/jxb/erab027.

Richards, D.E., Peng, J., and Harberd, N.P. (2000). Plant GRAS and metazoan STATs: one family? *BioEssays: News and Reviews in Molecular, Cellular and Developmental Biology* 22 (6): 573–577. https://doi.org/10.1002/(SICI)1521-1878(200006)22:6%3C573::AID-BIES10%3E3.0.CO;2-H.

Roca Paixão, J.F., Gillet, F.-X., Ribeiro, T.P. et al. (2019). Improved drought stress tolerance in *Arabidopsis* by CRISPR/dCas9 fusion with a histone Acetyl transferase. *Scientific Reports* 9 (1): https://doi.org/10.1038/s41598-019-44571-y.

Rubio, F., Nieves-Cordones, M., Horie, T., and Shabala, S. (2020). Doing 'business as usual' comes with a cost: evaluating energy cost of maintaining plant intracellular K^+ homeostasis under saline conditions. *New Phytologist* 225 (3): 1097–1104. https://doi.org/10.1111/nph.15852.

Saikia, B., Singh, S., Debbarma, J. et al. (2020). Multigene CRISPR/Cas9 genome editing of hybrid proline rich proteins (HyPRPs) for sustainable multi-stress tolerance in crops: the review of a promising approach. *Physiology and Molecular Biology of Plants* 26: 857–869. https://doi.org/10.1007/s12298-020-00782-6.

Sánchez-León, S., Gil-Humanes, J., Ozuna, C.V. et al. (2018). Low-gluten, nontransgenic wheat engineered with CRISPR/Cas9. *Plant Biotechnology Journal* 16 (4): 902–910. https://doi.org/10.1111/pbi.12837.

Sanghera, G.S., Wani, S.H., Hussain, W., and Singh, N.B. (2011). Engineering cold stress tolerance in crop plants. *Current Genomics* 12 (1): 30–43. https://doi.org/10.2174/138920211794520178.

Sarwar, M., Saleem, M.F., Ullah, N. et al. (2019). Role of mineral nutrition in alleviation of heat stress in cotton plants grown in glasshouse and field conditions. *Scientific Reports* 9 (1): https://doi.org/10.1038/s41598-019-49404-6.

Shabala, S. (2009). Salinity and programmed cell death: unravelling mechanisms for ion specific signalling. *Journal of Experimental Botany* 60: 709–711. https://doi.org/10.1093/jxb/erp013.

Shan, Q., Wang, Y., Li, J., and Gao, C. (2014). Genome editing in rice and wheat using the CRISPR/Cas system. *Nature Protocols* 9 (10): 2395–2410. https://doi.org/10.1038/nprot.2014.157.

Shen, C., Que, Z., Xia, Y. et al. (2017). Knock out of the annexin gene OsAnn3 via CRISPR/Cas9-mediated genome editing decreased cold tolerance in rice. *Journal of Plant Biology* 60 (6): 39–547. https://doi.org/10.1007/s12374-016-0400-1.

Shi, Y. (2017). Mechanistic insights into precursor messenger RNA splicing by the spliceosome. *Nature Reviews Molecular Cell Biology* 18 (11): 655–670. https://doi.org/10.1038/nrm.2017.86.

Shi, H., Quintero, F.J., Pardo, J.M., and Zhu, J.-K. (2002). The putative plasma membrane Na^+/H^+ antiporter SOS1 controls long-distance Na^+ transport in plants. *The Plant Cell* 14 (2): 65–477. https://doi.org/10.1105/tpc.010371.

Shi, J., Gao, H., Wang, H. et al. (2017). ARGOS8 variants generated by CRISPR-Cas9 improve maize grain yield under field drought stress conditions. *Plant Biotechnology Journal* 15 (2): 207–216. https://doi.org/10.1111/pbi.12603.

Shi, Y., Ding, Y., and Yang, S. (2018). Molecular regulation of CBF signaling in cold acclimation. *Trends in Plant Science* 23 (7): 23–637. https://doi.org/10.1016/j.tplants.2018.04.002.

Shkryl, Y., Yugay, Y., Avramenko, T. et al. (2021). CRISPR/Cas9-mediated Knock-out of HOS1 reveals its role in the regulation of secondary metabolism in *Arabidopsis thaliana*. *Plants* 10 (1): 104. https://doi.org/10.3390/plants10010104.

Shumayla, Tyagi, S., and Upadhyay, S.K. (2019). Receptor-like kinases and environmental stress in plants. In: *Molecular Approaches in Plant Biology and Environmental Challenges* (ed. S.P. Singh). Springer. https://doi.org/10.1007/978-981-15-0690-1_4.

Shumayla, Tyagi, S., Sharma, Y. et al. (2023). Expression of TaNCL2-A ameliorates cadmium toxicity by increasing calcium and enzymatic antioxidants activities in Arabidopsis. *Chemosphere* 29: 138636. https://doi.org/10.1016/j.chemosphere.2023.138636.

da Silva, D.C., da Silveira Falavigna, V., Fasoli, M. et al. (2016). Transcriptome analyses of the *Dof-like* gene family in grapevine reveal its involvement in berry, flower and seed development. *Horticulture Research* 3 (1): https://doi.org/10.1038/hortres.2016.42.

Silverstone, A.L., Ciampaglio, C.N., and Sun, T. (1998). The *Arabidopsis RGA* gene encodes a transcriptional regulator repressing the gibberellin signal transduction pathway. *The Plant Cell* 10 (2): 155–169. https://doi.org/10.1105/tpc.10.2.155.

Singh, A., Singh, P.K., Singh, R. et al. (2010). SNP haplotypes of the BADH1 gene and their association with aroma in rice (*Oryza sativa* L.). *Molecular Breeding* 26 (2): 325–338. https://doi.org/10.1007/s11032-010-9425-1.

Singh, S., Rahangdale, S., Pandita, S. et al. (2022). CRISPR/Cas9 for insect pests management: a comprehensive review of advances and applications. *Agriculture* 12: 1896. https://doi.org/10.3390/agriculture12111896.

Song, M., Linghu, B., Huang, S. et al. (2022). Identification of nuclear pore complexes (NPCs) and revealed outer-ring component BnHOS1 related to cold tolerance in B. napus. *International Journal of Biological Macromolecules* 223 (Pt A): 1450–1461. https://doi.org/10.1016/j.ijbiomac.2022.11.148.

Stone, S.L. (2014). The role of ubiquitin and the 26S proteasome in plant abiotic stress signaling. *Frontiers in Plant Science* 5. https://doi.org/10.3389/fpls.2014.00135.

Stone, S.L. (2019). Role of the ubiquitin proteasome system in plant response to abiotic stress. *International Review of Cell and Molecular Biology* 343: 65–110. https://doi.org/10.1016/bs.ircmb.2018.05.012.

Sun, J., Chen, S.-L., Dai, S.-X. et al. (2009). Ion flux profiles and plant ion homeostasis control under salt stress. *Plant Signaling & Behavior* 4 (4): 261–264. https://doi.org/10.4161/psb.4.4.7918.

Sun, X., Jones, W.T., and Rikkerink, E.H.A. (2012). GRAS proteins: the versatile roles of intrinsically disordered proteins in plant signalling. *Biochemical Journal* 442 (1): 1–12. https://doi.org/10.1042/bj20111766.

Sun, C., Chen, D., Fang, J. et al. (2014). Understanding the genetic and epigenetic architecture in complex network of rice flowering pathways. *Protein & Cell* 5 (12): 889–898. https://doi.org/10.1007/s13238-014-0068-6.

Sun, T.-J., Fan, L., Yang, J. et al. (2019). A glycine max sodium/hydrogen exchanger enhances salt tolerance through maintaining higher Na^+ efflux rate and K^+/Na^+ ratio in *Arabidopsis*. *BMC Plant Biology* 19 (1): https://doi.org/10.1186/s12870-019-2084-4.

Sun, C., Liu, L., Wang, L. et al. (2020). Melatonin: a master regulator of plant development and stress responses. *Journal of Integrative Plant Biology* 63 (1): 126–145. https://doi.org/10.1111/jipb.12993.

Sun, S., Hu, C., Qi, X. et al. (2021a). The AaCBF4-AaBAM3.1 module enhances freezing tolerance of kiwifruit (*Actinidia arguta*). *Horticulture Research* 8 (1): https://doi.org/10.1038/s41438-021-00530-1.

Sun, T., Ma, N., Wang, C. et al. (2021b). A golgi-localized sodium/hydrogen exchanger positively regulates salt tolerance by maintaining higher K^+/Na^+ ratio in soybean. *Frontiers in Plant Science* 12. https://doi.org/10.3389/fpls.2021.638340.

Sushmita, K.G., Upadhyay, S.K., and Verma, P.C. (2021). An overview of genome-engineering methods. In: *Genome Engineering for Crop Improvement*. Wiley. https://doi.org/10.1002/9781119672425.ch1.

Svitashev, S., Young, J.K., Schwartz, C. et al. (2015). Targeted mutagenesis, precise gene editing, and site-specific gene insertion in maize using Cas9 and Guide RNA. *Plant Physiology* 169 (2): 931–945. https://doi.org/10.1104/pp.15.00793.

Świeżawska, B., Duszyn, M., Jaworski, K., and Szmidt-Jaworska, A. (2018). Downstream targets of cyclic nucleotides in plants. *Frontiers in Plant Science* 9. https://doi.org/10.3389/fpls.2018.01428.

Tang, Y., Kuang, J., Wang, F. et al. (2013). Molecular characterization of PR and WRKY genes during SA- and MeJA-induced resistance against *Colletotrichum musae* in banana fruit. *Postharvest Biology and Technology* 79: 62–68. https://doi.org/10.1016/j.postharvbio.2013.01.004.

Tang, J., Tian, X., Mei, E. et al. (2022). WRKY53 negatively regulates rice cold tolerance at the booting stage by fine-tuning anther gibberellin levels. *The Plant Cell* 34 (11): 4495–4515. https://doi.org/10.1093/plcell/koac253.

Taoka, K., Ohki, I., Tsuji, H. et al. (2013). Structure and function of florigen and the receptor complex. *Trends in Plant Science* 18 (5): 287–294. https://doi.org/10.1016/j.tplants.2013.02.002.

Tarchoun, N., Saadaoui, W., Mezghani, N. et al. (2022). The effects of salt stress on germination, seedling growth and biochemical responses of tunisian squash (*Cucurbita maxima* Duchesne) germplasm. *Plants* 11 (6): 800. https://doi.org/10.3390/plants11060800.

Tarkowski, Ł.P. and Van den Ende, W. (2015). Cold tolerance triggered by soluble sugars: a multifaceted countermeasure. *Frontiers in Plant Science* 6: https://doi.org/10.3389/fpls.2015.00203.

Thomashow, M.F. (2010). Molecular basis of plant cold acclimation: insights gained from studying the CBF cold response pathway: figure 1. *Plant Physiology* 154 (2): 571–577. https://doi.org/10.1104/pp.110.161794.

Tian, H., Chen, S., Yang, W. et al. (2017). A novel family of transcription factors conserved in angiosperms is required for ABA signalling. *Plant, Cell & Environment* 40 (12): 2958–2971. https://doi.org/10.1111/pce.13058.

Tian, Q., Shen, L., Luan, J. et al. (2021). Rice shaker potassium channel OsAKT2 positively regulates salt tolerance and grain yield by mediating K^+ redistribution. *Plant, Cell & Environment* 44 (9): 2951–2965. https://doi.org/10.1111/pce.14101.

Tondelli, A., Francia, E., Barabaschi, D. et al. (2011). Inside the CBF locus in poaceae. *Plant Science* 180 (1): 39–45. https://doi.org/10.1016/j.plantsci.2010.08.012.

Torres, M.A. (2010). ROS in biotic interactions. *Physiologia Plantarum* 138 (4): 414–429. https://doi.org/10.1111/j.1399-3054.2009.01326.x.

Tran, M.T., Doan, D.T.H., Kim, J. et al. (2021). CRISPR/Cas9-based precise excision of SlHyPRP1 domain(s) to obtain salt stress-tolerant tomato. *Plant Cell Reports* 40 (6): 999–1011. https://doi.org/10.1007/s00299-020-02622-z.

Ueda, M., Matsui, A., Tanaka, M. et al. (2017). The distinct roles of class I and II RPD3-Like histone deacetylases in salinity stress response. *Plant Physiology* 175 (4): 1760–1773. https://doi.org/10.1104/pp.17.01332.

Uji, Y., Akimitsu, K., and Gomi, K. (2017). Identification of OsMYC2-regulated senescence-associated genes in rice. *Planta* 245 (6): 1241–1246. https://doi.org/10.1007/s00425-017-2697-5.

Ullah, A., Sun, H., Hakim et al. (2018). A novel cotton WRKY gene, *GhWRKY6*-like, improves salt tolerance by activating the ABA signaling pathway and scavenging of reactive oxygen species. *Physiologia Plantarum* 162 (4): 439–454. https://doi.org/10.1111/ppl.12651.

Upadhyay, S.K. (2021). *Genome Engineering for Crop Improvement*. Wiley. https://doi.org/10.1002/9781119672425.

Upadhyay, S.K. and Shumayla (2023). *Plant Receptor-Like Kinases: Role in Development and Stress*. Academic Press. https://doi.org/10.1016/C2020-0-03988-5.

Upadhyay, S.K., Kumar, J., Alok, A. et al. (2013). RNA guided genome editing for multiple target gene mutations in wheat. *G3: Genes|Genomes|Genetics* 3 (12): 2233–2238.

Van de Peer, Y., Mizrachi, E., and Marchal, K. (2017). The evolutionary significance of polyploidy. *Nature Reviews Genetics* 18 (7): 411–424. https://doi.org/10.1038/nrg.2017.26.

Verma, V., Ravindran, P., and Kumar, P.P. (2016). Plant hormone-mediated regulation of stress responses. *BMC Plant Biology* 16 (1): https://doi.org/10.1186/s12870-016-0771-y.

Viana, V.E., Carlos da Maia, L., Busanello, C. et al. (2021). When rice gets the chills: comparative transcriptome profiling at germination shows WRKY transcription factor responses. *Plant Biology* 23 (S1): 100–112. https://doi.org/10.1111/plb.13262.

Vieira Dos Santos, C. and Rey, P. (2006). Plant thioredoxins are key actors in the oxidative stress response. *Trends in Plant Science* 11 (7): 329–334. https://doi.org/10.1016/j.tplants.2006.05.005.

Vishwakarma, K., Upadhyay, N., Kumar, N. et al. (2017). Abscisic acid signaling and abiotic stress tolerance in plants: a review on current knowledge and future prospects. *Frontiers in Plant Science* 8. https://doi.org/10.3389/fpls.2017.00161.

Vlčko and Ohnoutková (2020). Allelic variants of CRISPR/Cas9 induced mutation in an *Inositol Trisphosphate 5/6 Kinase* gene manifest different phenotypes in barley. *Plants* 9 (2): 195. https://doi.org/10.3390/plants9020195.

Wahl, M.C., Will, C.L., and Lührmann, R. (2009). The Spliceosome: design principles of a dynamic RNP machine. *Cell* 136 (4): 701–718. https://doi.org/10.1016/j.cell.2009.02.009.

Wang, G., Kong, F., Zhang, S. et al. (2015). A tomato chloroplast-targeted DNAJ protein protects rubisco activity under heat stress. *Journal of Experimental Botany* 66 (11): 3027–3040. https://doi.org/10.1093/jxb/erv102.

Wang, F., Wang, C., Liu, P. et al. (2016). Enhanced rice blast resistance by CRISPR/Cas9-targeted mutagenesis of the ERF transcription factor gene *OsERF922*. *PLoS One* 11 (4): e0154027. https://doi.org/10.1371/journal.pone.0154027.

Wang, L., Chen, L., Li, R. et al. (2017). Reduced drought tolerance by CRISPR/Cas9-mediated SlMAPK3 mutagenesis in tomato plants. *Journal of Agricultural and Food Chemistry* 65 (39): 8674–8682. https://doi.org/10.1021/acs.jafc.7b02745.

Wang, H., Gong, M., Xin, H. et al. (2018). Effects of chilling stress on the accumulation of soluble sugars and their key enzymes in *Jatropha curcas* seedlings. *Physiology and Molecular Biology of Plants* 24 (5): 857–865. https://doi.org/10.1007/s12298-018-0568-6.

Wang, B., Zhong, Z., Wang, X. et al. (2020a). Knock-out of the OsNAC006 transcription factor causes drought and heat sensitivity in rice. *International Journal of Molecular Sciences* 21 (7): 2288. https://doi.org/10.3390/ijms21072288.

Wang, Z., Hong, Y., Li, Y. et al. (2020b). Natural variations in *SlSOS1* contribute to the loss of salt tolerance during tomato domestication. *Plant Biotechnology Journal* 19 (1): 20–22. https://doi.org/10.1111/pbi.13443.

Wang, T., Xun, H., Wang, W. et al. (2021a). Mutation of GmAITR genes by CRISPR/Cas9 genome editing results in enhanced salinity stress tolerance in soybean. *Frontiers in Plant Science* 12: https://doi.org/10.3389/fpls.2021.779598.

Wang, X., Ren, P., Ji, L. et al. (2021b). OsVDE, a xanthophyll cycle key enzyme, mediates abscisic acid biosynthesis and negatively regulates salinity tolerance in rice. *Planta* 255 (1): https://doi.org/10.1007/s00425-021-03802-1.

Wang, Z., Wang, Y., Tong, Q. et al. (2021c). Transcriptomic analysis of grapevine DOF transcription factor gene family in response to cold stress and functional analyses of the *VaDof17d* gene. *Planta* 253 (2): https://doi.org/10.1007/s00425-021-03574-8.

Wang, Z., Wong, D.C.J., Wang, Y. et al. (2021d). GRAS-domain transcription factor PAT1 regulates jasmonic acid biosynthesis in grape cold stress response. *Plant Physiology* 186 (3): 1660–1678. https://doi.org/10.1093/plphys/kiab142.

Wang, B., Wang, Y., Yu, W. et al. (2022a). Knocking out the transcription factor OsNAC092 promoted rice drought tolerance. *Biology* 11 (12): 1830. https://doi.org/10.3390/biology11121830.

Wang, Q., Bai, H., Zada, A., and Jiao, Q. (2022b). DORN1 is involved in drought stress tolerance through a Ca^{2+}-dependent pathway. *International Journal of Molecular Sciences* 23 (22): 14213. https://doi.org/10.3390/ijms232214213.

Waqas, M.A., Wang, X., Zafar, S.A. et al. (2021). Thermal stresses in maize: effects and management strategies. *Plants* 10 (2): 293. https://doi.org/10.3390/plants10020293.

Weeks, D.P., Spalding, M.H., and Yang, B. (2016). Use of designer nucleases for targeted gene and genome editing in plants. *Plant Biotechnology Journal* 14 (2): 483–495. https://doi.org/10.1111/pbi.12448.

Wen, L., Liu, T., Deng, Z. et al. (2022). Characterization of NAC transcription factor NtNAC028 as a regulator of leaf senescence and stress responses. *Frontiers in Plant Science* 13: https://doi.org/10.3389/fpls.2022.941026.

Wu, H., Ye, H., Yao, R. et al. (2015). OsJAZ9 acts as a transcriptional regulator in jasmonate signaling and modulates salt stress tolerance in rice. *Plant Science* 232: 1–12. https://doi.org/10.1016/j.plantsci.2014.12.010.

Wu, H., Zhang, X., Giraldo, J.P., and Shabala, S. (2018). It is not all about sodium: revealing tissue specificity and signalling roles of potassium in plant responses to salt stress. *Plant and Soil* 431 (1–2): 1–17. https://doi.org/10.1007/s11104-018-3770-y.

Wu, Z., Liang, J., Wang, C. et al. (2019). Alternative splicing provides a mechanism to regulate LlHSFA3 function in response to heat stress in lily. *Plant Physiology* 181 (4): 1651–1667. https://doi.org/10.1104/pp.19.00839.

Wu, S., Han, B., and Jiao, Y. (2020). Genetic contribution of paleopolyploidy to adaptive evolution in Angiosperms. *Molecular Plant* 13 (1): 59–71. https://doi.org/10.1016/j.molp.2019.10.012.

Wu, J., Gao, T., Hu, J. et al. (2022). Research advances in function and regulation mechanisms of plant small heat shock proteins (sHSPs) under environmental stresses. *Science of the Total Environment* 825: 154054. https://doi.org/10.1016/j.scitotenv.2022.154054.

Xing, B., Gu, C., Zhang, T. et al. (2021). Functional study of BpPP2C1 revealed its role in salt stress in *Betula platyphylla*. *Frontiers in Plant Science* 11: https://doi.org/10.3389/fpls.2020.617635.

Xiu-Yun, W. and Tian, L. (2017). A Casein Kinase II phosphorylation site in AtYY1 affects its activity, stability, and function in the ABA response. *Frontiers in Plant Science* 8: 323. https://doi.org/10.3389/fpls.2017.00323.

Xue, T., Wang, D., Zhang, S. et al. (2008). Genome-wide and expression analysis of protein phosphatase 2C in rice and *Arabidopsis*. *BMC Genomics* 9 (1): https://doi.org/10.1186/1471-2164-9-550.

Yadav, S., Modi, P., Dave, A. et al. (2020). *Effect of Abiotic Stress on Crops* (ed. M. Hasanuzzaman, M.C.M.T. Filho, M. Fujita, and T.A.R. Nogueira). IntechOpen.

Yang, C., Zhao, L., Zhang, H. et al. (2014). Evolution of physiological responses to salt stress in hexaploid wheat. *Proceedings of the National Academy of Sciences* 111 (32): 11882–11887. https://doi.org/10.1073/pnas.1412839111.

Yang, C., Huang, Y., Lv, P. et al. (2022). NAC transcription factor GmNAC12 improved drought stress tolerance in soybean. *International Journal of Molecular Sciences* 23 (19): 12029. https://doi.org/10.3390/ijms231912029.

Ye, N., Jia, L., and Zhang, J. (2012). ABA signal in rice under stress conditions. *Rice* 5 (1): 1. https://doi.org/10.1186/1939-8433-5-1.

Ye, C.-Y., Yang, X., Xia, X., and Yin, W. (2013). Comparative analysis of cation/proton antiporter superfamily in plants. *Gene* 521 (2): 245–251. https://doi.org/10.1016/j.gene.2013.03.104.

Yeom, S.-I., Seo, E., Oh, S.-K. et al. (2012). A common plant cell-wall protein HyPRP1 has dual roles as a positive regulator of cell death and a negative regulator of basal defense against pathogens. *The Plant Journal* 69 (5): 755–768. https://doi.org/10.1111/j.1365-313x.2011.04828.x.

Ying, S., Scheible, W.-R., and Lundquist, P.K. (2023). A stress-inducible protein regulates drought tolerance and flowering time in *Brachypodium* and *Arabidopsis*. *Plant Physiology* 191 (1): 643–659. https://doi.org/10.1093/plphys/kiac486.

Yokoi, S., Quintero, F.J., Cubero, B. et al. (2002). Differential expression and function of *Arabidopsis thaliana* NHX Na^{+}/H^{+} antiporters in the salt stress response. *The Plant Journal* 30 (5): 529–539. https://doi.org/10.1046/j.1365-313x.2002.01309.x.

Yoo, C.Y., Mano, N., Finkler, A. et al. (2019). A Ca^{2+}/CaM-regulated transcriptional switch modulates stomatal development in response to water deficit. *Scientific Reports* 9 (1): https://doi.org/10.1038/s41598-019-47529-2.

Yu, Q., Wang, B., Li, N. et al. (2017). CRISPR/Cas9-induced targeted mutagenesis and gene replacement to generate long-shelf life tomato lines. *Scientific Reports* 7 (1): https://doi.org/10.1038/s41598-017-12262-1.

Yu, W., Wang, L., Zhao, R. et al. (2019). Knock-out of SlMAPK3 enhances tolerance to heat stress involving ROS homeostasis in tomato plants. *BMC Plant Biology* 19 (1): https://doi.org/10.1186/s12870-019-1939-z.

Yuan, Y., Fang, L., Karungo, S.K. et al. (2016). Overexpression of VaPAT1, a GRAS transcription factor from *Vitis amurensis*, confers abiotic stress tolerance in *Arabidopsis*. *Plant Cell Reports* 35 (3): 655–666. https://doi.org/10.1007/s00299-015-1910-x.

Zhang, Y., Liang, Z., Zong, Y. et al. (2016). Efficient and transgene-free genome editing in wheat through transient expression of CRISPR/Cas9 DNA or RNA. *Nature. Communications* 7 (1): https://doi.org/10.1038/ncomms12617.

Zhang, Z., Li, J., Li, F. et al. (2017). OsMAPK3 phosphorylates OsbHLH002/OsICE1 and inhibits its ubiquitination to activate OsTPP1 and enhances rice chilling tolerance. *Developmental Cell* 43 (6): 731–743. e5. https://doi.org/10.1016/j.devcel.2017.11.016.

Zhang, M., Su, J., Zhang, Y. et al. (2018a). Conveying endogenous and exogenous signals: MAPK cascades in plant growth and defense. *Current Opinion in Plant Biology* 45 (Pt A): 1–10. https://doi.org/10.1016/j.pbi.2018.04.012.

Zhang, Z., Ge, X., Luo, X. et al. (2018b). Simultaneous editing of two copies of Gh14-3-3d confers enhanced transgene-clean plant defense against *Verticillium dahliae* in allotetraploid upland cotton. *Frontiers in Plant Science* 9: 842. https://doi.org/10.3389/fpls.2018.00842.

Zhang, Z., Yuan, L., Liu, X. et al. (2018c). Evolution analysis of Dof transcription factor family and their expression in response to multiple abiotic stresses in *Malus domestica*. *Gene* 639: 137–148. https://doi.org/10.1016/j.gene.2017.09.039.

Zhang, C., Srivastava, A.K., and Sadanandom, A. (2019a). Targeted mutagenesis of the SUMO protease, *Overly Tolerant to Salt1* in rice through CRISPR/Cas9-mediated genome editing reveals a major role of this SUMO protease in salt tolerance. *bioRxiv*. https://doi.org/10.1101/555706.

Zhang, A., Liu, Y., Wang, F. et al. (2019b). Enhanced rice salinity tolerance via CRISPR/Cas9-targeted mutagenesis of the OsRR22 gene. *Molecular Breeding* 39 (3): https://doi.org/10.1007/s11032-019-0954-y.

Zhang, Z., Dong, J., Ji, C. et al. (2019c). NAC-type transcription factors regulate accumulation of starch and protein in maize seeds. *Proceedings of the National Academy of Sciences* 116 (23): 11223–11228. https://doi.org/10.1073/pnas.1904995116.

Zhang, H., Li, G., Fu, C. et al. (2020). Genome-wide identification, transcriptome analysis and alternative splicing events of Hsf family genes in maize. *Scientific Reports* 10 (1): https://doi.org/10.1038/s41598-020-65068-z.

Zhang, T.Y., Li, Z.Q., Zhao, Y.D. et al. (2021). Ethylene-induced stomatal closure is mediated via MKK1/3-MPK3/6 cascade to EIN2 and EIN3. *Journal of Integrative Plant Biology* 63 (7): 1324–1340. https://doi.org/10.1111/jipb.13083.

Zhang, M., Cao, J., Zhang, T. et al. (2022a). A putative plasma membrane Na^+/H^+ antiporter *GmSOS1* is critical for salt stress tolerance in *Glycine max*. *Frontiers in Plant Science* 13. https://doi.org/10.3389/fpls.2022.870695.

Zhang, M., Zhao, R., Huang, K. et al. (2022b). The OsWRKY63-OsWRKY76-OsDREB1B module regulates chilling tolerance in rice. *The Plant Journal* 112 (2): 383–398. https://doi.org/10.1111/tpj.15950.

Zhang, Y., Li, P., Niu, Y. et al. (2023). Evolution of the WRKY66 gene family and its mutations generated by the CRISPR/Cas9 system increase the sensitivity to salt stress in *Arabidopsis*. *International Journal of Molecular Sciences* 24 (4): 3071. https://doi.org/10.3390/ijms24043071.

Zhao, J. and Blunck, R. (2016). The isolated voltage sensing domain of the Shaker potassium channel forms a voltage-gated cation channel. *eLife* 5: https://doi.org/10.7554/elife.18130.

Zhao, L., Yang, T., Xing, C. et al. (2019). The β-amylase PbrBAM3 from pear (*Pyrus betulaefolia*) regulates soluble sugar accumulation and ROS homeostasis in response to cold stress. *Plant Science* 287: 110184. https://doi.org/10.1016/j.plantsci.2019.110184.

Zhao, J., Lu, Z., Wang, L., and Jin, B. (2020). Plant responses to heat stress: physiology, transcription, noncoding RNAs, and epigenetics. *International Journal of Molecular Sciences* 22 (1): 117. https://doi.org/10.3390/ijms22010117.

Zheng, M., Lin, J., Liu, X. et al. (2021). Histone acetyltransferase TaHAG1 acts as a crucial regulator to strengthen salt tolerance of hexaploid wheat. *Plant Physiology* 186 (4): 1951–1969. https://doi.org/10.1093/plphys/kiab187.

Zhong, X., Hong, W., Shu, Y. et al. (2022). CRISPR/Cas9 mediated gene-editing of GmHdz4 transcription factor enhances drought tolerance in soybean (*Glycine max* [L.] Merr.). *Frontiers in Plant Science* 13. https://doi.org/10.3389/fpls.2022.988505.

Zhou, Z., Licklider, L.J., Gygi, S.P., and Reed, R. (2002). Comprehensive proteomic analysis of the human spliceosome. *Nature* 419 (6903): 182–185. https://doi.org/10.1038/nature01031.

Zhou, H., Li, J., Liu, X. et al. (2021). The divergent roles of the rice *BCL-2 associated athanogene* (*BAG*) Genes in plant development and environmental responses. *Plants* 10 (10): 2169. https://doi.org/10.3390/plants10102169.

Zhu, J.-K. (2016). Abiotic stress signaling and responses in plants. *Cell* 167 (2): 313–324. https://doi.org/10.1016/j.cell.2016.08.029.

11

Genome Editing for Improvement of Nutrition and Quality in Vegetable Crops

Payal Gupta[1], Suhas G. Karkute[2], Prasanta K. Dash[1], and Achuit K. Singh[2]

[1] *ICAR-National Institute for Plant Biotechnology, New Delhi, India*
[2] *Division of Vegetable Improvement, ICAR-Indian Institute of Vegetable Research, Varanasi, Uttar Pradesh, India*

11.1 Vegetables and Human Nutrition

Vegetables in general are the different parts such as fruits, leaves, stems and roots of herbaceous plants consumed by humans or other animals as part of meal. Vegetables are an essential component of human nutrition as they provide a wide range of nutrients that are absent in staple food like rice, wheat and corn but are important for overall health and well-being. Vegetables are low in calories but high in nutrients such as vitamins, minerals and fibre. They are rich in vitamins A, C and K, as well as micronutrients like magnesium, folate and potassium. The fibre in vegetables helps to promote regular bowel movements, helps in easing constipation and keep digestive system working properly (Ulger et al. 2018). Some vegetables, such as artichokes, chicory root, leeks and asparagus, contain prebiotics that promote the growth of healthy gut bacteria and modulates the microbiota in the gastrointestinal tract (Holscher 2017). Vegetables also protect the skin from damage caused by environmental factors and aging by providing antioxidants like vitamin C and beta-carotene (Wallace et al. 2020). The variety of phytochemicals present in different vegetables act as strong antioxidants which protect against free radicle damage and also carry out detoxification of carcinogens (Herrera et al. 2009; Upadhyay and Singh 2021, 2023). Furthermore, it is reported that vegetable-rich diets, in addition to being a source of nutrition; can reduce the chances of heart ailments, stroke and numerous cancers (Harvard school). The cruciferous vegetables like cabbage, cauliflower, kale, broccoli, turnip, etc., are rich in glucosinolates that are known to have anti-cancerous properties for lung, prostate and breast cancer (Dias 2012; Dixit et al. 2021). Based on research in China, it is been revealed that plant-based diets promote longevity (Campbell 2004) and the vegetables combat the 'Hidden Hunger' of malnutrition. Owing to the presence of plethora of nutritional elements, such as antioxidants, dietary fibre, minerals, vitamins and a diverse class of phytochemical compounds, there has been a surge in interest in the health benefits of eating vegetables (Septembre-Malaterre et al. 2018). Apart from delivering benefits to consumers, farmers

Applications of Genome Engineering in Plants, First Edition. Edited by Santosh Kumar Upadhyay.
© 2024 John Wiley & Sons Ltd. Published 2024 by John Wiley & Sons Ltd.

practicing vegetable cultivation typically earn significantly greater farm incomes than cereal growers, with per capita income as high as five times (Lumpkin et al. 2005).

Like all other crops, plants and vegetable crops are severely affected by various biotic and abiotic stresses, and as a result, insect pests, extreme temperatures and a limited water availability have become important obstacles in vegetable yield. Climate change will amplify these issues even further (Bisbis and Gruda 2018). All these environmental stresses affect the nutrition content as well as quality of vegetables.

The world population is increasing exponentially, and it is expected to further increase to 10 billion over 30 years in near future, leading to an increased demand for food crops (Hunter et al. 2017). Modern agriculture will soon enter saturation phase and experience significant hurdles in producing quality crops with higher yield that require minimum resources (Tilman et al. 2011). It has been advised to consume more than 400 g of fruits and vegetables every day for healthy life, and thus, demand of vegetables is going to enhance dramatically in near future. To accomplish this goal, scientific community is working hard to develop improved vegetable varieties with enhanced yield, improved nutritional content and increased tolerance to various stresses by employing both traditional and modern breeding approaches (Abdallah et al. 2015).

11.2 Important Quality Parameters of Vegetables

The quality of vegetable is an important criterion for the consumer point of view. The price of the vegetables in market depends on the quality of vegetables, and good quality vegetables help farmers to get the better remuneration for their products. There are several quality parameters such as sensory properties, nutritive values, chemical constituents and medicinal properties (Abbott 1999) to consider when evaluating vegetables. The most important parameter in assessing the quality is the appearance of vegetables. Consumers prefer vegetables with fresh and vibrant appearance with no signs of decay, damage, or discolouration. The appearance also includes the texture of the vegetables, which should be firm, with a crisp texture, and vegetables should not be wilted or soft. Size and shape of the vegetables is also considered as a quality parameter and vegetables or fruits with uniform size and shape are ideal. Quality is also judged by the flavour of the vegetables, and it is one of the most desired traits by consumers. High-quality vegetables should have a fresh and distinctive taste that is not bland, bitter, or off-flavour. Vegetables should be rich in nutrients such as vitamins, minerals and fibres. They should not have lost any of their nutritional value during transportation or storage. For ensuring that the vegetables have all the right quality parameters like maximum flavour, texture and nutritional value, it is necessary to harvest them at right time and right ripeness level. Overall, high-quality vegetables should be fresh, flavourful, nutritious and free from contaminants.

11.3 Approaches for Improving Nutrition Content in Vegetables

As the vegetables are primarily consumed for fulfilling the requirement of various nutrients, it is of utmost importance to improve the nutrition content in various vegetables. Enhancing the nutrition content or biofortification of vegetables can be accomplished by different approaches that have been thoroughly described by Garg et al. (2018). Some of the important approaches are described here.

1) **Selecting nutrient-dense varieties**: Choosing vegetables that are naturally high in nutrients such as vitamins and minerals can help increase their nutritional value.
2) **Breeding for improved nutrition**: Plant breeders can use breeding techniques to develop vegetables with enhanced nutrient profiles such as higher levels of vitamin A or iron through marker-assisted breeding. This is the sustainable, cost effective and most preferred method which requires genotypic variation in nutrient content. Anthocyanin-rich tomato 'Sun Black' is one of the examples where conventional breeding has been utilized for biofortification (Mazzucato et al. 2008).
3) **Biotechnology for biofortification**: Transgenic is one of the most potential approaches for biofortification of vegetable and other crops. It is not limited by the lack of genetic variation in nutrient content in the germplasm (Zhu et al. 2007).
4) **Agronomic approach for biofortification**: Growing vegetables in soil that is rich in nutrients can increase their nutritional content. This can be achieved through organic farming practices, using compost, or adding nutrients to the soil. For instance, application of chemical fertilizer in orange fleshed *Ipomoea* enhanced β-carotene levels (Laurie et al. 2012).
5) **Using appropriate growing conditions**: Providing vegetables with the right amount of sunlight, water and temperature can also help increase their nutrient content.
6) **Harvesting and storing vegetables properly**: Proper harvesting and storage of vegetables can help preserve their nutritional content. For example, harvesting vegetables at the perfect time and storing them in the optimum conditions can help prevent nutrient loss.

   Overall, by adopting these methods, it is possible to improve the nutritional content of vegetables and make them an even healthier part of a balanced diet.

11.3.1 Breeding for Improving Nutrition in Vegetable Crops

Plant breeding is an intricate procedure which is used to develop new crop varieties with desirable qualities, and it includes the approaches for combining these desirable attributes to produce superior varieties (Glenn et al. 2017). The initial stage in plant breeding is to identify and make use of genetic variability within a plant species. This has been accomplished for years by crossing best identified lines or by inducing physical or chemical mutations to crop plants (Holme et al. 2019; Kleter et al. 2019).

In mutation breeding, different mutations such as point mutations and deletions in the genomic DNA have been induced by various mutagens (Jankowicz-Cieslak et al. 2017), and so far, thousands of important varieties have been developed through these mutant populations (Ahloowalia et al. 2004). Although it is routinely utilized to generate variety in ornamentals, mutagenesis has mostly fallen short of its promise in vegetable breeding, (Bernardo 2016). Moreover, enhancing the nutritional content and reducing the antinutritional factors in vegetable crops is highly challenging due to several limitations like non-availability of desired genetic resources and inherent technological limitations of breeding process. Nevertheless, breeding technology has been successfully employed for developing the nutritionally improved cultivars of many vegetable crops such as β-carotene-rich orange cauliflower 'Pusa Beta Kesari', cassava with improved nutrition value (Maziya-Dixon et al. 2000), anthocyanin-rich tomato 'Sun Black' and 'Black Galaxy' (Mazzucato et al. 2008), iron- and zinc-rich potato 'INIA 321 Kawsay' (Garg et al. 2018).

11.3.2 Genome Editing Technologies

There have been numerous notable advancements in molecular biology methods for improving crop yield and quality during the last few decades. In the recent times, the field of genome editing has witnessed ample progress after the development of site-directed nucleases (SDN), which can modify crops in accelerated manner and show higher probability for introducing desired characteristics for the development of improved vegetable crops (Tian et al. 2021). SDNs allow insertion, deletion, or replacement of the particular DNA fragment at a specific intended position in the genome which helps to alter the desired genes and traits in a precise manner (Corte et al. 2019). There are a wide variety of genome editing techniques that can be employed in plants such as clustered regularly interspaced short palindromic repeats (CRISPR)/CRISPR-associated enzymes, transcription activator-like effector nucleases (TALENs) and zinc-finger nucleases (ZFNs) (Miller et al. 2007; Upadhyay 2021). Genome editing tools create the double strand breaks (DSBs) at the desired genomic location which subsequently activates the cellular DNA repair process either through non-homologous end joining (NHEJ) or homology-directed repair (HDR) (Zhu et al. 2020; Sushmita et al. 2021). The first-generation genome editing tools ZFN and TALEN have been successfully implemented in plants to create desired mutations. However, their applications were limited due to complexity of designing, low efficiency and high off target effects (Jaganathan et al. 2018). The most recent, CRISPR/Cas9 technology which is considered as a second-generation genome editing technique is simple in designing, cost effective and has very high efficiency of editing. It has been widely used in vegetable crops for creating novel and desired traits by targeting the specific genes (Abdallah et al. 2015; Karkute et al. 2017).

11.3.2.1 CRISPR/Cas9 and Advances in Genome Editing

CRISPR/Cas9 is a sophisticated genome-editing tool widely used by scientists to alter target/specific regions in the genome by changing, adding, or removing certain DNA sequences. On the basis of structure and function of Cas proteins, CRISPR/Cas systems are categorized into two classes (class I and class II) which are further differentiated into six sub-types (types I–VI) (Makarova and Koonin 2015). Sub-types I, III and IV are included in Class I, whereas sub-types II, V and VI are included in Class II. Sub-types I, II and V Cas proteins target DNA molecule, while types III and VI target DNA as well as RNA molecules. The mechanism of action of sub-type IV system is still unknown (Liu et al. 2020).

The most extensively used CRISPR nuclease in CRISPR/Cas technology is *Streptococcus pyogenes* type II Cas9 (SpCas9) (Doudna and Charpentier 2014). The Cas and guide RNA complex identifies the target region through protospacer adjacent motif (PAM) and Cas9 subsequently creates DSBs to induce cellular DNA repair pathways. CRISPR systems are differentiated by a set of *Cas* genes, non-coding RNAs and direct repeats or repetitive elements. These repetitive elements are separated by short variable sequences known as protospacers. The protospacers are acquired by bacteria from the viral DNA particles where the protospacer sequence is marked by a PAM sequence specific to Cas proteins and these protospacers are transcribed as CRISPR RNA array (Barrangou et al. 2007; Brouns et al. 2008; Bhalothia et al. 2020). For example, in case of *Streptococcus pyogenes,* Cas9 targets the protospacer having PAM sequence of 5′-NGG-3′. In order to widen umbrella of the target sites and reduce the off-target effects by Cas9, many variations have been introduced, which rely on modifications in the Cas9 protein structural properties. Among such examples are SpCas9-EQR (NGA-PAM), SpCas9-VQR (NGA-PAM) and SpCas9-VRER (NGCG-PAM), which have higher recognition flexibility but lower cleavage efficiency (Klienstiver et al. 2015; Hu et al. 2016, 2018; Alok et al. 2018, 2021; Yamamoto et al. 2019). SpCas9-NG version which recognizes PAM sequence of only two nucleotides, NG having wider sequence recognition ability induced mutations in Arabidopsis and rice with high efficiency (Hua et al. 2019; Zhong et al. 2019).

Alternatively, novel Cas9 proteins from several other bacteria can be utilized to widen the recognition sequence window. To date, several such Cas9 proteins have been characterized in *Neisseria meningitidis* (Hou et al. 2013), *Staphylococcus aureus* (Ran et al. 2015), *Streptococcus thermophilus* (Muller et al. 2016), *Francisella novicida* (Hirano et al. 2016) and *Campylobacter jejuni* (Kim et al. 2017), which are available to the user. Compared to SpCas9, these Cas9 variants have their own advantages like smaller size of SaCas9. Further, two or more variants of Cas9 can be easily utilized for multiplex editing to target multiple genomic locations as their PAM sites are different (Steinert et al. 2015).

Since the time of development of CRISPR/Cas system for genome editing, the discovery of new CRISPR system has gained pace, which resulted in identification of type VI CRISPR-Cas system which has Cas13 protein which can target RNA molecule (Abudayyeh et al. 2016; Cox et al. 2017). This has its own applications for knockdown of the gene where phenotypic effect of reduced expression can be studied without changing the genomic DNA. This technique has been demonstrated in tobacco and rice for RNA editing and in case of Arabidopsis for editing RNA viruses to induce viral disease resistance (Cox et al. 2017; Aman et al. 2018). Sub-type V CRISPR/Cas systems have variable roles including cleavage and creating nicks in dsDNA, ssDNA, dsRNA and ssRNA (Harrington et al. 2018). Thus, all these variants of CRISPR/Cas systems provide the opportunity to target single- and double-stranded DNA and RNA molecules and broaden the portfolio of genome editing techniques.

11.3.2.2 Mechanism of CRISPR/Cas-Mediated Genome Editing in Plants

In CRISPR/Cas9 system, guide RNA defines the specificity of the target site in the genome where Cas9 induces DSBs (Urnov et al. 2010). Following Cas9 cleavage, the target locus undergoes one of the two primary DNA damage repair processes: NHEJ which is error-prone and homology dependent repair which is a template-based repair mechanism. Genome editing basically exploits these two pathways for introduction of desired mutations or edits in the genomic DNA. During the process of NHEJ, often there is insertion or deletion of few nucleotides resulting in the frameshift mutation of the target gene and deletion of gene function (Perez et al. 2008). A bigger fragment deletion has also been achieved by employing two gRNAs targeting the locations flanking the desired deletion (Chen et al. 2011; Cong et al. 2013; Upadhyay et al. 2013). In contrast to

NHEJ, frequency of HDR is low and it normally occurs in dividing cells with variable efficiency in different cell types and states (Saleh-Gohari and Helleday 2004). However, HDR is very useful tool to induce precise, specified alterations at a target locus when an exogenously injected repair template is present. Here, the repair template is provided in the form of dsDNA or ssDNA having arms homologous to sequences flanking the insertion region.

11.4 Applications of Genome Editing for Improvement of Vegetable Nutrition and Quality

Vegetables are rich source of several essential nutrients in human diet; however, some of these crops may not have adequate levels of certain nutrients, or their nutritional content may be limited due to genetic factors. More importance to yield during the process of selection and breeding resulted in reduced nutrient content in available cultivars. Targeted modifications of a plant's DNA by genome editing has the potential to improve the nutritional content of vegetable crops. Genome editing can be employed to improve the nutritional content and quality of vegetable crops in a number of ways. It can be used to increase the levels of essential nutrients, such as vitamins and minerals, in vegetables or improving flavour by modifying the genes responsible for flavour compounds in vegetables. The modification of genes involved in ripening and decay in vegetables can result in extended shelf life. Apart from this, increasing the tolerance to various pests and diseases through genome editing reduces the need for chemical pesticides and improve the quality of the vegetables. Another way is to modify the composition of the crop to make it more beneficial to human health, such as reducing the level of harmful compounds or increasing the level of antioxidants. There are numerous reports where genome editing has been successfully utilized for improving quality and nutrition of vegetable crops (Karkute et al. 2023) and the present status has been given in Table 11.1.

Table 11.1 Status of genome edited vegetable crops for nutrition and quality improvement.

Plant	Target gene	Trait	References
Tomato	*Pectate lyase (PL), polygalacturonase 2a (PG2a,* and *beta-galactanase (TBG4)*	Development of cell wall and modification of fruit color and weight	Wang et al. (2019a)
	APETALA2a (AP2a), NON-RIPENING (NOR), and FRUITFULL (FUL1/TDR4 and *FUL2/MBP7)*	Development and ripening of fruit	Wang et al. (2019a)
	Psy1 and *CrtR-b2*	Change in carotenoid biosynthesis	D'Ambrosio et al. (2018)
	lncRNA1459	Alters fruit ripening, lycopene, ethylene and carotenoid biosynthesis	Li et al. (2018a)

(*Continued*)

Table 11.1 (Continued)

Plant	Target gene	Trait	References
	SGR1, Blc, LCY-E, LCY-B1, LCY-B2	Increased lycopene content	Li et al. (2018a)
	PDS and *GABA-TP1, GABA-TP2, GABA-TP3, CAT9* and *SSADH*	Increased γ-aminobutyric acid content	Li et al. (2018a)
	GAD2, GAD3	Higher γ-aminobutyric acid content	Nonaka et al., 2017
	SlMYB12	Pink fruit color	Deng et al. (2018)
	RIN	Fruit ripening	Jung et al. (2018)
	SP, MUILT, FAS, CyCb, OVUTE and *FW2.2*	Fruit size and lycopene accumulation	Zsogon et al. (2018)
	SP, SP5, CLV3 and *WUS, GGP1*	Plant structure, fruit ripening, day-length response, vitamin C, and fruit size	Li et al. (2018c)
	RIN and *ethylene*	Fruit ripening	Li et al. (2020)
	RIN	Ethylene production and fruit ripening	Ito et al. (2017)
	SlORRM4	Fruit ripening	Yang et al. (2017b)
	ALC	Delayed fruit ripening causing long-shelf life	Yu et al. (2017)
	CLAVATA-WUSICHEL	Altered locule number	Rodriguez-Leal et al. (2017)
	Solyc12g038510	Jointless mutant, abscission	Roldan et al. (2017)
	phytoene synthase (PSY)	Fruit color	Filler et al. (2017)
	LEAFYCOTYLEDON1-LIKE4	Fruit metabolism	Gago et al. (2017)
	SlIAA9	Parthenocarpy	Ueta et al. (2017)
	SlAGL6	Parthenocarpic fruits	Klap et al. (2017)
	ANT1	Increased contents of anthocyanin	Cermak et al. (2015)
Brinjal	*SmelPPO4, SmelPPO5, and SmelPPO6* genes	Reduced levels of flesh browning	Maioli et al. (2020)
Potato	*GBSS* genes	Starch biosynthesis	Andersson et al. (2018)
	GBSS1	Starch biosynthesis	Kusano et al. (2018)
	GBSS	Starch and tuber quality	Andersson et al. (2018)
	St16DOX	Glycoalkaloids metabolism	Nakayasu et al. (2018)

Table 11.1 (Continued)

Plant	Target gene	Trait	References
Carrot	*F3H*	Change in the anthocyanin biosynthesis pathway	Klimek-Chodacka et al. (2018)
	DcMYB113-like	Anthocyanin biosynthesis	Xu et al. (2019b)
Sweet potato	*GBSSI, SBEII*	Quality improvement, amylose, amylopectin	Wang et al. (2019b)
Chionese kale	*BoaCRTISO*	Carotenoid and chlorophyll biosynthesis, color	Sun et al. (2020)

Source: Adapted from Devi et al. (2022), Ku et al. (2020) under CC BY.

11.4.1 Improvement in the Appearance in Terms of Shape and Size

CRISPR/Cas9 technique is being employed to improve crop form and size based on the demands and preferences of the consumer. Several genes/QTLs responsible for crop appearance are well characterized, which can be the targets for genome editing. Tomato is the crop where maximum information related to shape and size of the fruits is available, making it an ideal crop for genome editing to target these traits. Researchers have manipulated tomato fruit form and size by manipulating several genes such as *OVATE, CLV, fas, lc and ENO* (Rodríguez-Leal et al. 2017; Zsogon et al. 2018; Yuste-Lisbona et al. 2020). It was found that *OVATE* is involved in asymmetric and *SUN* is involved in symmetric fruit elongation (Liu et al. 2002; Lin et al. 2014), whereas *SlWUS* and *SlCLV3* govern tomato locule quantity. Gain-of-function mutation in *CLV3* and partially loss-of-function mutation in *WUS* are identified as *fas* and *lc* loci, respectively, where fruit size is increased in both the mutants (Frary et al. 2000; Tanksley 2004; Munos et al. 2011). Disruption of *CLV-WUS* cis-regulatory areas/promoter also produced larger fruits confirming the role of these genes (Li et al. 2018c). Mutation in *SlAGAMOUS-LIKE6 (SlAGL6)* gene using CRISPR/Cas9 produced parthenocarpic fruits in high temperature without compromising fruit weight, shape, or pollen viability (Klap et al. 2017).

11.4.2 Improvement of the Shelf-Life

Texture is another important trait in fruits and vegetables defining their acceptability as well as price in the market. Plant breeders are constantly working to improve textural features in order to increase shelf life. Texture is governed by several enzymes like pectate lyase, polygalacturonase, β-galactosidase, pectin methylesterase, endo-b-(1,4)-glucanase, expansin and N-glycoprotein-modifying enzymes which are thus the target genes for genome editing to modify the texture. With respect to shelf life, there are several natural mutants available, which have enhanced shelf life with mutations in *alc, Cnr, Nr, nor* and *rin* genes. Although the mutations in these genes have other side effects like low nutritional value, change in colour, loss of flavour etc., they are the important targets for enhancing the shelf life using genome editing approaches (Wang et al. 2020). In contrast, mutation in *alc* gene preserved fruit colour as well as the smell of the fruits along with increased shelf life (Yu et al. 2017). Cell wall disintegrating enzymes can also lead to changes in fruit texture. Pectate lyase is one among such enzymes, which is involved in pectin degradation and thus can destabilize the cell wall and causes fruit softening (Marín-Rodríguez et al. 2002). Therefore, inhibition of the *PL* gene is expected to produce firmer fruits, which has been achieved earlier through RNAi (Yang et al. 2017a). Similar phenotype has been obtained recently by CRISPR/Cas9-mediated mutation of pectate lyase gene in tomato without compromising other fruit traits like nutrition and quality (Uluisik et al. 2016; Wang et al. 2019a). Polygalacturonase suppression by antisense technique in strawberry produces more firm fruits with enhanced shelf life (Quesada et al. 2009). CRISPR/Cas9-mediated knockout of this gene could develop strawberry cultivars for increased shelf life.

11.4.3 Improvement of the Ripening Time

Ripening regulation is one of the major concerns with respect to fleshy fruit/vegetable species. Tomato plants are a suitable model for the study of growth and ripening owing to their short life cycle, ease of transformation and effective replication. Fruit ripening during its maturation is an intricate and complex phenomenon that involves biochemical and physiological pathways governed by several genes. These genes are effective targets for controlling the ripening process using genome editing techniques (Martin-Pizarro and Posé 2018). Tomato fruits with significant delayed ripening have been developed through CRISPR/Cas9 technology by knocking out the *SlORRM4* gene (Yang et al. 2017b). Furthermore, *SlORRM4* deficiency had drastic impact on the mitochondrial functioning. In one of the experiments, where DNA demethylase gene *SlDML2* was knocked out using CRISPR/Cas9 tool, there was increased DNA methylation leading to identification of ripening induced as well as ripening suppressed genes (Lang et al. 2017). This also established the role of DNA methylation in tomato ripening process. *RIN* (*RIPENING INHIBITOR*) gene plays a significant role in controlling the ethylene production in tomato, making it a suitable target to manipulate ethylene level and hence ripening. RIN knockout tomato plants developed using CRISPR/Cas9 impacted the normal fruit ripening process and produced the fruits with incomplete ripening and reduced colour (Ito et al. 2015). Another study showed that CRISPR/Cas9-mediated modification of *lncRNA1459* gene dramatically reduced ethylene synthesis and lycopene accumulation ultimately altering fruit ripening (Li et al. 2018a).

11.4.4 Improvement in Colour of the Fruit/Vegetable

Colour of the fruits and vegetables is regulated by different pigments like anthocyanins, carotenoids and polyphenols. Colour is one of the most important factors influencing the preference of the consumers. For example, in case of tomatoes, pink fruits are preferred in Asian countries, whereas red fruits are consumed in European and American countries (Lin et al. 2014). Therefore, depending on the consumer's preference, the colour of the fruits and vegetables can be modified by manipulating the genes involved in the production of various pigments using genome editing technologies. For example, loss of flavonoid pigments in tomato results in pink colour of the fruits. Development of pink tomatoes has been accomplished by targeting the *SlMYB12* gene which is a transcription factor involved in flavonoid biosynthesis using CRISPR/Cas9 technique (Yang et al. 2019). Similarly, yellow and purple tomatoes have been successfully produced by mutating *PSY1* and *ANT1* genes, respectively (Filler et al. 2017; Chattopadhyay et al. 2021). Yellow phenotype was due to the reduction of overall lycopene content in *PSY1* mutant and purple colour was the result of higher accumulation of anthocyanin (Vu et al. 2020). CRISPR/Cas9-mediated metabolic engineering of carotenoid metabolism pathway has been carried out by targeting five genes which significantly enhanced the lycopene content in tomato (Li et al. 2018d). There was also the prevention of lycopene catabolism to β- and α-carotene. Lycopene content has also been enhanced in Chinese kale by knockout of *carotenoid isomerase* gene (*BoaCRTISO*) which resulted in yellow colour and improved market prospects (Sun et al. 2020). The solid purple carrot has been mutated to produce yellow carrots by using CRISPR/Cas9-mediated knockout of *DcMYB7* gene which is an *R2R3-MYB* (Xu et al. 2019a). On the other hand, knockout of *F3H* gene involved in anthocyanin biosynthesis resulted in callus discolouration in carrots due to reduced accumulation of anthocyanin (Klimek-Chodacka et al. 2018).

Lack of chlorophyll pigmentation due to disruption of *phytoene desaturase* (*PDS*) gene in certain plants impacts carotenoids and chlorophyll production, resulting in albino plant phenotypes. Apart from few papers in which albinism was used to standardize the gene-editing process using CRISPR/Cas-based technology, limited information is available in case of vegetable crops regarding the albino phenotypes. So far, phytoene desaturase gene in cabbage, carrot, melon, tomato and watermelon has been knocked out using CRISPR/Cas technology producing albino phenotype (Pan et al. 2016; Tian et al. 2017; Hooghvorst et al. 2019; Ma et al. 2019; Xu et al. 2019b).

11.4.5 Biofortification of Vegetable Crops Through Genome Editing

With the increasing awareness for health among consumers, there is increased demand for healthier and nutrition-rich foods. Fruits and vegetables have several important nutrients responsible for anti-cancer, anti-inflammatory and anti-oxidant properties. To fulfil the demand of consumers for healthier food, breeders are aiming to develop the fruits and vegetable varieties rich in nutrition and health beneficial compounds. Therefore, biofortification of fruits and vegetables to enhance the content of various nutrients like iron, zinc, anthocyanins, carotenoids, GABA, vitamins etc. is necessary and genome editing approach can be aptly used to achieve this goal.

11.4.5.1 Metabolic Engineering of Carotenoid Biosynthesis Pathway

Carotenoids are the source of antioxidant activities in cells and also prevent eye diseases in humans. Lycopene and phytoene are important for reducing the risk of cardiovascular disease. It is not possible for humans to synthesize carotenoid in their body and hence the requirement must be fulfilled through the diet. Earlier genetic engineering approach was used to produce β-carotene in rice by introducing *CrtI* and *PSY* genes. In tomato, *Psy1* and *CrtR-b2* genes were knocked out to alter the carotene content by manipulating the carotenoid biosynthesis pathway (D'Ambrosio et al. 2018). Lycopene-enriched tomatoes were developed by disrupting five carotenoid biosynthetic genes (*SGR1*, *LCYe*, *BLC*, *LCY*-B1 and *LCY*-B2), which dramatically enhanced the lycopene content to the level of fivefolds (Li et al. 2018d). In another study, mutation in *SlDDB1, SlDET1, SlCYC-B* genes using CRISPR-based Target-AID base editing technique enhanced carotenoid, lycopene and β-carotene content (Hunziker et al. 2020).

11.4.5.2 Increasing γ-Amino Butyric Acid and Vitamin D Content

GABA is an important non-protein amino acid which acts as a neurotransmitter and has inhibitory properties functioning in the anti-anxiety and blood pressure control systems (Nuss 2015). As a result, the food industry is now focussed on the development of novel GABA-rich foods. The glutamate decarboxylase (GAD) enzyme has essential role in biosynthesis of GABA and is responsible for the decarboxylation of glutamate to GABA. The action of GAD is inhibited by its own C-terminal autoinhibitory domain. Thus, the GABA content can be enhanced by removing this autoinhibitory domain from GAD. The CRISPR/Cas9 technique was successfully used to completely remove the C-terminus resulting in sevenfold GABA accumulation in edited tomato lines (Nonaka et al. 2017). Alternatively, multiplex editing of *SlGABA*-Ts and *SlSSADH* genes through CRISPR/Cas9 produced almost 20-fold higher GABA in tomato fruits but with reduced size of fruit and less production (Li et al. 2018b). GABA-rich vegetables are undeniably beneficial to human health. Nevertheless, chasing high GABA content indiscriminately may result in a reduced glutamate content along with faulty fruit phenotype (Gramazio et al. 2020).

Vitamin D is highly essential component for human health as its deficiency affects immune function. Plants are very poor source of Vitamin D and hence its deficiency is a global health problem. People with Vitamin D deficiency are at a greater risk of cancer, neurocognitive decline, dementia, depression and Parkinson's disease (Li et al. 2022). Undoubtedly, synthesis of Vitamin D in plant parts would be one of the best strategies to fulfil its requirement in human diet. With the CRISPR/Cas9 tool in hands, researchers have actually made plants to synthesize provitamin D3 in genome edited tomato (Li et al. 2022). This has been achieved by modifications in the duplicated section of pathway for phytosterol synthesis by mutating the *7-dehydrocholesterol reductase* (*Sl7-DR2*) gene which converts 7-dehydrocholesterol (7-DHC) or provitamin D3 into cholesterol. The green tomato fruits of these edited lines contain provitamin D3 to level of 0.3 μg/g dry weight, whereas in red fruit it was 0.2 μg/g dry weight which can be further increased by exposure to UVB.

11.4.6 Improvement of Starch Content

The quality of potato starch is important in many food production and processing industries. CRISPR-mediated genome editing improved starch quality in potato by knockout of different genes such as *starch-branching enzyme* genes (*SBE1* and *SBE2*) (Zhao et al. 2021), *granule-bound starch synthase* (*GBSS*) (Andersson et al. 2018; Johansen et al. 2019) and *starch synthase* (*SS6*) (Sevestre et al. 2020), which are involved in starch biosynthesis. CRISPR/Cas9 technology was employed to successfully improve the starch quality by targeting the *IbGBSSI* (*granule-bound starch synthase I*) and *IbSBEII* (*starch-branching enzyme II*) in sweet potato (Wang et al. 2019b). Malate build-up was achieved by CRISPR/Cas9-mediated modification of *SlALMT9* gene (Al-activated malate transporter 9) in tomatoes. (Ye et al. 2017).

11.4.7 Elimination of Anti-Nutritional Factors

The nutritional quality of fruits, vegetables and other crops is negatively impacted by several chemicals such as cadmium, phytic acid, gluten protein, etc. Here comes the role of genome editing to delete anything that is negatively regulated or impacted. The harmful chemicals, which are well known as antinutritional factors, can be systematically removed from plants through genome editing. For example, phytic acid present in several crops is critical antinutritional factor because it cannot be metabolized by humans which do not have phytic acid degrading enzymes. When individuals consume a large amount of phytic acid, absorption of minerals and protein is restricted due to the formation of complex of these nutrients with phytic acid (Oatway et al. 2001). ITPK gene encodes the enzyme required for catalysing the penultimate step of phytate synthesis. This ITPK gene mutation in rapeseeds using CRISPR/Cas9 resulted in 35% reduction in concentration of phytic acid without compromising plant growth and yield (Sashidhar et al. 2020).

Quality of fruits and vegetables may also be enhanced by decreasing undesirable chemical contents in plants. Potato includes two steroidal glycoalkaloids that are bitter and poisonous when consumed (α-solanine and α-chaconine). Using CRISPR/Cas9 technology, the *St16DOX* gene has been silenced to totally eliminate the accumulation of these steroidal glycoalkaloids in potato tubers (Nakayasu et al. 2018). Similar phenotype of reduced steroidal glycoalkaloids was obtained by knocking out of *Sterol Sidechain Reductase 2* (*SSR2*) using TALENs (Yasumoto et al. 2019). In vegetables like potato and brinjal, browning is caused by polyphenol oxidase (PPO) enzyme which mediates polyphenol oxidation. In order to reduce this browning, *PPO* gene was mutated using CRISPR/Cas9 technique in potato and brinjal. In genome-edited mutant potatoes, 69% reduction in PPO activity and 73% reduction in browning was observed in tubers (González et al. 2020). In case of brinjal, there are three *PPO* genes *SmelPPO4*, *SmelPPO5* and *SmelPPO6* involved in browning of cut fruits upon exposure to external environment. Simultaneous deletion of these three PPO genes was achieved by CRISPR/Cas9 tool which significantly prevented the browning of cut fruits (Maioli et al. 2020). In tomato, oxalic acid is considered as anti-nutritional factor as it is responsible for primary and secondary hyperoxaluria. To reduce the oxalic acid content in tomato fruits, *L1L4/NF-YB6* gene was mutated using ZFN, which resulted in successful reduction of oxalic acid levels in tomato (Gago et al. 2017).

11.5 Challenges and Future Prospects

Genome editing technologies, especially CRISPR/Cas-based tools are advanced to have several kinds of applications to enhance the quality, yield and resistance to various biotic and abiotic stresses (Doudna and Charpentier 2014; Langner et al. 2018). The technique is efficient in mutating the target gene or changing its function precisely through NHEJ to have desired traits in crops. However, like every technology has limitations, the CRISPR/Cas technique also has few hurdles that need to be taken care of while using it. One of the major limitations of all the genome editing techniques is off-target editing in genomes which can be avoided by judicious selection of gene and careful designing of gRNAs. Therefore, it is challenging to edit the plants without off-target effects in the absence of complete genome sequence information. It has limitations in alteration of various agronomic traits which are usually quantitative in nature and governed by more than one gene. Although multiplex editing approach can be used to target multiple genes at a time, the complexity of quantitative trait makes it difficult to achieve the desired phenotypes. Alternatively, such pleotropic traits can be modified by site-directed deletion, insertion or chromosomal recombination for creating desired alleles in the elite cultivars (Zhu et al. 2020).

Availability of standardized transformation protocol for delivering the constructs in plant cells and subsequent regeneration from transformed plant cells is another major hurdle for genome editing in vegetable crops (Lino et al. 2018; Chuang et al. 2021). These protocols are the preliminary requirements for carrying out the genome editing work. Several vegetable crops are difficult to transform, and it is more difficult to regenerate plants from tissues (Niazian et al. 2017). The worldwide approval of genome-edited crops for commercialization will speed up the development of these protocols in important vegetable crops to harness the benefit of this wonderful technology.

In addition to genome sequence information and transformation protocol which vary from crop to crop, there are some limitations which are inherent to the technology. Among them, editing efficiency of the gRNA is most critical factor. It is reported that all the gRNA does not have same efficiency of editing and therefore, selection of best gRNA among the available ones is critical for successful editing of the target gene. Secondary structure formation of gRNA alone and after pairing with target sequence and expression levels of gRNA and Cas9 also contributes in the success of target editing. Considering the enormous potential of the genome editing techniques for improvement of vegetable crops, we anticipate that all these limitations or hurdles will have solutions soon.

11.6 Conclusion

CRISPR/Cas9 is a promising technology for increased precision-based plant breeding that is widely used in plant molecular biology. However, there are some limitations associated with this technique which we believe can be taken care of in the near future. Various CRISPR/Cas9-based tools have already shown encouraging results in editing essential vegetable crops. Some of the genome-edited plant products are approved for commercial cultivation and human consumption. High GABA tomato is the first genome-edited product that has been approved for human consumption in Japan (Waltz 2022). This has led the path for adoption of several such vegetable crops due to dramatic increase in nutrition-rich food demand by humans worldwide. Many countries including US, China, India and recently European union have exempted biosafety regulation of CRISPR-based crops unlike GMOs which will allow the introduction of many genome-edited crops to the farmers' field and consumers' plate.

References

Abbott, J.A. (1999). Quality measurement of fruits and vegetables. *Postharvest Biology and Technology* 15 (3): 207–225.

Abdallah, N.A., Prakash, C.S., and McHughen, A.G. (2015). Genome editing for crop improvement: challenges and opportunities. *GM Crops Food* 6: 183–205. https://doi.org/10.1080/21645698.2015.1129937.

Abudayyeh, O.O., Gootenberg, J.S., Konermann, S. et al. (2016). C2c2 is a single-component programmable RNA-guided RNA-targeting CRISPR effector. *Science* 353 (6299): aaf5573.

Ahloowalia, B.S., Maluszynski, M., and Nichterlein, K. (2004). Global impact of mutation-derived varieties. *Euphytica* 135: 187–204.

Alok, A., Kumar, J., and Upadhyay, S.K. (2018). Engineering in hairy roots using CRISPR/Cas9-mediated editing. In: *Hairy Roots* (ed. V. Srivastava, S. Mehrotra, and S. Mishra). Singapore: Springer https://doi.org/10.1007/978-981-13-2562-5_15.

Alok, A., Chauhan, H., Upadhyay, S.K. et al. (2021). Compendium of plant-specific CRISPR vectors and their technical advantages. *Life* 11: 1021. https://doi.org/10.3390/life11101021.

Aman, R., Ali, Z., Butt, H. et al. (2018). RNA virus interference via CRISPR/Cas13a system in plants. *Genome Biology* 19 (1): 1–9.

Andersson, M., Turesson, H., Olsson, N. et al. (2018). Genome editing in potato via CRISPR-Cas9 ribonucleoprotein delivery. *Physiologia Plantarum* 164: 378–384. https://doi.org/10.1111/ppl.12731.

Barrangou, R., Fremaux, C., Deveau, H. et al. (2007). CRISPR provides acquired resistance against viruses in prokaryotes. *Science* 315 (5819): 1709–1712.

Bernardo, R. (2016). Bandwagons I, too, have known. *Theoretical and Applied Genetics* 129: 2323–2332.

Bhalothia, P., Yajnik, K., Alok, A., and Upadhayay, S.K. (2020). The current progress of CRISPR/Cas9 development in plants. In: *Genome Engineering via CRISPR-Cas9 Systems* (ed. S. Dhar). Elsevier https://doi.org/10.1016/B978-0-12-818140-9.00011-8.

Bisbis, M.B., Gruda, N., and Blanke, M. (2018). Potential impacts of climate change on vegetable production and product quality – a review. *Journal of Cleaner Production* 170: 1602–1620.

Brouns, S.J., Jore, M.M., Lundgren, M. et al. (2008). Small CRISPR RNAs guide antiviral defense in prokaryotes. *Science* 321 (5891): 960–964.

Campbell, T.M. II (2004). *The China Study: The Most Comprehensive Study of Nutrition ever Conducted and the Startling Implications for Diet, Weight Loss and Long-Term Health*. BenBella Books, Inc.

Čermak, T., Baltes, N.J., Čegan, R. et al. (2015). High frequency, precise modification of the tomato genome. *Genome Biology* 16: 1–15.

Chattopadhyay, T., Hazra, P., Akhtar, S. et al. (2021). Skin colour, carotenogenesis and chlorophyll degradation mutant alleles: genetic orchestration behind the fruit colour variation in tomato. *Plant Cell Reports* 40: 767–782.

Chen, F., Pruett-Miller, S.M., Huang, Y. et al. (2011). High-frequency genome editing using ssDNA oligonucleotides with zinc-finger nucleases. *Nature Methods* 8 (9): 753–755.

Chuang, Y.F., Phipps, A.J., Lin, F.L. et al. (2021). Approach for in vivo delivery of CRISPR/Cas system: a recent update and future prospect. *Cellular and Molecular Life Sciences* 78: 2683–2708.

Cong, L., Ran, F.A., Cox, D. et al. (2013). Multiplex genome engineering using CRISPR/Cas systems. *Science* 339 (6121): 819–823.

Cox, D.B., Gootenberg, J.S., Abudayyeh, O.O. et al. (2017). RNA editing with CRISPR-Cas13. *Science* 358 (6366): 1019–1027.

D'Ambrosio, C., Stigliani, A.L., and Giorio, G. (2018). CRISPR/Cas9 editing of carotenoid genes in tomato. *Transgenic Research* 27 (4): 367–378.

Deng, L., Wang, H., Sun, C. et al. (2018). Efficient generation of pink-fruited tomatoes using CRISPR/Cas9 system. *Journal of Genetics and Genomics* 45 (1): 51–54.

Devi, R., Chauhan, S., and Dhillon, T.S. (2022). Genome editing for vegetable crop improvement: challenges and future prospects. *Frontiers in Genetics* 13: 1037091.

Dias, J.S. (2012). Nutritional quality and health benefits of vegetables: a review. *Food and Nutrition Sciences* 3 (10): 1354–1374.

Dixit, S., Shukla, A., Singh, V. et al. (2021). Bioprospecting of natural compounds for industrial and medical applications; current scenario and bottleneck. In: *Bioprospecting of Plant Biodiversity for Industrial Molecules* (ed. S.K. Upadhyay), 53–57. Wiley https://doi.org/10.1002/9781119718017.ch3.

Doudna, J.A. and Charpentier, E. (2014). The new frontier of genome engineering with CRISPR-Cas9. *Science* 346 (6213): 1258096.

Erpen-Dalla Corte, L.M., Mahmoud, L.S., Moraes, T. et al. (2019). Development of improved fruit, vegetable, and ornamental crops using the CRISPR/Cas9 genome editing technique. *Plants* 8 (12): 601.

Filler, H.S., Melamed, B.C., and Levy, A.A. (2017). Targeted recombination between homologous chromosomes for precise breeding in tomato. *Nature Communications* 8: 15605. https://doi.org/10.1038/ncomms15605.

Frary, A., Nesbitt, T.C., Frary, A. et al. (2000). fw2. 2: a quantitative trait locus key to the evolution of tomato fruit size. *Science* 289 (5476): 85–88.

Gago, C., Drosou, V., Paschalidis, K. et al. (2017). Targeted gene disruption coupled with metabolic screen approach to uncover the LEAFY COTYLEDON1-LIKE4 (L1L4) function in tomato fruit metabolism. *Plant Cell Reports* 36: 1065–1082.

Garg, M., Sharma, N., Sharma, S. et al. (2018). Biofortified crops generated by breeding, agronomy, and transgenic approaches are improving lives of millions of people around the world. *Frontiers in Nutrition* 5: 12.

Glenn, K.C., Alsop, B., Bell, E. et al. (2017). Bringing new plant varieties to market: plant breeding and selection practices advance beneficial characteristics while minimizing unintended changes. *Crop Science* 57 (6): 2906–2921.

González, M.N., Massa, G.A., Andersson, M. et al. (2020). Reduced enzymatic browning in potato tubers by specific editing of a polyphenol oxidase gene via ribonucleoprotein complexes delivery of the CRISPR/Cas9 system. *Frontiers in Plant Science* 10: 1649.

Gramazio, P., Takayama, M., and Ezura, H. (2020). Challenges and prospects of new plant breeding techniques for GABA improvement in crops: tomato as an example. *Frontiers in Plant Science* 11: 577980.

Harrington, L.B., Burstein, D., Chen, J.S. et al. (2018). Programmed DNA destruction by miniature CRISPR-Cas14 enzymes. *Science* 362 (6416): 839–842.

Herrera, E., Jiménez, R., Aruoma, O.I. et al. (2009). Aspects of antioxidant foods and supplements in health and disease. *Nutrition Reviews* 67 (suppl_1): S140–S144.

Hirano, H., Gootenberg, J.S., Horii, T. et al. (2016). Structure and engineering of Francisella novicida Cas9. *Cell* 164 (5): 950–961.

Holme, I.B., Gregersen, P.L., and Brinch-Pedersen, H. (2019). Induced genetic variation in crop plants by random or targeted mutagenesis: convergence and differences. *Frontiers in Plant Science* 10: 1468.

Holscher, H.D. (2017). Dietary fiber and prebiotics and the gastrointestinal microbiota. *Gut Microbes* 8 (2): 172–184.

Hooghvorst, I., López-Cristoffanini, C., and Nogués, S. (2019). Efficient knockout of phytoene desaturase gene using CRISPR/Cas9 in melon. *Scientific Reports* 9 (1): 1–7.

Hou, Z., Zhang, Y., Propson, N.E. et al. (2013). Efficient genome engineering in human pluripotent stem cells using Cas9 from Neisseria meningitidis. *Proceedings of the National Academy of Sciences* 110 (39): 15644–15649.

Hu, X., Wang, C., Fu, Y. et al. (2016). Expanding the range of CRISPR/Cas9 genome editing in rice. *Molecular Plant* 9 (6): 943–945.

Hu, X., Meng, X., Liu, Q. et al. (2018). Increasing the efficiency of CRISPR-Cas9-VQR precise genome editing in rice. *Plant Biotechnology Journal* 16 (1): 292–297.

Hua, K., Tao, X., Han, P. et al. (2019). Genome engineering in rice using Cas9 variants that recognize NG PAM sequences. *Molecular Plant* 12 (7): 1003–1014.

Hunter, M.C., Smith, R.G., Schipanski, M.E. et al. (2017). Agriculture in 2050: recalibrating targets for sustainable intensification. *Bioscience* 67: 386–391. https://doi.org/10.1093/biosci/bix010.

Hunziker, J., Nishida, K., Kondo, A. et al. (2020). Multiple gene substitution by Target-AID base-editing technology in tomato. *Scientific Reports* 10 (1): 20471.

Ito, Y., Nishizawa-Yokoi, A., Endo, M. et al. (2015). CRISPR/Cas9-mediated mutagenesis of the RIN locus that regulates tomato fruit ripening. *Biochemical and Biophysical Research Communications* 467 (1): 76–82.

Ito, Y., Nishizawa-Yokoi, A., Endo, M. et al. (2017). Reevaluation of the RIN mutation and the role of RIN in the induction of tomato ripening. *Nature Plants* 3 (11): 866–874.

Jaganathan, D., Ramasamy, K., Sellamuthu, G. et al. (2018). CRISPR for crop improvement: an update review. *Frontiers in Plant Science* 9: 985.

Jankowicz-Cieslak, J., Mba, C., and Till, B.J. (2017). Mutagenesis for crop breeding and functional genomics. In: *Biotechnologies for Plant Mutation Breeding: Protocols* (ed. J. Jankowicz-Cieslak, T.H. Tai, J. Kumlehn, and B.J. Till), 3–18. Springer.

Johansen, I.E., Liu, Y., Jørgensen, B. et al. (2019). High efficacy full allelic CRISPR/Cas9 gene editing in tetraploid potato. *Scientific Reports* 9 (1): 1–7.

Jung, Y.J., Lee, G.J., Bae, S., and Kang, K.K. (2018). Reduced ethylene production in tomato fruits upon CRSPR/Cas9-mediated LeMADS-RIN mutagenesis. *Horticultural Science and Technology* 36 (3): 396–405.

Karkute, S.G., Singh, A.K., Gupta, O.P. et al. (2017). CRISPR/Cas9 mediated genome engineering for improvement of horticultural crops. *Frontiers in Plant Science* 8: 1635.

Karkute, S.G., Tribhuvan, K.U., and Yadav, S. (2023). Nutrition rich fruits and vegetables: status and opportunities in the era of genome editing. *Journal of Allbiosolution* 5 (1): 1–8.

Kim, E., Koo, T., Park, S.W. et al. (2017). In vivo genome editing with a small Cas9 orthologue derived from *Campylobacter jejuni*. *Nature Communications* 8 (1): 14500.

Klap, C., Yeshayahou, E., Bolger, A.M. et al. (2017). Tomato facultative parthenocarpy results from Sl AGAMOUS-LIKE 6 loss of function. *Plant Biotechnology Journal* 15 (5): 634–647.

Kleinstiver, B.P., Prew, M.S., Tsai, S.Q. et al. (2015). Engineered CRISPR-Cas9 nucleases with altered PAM specificities. *Nature* 523 (7561): 481–485.

Kleter, G.A., Kuiper, H.A., and Kok, E.J. (2019). Gene-edited crops: towards a harmonized safety assessment. *Trends in Biotechnology* 37 (5): 443–447.

Klimek-Chodacka, M., Oleszkiewicz, T., Lowder, L.G. et al. (2018). Efficient CRISPR/Cas9-based genome editing in carrot cells. *Plant Cell Reports* 37: 575–586.

Ku, H.K. and Ha, S.H. (2020). Improving nutritional and functional quality by genome editing of crops: status and perspectives. *Frontiers in Plant Science* 11: 577313.

Kusano, H., Ohnuma, M., Mutsuro-Aoki, H. et al. (2018). Establishment of a modified CRISPR/Cas9 system with increased mutagenesis frequency using the translational enhancer dMac3 and multiple guide RNAs in potato. *Scientific Reports* 8 (1): 13753.

Lang, Z., Wang, Y., Tang, K. et al. (2017). Critical roles of DNA demethylation in the activation of ripening-induced genes and inhibition of ripening-repressed genes in tomato fruit. *Proceedings of the National Academy of Sciences* 114 (22): E4511–E4519.

Langner, T., Kamoun, S., and Belhaj, K. (2018). CRISPR crops: plant genome editing toward disease resistance. *Annual Review of Phytopathology* 56: 479–512.

Laurie, S.M., Faber, M., Van Jaarsveld, P.J. et al. (2012). β-Carotene yield and productivity of orange-fleshed sweet potato (Ipomoea batatas L. Lam.) as influenced by irrigation and fertilizer application treatments. *Scientia Horticulturae* 142: 180–184.

Li, R., Fu, D., Zhu, B. et al. (2018a). CRISPR/Cas9-mediated mutagenesis of lncRNA1459 alters tomato fruit ripening. *The Plant Journal* 94 (3): 513–524.

Li, R., Li, R., Li, X. et al. (2018b). Multiplexed CRISPR/Cas9-mediated metabolic engineering of γ-aminobutyric acid levels in *Solanum lycopersicum*. *Plant Biotechnology Journal* 16 (2): 415–427.

Li, T., Yang, X., Yu, Y. et al. (2018c). Domestication of wild tomato is accelerated by genome editing. *Nature Biotechnology* 36 (12): 1160–1163.

Li, X., Wang, Y., Chen, S. et al. (2018d). Lycopene is enriched in tomato fruit by CRISPR/Cas9-mediated multiplex genome editing. *Frontiers in Plant Science* 9: 1–12.

Li, S., Zhu, B., Pirrello, J. et al. (2020). Roles of RIN and ethylene in tomato fruit ripening and ripening associated traits. *New Phytologist* 226 (2): 460–475.

Li, J., Scarano, A., Gonzalez, N.M. et al. (2022). Biofortified tomatoes provide a new route to vitamin D sufficiency. *Nature Plants* 8 (6): 611–616.

Lin, T., Zhu, G., Zhang, J. et al. (2014). Genomic analyses provide insights into the history of tomato breeding. *Nature Genetics* 46 (11): 1220–1226.

Lino, C.A., Harper, J.C., Carney, J.P., and Timlin, J.A. (2018). Delivering CRISPR: a review of the challenges and approaches. *Drug Delivery* 25 (1): 1234–1257.

Liu, J., Van Eck, J., Cong, B., and Tanksley, S.D. (2002). A new class of regulatory genes underlying the cause of pear-shaped tomato fruit. *Proceedings of the National Academy of Sciences* 99 (20): 13302–13306.

Liu, Z., Dong, H., Cui, Y. et al. (2020). Application of different types of CRISPR/Cas-based systems in bacteria. *Microbial Cell Factories* 19 (1): 1–14.

Lumpkin, T., Weinberger, K., and Moore, S. (2005). *Increasing Income Through Fruit and Vegetable Production Opportunities and Challenges*. CGIAR.

Ma, C., Liu, M., Li, Q. et al. (2019). Efficient BoPDS gene editing in cabbage by the CRISPR/Cas9 system. *Horticultural Plant Journal* 5 (4): 164–169.

Maioli, A., Gianoglio, S., Moglia, A. et al. (2020). Simultaneous CRISPR/Cas9 editing of three ppo genes reduces fruit flesh browning in Solanum melongena L. *Frontiers in Plant Science* 11: 607161.

Makarova, K.S. and Koonin, E.V. (2015). Annotation and classification of CRISPR-Cas systems. *CRISPR: Methods and Protocols* 1311: 47–75.

Marín-Rodríguez, M.C., Orchard, J., and Seymour, G.B. (2002). Pectate lyases, cell wall degradation and fruit softening. *Journal of Experimental Botany* 53 (377): 2115–2119.

Martín-Pizarro, C. and Posé, D. (2018). Genome editing as a tool for fruit ripening manipulation. *Frontiers in Plant Science* 9: 1415.

Maziya-Dixon, B., Kling, J.G., Menkir, A., and Dixon, A. (2000). Genetic variation in total carotene, iron, and zinc contents of maize and cassava genotypes. *Food and Nutrition Bulletin* 21 (4): 419–422.

Mazzucato, A., Papa, R., Bitocchi, E. et al. (2008). Genetic diversity, structure and marker-trait associations in a collection of Italian tomato (*Solanum lycopersicum* L.) landraces. *Theoretical and Applied Genetics* 116: 657–669.

Miller, J.C., Holmes, M.C., Wang, J. et al. (2007). An improved zinc-finger nuclease architecture for highly specific genome editing. *Nature Biotechnology* 25 (7): 778–785.

Müller, M., Lee, C.M., Gasiunas, G. et al. (2016). Streptococcus thermophilus CRISPR-Cas9 systems enable specific editing of the human genome. *Molecular Therapy* 24 (3): 636–644.

Muños, S., Ranc, N., Botton, E. et al. (2011). Increase in tomato locule number is controlled by two single-nucleotide polymorphisms located near WUSCHEL. *Plant Physiology* 156 (4): 2244–2254.

Nakayasu, M., Akiyama, R., Lee, H.J. et al. (2018). Generation of α-solanine-free hairy roots of potato by CRISPR/Cas9 mediated genome editing of the St16DOX gene. *Plant Physiology and Biochemistry* 131: 70–77.

Niazian, M., Noori, S.S., Galuszka, P., and Mortazavian, S.M.M. (2017). Tissue culture-based agrobacterium-mediated and in planta transformation methods. *Soil and Water Research* 53 (4): 133–143.

Nonaka, S., Arai, C., Takayama, M. et al. (2017). Efficient increase of γ-aminobutyric acid (GABA) content in tomato fruits by targeted mutagenesis. *Scientific Reports* 7 (1): 1–14.

Nuss, P. (2015). Anxiety disorders and GABA neurotransmission: a disturbance of modulation. *Neuropsychiatric Disease and Treatment* 165–175. https://doi.org/10.2147/NDT.S58841.

Oatway, L., Vasanthan, T., and Helm, J.H. (2001). Phytic acid. *Food Reviews International* 17 (4): 419–431.

Pan, C., Ye, L., Qin, L. et al. (2016). CRISPR/Cas9-mediated efficient and heritable targeted mutagenesis in tomato plants in the first and later generations. *Scientific Reports* 6 (1): 24765.

Perez, E.E., Wang, J., Miller, J.C. et al. (2008). Establishment of HIV-1 resistance in CD4+ T cells by genome editing using zinc-finger nucleases. *Nature Biotechnology* 26 (7): 808–816.

Quesada, M.A., Blanco-Portales, R., Posé, S. et al. (2009). Antisense down-regulation of the FaPG1 gene reveals an unexpected central role for polygalacturonase in strawberry fruit softening. *Plant Physiology* 150 (2): 1022–1032.

Ran, F.A., Cong, L., Yan, W.X. et al. (2015). In vivo genome editing using Staphylococcus aureus Cas9. *Nature* 520 (7546): 186–191.

Rodríguez-Leal, D., Lemmon, Z.H., Man, J. et al. (2017). Engineering quantitative trait variation for crop improvement by genome editing. *Cell* 171 (2): 470–480.

Roldan, M.V.G., Perilleux, C., Morin, H. et al. (2017). Natural and induced loss of function mutations in SlMBP21 MADS-box gene led to jointless-2 phenotype in tomato. *Scientific Reports* 7 (1): 4402.

Saleh-Gohari, N. and Helleday, T. (2004). Conservative homologous recombination preferentially repairs DNA double-strand breaks in the S phase of the cell cycle in human cells. *Nucleic Acids Research* 32 (12): 3683–3688.

Sashidhar, N., Harloff, H.J., Potgieter, L., and Jung, C. (2020). Gene editing of three BnITPK genes in tetraploid oilseed rape leads to significant reduction of phytic acid in seeds. *Plant Biotechnology Journal* 18 (11): 2241–2250.

Septembre-Malaterre, A., Remize, F., and Poucheret, P. (2018). Fruits and vegetables, as a source of nutritional compounds and phytochemicals: changes in bioactive compounds during lactic fermentation. *Food Research International* 104: 86–99. https://doi.org/10.1016/j.foodres.2017.09.031.

Sevestre, F., Facon, M., Wattebled, F., and Szydlowski, N. (2020). Facilitating gene editing in potato: a single-nucleotide polymorphism (SNP) map of the *Solanum tuberosum* L. cv. Desiree genome. *Scientific Reports* 10 (1): 2045.

Steinert, J., Schiml, S., Fauser, F., and Puchta, H. (2015). Highly efficient heritable plant genome engineering using Cas9 orthologues from *Streptococcus thermophilus* and *Staphylococcus aureus*. *The Plant Journal* 84 (6): 1295–1305.

Sun, B., Jiang, M., Zheng, H. et al. (2020). Color-related chlorophyll and carotenoid concentrations of Chinese kale can be altered through CRISPR/Cas9 targeted editing of the carotenoid isomerase gene BoaCRTISO. *Horticulture Research* 7 (1): 161.

Sushmita, K.G., Upadhyay, S.K., and Verma, P.C. (2021). An overview of genome-engineering methods. In: *Genome Engineering for Crop Improvement*. Wiley https://doi.org/10.1002/9781119672425.ch1.

Tanksley, S.D. (2004). The genetic, developmental, and molecular bases of fruit size and shape variation in tomato. *The Plant Cell* 16 (suppl_1): S181–S189.

Tian, S., Jiang, L., Gao, Q. et al. (2017). Efficient CRISPR/Cas9-based gene knockout in watermelon. *Plant Cell Reports* 36: 399–406.

Tian, S.W., Xing, S.N., and Xu, Y. (2021). Advances in CRISPR/Cas9-mediated genome editing on vegetable crops. *In Vitro Cellular & Developmental Biology. Plant* 57: 672–682.

Tilman, D., Balzer, C., Hill, J., and Befort, B.L. (2011). Global food demand and the sustainable intensification of agriculture. *Proceedings of the National Academy of Sciences of the United States of America* 108: 20260–20264. https://doi.org/10.1073/pnas.1116437108.

Ueta, R., Abe, C., Watanabe, T. et al. (2017). Rapid breeding of parthenocarpic tomato plants using CRISPR/Cas9. *Scientific Reports* 7 (1): 507.

Ülger, T.G., Songur, A.N., Çırak, O., and Çakıroğlu, F.P. (2018). Role of vegetables in human nutrition and disease prevention. In: *Vegetables – Importance of Quality Vegetables to Human Health* (ed. Md.Asaduzzaman and T. Asao), 7–32. IntechOpen https://doi.org/10.5772/intechopen.77038.

Uluisik, S., Chapman, N.H., Smith, R. et al. (2016). Genetic improvement of tomato by targeted control of fruit softening. *Nature Biotechnology* 34 (9): 950–952.

Upadhyay, S.K. (2021). *Genome Engineering for Crop Improvement*. Wiley https://doi.org/10.1002/9781119672425.

Upadhyay, S.K. and Singh, S.P. (2021). *Bioprospecting of Plant Biodiversity for Industrial Molecules*. Wiley https://doi.org/10.1002/9781119718017.

Upadhyay, S.K. and Singh, S.P. (2023). *Plants as Bioreactors for Industrial Molecules*. Wiley https://doi.org/10.1002/9781119875116.

Upadhyay, S.K., Kumar, J., Alok, A. et al. (2013). RNA guided genome editing for multiple target gene mutations in wheat. *G3: Genes|Genomes|Genetics* 3 (12): 2233–2238.

Urnov, F.D., Rebar, E.J., Holmes, M.C. et al. (2010). Genome editing with engineered zinc finger nucleases. *Nature Reviews Genetics* 11 (9): 636–646.

Vu, T.V., Sivankalyani, V., Kim, E.J. et al. (2020). Highly efficient homology-directed repair using CRISPR/Cpf1-geminiviral replicon in tomato. *Plant Biotechnology Journal* 18 (10): 2133–2143.

Wallace, T.C., Bailey, R.L., Blumberg, J.B. et al. (2020). Fruits, vegetables, and health: a comprehensive narrative, umbrella review of the science and recommendations for enhanced public policy to improve intake. *Critical Reviews in Food Science and Nutrition* 60 (13): 2174–2211.

Waltz, E. (2022). GABA-enriched tomato is first CRISPR-edited food to enter market. *Nature Biotechnology* 40 (1): 9–11.

Wang, D., Samsulrizal, N.H., Yan, C. et al. (2019a). Characterization of CRISPR mutants targeting genes modulating pectin degradation in ripening tomato. *Plant Physiology* 179 (2): 544–557.

Wang, H., Wu, Y., Zhang, Y. et al. (2019b). CRISPR/Cas9-based mutagenesis of starch biosynthetic genes in sweet potato (Ipomoea Batatas) for the improvement of starch quality. *International Journal of Molecular Sciences* 20 (19): 4702.

Wang, R., Lammers, M., Tikunov, Y. et al. (2020). The rin, nor and Cnr spontaneous mutations inhibit tomato fruit ripening in additive and epistatic manners. *Plant Science* 294: 110436.

Xu, Z.S., Yang, Q.Q., Feng, K., and Xiong, A.S. (2019a). Changing carrot color: insertions in DcMYB7 alter the regulation of anthocyanin biosynthesis and modification. *Plant Physiology* 181 (1): 195–207.

Xu, Z.S., Feng, K., and Xiong, A.S. (2019b). CRISPR/Cas9-mediated multiply targeted mutagenesis in orange and purple carrot plants. *Molecular Biotechnology* 61: 191–199. https://doi.org/10.1007/s12033-018-00150-6.

Yamamoto, A., Ishida, T., Yoshimura, M. et al. (2019). Developing heritable mutations in Arabidopsis thaliana using a modified CRISPR/Cas9 toolkit comprising PAM-altered Cas9 variants and gRNAs. *Plant and Cell Physiology* 60 (10): 2255–2262.

Yang, L., Huang, W., Xiong, F. et al. (2017a). Silencing of Sl PL, which encodes a pectate lyase in tomato, confers enhanced fruit firmness, prolonged shelf-life and reduced susceptibility to grey mould. *Plant Biotechnology Journal* 15 (12): 1544–1555.

Yang, Y., Zhu, G., Li, R. et al. (2017b). The RNA editing factor SlORRM4 is required for normal fruit ripening in tomato. *Plant Physiology* 175 (4): 1690–1702.

Yang, T., Deng, L., Zhao, W. et al. (2019). Rapid breeding of pink-fruited tomato hybrids using the CRISPR/Cas9 system. *Journal of genetics and genomics= Yi chuan xue bao* 46 (10): 505–508.

Yasumoto, S., Umemoto, N., Lee, H.J. et al. (2019). Efficient genome engineering using platinum TALEN in potato. *Plant Biotechnology* 36 (3): 167–173.

Ye, J., Wang, X., Hu, T. et al. (2017). An InDel in the promoter of Al-ACTIVATED MALATE TRANSPORTER9 selected during tomato domestication determines fruit malate contents and aluminum tolerance. *The Plant Cell* 29 (9): 2249–2268.

Yu, Q.H., Wang, B., Li, N. et al. (2017). CRISPR/Cas9-induced targeted mutagenesis and gene replacement to generate long-shelf life tomato lines. *Scientific Reports* 7 (1): 11874.

Yuste-Lisbona, F.J., Fernández-Lozano, A., Pineda, B. et al. (2020). ENO regulates tomato fruit size through the floral meristem development network. *Proceedings of the National Academy of Sciences* 117 (14): 8187–8195.

Zhao, X., Jayarathna, S., Turesson, H. et al. (2021). Amylose starch with no detectable branching developed through DNA-free CRISPR-Cas9 mediated mutagenesis of two starch branching enzymes in potato. *Scientific Reports* 11 (1): 1–13.

Zhong, Z., Sretenovic, S., Ren, Q. et al. (2019). Improving plant genome editing with high-fidelity xCas9 and non-canonical PAM-targeting Cas9-NG. *Molecular Plant* 12 (7): 1027–1036.

Zhu, C., Naqvi, S., Gomez-Galera, S. et al. (2007). Transgenic strategies for the nutritional enhancement of plants. *Trends in Plant Science* 12 (12): 548–555.

Zhu, H., Li, C., and Gao, C. (2020). Applications of CRISPR–Cas in agriculture and plant biotechnology. *Nature Reviews Molecular Cell Biology* 21 (11): 661–677.

Zsögön, A., Čermák, T., Naves, E.R. et al. (2018). De novo domestication of wild tomato using genome editing. *Nature Biotechnology* 36 (12): 1211–1216.

12

Insight into the Flavonoids Enrichment in Plants by Genome Engineering

Elena V. Mikhaylova

Institute of Biochemistry and Genetics, Ufa Federal Research Center RAS, Ufa, Russia

12.1 The Importance of Flavonoids

Flavonoids are a low-molecular-weight polyphenolic secondary metabolites, specific for plants. Thy have a C6—C3—C6 structure, represented by two benzene rings (ring A and ring B) connected by a C3-ring, labeled as C (Figure 12.1). More than 9000 flavonoids have been characterized in plants, which makes them one of the best studied classes of metabolites. Hydroxylation and oxidation provide a structural diversity of flavonoids. The most important groups are anthocyanidins, anthoxanthins, flavanones, flavanonols, isoflavonoids, and neoflavonoids (Havsteen 2002; Shah et al. 2019; Singla et al. 2019).

Flavonoids perform important functions in plants, such as UV filtration, floral pigmentation, attraction of pollinators, symbiotic nitrogen fixation, and physiological regulators, including stress response. They act as detoxifying agents and antimicrobial defensive compounds, and provide frost and drought resistance (Panche et al. 2016).

These compounds are indispensable in human nutrition due to antioxidant properties, associated with health-promoting effects. The ability of flavonoids to prevent damage caused by free radicals, combat diabetes and Alzheimer disease, their anti-cholinesterase, antitumorous, antibacterial, antiviral, and anti-inflammatory activities are being investigated worldwide (Havsteen 2002; Ullah et al. 2020).

Quercetin belongs to the most common and largest flavonol subclass, characterized by a hydroxyl group in position 3 of the C ring. Quercetin aglycone can interact with receptors responsible for the development of chemically induced cancers (Murakami et al. 2008). Quercetin also demonstrates anti-inflammatory, antiviral, antioxidant, and psychostimulant activities (Li et al. 2016).

Naringenin and hesperedin belong to flavanone (dihydroflavone) subclass, where double bond between positions 2 and 3 and the C ring are saturated. Naringenin is known for its potential health effects, such as activity against Alzheimer's disease and antitumor, antioxidant, anti-inflammatory, antiviral, antibacterial, antiadipogenic, and cardioprotective properties. Naringenin demonstrated inhibitory effect against dengue virus, chikungunya virus and hepatitis C virus, microbes and yeast, as well as multiple types of cancer. It also

Applications of Genome Engineering in Plants, First Edition. Edited by Santosh Kumar Upadhyay.
© 2024 John Wiley & Sons Ltd. Published 2024 by John Wiley & Sons Ltd.

improved memory and reduced amyloid and tau proteins in mice. Exposure of cells to naringenin reduced DNA hydroxyl damage (Havsteen 2002; Salehi et al. 2019a). Hesperidin is believed to play a role in plant defense against pathogens. It is also efficient against cardiovascular disease, neurodegeneration, and tumors due to the induction of apoptosis of cancer cells and cell cycle arrest (Devi et al. 2015; Aggarwal et al. 2020; Ullah et al. 2020).

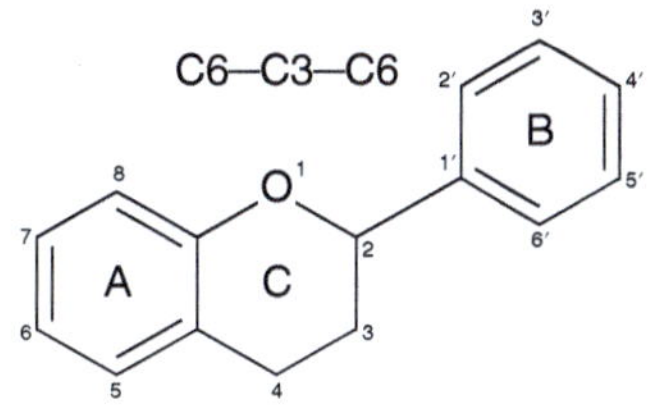

Figure 12.1 Basic skeleton structure of flavonoids.

Taxifolin and aromadendrin belong to flavanonol (dihydroflavonol) subclass. Taxifolin is known for its effect against inflammation, microbial infections, oxidative stress, cancer, diabetes, cardiovascular, and liver disorders. Ability to prevent angiogenesis, trigger apoptosis and inhibit cytochrome P450 enzymes, P-glycoprotein, cancer cell lipogenesis and reactive oxidative species (ROS) contributes to taxifolin antitumor activity. Taxifolin also enhanced the effect of antibiotics (Sunil and Xu 2019; Das et al. 2021). Aromadendrin 7-methyl ether possesses anti-inflammatory and antitumor activities and is capable of stimulating insulin-mediated glucose uptake in the treatment of type 2 diabetes (Malla et al. 2012).

Catechin and thearubigin belong to the flavan-3-ol (flavanol) subclass, characterized by the absence of double bond between positions 2 and 3 and the hydroxyl group bound to position 3. Thearubigin is the main polyphenol of the black tea with an anti-inflammatory and anti-oxidant activity (Maity et al. 2003). Catechins have beneficial effects on neurodegenerative diseases, oxidative stress, osteoporosis, and inflammation. However, in humans excessive doses of catechins can induce the formation of autoantibodies and autoimmune reaction or have hepatotoxic effect; these compounds are necessary for plants. They are involved in allelopathic interactions, being released in the environment to act as antibiotics and herbicides (Isemura 2019).

Luteolin, apigenin, and tangeritin belong to subclass of flavones, which is characterized by a double bond between positions 2 and 3 and a ketone in position 4 of the C ring. Luteolin is known for its antitumorous, anti-inflammatory, and neuroprotective activities. This molecule prevents the development of several types of cancer by inhibiting proliferation and inducing apoptosis of malignant cells and reversing the epithelial-mesenchymal transition. Luteolin also can induce lethal endoplasmic reticulum stress response and mitochondrial dysfunction in cancer cells (Imran et al. 2019; Yang et al. 2019). Apigenin has demonstrated positive effect on the treatment of cancer, diabetes, neurodegenerative diseases, depression, and insomnia (Salehi et al. 2019b).

Although these medical properties of flavonoids are observed *in vitro*, they are rarely confirmed in clinical trials, probably due to poor absorption in the human body. Proanthocyanidins, galloylated tea catechins, and anthocyanins are the least absorbable (Dixon and Pasinetti 2010). Nevertheless, flavonoids have numerous nonmedical applications.

Flavonoids are widely used as colorants (Giusti and Wallace 2009). Knowledge of an anthocyanin's chemistry allowed to stabilize these molecules, providing orange, red, blue, and violet pigmentation. Yellow flavonoid pigments, such as aurones and chalcones, co-occur with carotenoids. The flavonoid colorants are usually considered safe and approved for use in food.

Flavonoids are also applied in the industry and in agriculture as allelochemicals and biopesticides. They can act either inhibitory or stimulatory, depending on the concentration. It has been shown that quercetin and kaempferol released by plants such as legumes and walnut trees possess phytoinhibitory activity. However, in lower concentrations these flavonoids can stimulate seed germination (Palma-Tenango et al. 2017; Schnarr et al. 2022).

Tannins are oligomers or polymers derived from flavan-3-ols, flavan 3, 4-diols, and ellagic and gallic acids. The color of tannins range from yellow to brown. These flavonoids have been used for tanning leather since ancient times. They also find application as clarifying agent in the wine industry, as aroma ingredient, as anti-corrosive agents, and in the production of resins for 3D printing. Tannins also possess adhesive, biocide, and fungicide activity. Due to the antibacterial and antifungal effect, tannins can preserve fishing nets, wood, and iron materials from degradation (Pizzi 2019; Das et al. 2020).

Anthocyanins are an important part of the diet with antioxidant properties. They are often used as natural dyes. Due to the ability to change color depending on environmental conditions, they can also be used as pH indicators. Anthocyanins are accumulated in response to fungal and bacterial infections, heavy metals, pests, and other stress factors. For example, overexpression of anthocyanin rhamnosyltransferase genes *UGT79B2* and *UGT79B3* increased anthocyanin accumulation and tolerance to abiotic stress in *Arabidopsis*. Transcription factors Delila, HAT, and NAC regulate not only anthocyanin biosynthesis but also abiotic stress response (Amogne et al. 2020; Mikhaylova et al. 2022).

Isoflavonoids are the 3-phenylchromen-4-one backbone molecules, precursors of phytoalexins, involved in plant–microbe interactions. They are characteristic of legumes and can also be present in microorganisms. Like the 2-phenylchroman flavonoids, they are divided into several classes. Isoflavones can act as phytoestrogens (genistein, daidzein) and herbicides (isoschaftoside). *Desmodium uncinatum* can inhibit the growth of a parasitic plant Striga by synthesizing isoschaftoside in the roots. Rotenoids are used as pesticides against pests and alien fish species (Dixon and Pasinetti 2010; Hooper et al. 2010; Panche et al. 2016).

Due to the valuable qualities and prospects for practical application, flavonoids are one of the most well-studied molecules. The flavonoid biosynthesis pathway is well understood and described.

12.2 Flavonoid Biosynthesis Pathway

Flavonoids are synthesized through shikimate and arogenate pathways with the participation of enzymes. They are originally derived from phenylalanine and tyrosine, which are subjected to chemical modification by hydroxylation, methoxylation, or glycosylation (Shah et al. 2019). Phenylalanine is converted to *p*-coumaroyl-CoA by phenylalanine ammonia lyase (PAL), cinnamic acid 4-hydroxylase (C4H), and 4-coumarate: CoA ligase (4CL). Then chalcone synthase (CHS) initiates flavonoid biosynthesis. Chalcones and flavanones are the earliest synthesized compounds and universal flavonoid precursors. The activity of chalcone isomerase (CHI), chalcone reductase (CHR), isoflavone synthase (IFS), flavone synthase (FNS), flavonoid 3′-hydroxylase (F3′H), flavonoid 3′,5′-hydroxylase (F3′5′H), flavonol synthase (FLS), flavanone 3β-hydroxylase (F3H), flavanone 4-reductase (FNR), and dihydroflavonol 4-reductase (DFR) results in branching of the flavonoid biosynthesis pathway (Liu et al. 2021b). Anthocyanin biosynthesis require anthocyanidin synthase (ANS), leucoanthocyanidin dioxygenase (LDOX), and uridine diphosphate-dependent glucosyltransferase (UGT/UFGT). Leucoanthocyanidin reductase (LAR) and anthocyanidin reductase (ANR) catalyze the formation of flavan-3-ols. Methyltransferase (MT) and rhamnosyltransferase (RT) are involved in the transformation of flavonols and isoflavones. Competition between FLS and DFR contributes to either flavonol or anthocyanin synthesis, and F3H competes with FNS, IFS, and FNR during transformation of flavanones (Khusnutdinov et al. 2021). There are other phenylpropanoid metabolites except flavonoids, such as lignin, salicylic acid and caffeic acid (Liu et al. 2021b).

The general scheme for the flavonoid biosynthesis is represented in Figure 12.2.

The possibility to modulate flavonoid biosynthesis by genetic engineering has been confirmed multiple times. Transformation of tomato with genes encoding CHI, CHR, CHS, FNS and stilbene synthase promoted the accumulation of stilbenes, deoxychalcones, flavones, and flavonols in the fruit peel. Consumption of these fruits improved cardiovascular risk factors in mice (Rein et al. 2006). Overexpression of the *F3H* gene redirected flavonoid biosynthesis in sorghum, promoting the accumulation of kaempferol, isorhamnetin, and syringetin that were absent in nontransgenic plants (Wang et al. 2020). Flax overexpressing *CHS, CHI,* and *DFR* genes from *Petunia hybrida* was characterized by increased flavonoid (kaempferol, catechine), phenolic acid, and lignan content (Żuk et al. 2011.) Introduction of *PAL* and *CHS* genes contributed to accumulation of naringenin in rice, and addition of *F3H, IFS,* and *FNS* genes – to accumulation of kaempferol, genistein, and apigenin, respectively (Ogo et al. 2013). Therefore, structural flavonoid biosynthesis genes proved to be noteworthy targets in fundamental and applied research. Nevertheless, regulatory genes attract the attention even more often, since they are less studied.

A large number of MYB, bHLH (MYC), and WD40 transcription factors modulate biosynthesis of flavonoids by binding to specific cis-acting elements in the promoter regions of structural genes, encoding the above-listed enzymes (Khusnutdinov et al. 2021). These three types of transcription factors can act together for activation or repression as MBW complexes; however, *MYB* and *bHLH* can also act directly by binding to MYB motifs in structural gene promoters (Aharoni et al. 2001). The duplication and divergence of *MYB* and *bHLH* genes contribute to morphological and metabolic diversity in higher plants (Dubos et al. 2010).

MYB proteins comprise the largest family of transcription factors. Depending on the number of R motifs in a *MYB* gene (from one to four), they can be divided into subfamilies. Each motif is approximately 52 amino acids long, and adopts a helix-turn-helix conformation, where the second helix is capable of sequence-specific DNA binding (Feller et al. 2011). MYB transcription factors can contain activation or repression domain in the C-termini. Most plant MYB proteins contain two imperfect MYB repeats and belong to subfamily R2R3, but R3 MYBs are also important because they can act as repressors of flavonoid biosynthesis (Lloyd et al. 2017). Homologs PAP1, MYB113, ANT1, AN2, C1, MYB1, and LAP1 upregulated the transcription of 38 flavonoid biosynthesis genes. For example, overexpression of *AtPAP1* gene promoted purple pigmentation of a whole Arabisopsis plant. Overexpression of maize gene *ZmC1* in wheat promoted increased pigmentation in vegetative tissues (Riaz et al. 2019). PAP1, PAP2, and MYB10 also act as activators; however, the role of MYB1 and MYB6 appears to depend on plant species. Ectopic expression of tobacco

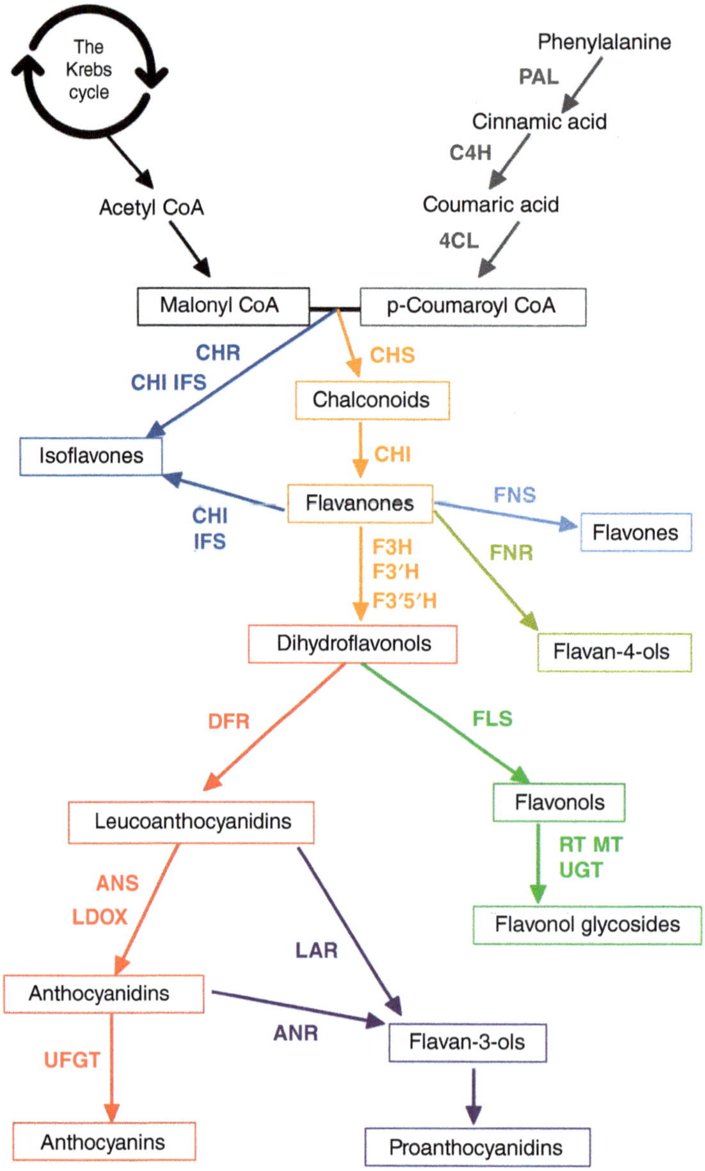

Figure 12.2 Flavonoid biosynthesis pathway.

PAP2 gene contributed to an increased anthocyanin accumulation in all organs of *A. thaliana* (Qi et al. 2020). Overexpression of *MYB10* gene upregulated the early flavonoid biosynthesis genes, but only in certain tissues (Laitinen et al. 2008; Feng et al. 2010).

R2R3 proteins MYB4 and MYB57, as well as R3 proteins MYBL2, MYBL1 and CPC, downregulate the synthesis of flavonoids (Khusnutdinov et al. 2021). Overexpression of *MYB57* and *MYB182* genes in poplar contributed to a lower anthocyanin and proanthocyanidin content and inhibited flavonoid and shikimate pathway genes (Yoshida et al. 2015;

Wan et al. 2017). Overexpression of *MYB4* gene in banana resulted in the inhibition of not only structural genes of flavonoid biosynthesis but also *bHLH* genes (Deng et al. 2021). Overexpression of *MYBL2* gene inhibited biosynthesis of anthocyanins and proanthocyanidins (Dubos et al. 2008; Gates et al. 2018).

Highly conserved bHLH domains comprise approximately 60 amino acids, which form two helices, connected by a loop. This structure can interact with bHLH and non-bHLH proteins. bHLH proteins are also capable of regulating transcription directly by recognizing protein-binding site (enhancer box) with a specific region at the N-terminus (Feller et al. 2011). TT8, R, GL3, Delila, Myc1, and bHLH1 are positive regulators of flavonoid biosynthesis. Negative regulators among bHLH have not been reported (Khusnutdinov et al. 2021). Overexpression of *ZmR* gene of maize in wheat and *Delila* gene of snapdragon in tobacco resulted in hyperaccumulation of anthocyanins in different tissues (Naing et al. 2017; Riaz et al. 2019). Naturally occurring mutations in the *TT8* gene have caused depigmentation of seeds in *B. juncea, B. rapa* and *A. thaliana* (Li et al. 2012; Padmaja et al. 2014; Lim et al. 2017).

WD40 proteins contain multiple tandem repeats of a WD motif, which comprises around 40 amino acids ending in tryptophan (W) and aspartic acid (D). However, this motif can be found in different proteins that do not belong to WD40 family. Unlike MYB and bHLH, WD40 does not bind to gene promoters, but only capable of interacting with other proteins through its WD domain and provides the formation of complexes (Miller et al. 2016; Lloyd et al. 2017). TTG1 and its homologs such as An11 participate in the formation of MBW complexes consisting of MYB, bHLH, and WD40 proteins (PAP1/2–GL3–TTG1 or TT2-TT8- TTG1) (Khusnutdinov et al. 2021). Loss of function of *WD40* gene in *Medicago truncatula* contributed to a reduced content of flavonols, anthocyanins, proanthocyanidins, and benzoic acid in the seeds (Pang et al. 2009). *WD40* gene of *Camelina sativa* upregulated the expression of flavonoid biosynthesis genes and promoted anthocyanin accumulation in green tea and tobacco (Liu et al. 2018).

It should be noted that these transcription factors participate in the regulation of other traits, such as trichome development, stomatal opening, and fatty acid content.

Other molecules such as LBD and NAC transcription factors, HD-ZIP proteins, and microRNAs can also be involved in flavonoid biosynthesis pathway (Khusnutdinov et al. 2021).

Therefore, flavonoids are secondary metabolites, and their biosynthesis depends on multiple enzymes and transcription factors. Due to the many beneficial properties of flavonoids, they often become targets of genetic engineering and gene editing. Knowledge of the flavonoid biosynthesis pathway and the large number of genes involved in this process allows for the numerous approaches to the flavonoid enrichment in plants and biotechnological production of flavonoids.

12.3 In Planta Flavonoid Enrichment via Genome Editing

Flavonoid biosynthesis genes can be used as targets not only in traditional genetic engineering but also in CRISPR gene editing. CRISPR/Cas allows for programmable and precise editing without off-target effects. Flavonoid enrichment does not only enhance stress tolerance and consumer qualities but can also serve as a visible marker. Therefore, editing of flavonoid biosynthesis genes allows for easier evaluation of the efficiency of new CRISPR techniques. For this reason, *DFR* gene has become one the most frequently used targets for gene editing (Table 12.1). Being an essential gene of anthocyanin biosynthesis, it provides stable phenotypic effects when up- or downregulated. For example, CRISPR-induced knockout of this gene promoted flower discoloration in *Ipomoea* and seed discoloration in rice (Watanabe et al. 2017; Jung et al. 2019). *F3′*H mutants of black rice generated via CRISPR demonstrated depigmented seed phenotype just like *DFR* mutants, and 10-fold reduced anthocyanin content (Jung et al. 2019).

The traditional CRISPR approach is to induce mutations and gene knockout. Knockin of a promoter, a gene, or repair of the sequence of a mutant gene is much more complicated

Table 12.1 Effects of the editing of flavonoid biosynthesis genes.

Edited gene	Method	Effect	References
DFR	Knockout	Flower discoloration in *Ipomoea*, seed discoloration in rice, reduced anthocyanin pigmentation in tomato	Watanabe et al. (2017), Jung et al. (2019), Danilo et al. (2018)
ANT1	Knockin	Hyperaccumulation of anthocyanins in tomato	Čermák et al. (2015)
*F3′*H	Knockout	Mutants of black rice generated via CRISPR had the same ocher seed phenotype as *DFR* mutants and 10-fold reduced anthocyanin content (Ryder et al. 2017; Qi 2019)	Jung et al. (2019)
C4H	CRISPRi	Accumulation of pinostrobin, naringenin, and chlorogenic acid	Karlson et al. (2022)
PAL2, *CHS*, *CHI*, and *FLS*	CRISPRa	Increased content of naringenin, eriodictyol, kaempferol, and quercetin in *N. benthamiana*	Selma et al. (2022)
F3H1, *F3H2*, and *FNSII-1*	Knockout	Improved isoflavone content and resistance to soybean mosaic virus	Zhang et al. (2020)
UGT	Knockout	Increased daidzein and formononetin contents in soybean, reduced content of genistein. Resistance to *Helicoverpa armigera* and *Spodoptera litura*	Zhang et al. (2022b)
UGT79B2 and *UGT79B3*	Knockout	Anthocyanin accumulation, tolerance to cold, drought, and salt stress in *Arabidopsis*	Li et al. (2017)
IFS1	Knockout	Reduced levels of isoflavones formononetin, biochanin A and genistein in red clover	Dinkins et al. (2021)
UDP	Knockout	Reduced content of phloridzin in apple	Miranda et al. (2023)
myb115	Knockout	Reduced proanthocyanidin content and greater sensitivity to fungal pathogen *Dothiorella gregaria* in poplar	Wang et al. (2017)
F3H	Knockout	Changed flower color, depigmentation of hypocotyls	Nishihara et al. (2018), Liu et al. (2021c)

Table 12.1 (Continued)

Edited gene	Method	Effect	References
F3′H	Knockout	Accumulation of pelargonidin instead of cyanidin in poinsettia; production of kaempferol instead of quercetin and lack of cyaniding-type anthocyanins in mice	Nitarska et al. (2021)
5/3′AT, 5GT, and *3′GT*	Knockout	Predominance of different delphinidin derivatives and new flower shades in gentian	Tasaki et al. (2019)
PSY1, *MYB12*, and *SGR1*	Knockout	Tomato genotypes with different fruit colors	Yang et al. (2023)
AN2	Knockout	Decreased anthocyanin content	Zhi et al. (2020), Yan et al. (2020)
TT2	Knockout	Reduced flavonoid content and improved fatty acid composition in oilseed rape	Xie et al. (2020)
TT8	Knockout	Depigmentation, elevated oil and protein content, and altered fatty acid composition in the seeds; depigmentation of seeds and flowers, elevated protein and lipid content in tobacco	Zhai et al. (2020), Tian et al. (2020)
MYB113	Knockout	Depigmentation in carrot	Xu et al. (2019)
MYB45	Knockout	Increased content of rutin, catechin, and other flavonoids in tartary buckwheat	Wen et al. (2022)
MYB57	Knockout	Increased anthocyanin and proanthocyanidin content in poplar	Wan et al. (2017)
ZIP36	Knockout	Increased accumulation of naringenin chalcone, naringenin, dihydroflavonols, and cyanidin-3-O-glucoside and decreased content of stilbenes, lignans, and some flavonols	Tu et al. (2022)
MYB4	Knockout	Accumulation of sinapate esters instead of flavonoids in Arabidopsis	Jin (2000)
GLABRA2	Knockout	Elevated anthocyanin accumulation in Arabidopsis	Wang et al. (2015)
PAP1	CRISPRa	No effect in *Arabidopsis*	Lowder et al. (2015), Park et al. (2017)
UFGT	CRISPRa	No effect in grape	Ren et al. (2022)

and requires donor DNA (Alok et al. 2021; Sushmita et al. 2021; Upadhyay 2021). However, it has been successfully accomplished on the same *DFR* gene. CRISPR-induced deletion in its sequence contributed to reduced anthocyanin accumulation in tomato; however, knock-in of the previously deleted sequence restored pigmentation (Danilo et al. 2018). Insertion of the 35S promoter upstream of the tomato *ANT1* gene resulted in hyperaccumulation of anthocyanins (Čermák et al. 2015). However, we were unable to find more information on other successful knockin cases.

There is also an option to modulate gene expression without the induction of mutations using Cas fused to activator or repressor protein, also known as CRISPRa and CRISPRi, respectively (Mikhaylova et al. 2022). For example, activation of *DFR* gene by dCas9:EDLL-S2:VPR system increased gene expression level in tomato and tobacco up to 10,000-fold (Selma et al. 2019). CRISPR-mediated downregulation of the *C4H* gene contributed to activation of *4CL* and *CHS* genes and accumulation of pinostrobin, naringenin, and chlorogenic acid (Karlson et al. 2022). Targeting of *PAL2*, *CHS*, *CHI*, and *FLS* genes using CRISPR activators increased the content of naringenin, eriodictyol, kaempferol, and quercetin in leaves of *N. benthamiana* (Selma et al. 2022). This approach is also rarely used and requires an advanced CRISPR/Cas toolkit.

Therefore, knockout of structural flavonoid biosynthesis genes is the most common and reliable way of metabolic engineering. The shift in flavonoid biosynthesis pathway can promote not only phenotypic effect, but also stress tolerance. For example, editing of *F3H1*, *F3H2,* and *FNSII-1* genes of soybean significantly improved isoflavone content and resistance to soybean mosaic virus (Zhang et al. 2020). *UGT* loss of function mutations in soybean contributed to an increased daidzein and formononetin contents, while genistein content was reduced. The knockout mutants were also resistant to *Helicoverpa armigera* and *Spodoptera litura*; however, overexpression of the *UGT* gene decreased resistance to these pests (Zhang et al. 2022b). CRISPR-induced knockout of *UGT79B2* and *UGT79B3* genes in *Arabidopsis* reduced not only anthocyanin accumulation but also tolerance to abiotic stress, such as cold, drought, and salinity. Plants overexpressing these genes had opposite characteristics (Li et al. 2017). The CRISPR knockout of the *IFS1* gene in red clover decreased the content of isoflavones (formononetin, biochanin A, and genistein), but there was no effect on nodulation (Dinkins et al. 2021). CRISPR-induced mutation in phloretin-specific UDP-2′-O-glucosyltransferase (*PGT1*) gene in apple did not affect normal growth despite reduced content of phloridzin; however, RNAi knockdown lines demonstrated dwarf phenotype (Miranda et al. 2023). Mutations induced in *myb115* gene via CRISPR contributed to a reduction of proanthocyanidin content and greater sensitivity to fungal pathogen *Dothiorella gregaria* in poplar (Wang et al. 2017).

Regulatory genes also can affect stress tolerance. For example, CRISPR-induced knockout of the *microRNA171f* gene in rice contributed to downregulation of *CHS, CHI, F3H, F3′H, FLAVONE SYNTHASE II* (*FNSII*), *4-COUMARATE-COA LIGASE 2* (*4CL2*), and *FLAVANONE 2-HYDROXYLASE* (*F2H*) and reduced drought tolerance. The opposite effect was observed in plants overexpressing *MIR171f* (Um et al. 2022).

CRISPR has been successfully used to alter ornamental characteristics of flowers. In *Torenia fournieri* flower, color changed to pale blue due to the CRISPR-induced mutagenesis of the *F3H* gene (Nishihara et al. 2018). The knockout of *F3′H* gene in poinsettia by CRISPR/Cas9 promoted accumulation of pelargonidin instead of cyanidin, resulting in a depigmentation of bract (Nitarska et al. 2021). The knockout of acyltransferase gene *5/3′AT* and glycosyltransferase genes *5GT* and *3′GT* in blue-flowered gentian via CRISPR contributed to predominance of different delphinidin derivatives and new flower shades (Tasaki et al. 2019).

Color modification was also carried out on agricultural plants and their relatives. CRISPR/Cas9-mediated multiplex editing of *PSY1*, *MYB12*, and *SGR1* genes allowed for generation of new tomato genotypes with different fruit colors (Yang et al. 2023). Knockout of the *AN2* gene in purple tomato reduced the content of anthocyanins in vegetative tissues, but not in fruits (Zhi et al. 2020). However, in a similar study, anthocyanin content was reduced in fruits (Yan et al. 2020). CRISPR-induced knockout of a *F3H* gene in tomato with purple hypocotyls resulted in their depigmentation (Liu et al. 2021a).The knockout of *TT2* homologues in oilseed rape decreased flavonoid content, but improved fatty acid composition (Zie et al. 2020). CRISPR-induced mutations in two *TT8* gene copies in oilseed rape resulted in depigmentation, elevated oil and protein content, and altered fatty acid composition in the seeds (Zhai et al. 2020). Editing of both copies of the *TT8* gene contributed to depigmentation of seeds and flowers, elevated protein and lipid content in tobacco. *ANR* and *LAR* gene expression was downregulated in these mutants (Tian et al. 2020). CRISPR-induced knockout of *MYB113* gene in purple carrot also resulted in depigmentation (Xu et al. 2019). CRISPR editing of *CHS*, *CHI*, and *CHIL* genes in rice reduced the content of most flavonoids, especially flavone and flavone C-glycoside. However, the content of vitexinn and naringenin chalcone increased (Lam et al. 2022). CRISPR-generated *TTG1* knockout mutants of rice were characterized by decreased anthocyanin content and reduction of trichomes (Yang et al. 2021b). In a model plant *A. thaliana,* the knockout of *TTG1* had the same effect; however, *F3′H* knockout resulted in the formation of kaempferol instead of quercetin and cyanidin-type anthocyanins deficiency (Ryder et al. 2017; Qi 2019).

Unfortunately, at the moment CRISPR is mainly used to reduce the content of flavonoids by knockout of the corresponding genes, while the increase in their content is more important for agriculture. There are several examples of editing the genes of negative regulators of flavonoid biosynthesis. Mutagenesis of *MYB45* gene of tartary buckwheat via CRISPR increased the content of rutin, catechin, and other flavonoids (Wen et al. 2022). Knockout of *MYB57* gene via CRISPR increased the content of anthocyanins and proanthocyanidins in poplar; however, overexpression of this gene negatively affected accumulation of these flavonoids and suppressed structural genes *PAL4*, *CHI*, *4CL5*, *CHS*, *F3H*, *DFR1*, *ANR1*, *ANS1*, and *LAR1* (Wan et al. 2017). CRISPR-mediated knockout of *ZIP36* gene of grapevine activated a range of anthocyanin biosynthesis genes, increased accumulation of naringenin chalcone, naringenin, dihydroflavonols, and cyanidin-3-O-glucoside, and decreased the content of stilbenes, lignans, and some flavonols (Tu et al. 2022).

However, manipulating the expression of repressors can be rather tricky. For example, CRISPR-induced knockout of *MYB4* gene in *A. thaliana* contributed to accumulated sinapate esters (phenylpropanoid derivatives) instead of flavonoids (Jin 2000). Natural *GLABRA2* mutants of *Arabidopsis* were characterized by elevated anthocyanin accumulation (Wang et al. 2015). CRISPR mutants had trichomeless phenotype; however, changes in pigmentation were not reported (Malzahn et al. 2019; Kong et al. 2021).

Transcriptional activation via CRISPR can also produce various non-correlated results. For example, two- to sevenfold activation of *PAP1* gene in *Arabidopsis* did not contribute to changes in leaf color; however, two- to threefold activation contributed to the purple pigmentation of the leaves (Lowder et al. 2015; Park et al. 2017). Direct 10-fold transcriptional activation of the *UFGT* gene in grape cells using dCas9-VP64 and dCas9-TV systems did not contribute to pigmentation changes; however, introduction of pCAMBIA2300-VvMYBA1 overexpression vector increased the *UFGT* gene expression up to 25-fold and promoted anthocyanin accumulation (Ren et al. 2022).

In general, approaches for flavonoid enrichment via CRISPR editing require further development. At the moment, we lack the knowledge to accurately predict the consequences of editing a particular gene and its effect on flavonoid content. The regulation of any of the above-mentioned genes does not allow to change the content of a particular flavonoid, but can shift the pathway towards a certain group of flavonoids. There have been no reports of increased production of medically valuable compounds in gene-edited plants, also due to the difficulty of plant genetic transformation and *in vitro* regeneration.

The high frequency of use of *DFR* gene as a target may be due to a small copy number and the stability of effect. Other structural genes, such as *PAL*, *4Cl*, and *FLS* have multiple copies even in *A. thaliana*. The role of regulatory genes differ between plant species. Editing of these genes is also complicated by the presence of other transcription factors that regulate the expression of the same structural gene. Nevertheless, knockout of negative regulators of flavonoid biosynthesis can be considered a promising approach. Further development of CRISPR-induced knockin and transcriptional regulation techniques may also allow for flavonoid enrichment in plants.

12.4 Biotechnological Production of Flavonoids

For the production of certain flavonoids, CRISPR editing of bacterial cultures is a promising approach. Modification of bacteria is easier than plant genetic transformation, and desired flavonoids may be produced in bioreactors. *Escherichia coli*, *Saccharomyces sp.*, and *Streptomyces* sp. are well studied organisms that lack phenylpropanoid pathway; therefore a heterologous flavonoid pathway can be engineered in these microbes. It has been achieved in early 2000s via cloning of structural biosynthetic genes such as *PAL*, *4CL*, and *CHS* (Hwang et al. 2003). CRISPR approach in bacteria proved to be effective in eliminating of competing enzymes. CRISPRi repression of competing genes is superior to traditional gene knockout strategies, which often lead to undesirable effects and even cell death (Shah et al. 2019).

For example, downregulation of competing genes *sucABCD*, *fumC*, and *scpC* in *E. coli*, expressing three genes from the flavonoid pathway (*4CL*, *CHS*, and *CHI*), contributed to a 2.5-fold increase in naringenin content (Cress et al. 2015). Targeting of the *fabB*, *fabF*, *fumC*, *sucC*, and *adhE* genes allowed to increase naringenin concentration up to 7.4-fold (Wu et al. 2015). Downregulation of *eno*, *adhE*, *fabB*, *sucC*, *fumC*, and *fabF* in *E. coli* overexpressing *PAL*, *4CL*, *CHS*, and *CHI* flavonoid biosynthesis genes contributed to a 20-fold increase in accumulation of pinocembrin. This flavonoid compound is considered a therapeutic agent to reduce cerebral ischemia/reperfusion and blood-brain injury (Wu et al. 2016). Similar approach was used to increase production of peonidin 3-O-glucoside and naringenin 7-sulfate in *E. coli* (Shah et al. 2019).

Yeast (*S. cerevisiae*) naringenin cell factory was constructed via genetic engineering and pathway control CRISPR tools (Vanegas et al. 2017). In another study, daidzein, puerarin, and daidzin biosynthesis was rebuilt in yeast using CRISPR-induced knockin of *At4CL1*, *GmCHR5*, *GmCHS8*, and *GmCHI1B2* genes (Liu et al. 2021a). Biosynthesis of isorhamnetin and acacetin in yeast was achieved by integration of *FnsA*, *ubiC*, and *aroL* genes and deletion of the endogenous genes *ARO7* and *TRP3* by means of the CRISPR system (Zhang et al. 2022a; Zhang et al. 2022b).

Some endophytic fungi such as *Phomopsis liquidambaris* do not produce flavonoids, but they share flavonoid pathway with plants and do not require introduction of heterogeneous flavonoid genes. The knockout of the mitogen-activated protein kinase kinase (MAPKK) gene in *P. liquidambaris* using CRISPR-Cas9 activated the flavonoid biosynthesis and contributed to yellow pigmentation. Cinnamic acid, p-coumaric acid, kaempferol, quercetin and naringenin were synthesized in these fungi and released to fermentation broth (Yang et al. 2021a)

12.5 Conclusions

CRISPR/Cas approach allowed to modulate biosynthesis of anthocyanins and anthocyanidins, pinostrobin, naringenin, eriodictyol, kaempferol, quercetin, daidzein, formononetin, genistein, biochanin A, rutin, catechin, and many other flavonoids. At the moment CRISPR is mainly used to reduce the content of flavonoids by knockout of the corresponding genes. Targeting negative regulators of flavonoid biosynthesis contributes to accumulation of flavonoids; however, the role and functions of such genes differ between plant species. Engineering of heterologous flavonoid pathways in microorganism via CRISPR can be considered an alternative approach. Further development of CRISPR-induced knockin and transcriptional regulation techniques may also allow for flavonoid enrichment in plants.

References

Aggarwal, V., Tuli, H.S., Thakral, F. et al. (2020). Molecular mechanisms of action of hesperidin in cancer: recent trends and advancements. *Experimental Biology and Medicine (Maywood, N.J.)* 245 (5): 486–497.

Aharoni, A., De Vos, C.H.R., Wein, M. et al. (2001). The strawberry FaMYB1 transcription factor suppresses anthocyanin and flavonol accumulation in transgenic tobacco: strawberry Myb represses flavonoid biosynthesis. *The Plant Journal* 28 (3): 319–332.

Alok, A., Chauhan, H., Upadhyay, S.K. et al. (2021). Compendium of plant-specific CRISPR vectors and their technical advantages. *Life* 11: 1021. https://doi.org/10.3390/life11101021.

Amogne, N.Y., Ayele, D.W., and Tsigie, Y.A. (2020). Recent advances in anthocyanin dyes extracted from plants for dye sensitized solar cell. *Materials for Renewable and Sustainable Energy* 9 (4): 23.

Čermák, T., Baltes, N.J., Čegan, R. et al. (2015). High-frequency, precise modification of the tomato genome. *Genome Biology* 16 (1): 232.

Cress, B.F., Toparlak, Ö.D., Guleria, S. et al. (2015). CRISPathBrick: modular combinatorial assembly of type II-A CRISPR arrays for dCas9-mediated multiplex transcriptional repression in *E. coli*. *ACS Synthetic Biology* 4 (9): 987–1000.

Danilo, B., Perrot, L., Botton, E. et al. (2018). The DFR locus: a smart landing pad for targeted transgene insertion in tomato. *PLoS One* 13 (12): e0208395.

Das, A.K., Islam, M.N., Faruk, M.O. et al. (2020). Review on tannins: extraction processes, applications and possibilities. *South African Journal of Botany* 135: 58–70.

Das, A., Baidya, R., Chakraborty, T. et al. (2021). Pharmacological basis and new insights of taxifolin: a comprehensive review. *Biomedicine & Pharmacotherapy* 142: 112004.

Deng, G.-M., Zhang, S., Yang, Q.-S. et al. (2021). MaMYB4, an R2R3-MYB repressor transcription factor, negatively regulates the biosynthesis of anthocyanin in banana. *Frontiers in Plant Science* 11: 600704.

Devi, K.P., Rajavel, T., Nabavi, S.F. et al. (2015). Hesperidin: a promising anticancer agent from nature. *Industrial Crops and Products* 76: 582–589.

Dinkins, R.D., Hancock, J., Coe, B.L. et al. (2021). Isoflavone levels, nodulation and gene expression profiles of a CRISPR/Cas9 deletion mutant in the isoflavone synthase gene of red clover. *Plant Cell Reports* 40 (3): 517–528.

Dixon, R.A. and Pasinetti, G.M. (2010). Flavonoids and isoflavonoids: from plant biology to agriculture and neuroscience. *Plant Physiology* 154 (2): 453–457.

Dubos, C., Le Gourrierec, J., Baudry, A. et al. (2008). MYBL2 is a new regulator of flavonoid biosynthesis in *Arabidopsis thaliana*. *The Plant Journal* 55 (6): 940–953.

Dubos, C., Stracke, R., Grotewold, E. et al. (2010). MYB transcription factors in *Arabidopsis*. *Trends in Plant Science* 15 (10): 573–581.

Feller, A., Machemer, K., Braun, E.L., and Grotewold, E. (2011). Evolutionary and comparative analysis of MYB and bHLH plant transcription factors: plant MYB and bHLH factors. *The Plant Journal* 66 (1): 94–116.

Feng, S., Wang, Y., Yang, S. et al. (2010). Anthocyanin biosynthesis in pears is regulated by a R2R3-MYB transcription factor PyMYB10. *Planta* 232 (1): 245–255.

Gates, D.J., Olson, B.J.S.C., Clemente, T.E., and Smith, S.D. (2018). A novel R3 MYB transcriptional repressor associated with the loss of floral pigmentation in *Iochroma*. *The New Phytologist* 217 (3): 1346–1356.

Giusti, M.M. and Wallace, T.C. (2009). Flavonoids as natural pigments. In: *Handbook of Natural Colorants* (ed. T. Bechtold and R. Mussak), 255–275. Chichester: Wiley.

Havsteen, B.H. (2002). The biochemistry and medical significance of the flavonoids. *Pharmacology & Therapeutics* 96 (2–3): 67–202.

Hooper, A.M., Tsanuo, M.K., Chamberlain, K. et al. (2010). Isoschaftoside, a C-glycosylflavonoid from *Desmodium uncinatum* root exudate, is an allelochemical against the development of Striga. *Phytochemistry* 71 (8–9): 904–908.

Hwang, E.I., Kaneko, M., Ohnishi, Y., and Horinouchi, S. (2003). Production of plant-specific flavanones by *Escherichia coli* containing an artificial gene cluster. *Applied and Environmental Microbiology* 69 (5): 2699–2706.

Imran, M., Rauf, A., Abu-Izneid, T. et al. (2019). Luteolin, a flavonoid, as an anticancer agent: a review. *Biomedicine & Pharmacotherapy* 112: 108612.

Isemura, M. (2019). Catechin in human health and disease. *Molecules* 24 (3): 528.

Jin, H. (2000). Transcriptional repression by AtMYB4 controls production of UV-protecting sunscreens in *Arabidopsis*. *The EMBO Journal* 19 (22): 6150–6161.

Jung, Y.J., Lee, H.J., Kim, J.H. et al. (2019). CRISPR/Cas9-targeted mutagenesis of F3′H, DFR and LDOX, genes related to anthocyanin biosynthesis in black rice (*Oryza sativa* L.). *Plant Biotechnology Reports* 13 (5): 521–531.

Karlson, C.K.S., Mohd Noor, S.N., Khalid, N., and Tan, B.C. (2022). CRISPRi-mediated down-regulation of the cinnamate-4-hydroxylase (C4H) gene enhances the flavonoid biosynthesis in nicotiana tabacum. *Biology* 11 (8): 1127.

Khusnutdinov, E., Sukhareva, A., Panfilova, M., and Mikhaylova, E. (2021). Anthocyanin biosynthesis genes as model genes for genome editing in plants. *International Journal of Molecular Sciences* 22 (16): 8752.

Kong, X., Pan, W., Sun, N. et al. (2021). *GLABRA2* -based selection efficiently enriches Cas9-generated nonchimeric mutants in the T1 generation. *Plant Physiology* 187 (2): 758–768.

Laitinen, R.A.E., Ainasoja, M., Broholm, S.K. et al. (2008). Identification of target genes for a MYB-type anthocyanin regulator in gerbera hybrida. *Journal of Experimental Botany* 59 (13): 3691–3703.

Lam, P.Y., Wang, L., Lui, A.C.W. et al. (2022). Deficiency in flavonoid biosynthesis genes *CHS*, *CHI*, and *CHIL* alters rice flavonoid and lignin profiles. *Plant Physiology* 188 (4): 1993–2011.

Li, X., Chen, L., Hong, M. et al. (2012). A large insertion in bHLH transcription factor BrTT8 resulting in yellow seed coat in brassica rapa. *PLoS One* 7 (9): e44145.

Li, Y., Yao, J., Han, C. et al. (2016). Quercetin, inflammation and immunity. *Nutrients* 8 (3): 167.

Li, P., Li, Y.-J., Zhang, F.-J. et al. (2017). The *Arabidopsis* UDP-glycosyltransferases UGT79B2 and UGT79B3, contribute to cold, salt and drought stress tolerance via modulating anthocyanin accumulation. *The Plant Journal* 89 (1): 85–103.

Lim, S.-H., Kim, D.-H., Kim, J.K. et al. (2017). A radish basic helix-loop-helix transcription factor, RsTT8 acts a positive regulator for anthocyanin biosynthesis. *Frontiers in Plant Science* 8: 1917.

Liu, Y., Hou, H., Jiang, X. et al. (2018). A WD40 repeat protein from camellia sinensis regulates anthocyanin and proanthocyanidin accumulation through the formation of MYB–bHLH–WD40 ternary complexes. *International Journal of Molecular Sciences* 19 (6): 1686.

Liu, J., Wang, S., Wang, H. et al. (2021a). Rapid generation of tomato male-sterile lines with a marker use for hybrid seed production by CRISPR/Cas9 system. *Molecular Breeding* 41 (3): 25.

Liu, Q., Liu, Y., Li, G. et al. (2021b). De novo biosynthesis of bioactive isoflavonoids by engineered yeast cell factories. *Nature Communications* 12 (1): 6085.

Liu, W., Feng, Y., Yu, S. et al. (2021c). The flavonoid biosynthesis network in plants. *International Journal of Molecular Sciences* 22 (23): 12824.

Lloyd, A., Brockman, A., Aguirre, L. et al. (2017). Advances in the MYB–bHLH–WD repeat (MBW) pigment regulatory model: addition of a WRKY factor and co-option of an anthocyanin MYB for betalain regulation. *Plant and Cell Physiology* 58 (9): 1431–1441.

Lowder, L.G., Zhang, D., Baltes, N.J. et al. (2015). A CRISPR/Cas9 toolbox for multiplexed plant genome editing and transcriptional regulation. *Plant Physiology* 169 (2): 971–985.

Maity, S., Ukil, A., Karmakar, S. et al. (2003). Thearubigin, the major polyphenol of black tea, ameliorates mucosal injury in trinitrobenzene sulfonic acid-induced colitis. *European Journal of Pharmacology* 470 (1–2, 112): 103.

Malla, S., Koffas, M.A., Kazlauskas, R.J., and Kim, B.G. (2012). Production of 7-O-methyl aromadendrin, a medicinally valuable flavonoid, in Escherichia coli. *Applied and Environmental Microbiology* 78 (3): 684–694. https://doi.org/10.1128/AEM.06274-11.

Malzahn, A.A., Tang, X., Lee, K. et al. (2019). Application of CRISPR-Cas12a temperature sensitivity for improved genome editing in rice, maize, and *Arabidopsis*. *BMC Biology* 17 (1): 9.

Mikhaylova, E.V., Khusnutdinov, E.A., Chemeris, A.V., and Kuluev, B.R. (2022). Available toolkits for CRISPR/CAS genome editing in plants. *Russian Journal of Plant Physiology* 69 (1): 3.

Miller, J.C., Chezem, W.R., and Clay, N.K. (2016). Ternary WD40 repeat-containing protein complexes: evolution, composition and roles in plant immunity. *Frontiers in Plant Science* 6: 1108.

Miranda, S., Piazza, S., Nuzzo, F. et al. (2023). CRISPR /Cas9 genome-editing applied to *MdPGT1* in apple results in reduced foliar phloridzin without impacting plant growth. *The Plant Journal* 113 (1): 92–105.

Murakami, A., Ashida, H., and Terao, J. (2008). Multitargeted cancer prevention by quercetin. *Cancer Letters* 269 (2): 315–325.

Naing, A.H., Park, K.I., Ai, T.N. et al. (2017). Overexpression of snapdragon delila (Del) gene in tobacco enhances anthocyanin accumulation and abiotic stress tolerance. *BMC Plant Biology* 17 (1): 65.

Nishihara, M., Higuchi, A., Watanabe, A., and Tasaki, K. (2018). Application of the CRISPR/Cas9 system for modification of flower color in *Torenia fournieri*. *BMC Plant Biology* 18 (1): 331.

Nitarska, D., Boehm, R., Debener, T. et al. (2021). First genome edited poinsettias: targeted mutagenesis of flavonoid 3′-hydroxylase using CRISPR/Cas9 results in a colour shift. *Plant Cell, Tissue and Organ Culture* 147 (1): 49–60.

Ogo, Y., Ozawa, K., Ishimaru, T. et al. (2013). Transgenic rice seed synthesizing diverse flavonoids at high levels: a new platform for flavonoid production with associated health benefits. *Plant Biotechnology Journal* 11 (6): 734–746.

Padmaja, L.K., Agarwal, P., Gupta, V. et al. (2014). Natural mutations in two homoeologous TT8 genes control yellow seed coat trait in allotetraploid brassica juncea (AABB). *Theoretical and Applied Genetics* 127 (2): 339–347.

Palma-Tenango, M., Soto-Hernández, M., and Aguirre-Hernández, E. (2017). Flavonoids in agriculture. In: *Flavonoids – From Biosynthesis to Human Health* (ed. G.C. Justino), 189–201. InTech.

Panche, A.N., Diwan, A.D., and Chandra, S.R. (2016). Flavonoids: an overview. *Journal of Nutritional Science* 5: e47.

Pang, Y., Wenger, J.P., Saathoff, K. et al. (2009). A WD40 repeat protein from *Medicago truncatula* is necessary for tissue-specific anthocyanin and proanthocyanidin biosynthesis but not for trichome development. *Plant Physiology* 151 (3): 1114–1129.

Park, J.-J., Dempewolf, E., Zhang, W., and Wang, Z.-Y. (2017). RNA-guided transcriptional activation via CRISPR/dCas9 mimics overexpression phenotypes in *Arabidopsis*. *PLoS One* 12 (6): e0179410.

Pizzi, A. (2019). Tannins: prospectives and actual industrial applications. *Biomolecules* 9 (8): 344.

Qi, Y. (2019). *Plant Genome Editing with CRISPR Systems: Methods and Protocols*. New York: Springer.

Qi, Y., Gu, C., Wang, X. et al. (2020). Identification of the *Eutrema salsugineum* EsMYB90 gene important for anthocyanin biosynthesis. *BMC Plant Biology* 20 (1): 186.

Rein, D., Schijlen, E., Kooistra, T. et al. (2006). Transgenic flavonoid tomato intake reduces C-reactive protein in human C-reactive protein transgenic mice more than wild-type tomato. *The Journal of Nutrition* 136 (9): 2331–2337.

Ren, C., Li, H., Liu, Y. et al. (2022). Highly efficient activation of endogenous gene in grape using CRISPR/dCas9-based transcriptional activators. *Horticulture Research* 9: uhab037.

Riaz, B., Chen, H., Wang, J. et al. (2019). Overexpression of maize ZmC1 and ZmR transcription factors in wheat regulates anthocyanin biosynthesis in a tissue-specific manner. *International Journal of Molecular Sciences* 20 (22): 5806.

Ryder, P., McHale, M., Fort, A., and Spillane, C. (2017). Generation of stable nulliplex autopolyploid lines of *Arabidopsis thaliana* using CRISPR/Cas9 genome editing. *Plant Cell Reports* 36 (6): 1005–1008.

Salehi, B., Fokou, P., Sharifi-Rad, M. et al. (2019a). The therapeutic potential of naringenin: a review of clinical trials. *Pharmaceuticals* 12 (1): 11.

Salehi, B., Venditti, A., Sharifi-Rad, M. et al. (2019b). The therapeutic potential of apigenin. *International Journal of Molecular Sciences* 20 (6): 1305.

Schnarr, L., Segatto, M.L., Olsson, O. et al. (2022). Flavonoids as biopesticides – systematic assessment of sources, structures, activities and environmental fate. *Science of the Total Environment* 824: 153781.

Selma, S., Bernabé-Orts, J.M., Vazquez-Vilar, M. et al. (2019). Strong gene activation in plants with genome-wide specificity using a new orthogonal CRISPR /Cas9-based programmable transcriptional activator. *Plant Biotechnology Journal* 17 (9): 1703–1705.

Selma, S., Sanmartín, N., Espinosa-Ruiz, A. et al. (2022). Custom-made design of metabolite composition in *N. benthamiana* leaves using CRISPR activators. *Plant Biotechnology Journal* 20 (8): 1578–1590.

Shah, F.L.A., Ramzi, A.B., Baharum, S.N. et al. (2019). Recent advancement of engineering microbial hosts for the biotechnological production of flavonoids. *Molecular Biology Reports* 46 (6): 6647–6659.

Singla, R.K., Dubey, A.K., Garg, A. et al. (2019). Natural polyphenols: chemical classification, definition of classes, subcategories, and structures. *Journal of AOAC International* 102 (5): 1397–1400.

Sunil, C. and Xu, B. (2019). An insight into the health-promoting effects of taxifolin (dihydroquercetin). *Phytochemistry* 166: 112066.

Sushmita, Kaur, G., Upadhyay, S.K., and Verma, P.C. (2021). An overview of genome-engineering methods. *Wiley*. https://doi.org/10.1002/9781119672425.ch14.

Tasaki, K., Higuchi, A., Watanabe, A. et al. (2019). Effects of knocking out three anthocyanin modification genes on the blue pigmentation of gentian flowers. *Scientific Reports* 9 (1): 15831.

Tian, Y., Liu, X., Li, T., Qin, H., Li, X., Chen, K., Chen, F., Xu, Y. (2020) CRISPR-Cas9 mediated knockout of NtAn1 to enhance the lipid accumulation in tobacco seed for biodiesel production, Preprint (Version 1) available at Research Square. https://doi.org/10.21203/rs.3.rs-37669/v1.

Tu, M., Fang, J., Zhao, R. et al. (2022). CRISPR/Cas9-mediated mutagenesis of *VvbZIP36* promotes anthocyanin accumulation in grapevine (*Vitis vinifera*). *Horticulture Research* 9: uhac022.

Ullah, A., Munir, S., Badshah, S.L. et al. (2020). Important flavonoids and their role as a therapeutic agent. *Molecules* 25 (22): 5243.

Um, T., Choi, J., Park, T. et al. (2022). Rice *microRNA171f/SCL6* module enhances drought tolerance by regulation of flavonoid biosynthesis genes. *Plant Direct* 6 (1).

Upadhyay, S.K. (2021). Genome engineering for crop improvement. *Wiley*. https://doi.org/10.1002/9781119672425.

Vanegas, K.G., Lehka, B.J., and Mortensen, U.H. (2017). SWITCH: a dynamic CRISPR tool for genome engineering and metabolic pathway control for cell factory construction in Saccharomyces cerevisiae. *Microbial Cell Factories* 16 (1): 25.

Wan, S., Li, C., Ma, X., and Luo, K. (2017). PtrMYB57 contributes to the negative regulation of anthocyanin and proanthocyanidin biosynthesis in poplar. *Plant Cell Reports* 36 (8): 1263–1276.

Wang, X., Wang, X., Hu, Q. et al. (2015). Characterization of an activation-tagged mutant uncovers a role of GLABRA2 in anthocyanin biosynthesis in *Arabidopsis*. *The Plant Journal* 83 (2): 300–311.

Wang, L., Ran, L., Hou, Y. et al. (2017). The transcription factor MYB115 contributes to the regulation of proanthocyanidin biosynthesis and enhances fungal resistance in poplar. *The New Phytologist* 215 (1): 351–367.

Wang, L., Lui, A.C.W., Lam, P.Y. et al. (2020). Transgenic expression of flavanone 3-hydroxylase redirects flavonoid biosynthesis and alleviates anthracnose susceptibility in sorghum. *Plant Biotechnology Journal* 18 (11): 2170–2172.

Watanabe, K., Kobayashi, A., Endo, M. et al. (2017). CRISPR/Cas9-mediated mutagenesis of the dihydroflavonol-4-reductase-B (DFR-B) locus in the Japanese morning glory ipomoea (*Pharbitis*) nil. *Scientific Reports* 7 (1): 10028.

Wen, D., Wu, L., Wang, M. et al. (2022). CRISPR/Cas9-mediated targeted mutagenesis of FtMYB45 promotes flavonoid biosynthesis in tartary buckwheat (*Fagopyrum tataricum*). *Frontiers in Plant Science* 13: 879390.

Wu, J., Du, G., Chen, J., and Zhou, J. (2015). Enhancing flavonoid production by systematically tuning the central metabolic pathways based on a CRISPR interference system in *Escherichia coli*. *Scientific Reports* 5 (1): 13477.

Wu, J., Zhang, X., Zhou, J., and Dong, M. (2016). Efficient biosynthesis of (2S)-pinocembrin from d-glucose by integrating engineering central metabolic pathways with a pH-shift control strategy. *Bioresource Technology* 218: 999–1007.

Xie, T., Chen, X., Guo, T. et al. (2020). Targeted knockout of *BnTT2* homologues for yellow-seeded *Brassica napus* with reduced flavonoids and improved fatty acid composition. *Journal of Agricultural and Food Chemistry* 68 (20): 5676–5690.

Xu, Z.-S., Feng, K., and Xiong, A.-S. (2019). CRISPR/Cas9-mediated multiply targeted mutagenesis in orange and purple carrot plants. *Molecular Biotechnology* 61 (3): 191–199.

Yan, S., Chen, N., Huang, Z. et al. (2020). *Anthocyanin Fruit* encodes an R2R3-MYB transcription factor, SlAN2-like, activating the transcription of *SlMYBATV* to fine-tune anthocyanin content in tomato fruit. *The New Phytologist* 225 (5): 2048–2063.

Yang, P.-W., Lu, Z.-Y., Pan, Q. et al. (2019). MicroRNA-6809-5p mediates luteolin-induced anticancer effects against hepatoma by targeting flotillin 1. *Phytomedicine* 57: 18–29.

Yang, Q., Wu, M., Zhu, Y.-L. et al. (2021a). The disruption of the MAPKK gene triggering the synthesis of flavonoids in endophytic fungus *Phomopsis liquidambaris*. *Biotechnology Letters* 43 (1): 119–132.

Yang, X., Wang, J., Xia, X. et al. (2021b). *OsTTG1*, a WD40 repeat gene, regulates anthocyanin biosynthesis in rice. *The Plant Journal* 107 (1): 198–214.

Yang, T., Ali, M., Lin, L. et al. (2023). Recoloring tomato fruit by CRISPR/Cas9-mediated multiplex gene editing. *Horticulture Research* 10 (1): uhac214.

Yoshida, K., Ma, D., and Constabel, C.P. (2015). The MYB182 protein down-regulates proanthocyanidin and anthocyanin biosynthesis in poplar by repressing both structural and regulatory flavonoid genes. *Plant Physiology* 167 (3): 693–710.

Zhai, Y., Yu, K., Cai, S. et al. (2020). Targeted mutagenesis of *BnTT8* homologs controls yellow seed coat development for effective oil production in *Brassica napus* L. *Plant Biotechnology Journal* 18 (5): 1153–1168.

Zhang, P., Du, H., Wang, J. et al. (2020). Multiplex CRISPR/Cas9-mediated metabolic engineering increases soya bean isoflavone content and resistance to soya bean mosaic virus. *Plant Biotechnology Journal* 18 (6): 1384–1395.

Zhang, H., Li, Z., Zhou, S. et al. (2022a). A fungal NRPS-PKS enzyme catalyses the formation of the flavonoid naringenin. *Nature Communications* 13 (1): 6361.

Zhang, Y., Guo, W., Chen, L. et al. (2022b). CRISPR/Cas9-mediated targeted mutagenesis of GmUGT enhanced soybean resistance against leaf-chewing insects through flavonoids biosynthesis. *Frontiers in Plant Science* 13: 802716.

Zhi, J., Liu, X., Li, D. et al. (2020). CRISPR/Cas9-mediated SlAN2 mutants reveal various regulatory models of anthocyanin biosynthesis in tomato plant. *Plant Cell Reports* 39 (6): 799–809.

Żuk, M., Kulma, A., Dymińska, L. et al. (2011). Flavonoid engineering of flax potentiate its biotechnological application. *BMC Biotechnology* 11 (1): 10.

13

Genome Engineering in Medicinal Plants for Improved Therapeutics: Current Scenario and Future Perspective

Buket Çakmak Güner

Department of Medical Biology, School of Medicine, Ankara University, Ankara, Turkey

13.1 Introduction

It is becoming increasingly common for plant extracts, in particular herbal medicines, to play an important role in pharmaceutical industry because they have been regarded as alternative to chemical medicines because they are more traditional (Gibbons 2003). Plants have had a long history in medicine, notably in the Chinese and Ayurvedic systems (Gibbons 2003). Approximately 500,000 species of plants are used as medicines (Joppa et al. 2011; Salmeron-Manzano et al. 2020). In the beginning of using of traditional medicines, trial-and-error method was used to identify plants that were useful in treating illness or feeling better, and in this way, plants that were beneficial to health were discovered (Grover et al. 2002; Oluyemisi et al. 2012). Over the generations, their use has been refined, which has become to known as "traditional medicine" in many contexts.

It was known that many plants had therapeutic properties, but bioactive molecules that treat the diseases were unknown until the eighteenth century (Faridi et al. 2010). Most of the medicinal properties of plants come from secondary metabolites, which are major phytochemicals of pharmaceuticals. With the beginning of the modern science, it is possible to detect, isolate, and improve the bioactive molecules of medicinal plants.

The direct use of medicinal plants appears to have been displaced in modern medicine until recently (Atasanov et al. 2015). Medicine today relies heavily on pharmaceutical drugs, which are largely derived from natural ingredients, which are often utilized as raw materials (Arceusz et al. 2010). Despite this, much of the underdeveloped world does not have access to modern medicine, so many people depend on direct use of medicinal plants as traditional medicine (Salmeron-Manzano and Manzano-Agugliaro 2020). Despite its potential benefits, there are two major drawbacks that may accompany the potential tendency in using traditional medicines. In the first place, medicinal plants are used without sanitary regulation, without considering their potential health hazards that could include dangerous bioactive molecules (Chan 2003; Adoym 2009). For example,

Applications of Genome Engineering in Plants, First Edition. Edited by Santosh Kumar Upadhyay.
© 2024 John Wiley & Sons Ltd. Published 2024 by John Wiley & Sons Ltd.

Momordica charantina L. used in treating fevers and malaria can use severe blood sugar drops and induce comas in patients (Basch et al. 2003; Li et al. 2004; Street et al. 2008). Secondly, there has been an increase in the number of unresearched products, which has led false perspective.

Biotechnological techniques are very functional to eliminate the disadvantages arising from direct use of medicinal plants. By using different high-throughput methodologies, scientist can get the complete chemical and genomic information of medicinal plants and identify unknown natural molecules (Zhao et al. 2018; Niazian 2019). Next-generation sequencing (NGS) gives information in genus and species level using several molecular identification and classification techniques (Techen et al. 2014). Following NGS analysis, TALENs, ZFNs, and CRISPR-Cas9 technologies can be used for targeted genome editing, enabling genetic and metabolic engineering in medicinal plants. In addition, NGS is a promising technology for identification of new medicinal plants and also developing of new phytochemicals (Pouvreau et al. 2018).

In this chapter, genome engineering methods used in medicinal plants for producing therapeuticals are highlighted. The principals, advantages, disadvantages, concerns, and future perspectives are presented.

13.2 Genome Engineering in Plants

The development of transgenic plants and transformation techniques in the early 1980s sped up the development of plant genetic engineering technology. In the following years, transformation vectors, DNA delivery methods, chimeric genes, and plant regeneration systems were developed (Babich et al. 2020).

Transgenic plants are genetically modified plants that have been added new gene from other plants with the specific transformation techniques. As compared to traditional breeding, this technology has the advantage that it transfers only one gene and total genome of the plant does not change (Lima et al. 2014).

There are several methods for DNA delivery to the cells. Generally, DNA transfer methods can be categorized into several large categories (Martin et al. 2016). The first group is Agrobacterial and viral vectors and they belong to the natural gene transfer systems. There are two subgroups of artificial transfer systems in the second group: direct injections of DNA by chemicals, physical methods or mechanical methods into cells and protoplasts; and using artificial vector systems such as liposomes, microinjections, or microparticles (Diaz et al. 2019).

13.2.1 *Agrobacterium*-Mediated Transformation

Agrobacterium-mediated gene transfer involves the incorporation of a foreign gene into the bacterial plasmid. Plant cells are infected with bacteria carrying chimeric plasmids and required DNA is transferred (Gelvin 2003). A. *tumefaciens* is generally used in agrobacterial plant transformation, while A. *rhizogenes* is rarely used as bacterial vectors (Hellens et al. 2000). In this process, Agrobacterium's Ti plasmid is used to transfer small DNA fragments into plant cells. Transferred genes induce the formation of tumors or overgrown roots in which opines are synthesized, which are nutrients for *Agrobacterium* but not absorbed by plant cells. Ti plasmids have also been shown to function as a biological vector for transferring large DNA fragments since they contain all the necessary components (Radziejewska et al. 2018).

13.2.2 Biolistic or Particle Bombardment-Mediated Transformation

The principle of particle bombardment or biolistic method is forcing DNA molecules physically into the plant cells (Sanford 2000). Particle bombardment method does not depend on the genotype or receptor of the host when compared to *A. tumefaciens*-mediated transformation (Davies et al. 2013). In addition to this, biolistic plant transformation often results in multiple integrations that are scrambled (Altpeter et al. 2005). The biolistic method allows delivery of large fragments of DNA, such as bacterial artificial genomes (Ananiev et al. 2009).

The particle bombardment can deliver exogenous DNA both nuclear genomes and organelle genomes (Montanari et al. 2015). In the biolistic machinery, high pressure that delivers the exogenous DNA to the recipient target tissue by triggering it onto gold/tungsten carriers (microprojectile) with high velocity (Sanford 2000). During the movement of the microprojectile, the coated DNA remains inside the cell while the microprojectile passes through.

13.2.3 Electroporation-Mediated Transformation

Electroporation is the process of creating transient pores with electricity in prokaryotic and eukaryotic plasma membranes (Yarmush et al. 2014). Electricity causes creating of electropores in the cell membrane and these pores allow micro and micromolecules getting in or outside of the cell (Ozyigit 2020). Electropores can close and cells can recover under the optimal electrical pulse. On the other hand, unsuitable electrical currents such as very high field strength can cause overheating of the cells and resulting in cell deaths (Kotnik et al. 2012).

13.2.4 Chemical-Mediated Transformation

Calcium phosphate-mediated transfer, polyethylene glycol (PEG)-mediated transfer, DEAE-dextran-mediated transfer, and liposome-mediated transfer are examples of chemical transformation techniques. (Ozyigit 2020). In the calcium-mediated transformation, DNA is mixed with calcium chloride and potassium phosphate solution and obtained calcium phosphate precipitate. Then precipitated DNA and cells incubated and DNA enter the cells by endocytosis (Lee et al. 2019). PEG-mediated transfer is used for transfer genes into the protoplasts. In the DEAE-Dextran-mediated transfer, DEAE-Dextran is a low cost, simple, and effective transfer medium for transient cells that are commercially available at low prices (Uchiumi et al. 2014). Liposome-mediated transfer involves artificial lipid vesicles that act as delivery agents for foreign genome. Liposome surrounds the molecule, which is aimed to transfer and fuse with the cell membrane to transfer it (Allen and Cullis 2013).

13.3 Genome Editing in Plants

Techniques for genetic engineering cover a broad range of studies, such as producing transgenic animals, conducting functional gene analyses, creating diseases models, and developing drugs. Because of their large size, the delivery proteins such as enzymes, antibodies, and growth hormones have faced several obstacle. In order to overcome these difficulties, scientists explored alternative approaches, which resulted in the advancement of genome editing technology (Khalil 2020). A remarkable effort and a great deal of experimentation have now led to methods for changing the genetic structure of living organisms. Genome editing technologies have advanced recent years, and now three of "programmable" nucleases are frequently used in genome editing: zinc finger nucleases (ZFNs), transcription activator-like effector nucleases (TALEN), and the clustered regularly interspaced short palindromic repeats (CRISPR)/CRISPR associated protein Cas9 (CRISPR-Cas9) System (Gaj et al. 2016; Sushmita et al. 2021; Upadhyay 2021). With the discovery of genome engineering/editing technology, it will become possible to precisely manipulate the genomes or organisms to alter their particular characteristics.

Engineered ZFNs are generally used as production of animal models for disease treatment studies (Urnov et al. 2010). They are fusion proteins that include multiple site-specific DNA binding domains, and they are adapted from zinc finger that contains transcription factors. These transcription factors are binded to the endonuclease domain of the bacterial FokI restriction enzyme (Rehman et al. 2021). ZFNs act as a pair for cutting a specific region in the genome by identifying two flanking sequence sites, one on each strand. The FokI domains dimerize when ZFNs bind to the sequence site, cutting the DNA at the specific site, causing a double strand break (DSB). Then, cells repair double break strands in two different ways: homology-directed repair and nonhomologous end joining (Urnov et al. 2010). While ZFNs have many advantageous in genome editing, they also have a lot of potential disadvantages. For example, assembling of zinc-finger domains for binding nucleotides with high affinity is very complex process. Another example is limitation of the binding sites that ZFNs can only bind ZFN open sites (Khandagale and Nadaf 2016).

TALENs are transcription activator-like effectors (TALEs) that bind DNA by binding to the FokI enzyme's non-specific cleavage domain. This type of protein is secreted by phytopathogenic bacteria known as Xanthomonas spp (Khandagale and Nadaf 2016). In terms of its DNA-binding domain, it attaches to DNA and is capable of altering the expression pattern of genes in the host. In these process, DNA binding domain has 34 aminoacid repeats and each repeat could recognize a single nucleotide in the target genome while repeats of ZFNs could recognize three nucleotides (Rehman et al. 2021). Thus, designing of TALENs are easier than ZFNs.

CRISPR-Cas9 technology has emerged recently and become the most preferred genome editing tool in a short time. This technology has high specificity and activity in the target genome (Afzal et al. 2020). CRISPR-Cas9 systems combine the short RNA molecules and proteins to cleavage the specific target DNA molecules (Guo et al. 2022). CRISPR RNA and transactivating CRISPR RNA (tracrRNA) are RNA parts that guide protein part of the system; Cas9 is endonuclease and protein part of the system (Khandagale and Nadaf 2016). CRISPR technology has been got interest since its discovery and still continued. Many scientific researches demonstrated that CRISPR-Cas9 genome editing is applicable from

Table 13.1 Comparison of genome editing techniques.

ZFNs	TALENs	CRISPR-Cas9
Zinc finger domain include FokI nuclease domain	TALE-DNA domain includes FokI nuclease domain	CrRNA, tracrRNA, and Cas9 protein domains
Constitute DSBs in the target genome	Constitute DSBs in the target genome	Constitute DSBs or single-strand breaks (SSBs) in target genome
24–36 bp length of target sequence	24–59 bp length of target sequence	20–22 bp length of target sequence
Cloning and protein design steps are essential	Cloning and protein design steps are essential	Cloning step is not essential, protein design step is not complex
Multiplex editing is difficult	Multiplex editing is difficult	Multiplex editing is possible and easy
Expensive	Expensive	Cost effective

Source: Adapted from Khandagale and Nadaf (2016).

microorgansims to plant and human genome editing (Jiang et al. 2013, Shan et al. 2013; Feng et al. 2014; Alok et al. 2018, 2021; Bhalothia et al. 2020; Singh et al. 2022) (Table 13.1).

13.3.1 Applications in Medicinal Plants

Although there are several genome editing techniques in plants, CRISPR-Cas9 technology is the most used one (Guo et al. 2022). These technology has focused on manipulating genes involved in the biosynthesis pathways of secondary metabolites and other bioactive molecules. There are several examples of genome-edited medicinal plants such as *Salvia miltiorrhiza, Dendrobium officinale, Cannabis sativa, Camelina sativa, Dioscorea zingiberensis, Nicotiana tabacum, Papaver somniferum* (Dixit et al. 2021; Guo et al. 2022; Mishra et al. 2023).

Camelina sativa, which belongs to Brassicaceae family, has been used in many areas such as pharmaceutical, cosmetic, and bio-fuel industries (Mishra et al. 2023). It is rich in bioactive molecules such as phenolics and linoleic, α-linoleic fatty acid (Ergönül and Özbek 2020). Studies have been showed that these molecules have antioxidant, anticancer, and anti-inflammatory properties (Ergönül and Özbek 2020; Dey 2021). Genome editing studies in this medicinal plant focus on knocking out of *FAD2* (*Fatty Acid Desaturase 2*) and *FAE1* (*Fatty Acid Elongase 1*) genes for enhancing linoleic, α- linoleic fatty acid content (Guo et al. 2022).

Dioscorea zingiberensis, which belongs to Dioscoreaceae family, is considered as important plant for medicine. Diosgenin molecule of *D. zingiberensis* has been used in the treatment of several diseases in traditional Chinese medicine like cardiovascular disease, cough, stomach distension (Zhang et al. 2018). Diosgenin biosynthesis pathway requires E-isomer farnesyl pyrophosphate (FFP) synthase enzyme for production of diosgenin. For this purpose, FFP encoding gene *Dzfps* of *D. zingiberensis* is mutated with CRISPR-Cas9 genome editing technique (Feng et al. 2018). As a result of mutation, squalene molecule and *Dzfps* levels decrease.

Nicotiana tabacum, which belongs to Solanaceae family, is a main source of tobacco. It is rich in secondary metabolites such as alkaloids, flavonoids, and terpenoids. *Nicotiana tabacum* is also used in recombinant protein production such as xylanase, proteases and has important role in pharmaceutical and industrial areas (Dey 2021). CRISPR-Cas9 system has been applied for inactivating plant-specific glycans to produce glycoproteins and nicotine-free tobacco (Mercx et al. 2017; Schachtsiek and Stehle 2019).

P*apaver somniferum,* which belongs to Papaveraceae family, has been frequently used as analgesic, sedative purpose in different diseases (Masihuddin et al. 2018). One of the main chemical molecule for analgesic actions is benzylisoquinoline alkaloids (BIAs). *4′OMT* gene that regulates BIA biosynthesis in *P. somniferum* were knocked out with CRISPR-Cas9 editing technique and declining of BIA production levels were observed (Alagoz et al. 2016; Mishra et al. 2023).

Salvia miltiorrhiza is a medicinal plant, belongs to the Lamiaceae family. The main bioactive chemical molecules in this medicinal plant are diterpenoids and phenolic acids. These molecules have several pharmaceutical activities like cardiovascular, antioxidant, antifibrotic, etc. (Shi et al. 2021). The genes that encode molecules that have role in tanshionine biosynthetic pathway have been mutated with CRISPR-Cas9 genome editing technique, and it has been shown that diterpenoids and phenolic acid levels are reduced (Guo et al. 2022).

Dendrobium officinale is a medicinal plant that belongs to the Orchidaceae family. This medicinal plant has shown antimicrobial, anticancer, antifungal, anti-inflammatory properties thanks to chemical constituents like alkaloids, bibenzyls, and polysaccharides (Liang et al. 2018). Genome editing of *D. officinale* has suggested that CRISPR-Cas9 system is suitable method for deletions, insertions, and substitutions in the target genes (Kui et al. 2017) (Table 13.2).

Table 13.2 Medicinal plants and edited genes with CRISPR-Cas9 delivery system.

Medicinal plant	Target genes	Result of editing
Camelina sativa	*FAD2*	Enhancing of produced oleic acid content
	FAE1	Enhancing of produced oleic and α-linoleic acid content
Dioscorea zingiberensis	*Dzfps*	Decrease in squalene levels
Nicotiana tabacum	*XyIT-β*	Decrease in β-xylose and α-fructose contents
	FucT-α	Nicotine-free plant production
	BBL	
papaver somniferum	*4′OMT2*	Decrease quantity of alkaloids
Salvia miltiorrhiza	*SmCPS1*	Decrease in tanshinone I/IIA level
	SmLAC	Decrease in RA and LAB levels
	SmRAS	Decrease in RA and LAB levels
Dendrobium officinale	C3H	Generating of deletions, insertions, and substitutions
	CCR	
	C4H	
	IRX	
	4CL	

Source: Adapted from Guo et al. (2022) and Mishra et al. (2023).

13.4 Medicinal Plants: Comparison of Traditional and Scientific Use

It is becoming increasingly important to use plant extracts in pharmacy, especially herbal medicinal products, and these products are gaining in popularity from the perspective of the general public, who see these products as viable alternatives to traditional chemical entity medicines (Firenzuoli and Gori 2007). Due to their complex nature, extracts have dozens to hundreds of discrete components that can act on numerous pharmacological targets as they are complex mixtures (Firenzuoli et al. 2004). Herbal practitioners often regard this as a distinct advantage, as it allows for a variety of actions to be achieved, but also because inactive ingredients may increase the activity of certain components (Firenzuoli and Gori 2007). However, herbal medicines carry risks. For this reason, research in this area needs to be intensified, as the public often believes that all natural treatments are inherently safe.

In the traditional use of herbal medicine; people could use herbal plants or parts of plants for treatment, they could use mixture of many plants to get better results. People generally believe that medicinal plants are safe without any adverse reaction. On the other hand, in the scientific use, scientists use suitable extracts and pharmaceutical preparation of plants. They generally use standardized and purified chemical ingredients that have a pharmacological activity. They also try to eliminate possible side effects, drug interactions, contraindications, etc. (Firenzuoli and Gori 2007).

13.5 Chemical Components of Medicinal Plants

Two types of compounds named primary metabolites and secondary metabolites include phenolics, flavonoids, alkaloids, steroids, terpenes, and tannins have been produced by plants (Farnsworth 1966). Any modern drugs are created from these natural metabolites or combination of such substances. Several medicinal properties are present in plants, including antifungal, antiviral, antibacterial, anticancer, sedative, laxative, diuretic, antihelminthic, etc. (Pant 2014).

One of the most important issue of medicinal plant is the scanty information about the total composition of the plants. Additionally, chemical composition of medicinal plants can be changed depending on the factors like part of the plant, storage conditions, humidity, harvest time, geographical location, etc. (Sethi et al. 2021). Merchandized products may also vary from batch to another batch in terms of chemical constituent content and concentration (Chavan et al. 2006). It is also possible for the same manufacturer to produce products containing different substances over different periods of time, even when they use standardized conditions to achieve high pharmaceutical quality (Firenzuoli and Gori 2007).

Through high-throughput experiments, massive databases of genomic, proteomic, and chemical data have been created, and these databases can now be accessed with efficient separation methods and powerful spectrometric techniques for identifying active compounds and elucidating their structures (Chavan et al. 2006). To create effective, safe, and personalized medicines, postgenomic era information, computational biology, and synthetic biology approaches merge to create novel drugs (Singh et al. 2019).

13.6 Using Biotechnological Techniques in Medicinal Plant Production

Mass production of the desired product with minimal inputs is necessary to meet the growing needs in every field. In the same way, the growing demand for pharmaceuticals has led us to switch to more efficient processes such as plants. Our basic needs are mostly fulfilled by plants since ancient times. Today, developments in genetic engineering area are allowing plants to be used as factories to produce proteins, pharmaceuticals, vaccines, industrial enzymes, etc. (Georgiev et al. 2018). These systems are the most cost-effective for producing recombinant proteins. There are many factors to consider when choosing a system, including the product safety, storage, cost-effectiveness, the purification process, the environmental concerns, and the public's acceptance (Niazian 2019).

It is crucial to assess the genetic variability of unexplored genetic backgrounds in order to develop an efficient plant production program (Niazian et al. 2017). RFLP (restriction fragment length polymorphism), RAPD (randomly amplified polymorphic DNA), and ISSR (Inter-Simple Sequence Repeat) are efficient morphological and molecular markers for this purpose (Niazian 2019). However, these methods are costly, slow, and time consuming. Therefore, high-throughput techniques have been developed for faster and more effective methods for identifying genetic diversity. Using microarray technology in medicinal plants is very effective since it allows identification of differences in expression patterns under different growth conditions (Ishkanian et al. 2004). Therefore, it is possible to track and find genomic fragments that are expressed in different growing environments, whether *in vitro* or in the field.

With the next-generation sequencing (NGS), several genomic markers can be discovered, sequenced, and genotyped quickly and cost effectively (Davey et al. 2011). NGS can also be used to investigate the metabolisms of plants that have not been studied and are unexplored, such as medicinal plants (Liu et al. 2017). The NGS technology can be used for cDNA sequencing (RNA-seq) or transcriptome profiling. RNA-seq consists of sequencing the entire transcriptome of a population. Compared with microarrays, NGS-based transcriptome analysis is more effective at distinguishing unknown genes. This method has been successfully used in order to identify the unigenes responsible for biosynthesis of monoterpenoids in ajowan (*Trachyspermum ammi L.*) (Soltani Howyzeh et al. 2018). A functional breeding program can be developed from the results of such researches.

In addition to all the applications outlined above, NGS has other potential applications in the field of developing of faster production of medicinal plants (Figure 13.1). Molecular taxon identifying of plants can be achieved with DNA barcoding through NGS technologies. The restriction-site-assisted DNA sequencing (RAD-seq) method can be implemented to wild-type medicinal plants without reference genome information (Davey et al. 2011). Single nucleotide polymorphism (SNP) markers serve as a link between genome-wide association studies (GWAS) and NGS. GWAS can detect QTLs in plants using high-density SNP markers to cover their entire genome (Edwards and Batley 2010).

A reverse genetic screen method named as targeting induced local lesions in genomes (TILLING) can identify mutation synthesized isogens by analyzing their function. A targeted mutagenesis technique is used in this method to create one isogen, while a natural isogen is used for the other. Thus, TILLING can be used to compare natural and induced alterations in genome (Varshney et al. 2009).

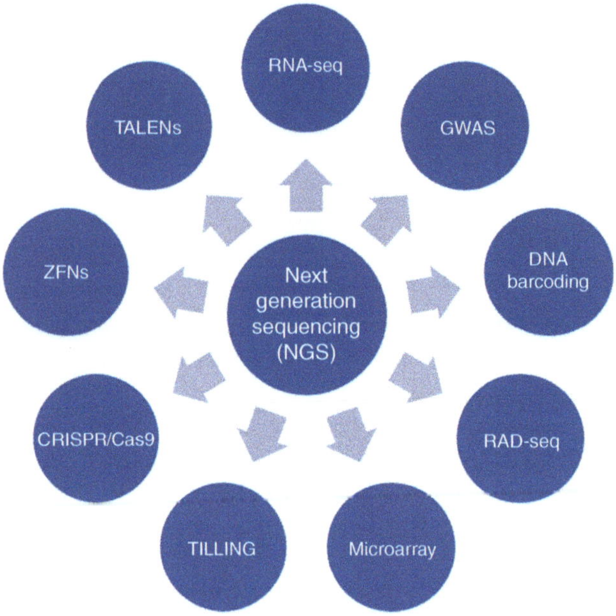

Figure 13.1 NGS and different biotechnological applications. *Source:* Adapted from Niazian (2019).

It is also possible to use NGS information to apply new technologies in biotechnology such as TALENs, ZFNs, and CRISPR-Cas9 genome editing to connect another study area such as synthetic biology. Functional genomic combines high-throughput genome sequencing and omic technologies to identify chemical molecules produced by genes involves in specific processes (Yamazaki et al. 2018).

13.7 *In Vitro* Culture Techniques in Herbal Medicine

"Plant tissue culture" is the cultivation of a single cell, tissue, or organin plants *in vitro* and aseptically. The main rule initiating *in vitro* culture is surface sterilization to prevent adhering microorganisms to the culture.

Plant tissue cultures could be developed in solid or liquid area containing growth nutrients. Today, the methodology for initiating *in vitro* cultures have been extensively studied, and the most used procedures is illustrated in Figure 13.2.

13.7.1 Plant Tissue Culture in Herbal Medicine

Plant tissue culture is a useful biotechnological method for medicinal plant production. This technique also offers controlled production of phytochemicals used as pharmaceuticals in medicinal plants. There are several ways to regenerate a plant species *in vitro*, including using a single cell, a tissue, an organ, or an explant derived from differentiated tissue. Production of unlimited plants and metabolites, various vegetative plant parts or organs may be used in these technologies such as nodes, meristems, stems, leaves, roots, endosperm, embryos, etc. (Pant 2014).

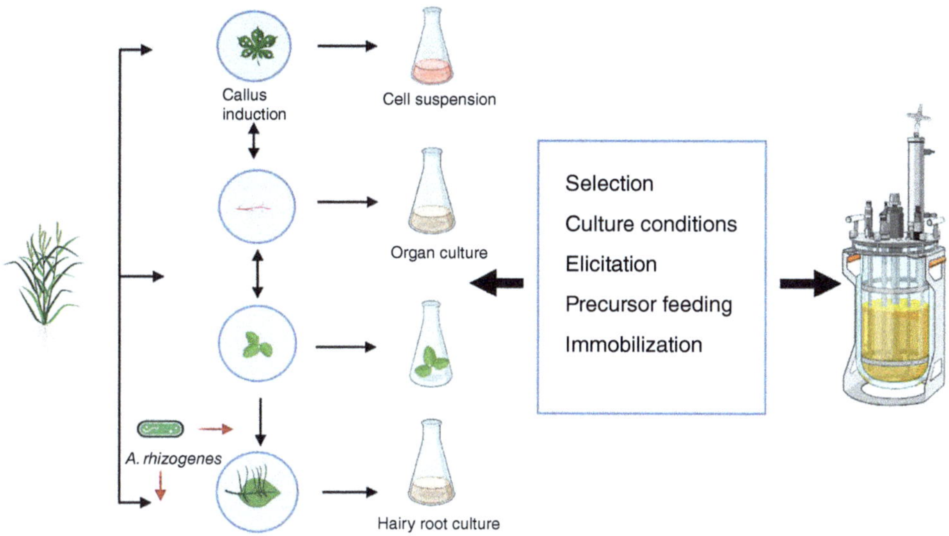

Figure 13.2 Bioactive molecule production via *in vitro* culture techniques. *Source:* Modified from Wawrosch and Zotchev (2021).

Biotic stress and low yield in medicinal plant production can surpass with plant tissue culture technique. Plant tissue culture has success in many different areas of medicinal plants such as conservation, micropagation, gene transformation, large-scale production with bioreactors (Grzegorczyk-Karolak et al. 2018). Plant tissue culture is useful method to comparing and evaluating the experiment conditions and effects of these conditions on the secondary metabolite production in medicinal plants (Lalaleo et al. 2018). Plant tissue culture technique both saves medicinal plants and active secondary metabolites and gives opportunity to product engineered molecules with initiating synthetic biology studies (Espinosa-Leal et al. 2018). Thus, plant tissue cultures are used for production of unidentified phytomolecules and mass production of medicinal plants, which are valuable for pharmaceutical industry. *In vitro* production of valuable phytochemicals are directly affected from culture conditions such as temperature, growth medium, transformation (Matkowski 2008). Medicinal plants can undergo *in vitro* chromosome duplication and artificial polyploidy induction, with many beneficial effects on their qualitative and quantitative properties. Additionally, overexpression of responsible genes in secondary metabolite biosynthesis pathways and the modification of the genes regulating to produce high-quality bioactive compounds depend on tissue culture (Pant 2014). Additionally, plant tissue culture enables hairy root cultures that have high capacity for production of secondary metabolites with higher biochemical and genetic stability compared to conventional cultures (Li et al. 2009).

13.7.2 Hairy Root Cultures in Herbal Medicine

Hairy roots are transformed by T-DNA plasmid of *Agrobacterium rhizogenes*. They also grow faster than other parts of plants in culture without requirement of additional growth regulators (Guillon et al. 2006). Thus, plant-based pharmaceutical production with hairy root culture is more advantageous than other cell and organ cultures. This system also offers large-scale production. Moreover, the *A. rhizogenes*-induced hairy root formation offers the possibility of introducing foreign genes into medicinal plants via transgenic plant technology using the Ri plasmid (Giri et al. 2001). Thus, hairy root cultures are an ideal tool to study biochemical pathways and metabolic engineering medicinal plants from a biochemical and gene expression perspective (Zhou and Wu 2006). Scientists have evaluated the capacity of hairy roots to produce pharmaceutical compounds in a variety of medicinal plants.

13.7.3 Callus and Cell Suspension Culture in Herbal Medicine

Calluses are undifferentiated cells that develop from any kind of explant in culture. In liquid nutrient media (suspension cultures), cells are suspended in medium and grown homogeneously (Pant 2014). Callus is suspended in a cell suspension in the liquid medium. Afterward, the suspension is shaken so that aggregates of cells disperse into single cells that are evenly distributed and grow continuously throughout the liquid medium (Babich et al. 2020). Therefore, suspension cultures is suitable system for production of high-quality secondary metabolites and other valuable chemical of plants.

13.7.4 Micropropagation in Herbal Medicine

Micropagation is a useful plant propagation process in agriculture, forestry, and horticulture in terms of mass-producing and proliferating plantlets in culture. Micropropagation can be used to produce high-quality, virus-free, disease-resistant seedlings and high yields of medicinal components. It is also possible to plant the seedlings on a large scale in the field. Micropropagation techniques are important for rapid production, preservation, and study of valuable medicinal plant species (Pant 2014). Additionally, micropropagation can be combined with other biotechnological methods to improve properties of medicinal plants by changing genotype by transformation and somaclonal variation.

13.7.5 Elicitation

An elicitor is a substance found in plants that stimulates the synthesis of secondary metabolites and phytoalexins (Rao and Ravinshankar 2002). In order to induce metabolite production, cell cultures are exposed to elicitors such as chitin or mannitol. Researches have shown that plant cell and tissue culture systems produce secondary metabolites more efficiently with elicitors (Zhou and Wu 2006).

13.7.6 Bioreactors for Large Scale Up

Bioreactors are indispensable for the mass production of pharmaceuticals when they are designed and operated efficiently. Cultured plant cells exhibit different morphologic and growth characteristics from those of microorganisms, including their relatively large size, tendency to aggregate, sensitivity to shear stress, and slow growth rate (Valdani et al. 2019). Microorganisms are typically grown in conventional fermentors that are suboptimal or even unsuitable for production of plant cells. A second consideration in the design of bioreactors for plant tissue cultures is to avoid product degradation and feedback inhibition, which is achieved by removing the product from the cells or culture medium (Niazian 2019). In bioreactors, oxygen transfer is another factor that affects secondary metabolite production and plant growth (Valdani et al. 2019). In spite of the fact that plant cells require relatively little oxygen for cell growth (compared to microbial cells), their need for oxygen increases significantly during metabolite synthesis.

13.8 Pharmaceutical Products from Medicinal Plants: Current Situation

Ultimate goal of tissue culture researches is production of valuable pharmaceuticals and medicines. In addition, cell and tissue cultures are useful for ensuring the continuity of biomass supply of phytomedicines, since breeders can select crops including phytomolecules with the objective of maintaining genetic consistency and the sustainability of wild plant populations. There have been a great number of secondary products produced in plant tissue cultures that exhibit a large quantity of bioactivities, as shown in Table 13.3. It is possible to extract or use some natural compounds (artemisin, paclitaxel, cannabinoids, atropine, ephedrine, digoxin, morphine, etc.) as templates or as precursors for synthesis or semi-synthesis. There are several major diseases that can be treated with these compounds, including Alzheimer's, cancer, microbial diseases, depression, and cardiovascular diseases (Babich et al. 2020).

13.8.1 Antimicrobial Molecules

It is becoming increasingly popular to utilize natural materials as treatment for microbial infections as they contain antimicrobial properties. Callus and cell cultures of *Arnebia euchroma, Lithosperum erythorhizon,* and *Onosma paniculatum* have been researched for shikonin production by the researchers. Shikonin and derivatives are used as antimicrobial drugs (Zhou and Wu 2006). *Syzygium aromaticum* (clove) is also one of the most effective antimicrobial herbs. Eugenol, 2-heptanol, eugenyl acetate are some of the active components of clove (Parham et al. 2020). These molecules have shown many pharmaceutical properties like antimicrobial, antiviral, anticancer, and antiallergic. *Cinnamomum verum* and *Cinnamomum zeylanicum* belong to *Lauraceae* family. These herbs have been traditionally used antimicrobial and antioxidant for years. Studies carried out recent years have been shown that components of these herbs have antimicrobial activity against different bacteria species.

13.8.2 Antioxidant Molecules

Rosmarinic acid and lithospermic acid are water soluble phenolic acids in the *Salvia* plants that have antioxidant activity in different diseases. Another antioxidant molecule is lycopene derived from carrots with cell culture technique (Zhou and Wu 2006). Garlic is also another antioxidant herbs. Phenolic and polysaccharide compounds and Allicin

Table 13.3 Active secondary metabolites isolated from medicinal plants via *in vitro* culture techniques.

Bioactive molecule	Plant species	Culture type
Alkaloids		
Berberine	*Aragemone mexicana*	Shoot culture
Camptothecin	*Ophiorrhiza mungos*	Root culture
Capsaicin & dihydrocapsaicin	*Capsium chinense*	Cell suspension culture
Indole alkaloids	*Isatis tinctoria*	Hairy root culture
Phenylpropanoids		
Chlorogenic acid derivatives	*Gardenia jasminoides*	Cell suspension culture
Cichoric acid	*Ocimum basilicum*	Cell suspension culture
Flavonoids (total)		
Flavonoids (Chrysin, wogonin)	*Ficus deltoidea*	Cell suspension culture
Flavonoids, phenolics, polysaccharides	*Oplopanax elatus*	Root culturc
Isoflavones	*Trifolium pratense*	*In vitro* cultivated plant
Isoquercetin	*Ocimum basilicum*	Cell suspension culture
Rosmarinic acid	*Ocimum basilicum/ Salivia lerifolia*	Cell suspension culture
Lignans		
Lignans (total)	*Linum albüm*	Hairy root culture
Dehydrodiconiferyl alcohol glucoside	*Linum usitatissimum*	Root culture
Guaiacylglycerol-β-coniferyl alcohol ether glucoside	*Linum usitatissimum*	Root culture
Lariciresinol diglucoside	*Linum usitatissimum*	Root culture
Secoisolariciresinol diglucoside	*Linum usitatissimum*	Root culture
Stilbenes		
Mulberroside A	*Morus alba*	Cell suspension culture
Terpenoids		
Artemisinin	*Artemisia annua*	Cell suspension culture
Betulinic acid	*Lantana camara*	Cell suspension culture
Bilobalide	*Ginkgo biloba*	Cell suspension culture
Ginkgolide A/B/C	*Ginkgo biloba*	Cell suspension culture
Ginsenosides	*Panax ginseng*	Cell suspension culture
Oleanolic acid	*Lantana camara*	Cell suspension culture
Oleanolic acid glycosides	*Calendula officinalis*	Hairy root culture
Paclitaxel	*Corylus avellana*	Cell suspension culture
	Taxus x media	Hairy root culture
	Stevia rebaudiana	Root culture
Steviol glycosides	*Lantana. Camara*	Cell suspension culture
Ursolic acid	*Withania somnifera*	Cell suspension culture
Withanolides	*Hybanthus enneaspermus*	Root culture
L- dopa		

Source: Modified from Wawrosch and Zotchev (2021).

molecules of garlic display antioxidant activity. Flavonols, hydrolyzable tannins, and hydrobenzoic acids of eucalyptus have also strong antioxidant activity (Parham et al. 2020).

13.8.3 Anticancer Molecules

To date, it has been reported that over a thousand plant species possess anticancer properties (Parham et al. 2020). One of the important examples of anticancer medication is isolation of indole alkaloids, such as vinblastine and vincristine from the *Catharantus roseus* (Zhou and Wu 2006). These molecules have been used in the treatment of different types of cancer such as breast and lung cancers, leukemias, lymphomas, advanced testicular cancer.

Another major success is discovery of Paclitaxel molecule (Taxol), which is a diterpenoid alkaloid and originally isolated from *Taxus brevifoila*. This molecule has been used for treatment of breast, ovarian, and lung cancers (Chandran et al. 2020). Because of the rareness of Taxus yew trees and slow growth and also low yield of Taxol obtaining, the demand of Taxol for pharmaceutical production cannot be met by extracting the drug from the trees. Over the years, experts have investigated alternative cultivation methods for various Taxus species for producing Taxol and other useful taxane compounds (Wawrosch and Zotchev 2021). Researchers discovered that cephalotaxine isolated from *Cephalotaxus fortunei* (plum yew) and its ester derivatives like isoharrringtonine, homoharringtonine, deoxyharringtonine, and harringtonine are effective molecules for the treatment of malignant lymphoma and leukemia (Zhou and Wu 2006).

These molecules have been produced by basic culture techniques such as cell suspension, tissue, callus cultures. Researchers changed the culture conditions and ingredients such as changing temperature or adding stimulants in different concentrations to find out optimum production condition.

13.8.4 Cardiovascular Molecules

There are many medicinal plant derived molecules used in the treatment of cardiovascular diseases. The antihypertensive alkaloid ajmalicine derived from *Catharanthus roseus* is one of these molecules. Tanshinone and its derivatives derived from *Salvia miltiorrhiza* (Wang et al. 2020), and forskolin derivative derived from *Coleus forskohli* (Ju et al. 2021) have also used in the cardiovascular disease treatment. It is also known that some medicinal plant molecules with antioxidant activity play a role in the treatment of cardiovascular diseases. Flavonoids, tannins, and phenolic acids of Tribulus (Tian et al. 2019); Cinnamaldehyde and eugenol of cinnamon (Willis et al. 2019); thymol and carvacrol of lemon balm are several important examples of antioxidant molecules that effect cardiovascular health (Pahram et al. 2020).

Many researchers have reported production of pharmaceutical molecules of medicinal plants. Some of these molecules have been approved by Food and Drug Administration (FDA), while clinical trials of most of bioactive molecules have not yet been conducted. A brief list of commercial pharmaceuticals produced by medicinal plants are given in Table 13.4.

Table 13.4 Some of commercial molecules produced by plant tissue cultures used in pharmaceuticals.

Plant species	Bioactive molecule	Major activity
Artemisia annua L.	Artemisinin	Antimalarial
Catharantus roseus L. G. Don	Vicristine, Vinblastine	Anticancer
Campotheca acuminata	Camptothecin	Anticancer, antiviral
Leucojum aestivum L.	Galantamine	Anti-Alzheimer
Narcissus sp. L.	Galantamine	Anti-Alzheimer
Hyoscyamus niger L.	Scopolamine	Parasympatholytic
*Pilocarpus sp.*Vahl	Pilocarpine	Anti-glaucoma
Podpphyllum emodi	Podophyllotoxin	Anticancer, antiviral
Podophyllum peltatum L.	Podophyllotoxin	Anticancer, antiviral
Rauwolfia serpentina L.	Reserpine	Hypotensive, sedative
Taxus sp. L.	Paclitaxel	Anticancer
Filipendula ulmaria L.	Aspirin	Cardiovascular
Colchicum autumnale L.	Colcicine	Anticancer
Digitalis purpurea L.	Digoxin	Cardiovascular
Papaver somniferum L.	Apomorphine hydrochloride	Neurological diseases
Atropa belladonna L.	Tiotropium bromide	Respiratory tract diseases
Allium sativum L.	Allicin	Antimicrobial

Source: Adapted from Moraes et al. (2021).

13.9 Future Perspective and Conclusion

There is no doubt that demand for plant-derived products has increased throughout the World. In light of this, it is essential that quality awareness be developed in order to ensure herbal drug efficacy. This perspective emphasizes the importance of clinical trials for determining the safety and effectiveness of herbal medicines. Standardization of herbal medicines become difficult because of the presence of complex and diverse secondary metabolites. In addition, therapeutic properties can change depending on plant age, geographical location and used part of plant (Giannenas et al. 2020). There are drastic differences in pharmacological activity between herbal products derived from the same species of plant due to the variation in phytochemical constituents. Furthermore, the timing of harvesting and the occurrence of microorganism contamination influence the global standards of medicinal plants. Biotechnology-based production methods have allowed to speed production of medicinal plant and bioactive compounds.

Plant tissue culture is core biotechnological technique in medicinal plant production. Furthermore, this technique helps the conservation of medicinal plant and contributes to improvement of chemical profile of plants. Because tissue culture technique provides controllable, sustainable, and environmentally friendly production, it can be alternative to traditional field cultivation (Wawrosch and Zotchev 2021). Tissue cultures are useful for application of genome editing technologies. Bioreactors enable large-scale production of the cultures and have a huge contribution on the production of bioactive molecules from medicinal plants. After decades of research, many high yielding cell suspensions and tissue and organ culture lines have been developed, but only a few process and produced molecule has been commercially deployed. *In vitro* systems with highly productive properties have been published decades ago, yet a number of studies published recently focus on secondary metabolites that have been widely studied for nearly 50 years. Yet, despite their obvious superiority, most of these processes have never been commercialized. Hence, plant-based *in vitro* systems lag behind while they achieve increased productivity in most cases.

There are many factors that determine the commercialization of *in vitro* plant production systems, such as economically production of molecules, regulatory requirements, market prices and consumer acceptance (Wawrosch and Zotchev 2021). But recently, there are changes in both markets and the clinical demand for certain secondary metabolites. Based on the research reviewed here, plant-based in vitro cultures have potential to produce bioactive metabolites to improve pharmaceuticals (Kreis 2019). Extensive researches for producing and controlling the quality of raw herbal drugs and formulation are required to achieve this purpose. It is also required that these plant based drugs studied in animals and tested in clinical trials before they are commercialized. Additionally, a practical plan should be developed to conserve these medicines from medicinal plants.

Genome editing is a promising technology with the ability of contributed elucidation of biosynthetic pathways, progress in genome engineering, and synthetic biology tools. The medicine industry can advance with scalable, efficient, and inexpensive methodologies to make easer the extraction and purification from plant biomass. Many technological approaches such as CRISPR-Cas9, TALENs, ZFNs, RNA*i*, and other genomic techniques have been used to develop pharmaceutical products (Sethi et al. 2021). Advance in high-throughput platforms, plant genomics, and computational approaches, combined with biotechnological techniques, have speeded up for novel pharmaceuticals production strategies.

In conclusion, properties of medicinal plants will be continued to investigate as therapeutic molecules in pharmaceutical industry. *In vivo* and *in vitro* pharmacological studies should be concentrated transforming medicinal plants and their bioactive molecules into alternative medicinal products. It appears that medicinal plants are a huge field needed discovery.

References

Adoum, O.A. (2009). Determination of toxicity levels of some savannah plants using brine shrimp test (BST). *Bayero Journal of Pure and Applied Sciences* 2 (1): 135–138.

Afzal, S., Sirohi, P., and Singh, N.K. (2020). A review of CRISPR associated genome engineering: application, advances and future prospects of genome targeting tool for crop improvement. *Biotechnology Letters* 42: 1611–1632.

Alagoz, Y., Gurkok, T., Zhang, B., and Unver, T. (2016). Manipulating the biosynthesis of bioactive compound alkaloids for next-generation metabolic engineering in opium poppy using CRISPR-Cas 9 genome editing technology. *Scientific Reports* 6 (1): 1–9.

Allen, T.M. and Cullis, P.R. (2013). Liposomal drug delivery systems: from concept to clinical applications. *Advanced Drug Delivery Reviews* 65 (1): 36–48.

Alok, A., Kumar, J., and Upadhyay, S.K. (2018). Engineering in hairy roots using CRISPR/Cas9-mediated editing. In: *Hairy Roots* (ed. V. Srivastava, S. Mehrotra, and S. Mishra). Singapore: Springer. https://doi.org/10.1007/978-981-13-2562-5_15.

Alok, A., Chauhan, H., Upadhyay, S.K. et al. (2021). Compendium of plant-specific CRISPR vectors and their technical advantages. *Life* 11: 1021. https://doi.org/10.3390/life11101021.

Altpeter, F., Baisakh, N., Beachy, R. et al. (2005). Particle bombardment and the genetic enhancement of crops: myths and realities. *Molecular Breeding* 15: 305–327.

Ananiev, E.V., Wu, C., Chamberlin, M.A. et al. (2009). Artificial chromosome formation in maize (*Zea mays* L.). *Chromosoma* 118: 157–177.

Arceusz, A., Radecka, I., and Wesolowski, M. (2010). Identification of diversity in elements content in medicinal plants belonging to different plant families. *Food Chemistry* 120 (1): 52–58.

Atanasov, A.G., Waltenberger, B., Pferschy-Wenzig, E.M. et al. (2015). Discovery and resupply of pharmacologically active plant-derived natural products: a review. *Biotechnology Advances* 33 (8): 1582–1614.

Babich, O., Sukhikh, S., Pungin, A. et al. (2020). Modern trends in the in vitro production and use of callus, suspension cells and root cultures of medicinal plants. *Molecules* 25 (24): 5805.

Basch, E., Gabardi, S., and Ulbricht, C. (2003). Bitter melon (*Momordica charantia*): a review of efficacy and safety. *American Journal of Health-System Pharmacy* 60 (4): 356–359.

Bhalothia, P., Yajnik, K., Alok, A., and Upadhayay, S.K. (2020). The current progress of CRISPR/Cas9 development in plants. In: *Genome Engineering via CRISPR-Cas9 Systems* (ed. S. Dhar). Elsevier. https://doi.org/10.1016/B978-0-12-818140-9.00011-8.

Chan, K. (2003). Some aspects of toxic contaminants in herbal medicines. *Chemosphere* 52 (9): 1361–1371.

Chandran, H., Meena, M., Barupal, T., and Sharma, K. (2020). Plant tissue culture as a perpetual source for production of industrially important bioactive compounds. *Biotechnology Reports* 26: e00450.

Chavan, P., Joshi, K., and Patwardhan, B. (2006). DNA microarrays in herbal drug research. *Evidence-based Complementary and Alternative Medicine* 3 (4): 447–457.

Davey, J.W., Hohenlohe, P.A., Etter, P.D. et al. (2011). Genome-wide genetic marker discovery and genotyping using next-generation sequencing. *Nature Reviews Genetics* 12 (7): 499–510.

Davies, K.M., Deroles, S.C., Boase, M.R. et al. (2013). Biolistics-based gene silencing in plants using a modified particle inflow gun. In: *Biolistic DNA Delivery: Methods and Protocols* (ed. S. Sudowe and A.B. Reske-Kunz), 63–74. Totowa, NJ: Humana Press.

Dey, A. (2021). CRISPR/Cas genome editing to optimize pharmacologically active plant natural products. *Pharmacological Research* 164: 105359.

Díaz, A., Villanueva, P., Oliva, V. et al. (2019). Genetic transformation of the filamentous fungus Pseudogymnoascus verrucosus of Antarctic origin. *Frontiers in Microbiology* 10: 2675.

Dixit, S., Shukla, A., Singh, V. et al. (2021). Engineering of plant metabolic pathway for nutritional improvement; recent advances and challenges. In: *Genome Engineering for Crop Improvement* (ed. S.K. Upadhyay), 351–379. https://doi.org/10.1002/9781119672425.ch20.

Edwards, D. and Batley, J. (2010). Plant genome sequencing: applications for crop improvement. *Plant Biotechnology Journal* 8 (1): 2–9.

Ergönül, P.G. and Özbek, Z.A. (2020). Cold pressed camelina (*Camelina sativa* L.) seed oil. In: *Cold Pressed Oils* (ed. M.F. Ramadan), 255–266. Academic Press.

Espinosa-Leal, C.A., Puente-Garza, C.A., and García-Lara, S. (2018). In vitro plant tissue culture: means for production of biological active compounds. *Planta* 248: 1–18.

Faridi, P., Zarshenas, M.M., Abolhassanzadeh, Z., and Mohagheghzadeh, A. (2010). Collection and storage of medicinal plants in the canon of medicine. *Pharmacognosy Journal* 2 (8): 216–218.

Farnsworth, N.R. (1966). Biological and phytochemical screening of plants. *Journal of Pharmaceutical Sciences* 55 (3): 225–276.

Feng, Z., Mao, Y., Xu, N. et al. (2014). Multigeneration analysis reveals the inheritance, specificity, and patterns of CRISPR/Cas-induced gene modifications in *Arabidopsis*. *Proceedings of the National Academy of Sciences* 111 (12): 4632–4637.

Feng, S., Song, W., Fu, R. et al. (2018). Application of the CRISPR/Cas9 system in *Dioscorea zingiberensis*. *Plant Cell, Tissue and Organ Culture (PCTOC)* 135: 133–141.

Firenzuoli, F. and Gori, L. (2007). Herbal medicine today: clinical and research issues. *Evidence-based Complementary and Alternative Medicine* 4 (S1): 37–40.

Firenzuoli, F., Gori, L., Crupi, A., and Neri, D. (2004). Flavonoids: risks or therapeutic opportunities? *Recenti Progressi in Medicina* 95 (7–8): 345–351.

Gaj, T., Sirk, S.J., Shui, S.L., and Liu, J. (2016). Genome-editing technologies: principles and applications. *Cold Spring Harbor Perspectives in Biology* 8 (12): a023754.

Gelvin, S.B. (2003). Agrobacterium-mediated plant transformation: the biology behind the “gene-jockeying” tool. *Microbiology and Molecular Biology Reviews* 67 (1): 16–37.

Georgiev, V., Slavov, A., Vasileva, I., and Pavlov, A. (2018). Plant cell culture as emerging technology for production of active cosmetic ingredients. *Engineering in Life Sciences* 18 (11): 779–798.

Giannenas, I., Sidiropoulou, E., Bonos, E. et al. (2020). The history of herbs, medicinal and aromatic plants, and their extracts: past, current situation and future perspectives. In: *Feed additives* (ed. P. Florou-Paneri, E. Christaki, and I. Giannenas), 1–18. Academic Press.

Gibbons, S. (2003). An overview of plant extracts as potential therapeutics. *Expert Opinion on Therapeutic Patents* 13 (4): 489–497.

Giri, A., Dhingra, V., Giri, C.C. et al. (2001). Biotransformations using plant cells, organ cultures and enzyme systems: current trends and future prospects. *Biotechnology Advances* 19 (3): 175–199.

Grover, J.K., Yadav, S., and Vats, V. (2002). Medicinal plants of India with anti-diabetic potential. *Journal of Ethnopharmacology* 81 (1): 81–100.

Grzegorczyk-Karolak, I., Wiktorek-Smagur, A., and Hnatuszko-Konka, K. (2018). An untapped resource in the spotlight of medicinal biotechnology: the genus *Scutellaria*. *Current Pharmaceutical Biotechnology* 19 (5): 358–371.

Guillon, S., Tremouillaux-Guiller, J., Pati, P.K. et al. (2006). Hairy root research: recent scenario and exciting prospects. *Current Opinion in Plant Biology* 9 (3): 341–346.

Guo, M., Chen, H., Dong, S. et al. (2022). CRISPR-Cas gene editing technology and its application prospect in medicinal plants. *Chinese Medicine* 17 (1): 33.

Hellens, R., Mullineaux, P., and Klee, H. (2000). Technical focus: a guide to *Agrobacterium* binary Ti vectors. *Trends in Plant Science* 5 (10): 446–451.

Ishkanian, A.S., Malloff, C.A., Watson, S.K. et al. (2004). A tiling resolution DNA microarray with complete coverage of the human genome. *Nature Genetics* 36 (3): 299–303.

Jiang, W., Zhou, H., Bi, H. et al. (2013). Demonstration of CRISPR/Cas9/sgRNA-mediated targeted gene modification in *Arabidopsis*, tobacco, sorghum and rice. *Nucleic Acids Research* 41 (20): e188–e188.

Joppa, L.N., Roberts, D.L., Myers, N., and Pimm, S.L. (2011). Biodiversity hotspots house most undiscovered plant species. *Proceedings of the National Academy of Sciences* 108 (32): 13171–13176.

Ju, H., Zhang, C., and Lu, W. (2021). Progress in heterologous biosynthesis of forskolin. *Journal of Industrial Microbiology and Biotechnology* 48 (1–2): kuab009.

Khalil, A.M. (2020). The genome editing revolution. *Journal, Genetic Engineering & Biotechnology* 18 (1): 1–16.

Khandagale, K. and Nadaf, A. (2016). Genome editing for targeted improvement of plants. *Plant Biotechnology Reports* 10: 327–343.

Kotnik, T., Kramar, P., Pucihar, G. et al. (2012). Cell membrane electroporation-part 1: the phenomenon. *IEEE Electrical Insulation Magazine* 28 (5): 14–23.

Kreis, W. (2019). Exploiting plant cell culture for natural product formation. *Journal of Applied Botany and Food Quality* 92: 216–225.

Kui, L., Chen, H., Zhang, W. et al. (2017). Building a genetic manipulation tool box for orchid biology: identification of constitutive promoters and application of CRISPR/Cas9 in the orchid, *Dendrobium officinale. Frontiers in Plant Science* 7: 2036.

Lalaleo, L., Testillano, P., Risueño, M.C. et al. (2018). Effect of in vitro morphogenesis on the production of podophyllotoxin derivatives in callus cultures of Linum album. *Journal of Plant Physiology* 228: 47–58.

Lee, J.E., Yin, Y., Lim, S.Y. et al. (2019). Enhanced transfection of human mesenchymal stem cells using a hyaluronic acid/calcium phosphate hybrid gene delivery system. *Polymers* 11 (5): 798.

Li, S.Y., Zhang, B.L., and Tan, F.P. (2004). Glycopenia coma caused by taking both dimethylbiguanide and *Momordica charantia* buccal tablet in one patient. *Chinese Journal of New Drugs and Clinical Remedies* 23 (3): 189–190.

Li, W., Li, M.F., Yang, D.L. et al. (2009). Production of podophyllotaxin by root culture of *Podophyllum hexandrum* Royle. *Electronic Journal of Biology* 5 (2): 34–39.

Liang, J., Chen, S., Hu, Y. et al. (2018). Protective roles and mechanisms of *Dendrobium* officinal polysaccharides on secondary liver injury in acute colitis. *International Journal of Biological Macromolecules* 107: 2201–2210.

Lima, G.P.P., Vianello, F., Corrêa, C.R. et al. (2014). Polyphenols in fruits and vegetables and its effect on human health. *Food and Nutrition Sciences* 5: 1065–1082.

Liu, X., Liu, Y., Huang, P. et al. (2017). The genome of medicinal plant *Macleaya cordata* provides new insights into benzylisoquinoline alkaloids metabolism. *Molecular Plant* 10 (7): 975–989.

Martin, M.Á., Goya, L., and Ramos, S. (2016). Antidiabetic actions of cocoa flavanols. *Molecular Nutrition & Food Research* 60 (8): 1756–1769.

Masihuddin, M., Jafri, M.A., Siddiqui, A., and Chaudhary, S. (2018). Traditional uses, phytochemistry and pharmacological activities of *Papaver somniferum* with special reference of unani medicine an updated review. *Journal of Drug Delivery and Therapeutics* 8 (5-s): 110–114.

Matkowski, A. (2008). Plant in vitro culture for the production of antioxidants—a review. *Biotechnology advances* 26 (6): 548–560.

Mercx, S., Smargiasso, N., Chaumont, F. et al. (2017). Inactivation of the β (1, 2)-xylosyltransferase and the α (1, 3)-fucosyltransferase genes in Nicotiana tabacum BY-2 cells by a multiplex CRISPR/Cas9 strategy results in glycoproteins without plant-specific glycans. *Frontiers in Plant Science* 8: 403.

Mishra, R., Agarwal, P., and Mohanty, A. (2023). Applications of genome editing techniques for the improvement of medicinal plants. In: *Phytochemical Genomics: Plant Metabolomics and Medicinal Plant Genomics*, 545–569. Singapore: Springer Nature Singapore.

Montanari, A., Bolotin-Fukuhara, M., D'Orsi, M.F. et al. (2015). Biolistic transformation for delivering DNA into the mitochondria. *Genetic Transformation Systems in Fungi* 1: 101–117.

Moraes, R.M., Cerdeira, A.L., and Lourenço, M.V. (2021). Using micropropagation to develop medicinal plants into crops. *Molecules* 26 (6): 1752.

Niazian, M. (2019). Application of genetics and biotechnology for improving medicinal plants. *Planta* 249: 953–973.

Niazian, M., Sadat Noori, S.A., Tohidfar, M., and Mortazavian, S.M.M. (2017). Essential oil yield and agro-morphological traits in some Iranian ecotypes of ajowan (*Carum copticum* L.). *Journal of Essential Oil-Bearing Plants* 20 (4): 1151–1156.

Oluyemisi, F., Henry, O., and Peter, O. (2012). Standardization of herbal medicines-A review. *International Journal of Biodiversity and Conservation* 4 (3): 101–112.

Ozyigit, I.I. (2020). Gene transfer to plants by electroporation: methods and applications. *Molecular Biology Reports* 47 (4): 3195–3210.

Pant, B. (2014). Application of plant cell and tissue culture for the production of phytochemicals in medicinal plants. *Infectious Diseases and Nanomedicine II: First International Conference (ICIDN–2012)*, Kathmandu, Nepal (15–18 December 2012), pp. 25–39. India: Springer.

Parham, S., Kharazi, A.Z., Bakhsheshi-Rad, H.R. et al. (2020). Antioxidant, antimicrobial and antiviral properties of herbal materials. *Antioxidants* 9 (12): 1309.

Pouvreau, B., Vanhercke, T., and Singh, S. (2018). From plant metabolic engineering to plant synthetic biology: the evolution of the design/build/test/learn cycle. *Plant Science* 273: 3–12.

Radziejewska, I., Borzym-Kluczyk, M., and Leszczyńska, K. (2018). Lotus tetragonolobus, *Ulex europaeus*, *Maackia amurensis*, and *Arachis hypogaea* (peanut) lectins influence the binding of *Helicobacter pylori* to gastric carbohydrates. *Advances in Clinical and Experimental Medicine* 27 (6): 807–811.

Rao, S.R. and Ravishankar, G.A. (2002). Plant cell cultures: chemical factories of secondary metabolites. *Biotechnology Advances* 20 (2): 101–153.

Rehman, S., Rehman, I.U., Jan, B. et al. (2021). Genome editing: applications for medicinal and aromatic plants. In: *Medicinal and Aromatic Plants* (ed. T. Aftab and K.R. Hakeem), 119–144. Academic Press.

Salmerón-Manzano, E. and Manzano-Agugliaro, F. (2020). Worldwide research on low cost technologies through bibliometric analysis. *Inventions* 5 (1): 9.

Salmerón-Manzano, E., Garrido-Cardenas, J.A., and Manzano-Agugliaro, F. (2020). Worldwide research trends on medicinal plants. *International Journal of Environmental Research and Public Health* 17 (10): 3376.

Sanford, J.C. (2000). The development of the biolistic process. *In Vitro Cellular & Developmental Biology-Plant* 36 (5): 303–308.

Schachtsiek, J. and Stehle, F. (2019). Nicotine-free, nontransgenic tobacco (*Nicotiana tabacum* l.) edited by CRISPR-Cas9. *Plant Biotechnology Journal* 17 (12): 2228.

Sethi, L., Kumari, K., and Dey, N. (2021). Engineering of plants for efficient production of therapeutics. *Molecular Biotechnology* 63 (12): 1125–1137.

Shan, Q., Wang, Y., Li, J. et al. (2013). Targeted genome modification of crop plants using a CRISPR-Cas system. *Nature Biotechnology* 31 (8): 686–688.

Shi, M., Du, Z., Hua, Q., and Kai, G. (2021). CRISPR/Cas9-mediated targeted mutagenesis of bZIP2 in *Salvia miltiorrhiza* leads to promoted phenolic acid biosynthesis. *Industrial Crops and Products* 167: 113560.

Singh, S., Singh, D.B., Singh, S. et al. (2019). Exploring medicinal plant legacy for drug discovery in post-genomic era. *Proceedings of the National Academy of Sciences, India Section B: Biological Sciences* 89: 1141–1151.

Singh, S., Rahangdale, S., Pandita, S. et al. (2022). CRISPR/Cas9 for insect pests management: a comprehensive review of advances and applications. *Agriculture* 12: 1896. https://doi.org/10.3390/agriculture12111896.

Soltani Howyzeh, M., Sadat Noori, S.A., Shariati, J.V., and Amiripour, M. (2018). Comparative transcriptome analysis to identify putative genes involved in thymol biosynthesis pathway in medicinal plant *Trachyspermum ammi* L. *Scientific Reports* 8 (1): 13405.

Street, R.A., Stirk, W.A., and Van Staden, J. (2008). South African traditional medicinal plant trade – challenges in regulating quality, safety and efficacy. *Journal of Ethnopharmacology* 119 (3): 705–710.

Sushmita, K.G., Upadhyay, S.K., and Verma, P.C. (2021). An overview of genome-engineering methods. In: *Genome Engineering for Crop Improvement* (ed. S.K. Upadhyay). Wiley. https://doi.org/10.1002/9781119672425.ch1.

Techen, N., Parveen, I., Pan, Z., and Khan, I.A. (2014). DNA barcoding of medicinal plant material for identification. *Current Opinion in Biotechnology* 25: 103–110.

Tian, C., Chang, Y., Zhang, Z. et al. (2019). Extraction technology, component analysis, antioxidant, antibacterial, analgesic and anti-inflammatory activities of flavonoids fraction from *Tribulus terrestris* L. leaves. *Heliyon* 5 (8): e02234.

Uchiumi, F., Larsen, S., Tanuma, S.I., and Figgs, G.P. (2014). Application of DEAE-dextran to an efficient gene transfer system. *Seikagaku* 86: 532–537.

Upadhyay, S.K. (2021). *Genome Engineering for Crop Improvement*. Wiley. https://doi.org/10.1002/9781119672425.

Urnov, F.D., Rebar, E.J., Holmes, M.C. et al. (2010). Genome editing with engineered zinc finger nucleases. *Nature Reviews Genetics* 11 (9): 636–646.

Valdiani, A., Hansen, O.K., Nielsen, U.B. et al. (2019). Bioreactor-based advances in plant tissue and cell culture: challenges and prospects. *Critical Reviews in Biotechnology* 39 (1): 20–34.

Varshney, R.K., Nayak, S.N., May, G.D., and Jackson, S.A. (2009). Next-generation sequencing technologies and their implications for crop genetics and breeding. *Trends in Biotechnology* 27 (9): 522–530.

Wang, D., Yu, W., Cao, L. et al. (2020). Comparative pharmacokinetics and tissue distribution of cryptotanshinone, tanshinone IIA, dihydrotanshinone I, and tanshinone I after oral administration of pure tanshinones and liposoluble extract of *Salvia miltiorrhiza* to rats. *Biopharmaceutics & Drug Disposition* 41 (1–2): 54–63.

Wawrosch, C. and Zotchev, S.B. (2021). Production of bioactive plant secondary metabolites through in vitro technologies—status and outlook. *Applied Microbiology and Biotechnology* 105 (18): 6649–6668.

Willis, S., Sunkara, R., Hester, F. et al. (2019). Chemopreventive and anti-inflammatory potential of select herbal teas and cinnamon in an in-vitro cell model. *Food and Nutrition Sciences* 10 (9): 1142–1156.

Yamazaki, M., Rai, A., Yoshimoto, N., and Saito, K. (2018). Perspective: functional genomics towards new biotechnology in medicinal plants. *Plant Biotechnology Reports* 12: 69–75.

Yarmush, M.L., Golberg, A., Serša, G. et al. (2014). Electroporation-based technologies for medicine: principles, applications, and challenges. *Annual Review of Biomedical Engineering* 16: 295–320.

Zhang, X., Jin, M., Tadesse, N. et al. (2018). Dioscorea zingiberensis CH Wright: an overview on its traditional use, phytochemistry, pharmacology, clinical applications, quality control, and toxicity. *Journal of Ethnopharmacology* 220: 283–293.

Zhao, X., Park, S.Y., Yang, D., and Lee, S.Y. (2018). Synthetic biology for natural compounds. *Biochemistry* 58 (11): 1454–1456.

Zhou, L.G. and Wu, J.Y. (2006). Development and application of medicinal plant tissue cultures for production of drugs and herbal medicinals in China. *Natural Product Reports* 23 (5): 789–810.

14

Nutraceuticals Enrichment by Genome Editing in Plants

Luis Alfonso Jiménez-Ortega, Jesus Christian Grimaldi-Olivas, Brandon Estefano Morales-Merida, and J. Basilio Heredia

Ciencia y Tecnología de Alimentos (Food Science and Technology), Alimentos Funcionales y Nutracéuticos (Nutraceuticals and Functional Foods), Centro de Investigación en Alimentación y Desarrollo, AC (CIAD), (Research Center for Food and Development), Culiacán, Sinaloa, México

14.1 Introduction

Food security is one of the issues that most concern the scientific community. It is an enormous challenge to reach the nutritional needs of the entire world population. By 2050, the world population is expected to be 9.8 billion people (UN 2017). In 2020, more than 811 million people went hungry worldwide, representing a serious public health problem (UN 2023). Therefore, the tools of biotechnology, in particular synthetic biology, are essential to contribute to food security in developing countries. The application of emerging technologies such as the CRISPR-Cas9 engineering has allowed science to design new varieties of crops with specific qualities that would enable the evaluation of the nutritional and nutraceutical needs of the target population to determine the appropriate experimental models for their correct development (Karkute et al. 2017; Fiaz et al. 2021; Upadhyay 2021a; Kumar et al. 2022).

Instead, a higher incidence and prevalence of chronic degenerative pathologies namely diabetes, cancer, obesity, and hypertension have been observed, which are related to inflammation and biological oxidative stress caused by endogenous and exogenous factors (Azzi 2022). That is why there is currently a great demand for fortified, nutraceutical, and functional food products, since it is well known that consuming phytochemicals is beneficial for health and can contribute to the treatment and/or prevention of chronic-degenerative pathologies (Pandita and Pandita 2022).

This chapter discusses and captures the use of genome editing technologies such as CRISPR-Cas9 for biofortification and improving the expression of metabolic pathways convoluted in synthesizing phytochemicals. And for upgrade, the nutritional and functional fulfilled of crops, e.g. to improve the biosynthesis of secondary metabolites and other micro and macro nutrients.

Applications of Genome Engineering in Plants, First Edition. Edited by Santosh Kumar Upadhyay.
© 2024 John Wiley & Sons Ltd. Published 2024 by John Wiley & Sons Ltd.

14.2 Functional and Biofortified Foods: Phytochemicals, Nutraceuticals, and Micronutrients

The phytochemical term refers to low-molecular-weight compounds characterized by being biosynthesized from the secondary metabolism of plants and other organisms, commonly called phytochemicals. These are classified into alkaloids, terpenes, saponins, phenolics, and sulfur-containing compounds. The turn of phrase nutraceutical indicates to a diversity of molecules such as polyunsaturated fatty acids, bioactive peptides, dietary fiber, plant extracts, and phytochemicals, which are safe, and exert a scientifically evaluated health benefit both in the blockage and treatment of most noncommunicable diseases. That is why they are a fundamental piece in developing functional foods, which are increasing in market size and demand (Zhu et al. 2013; Upadhyay and Singh 2021; Pandita and Pandita 2022; Upadhyay and Singh 2023).

On the other hand, biofortified foods refer to crops such as grains, seeds, and vegetables, which are modified to enhance the production of vitamins and minerals, among others (Zhu et al. 2013). However, the improvement approach must be comprehensive, enriching the nutraceutical and nutritional content of foods.

14.3 Metabolic Engineering to Enhance the Production of Phenolic Compounds

Phenolic compounds are one of the phytochemicals eminently appreciated by the food, agricultural, and pharmaceutical industries. They are characterized by presenting biological activities, mainly antioxidant (Zeb 2020), hypoglycemic (Leyva-López et al. 2016; Gutiérrez-Grijalva et al. 2019), anti-inflammatory (Leyva-López et al. 2016), antiobesogenic (Das et al. 2022), anticancer (Criollo-Mendoza et al. 2022), antibacterial (Yuan et al. 2021), and antifungal activity (Sudheeran et al. 2020), among others. Therefore, in recent years its consumption has become popular, increasing demand and production.

14.3.1 Biosynthetic Pathway of Phenolic Compounds

14.3.1.1 Phenolic Acids

The plants synthesize from primary metabolites such as phenylalanine or tyrosine, a large class of secondary metabolites called phenylpropanoids (Figure 14.1). Based on chemical structures, phenylpropanoids are divided into five groups: stilbenes, coumarins, monolignols, flavonoids, and phenolic acids.

Phenolic acids contain one or more hydroxyl or methoxy attached to an aromatic ring. They can be classified into hydroxycinnamic and hydroxybenzoic acid groups (Ghasemzadeh and Ghasemzadeh 2011; Marchiosi et al. 2020). The shikimate pathway is intricate in the biosynthesis of phenolic acids; they gain from phenylalanine, chorismic acid, tyrosine, and 3-dehydroquinate (Widhalm and Dudareva 2015).

Various studies have reported the effect of MYB family transcription factors, which operate as repressors or activators in the biosynthetic pathway of phenolic acids in

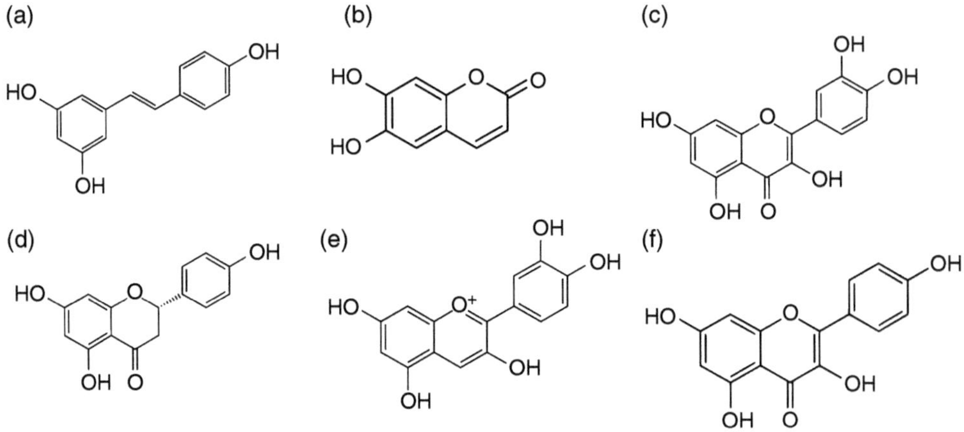

Figure 14.1 Main phenylpropanoid compounds. (a) Resveratrol. (b) Esculetin. *Source:* MarkusHagenlocher/Wikimedia Commons/CC BY-SA 3.0. (c) *p*-Coumaryl alcohol. *Source:* byrev/Pixabay. (d) Quercetin. *Source:* Kasyap/Wikimedia Commons/CC BY-SA 4.0. (e) Gallic acid. *Source:* Jonathan Billinger/Wikimedia Commons/CC BY-SA 2.0. (f) Kaempferol. *Source:* Yikrazuul, Public domain, via Wikimedia Commons.

Salvia miltiorrhiza. For instance, Zhang et al. (2013) notice that the downregulation of SmMYB39 increases 4-coumaric acid, rosmarinic acid, salvianolic acid B, and salvianolic acid A. In contrast, the content of phenolic acids is decreased in transgenic lines overexpressing SmMYB39.

Further, several phenolic acids biosynthetic pathway genes, such as *SmPAL2, SmC4H, Sm4CL2, SmHPPR*, and *SmRAS*, are activated by AtMYB75/PAP1 in transgenic *S. miltiorrhiza* (Zhang et al. 2010). Furthermore, co-expression of a bHLH TF (SmDEL) and MYB TF (SmROS) stimulates the expression of SmCHS and SmRAS, and this intensifies the accumulation of rosmarinic acid and salvianolic acid B (Wang et al. 2013).

14.3.1.2 Flavonoids

Flavonoids are the most generous polyphenol-type compounds in the plant kingdom. The chemical structure comprises a 15C backbone, two phenyl rings (A- and B-rings), and a heterocyclic pyrene ring (C).

They can be classified into different subgroups related status of the heterocyclic ring (saturation, substitution, and oxidation) as flavones, flavanones, flavanonols, isoflavones, aurones, anthocyanidins, and proanthocyanidins (Iwashina 2000; Grotewold 2006; Kumar and Pandey 2013; Agati et al. 2020).

Flavonoid biosynthesis begins with one *p*-coumaroyl-CoA and three malonyl-CoA molecules. The first step, chalcone synthase (CHS) catalyzes the condensation and cyclization of one *p*-coumaroyl-CoA and three malonyl-CoA molecules, yielding naringenin chalcone. Subsequently, chalcone is isomerized by chalcone isomerase (CHI) to yield a flavanone (naringenin). Flavanone is an intermediate of several flavonoid subclasses. The hydroxylation of naringenin by F3H (flavanone-3-hydroxylase) produces dihydro flavonols. Dihydro flavonol-4-reductase (DFR) reduces dihydroflavonols and gives rise to leucoanthocyanidins, while flavonol synthase (FLS) catalyzes the oxidation to flavonols. Flavanones are transformed into flavones by means of the enzymatic action of flavone synthase (FNS). Further, flavan-3-ols are generated by leucoanthocyanidin reductase (LAR), which reduces leucoanthocyanidins to flavan-3-ols, while anthocyanidin synthase (ANS) transforms leucoanthocyanidins to anthocyanidins. Ultimately, anthocyanidin reductase (ANR) reduced anthocyanidins to anthocyanins (Koes et al. 1994; Ngaki et al. 2012; Saito et al. 2013; Mohammad et al. 2020).

Diverse articles have reported various genes to play a part in regulating flavonoid biosynthesis, and the principal modifications are associated with transcription factors (TFs) like bHLH and MYB type.

In *Arabidopsis,* AtMYB12/PFG1, AtMYB11/PFG2, and AtMYB111/PFG3 activate to AtFLS, AtCHS, AtCHI, AtF3H, and AtF3′H, genes related to the biosynthesis of flavonol glycosides (Stracke et al. 2010). Additionally, MYB21 and its homologs MYB24 and MYB57 also manage the flavonol accumulation in *Arabidopsis* stamen through direct binding to the cis-element in the FLS (flavonol synthase) promoter (Zhang et al. 2021).

Moreover, research with *Arabidopsis* transgenic plants showed that the overexpression of AtMYB4 suppress the expression of AtC4H, AtCHS, and At4CL (Jin et al. 2000). Another study reported that the interruption of AtMYB7 permits the upregulation of PAL, C4H, 4CL, F3′H, DFR, and UGT and increases flavonol accumulation (Fornalé et al. 2013).

It has been observed that in fruits of *Vitis vinifera* the transcription factor VvMYC1 (bHLH TF) interacts with MYB5a, MYB5b, MYBA1/A2, and MYBPA1, which induce the promoters of anthocyanin and/or proanthocyanidin (PA) synthesis (Hichri et al. 2010). In chrysanthemums (*Chrysanthemum morifolium* Ramat.), CmbHLH2 induced CmDFR and activated anthocyanin accumulation (Xiang et al. 2015). Also, it has been observed in onion bulbs, where AcB2 (bHLH TF) increases the expression of AcANS and AcF3H1, improving anthocyanin production (Li et al. 2022).

14.3.2 Tools to Increase the Production of Phenolic Compounds in Plants and Crops

There are different tools for genetic improvements, such as homologous recombination, transcription activator-like effector nucleases, zinc finger nucleases, pentatricopeptide repeat proteins, RNA interference, the CRISPR/Cas9 system, cisgenesis, and intragenesis (Limera et al. 2017; Mohanta et al. 2017; Sushmita et al. 2021).

In tomatoes using homologous recombination, glycosylated and aglycone flavonoids such as deoxychalcones (butein and isoliquiritigenin), flavones (luteolin-7-glucoside and luteolin aglycon), and flavonols (quercetin glycosides and kaempferol glycosides) were produced, thereby total antioxidant capacity increased in tomato fruit peel (Schijlen et al. 2006).

Conversely, RNA interference also has been used, such as strategies for modifying flavonoid biosynthesis in *A. thaliana*, *V. vinifera*, *Prunus persica*, *Malus domestica*, and *Solanum lycopersicum* (Zhu et al. 2012; Wang et al. 2013; Yang et al. 2013; Qu et al. 2016; Singh et al. 2016; Dixit et al. 2021). Further, several studies based on CRISPR-Cas9 have reported the genetic upgrade of plants related to flavonoids and phenolic acids (Table 14.1).

Table 14.1 Improved plants and crops for the accumulation and production of phenolic compounds.

Plant material	Effect	Reference
Ipomoea [Pharbitis] nil	Anthocyanin-less white flowers	Watanabe et al. (2017)
Torenia fourneri L.	Anthocyanin-less pale blue (almost white) flowers	Nishihara et al. (2018)
Petunia hybrida cv. Madness midnight	Pale purplish-pink flower color	Yu et al. (2021)
Euphorbia pulcherrima	Changes color from vivid red to vivid reddish orange of bract	Nitarska et al. (2021)
Salvia miltiorrhiza	Changes in the content of phenolic acids	Zhou et al. (2018)
Daucus carota L.	Non-purple-colored calli	Klimek et al. (2018)
Oryza sativa L. cv Heugseonchal and Sinmyungh	Reduction of anthocyanin accumulation	Jung et al. (2019)
Solanum lycopersicum L.	Purple-color phenotype seedling	Danilo et al. (2018)
	Anthocyanin accumulation in the whole plant	Čermák et al. (2015)
	Pink-colored tomato fruits	Deng et al. (2018)
	Anthocyanin accumulation in tomato fruits	Zhi et al. (2020)
Punica granatum L.	3-O- and 4-O-glucosides of gallic acid accumulation	Chang et al. (2019)
Vitis vinifera L. rootstock 101–114	Lack of visible pigment	Sunitha and Rock (2020)
V. vinifera L. cv. Thompson Seedless	Anthocyanin accumulation	Tu et al. (2022)
Agaricus bisporus	Reduced fruit browning	Waltz, (2016)
Solanum tuberosum L.	Reduced fruit browning	González et al. (2020)
Solanum melongena L.	Reduced fruit browning	Maioli et al. (2020)
Fagopyrum tataricum	Flavonoids accumulation	Wen et al. (2022)

14.4 Metabolic Engineering to Enhance the Production of Terpenes

Terpenes are a wide range of hydrophobic compounds used as essential oils since ancient times. They are highly demanded by the food and pharmaceutical industries. They have powerful aromas and bioactive properties, which can be used as additives in certain products (Masyita et al. 2022). Likewise, biopesticides based on antibacterial and antifungal terpenes have recently been developed, making it an area in constant growth (Ninkuu et al. 2021).

14.4.1 Biosynthetic Pathway of Terpenes

Terpenes are hydrocarbons that have a chemical formula $(C5H8)_n$. They are classified into hemiterpene (C5, $n = 1$), monoterpene (C10, $n = 2$), sesquiterpene (C15, $n = 3$), diterpene (C20, $n = 4$), sesterterpene (C25, $n = 5$), triterpene (C30, $n = 6$), tetraterpene (C40, $n = 8$), and polyterpene (Figure 14.2). Further, there are primary terpenoid metabolites (carotenoids, sterols, and hormones) and specialized terpenoids, which are frequently specific to individual plant species (Ashour et al. 2010; Nagegowda 2010; Gutensohn et al. 2013).

The biosynthesis of terpenes is regulated by the mevalonate (MVA) pathway and the 2-C-methyl-D-erythritol-4-phosphate (MEP) pathway (Oldfield and Lin 2012). These reactions are produced in cytosol and plastid; the MVA pathway synthesizes isopentenyl diphosphate (IPP) from acetyl-CoA in the cytosol, while in the plastids MEP pathway permits the generation of IPP and DMAPP from pyruvate and glyceraldehyde-3-phosphate (Lange and Croteau 1999; Nagegowda 2010; Xiao and Zhong 2016).

Besides, terpene structures may be modified through acylation, dehydrogenation, hydroxylation, or glycosylation, giving way to chemically diverse terpenoids (Chen et al. 2011). There are various key enzymes in terpenes biosynthesis such as HMG-CoA reductase (HMGR), 1-deoxy-D-xylulose-5-phosphate synthase (DXS), 1-deoxy-D-xylulose-5-phosphate reductoisomerase (DXR), and terpene synthase (TPS) (Huang et al. 2021).

14.4.2 Tools to Increase the Production of Terpenes in Plants and Crops

Plants with innovative traits can be developed through the CRISPR/Cas9 system, which facilitates adding, deleting, or silencing multiple genes in the plant genome (Naik et al. 2022). In addition, it can increase the production of phytochemicals with improved quality and modulate the chemical profile for commercial production (Jan et al. 2021; Tang et al. 2022). Some examples of the improvement in genes related to terpenes are shown in Table 14.2.

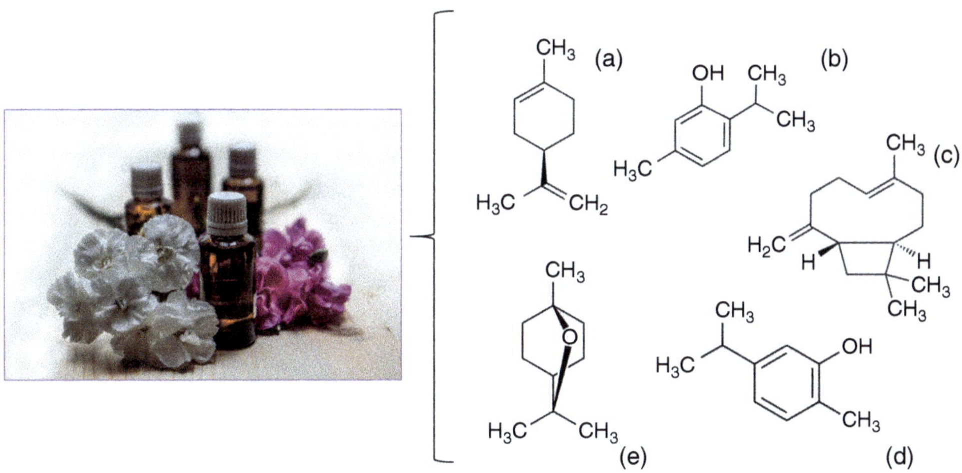

Figure 14.2 Some monoterpenes and aromatic terpenes molecules are used in the production of pharmacological products. (a) Limonene. *Source:* LiliGraphie/Adobe Stock. (b) Thymol. (c) β-caryophyllene. (d) Carvacrol. (e) Cineole.

Table 14.2 Plants and crops bred for terpene accumulation and production.

Plant material	Target gene	Effect	Reference
Taraxacum kok-saghyz	1-FTT	Increase in rubber content	Iaffaldano et al. (2016)
Salvia miltiorrhiza	SmCPS1	Changes in the accumulation of cryptotanshinone, tanshinone IIA, and tanshinone I	Li et al. (2017)
Solanum lycopersicum L.	Psy1 and CrtR-b2	Changes in the accumulation of carotenoids in flowers and fruits	D'Ambrosio et al. (2018)
Dioscorea zingiberensis	Dzfps	Decrease in squalene content	Feng et al. (2018)
Nicotiana tabacum	NtCPS2	The decrease of cis-abienol and labdene diol contents and gibberellin and abscisic acid (ABA) contents increased	He et al. (2021)
Zea mays	fps1-1	Production of zealexin and kauralexin	Tang et al. (2022)
N. tabacum	NtGGPPS1	Increased carotenoid levels in leaves	Dong et al. (2022)
Populus × euramericana cv.	PtHMGR	Increased ABA, GA, carotene, and lycopene content	Wei et al. (2019)
Panax ginseng	PgHMGR1	Enhanced the production of sterols and triterpenes	Kim et al. (2014)
A. thaliana	AtHMG1	Reduction of sterols and/or triterpenes	Suzuki et al. (2004)
	AtDXS	Reduction of neoxanthin, lutein, β-carotene, and violaxanthin	Araki et al. (2000)
Lavandula latifolia	SlDXS	Changes in α- and β-pinene, cineole, camphor, linalool, α-terpineol, and borneol content	Muñoz-Bertomeu et al. (2006)
V. vinifera L.	VvDXS	Accumulation of linalool, geraniol, and nerol	Battilana et al. (2011)
Ginkgo biloba	GbDXR	Changes in ginkgolides biosynthesis	Gong et al. (2005)
Coleus forskohlii	CfDXR	Decreased of forskolin production in roots	Engprasert et al. (2005)
Prunus persica L.	PpTPS1 and PpTPS3	Accumulation of linalool in peach fruit	Wei et al. (2022)
Morella rubra	MrTPS3 and MrTPS20	Accumulation of β-Caryophyllene and α-Pinene	Wang et al. (2022)

14.5 Metabolic Engineering to Enhance the Production of Alkaloids

Alkaloids become part of human beings since time immemorial. Used for multiple purposes and with potent bioactivities, alkaloids are found in 20% of plant species. Their main uses are as painkillers (Jiang et al. 2022), treatment of pathologies related to the central nervous system (Lin et al. 2020), antibacterial (Cushnie et al. 2014), anticancer (Mondal et al. 2019), and antidepressants (Ferraz et al. 2019), among others, highlighting that multiple FDA-approved drugs are alkaloid-based. By uniting the chemistry of natural products with metabolic engineering, alkaloids can be produced, isolated, and purified on an extensive to meet the demand for new, safer, and more efficient drugs. Some of the most used alkaloids globally are caffeine, capsaicin, piperine, morphine, vincristine, atropine, pilocarpine, and vincamine (Figure 14.3) (Daley and Cordell 2021; Heinrich et al. 2021).

14.5.1 Biosynthetic Pathway of Alkaloids

A useful and practical classification of alkaloids is by their biosynthetic origin. The protoalkaloids are derived from amino acids, while the pseudoalkaloids are synthesized through transamination reactions. Protoalkaloids are mainly derived from aromatic amino acids such as phenylalanine, tyrosine, lysine, and tryptophan. In comparison, pseudoalkaloids originated from acetate, pyruvic acid, ferulic acid, geraniol, saponins, adenine, and guanine. Its structure is characterized by the presence of an N atom, which is derived from amino acids (true alkaloids) or other molecules (pseudo alkaloids) (Aniszewski 2015a, b).

Its biosynthetic pathway is based on the main building blocks of the acetyl coenzyme A, shikimic acid, mevalonic acid, and 1-deoxyxylulose 5-phosphate pathways. For the synthesis of alkaloids, an accumulation of amine and aldehyde precursors must occur to form an iminium cation and, finally, a Mannich-type reaction (Lichman 2021). The classification and biosynthesis of alkaloids are complex because several types of alkaloids can be synthesized from a single amino acid. For example, from L-lysine, four different skeletons with different nuclei can be synthesized (piperidine, indolizine, quinolizidine, and pyridone) (Aniszewski 2015b).

The alkaloids that come from ornithine contain a pyrrolidine nucleus (pyrrolizidine alkaloids). Their synthesis begins from L-ornithine, where a decarboxylation occurs by pyridoxal phosphate giving rise to putrescine, which S-adenosylmethionine methylates to N-methyl putrescine. Hygrine is the first alkaloid synthesized from this route, from which other pyrrolidine and tropane alkaloids are derived.

Alkaloids derived from tyrosine are characterized by containing phenyl and phenyl propyl nuclei. Within this group, four main metabolic pathways serve as a precursor for other alkaloids: mescaline, colchicine, dopamine, and galanthamine. Alkaloids derived from tryptophan are characterized by holding the indole nucleus in a ring system, the latter originating from the anthranilic acid metabolic pathway and the building blocks of the shikimate pathway. Within this group highlight the metabolic pathways of psilocybin, harman, harmine, ajmalicine, tabersonine, catharanthine, vindoline, vinblastine,

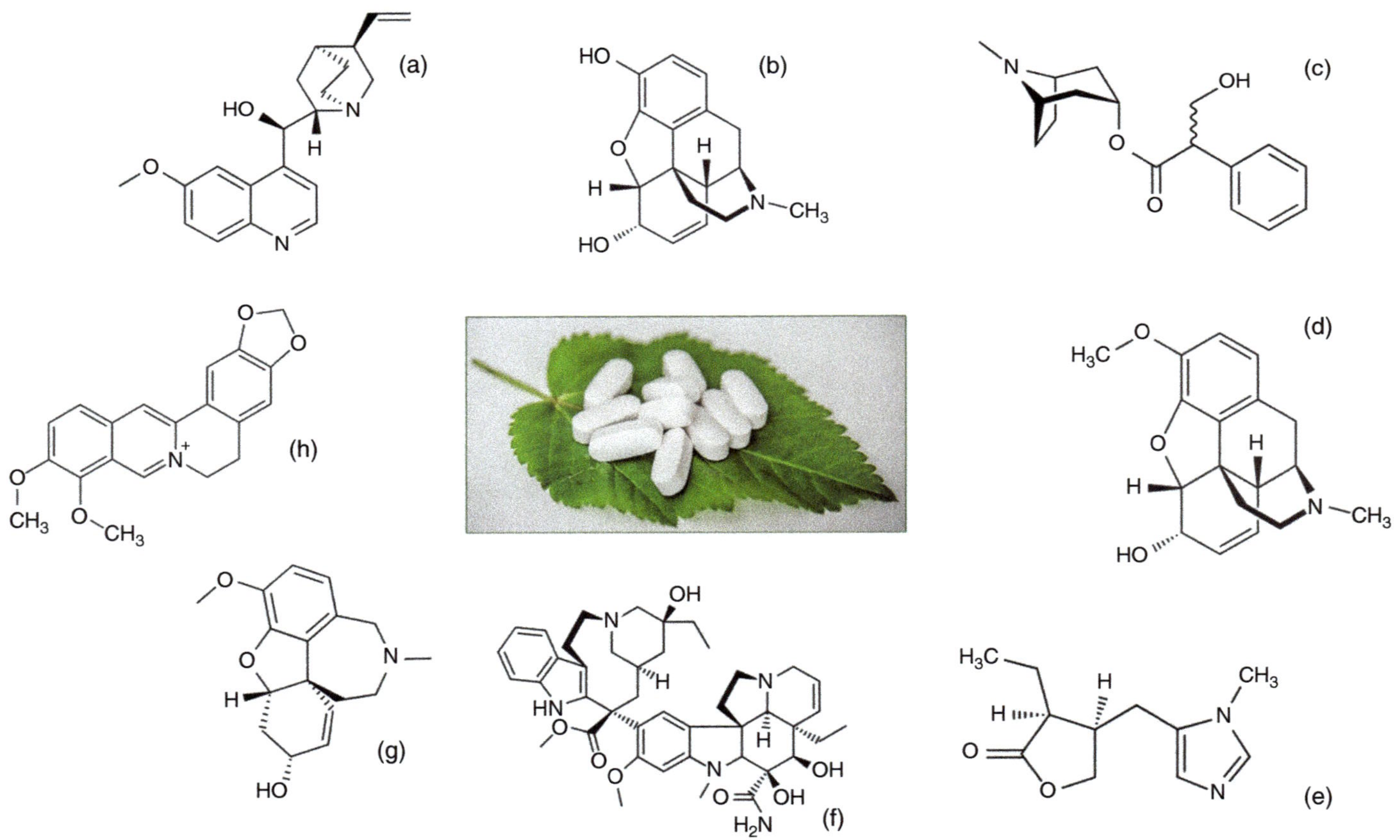

Figure 14.3 Chemical structures of pharmacologically potent alkaloids. (a) Quinine. (b) Morphine. (c) Atropine. (d) Codein. (e) Pilocarpine. (f) Vindesine. (g) Galanthamine (h) Berberine.

vincristine, strychnine, brucine, quinine, quinidine, eserine, and ergotamine, which originate numerous bioactive alkaloids.

Another important group is the alkaloids originated from nicotinic acid, which contain a pyridine nucleus that originates through the dihydro nicotinic acid pathway, transferring from dihydropyridine to nicotine and nornicotine.

The alkaloids derived from lysine contain at least four different nuclei (piperidine, indolizidine, quinolizidine, and pyridone). Within this group are the metabolic pathways of piperine, pelletierine, lobelanine, swainsonine, castanospermine, lupine, lupanine, sparteine, and cytisine, giving rise to a great structural diversity of alkaloids (Aniszewski 2015b).

14.5.2 Tools to Increase the Production of Alkaloids in Plants and Crops

The medical and industrial importance of alkaloids has led to the developing of technologies capable of increasing their biosynthetic performance. Some of the most widely used tools are hairy root transformation, whole plant transformation, and transient and cell culture using *Agrobacterium tumefaciens* and *A. rhizogenes*. Other techniques used are gene overexpression or silencing, enzyme engineering (endogenous and foreign enzyme overexpression), and expressing heterologous genes. Gene modification using the CRISPR/Cas9 system has been a promising technology for developing new and improved plant varieties (Mora-Vásquez et al. 2022). Table 14.3 shows some modified genes for the overexpression of alkaloids in plants.

Table 14.3 Metabolic engineering to enhance de production of alkaloids.

Plant material	Target gene	Effect	Reference
Catharanthus roseus	ORCA, ORCA3, ORCA4, ORCA5, AP2/ERF.	Overexpression of terpenoid indole alkaloids in flowers petals	Singh et al. (2020)
	HPTII, GUS, GFP	Enhance the production of indole alkaloids	Hanafy et al. (2016)
	BIS1, ORCA3 MYC2a,	Enhance the production of indole alkaloids	Schweizer et al. (2018)
Atropa belladonna	AbSAUR1	Enhance the production of tropane alkaloids	Bai et al. (2019)
	CYP82M3, UGT1, PYKS, LS		Zeng et al. (2022)
	AbCaM1, G2-EPSPS		Zhang et al. (2022b)
	AbODC		Zhao et al. (2020)
	AbH6H		Zeng et al. (2021)
	PYKS	Enhance the production of pyrrolidine alkaloid	Hasebe et al. (2021)
	PYKS AbPyKS	Enhance the production of pyrrolidine alkaloid	Parks et al. (2023)
	AbP450-5021, AbP450-116623	Enhance the production of tropane alkaloids	Sadre et al. (2022)

(*Continued*)

Table 14.3 (Continued)

Plant material	Target gene	Effect	Reference
Papaver somniferum	4′OMT1	Enhance the production of benzylisoquinoline alkaloids	Alagoz et al. (2016)
Solanum lycopersicum	MYC1, MYC2	Enhance the production of steroidal glycoalkaloids	Swinnen et al. (2022)
Nicotiana benthamiana	Glycosyltransferases	Enhance the production of monoterpene indole alkaloids	Dudley et al. (2022)
Nicotiana tabacum	BBL	Enhance the production of myosmine	Zhang et al. (2022a)

14.6 Metabolic Engineering to Enhance the Production of Vitamins and Minerals

Traditionally, biofortification has been one of the most widely used techniques to enhance the nutritional characteristics of crops of industrial interest. This can be reach by adding nutrients to fertilizers; however, this is not possible for all minerals, such as iron, as well as for vitamins, essential amino acids, and fatty acids because these are products of plant metabolism (Pérez-Massot et al. 2013). These micronutrients are essential for nutrition and the correct functioning of the organism. Figure 14.4 shows some crops abundant in vitamins and minerals.

Metabolic engineering is one of the technologies with which it is feasible to upgrade the content of vitamins and minerals by inserting genes from any source directly into the various local varieties and thus increase the levels of organic nutrients, in a faster way, compared to conventional breeding, which is limited by the availability of existing germplasm (Welch and Graham 2005). Table 14.4 summarizes some crops biofortified through metabolic and genetic engineering.

14.6.1 Tools to Increase the Production of Vitamins in Plants and Crops

Vitamin A belongs to the class of unsaturated organic compounds, which includes preformed retinoids including retinol and its derivatives (retinal and retinoic acid); it also contains several provitamin-A carotenoids, and β-carotene one of the main ones (Foyer et al. 2020). In plants, the way to increase β-carotene content is by altering the carotenoid pathway, either by increasing the availability of the precursors required in the pathway or by overexpressing enzymes in the early stages of the pathway between GGPP and lycopene, thus favoring β-branch metabolic flux by selecting LYCB activity at the expense of LYCE; another strategy is to increase the carotenoid sink to eliminate feedback constraints (Zhu et al. 2009; Farré et al. 2010; Bai et al. 2011).

Some studies have focused on producing carotenoid precursors that improve the carotenoid content, such as the overexpression of the enzyme DXP synthase in tomato plants (Enfissi et al. 2005). For example, the bacterial gene CrtB has been expressed in cassava roots, accumulating up to 21 μg/g of carotenoids, a more than 30-fold increase over the wild

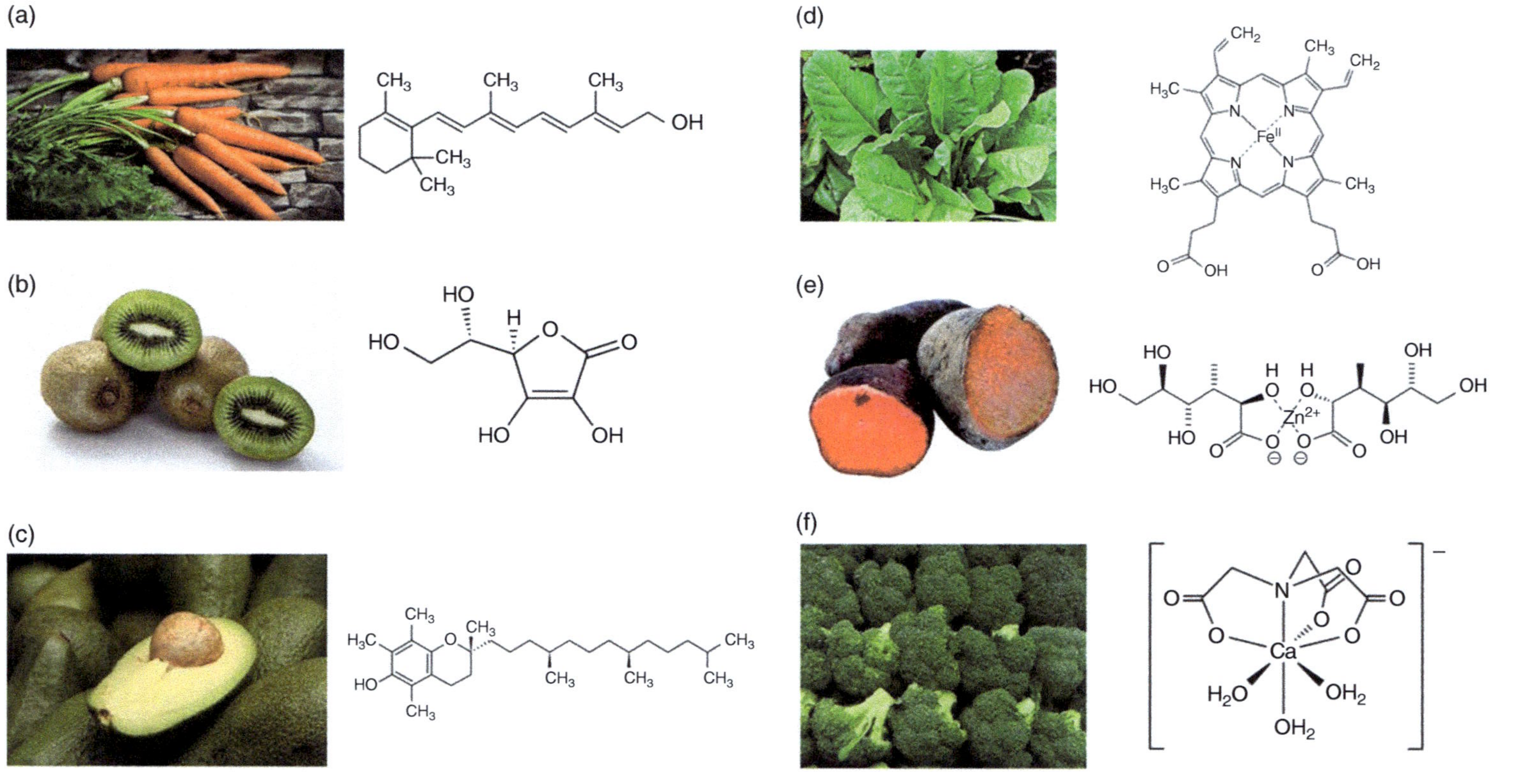

Figure 14.4 Fruits and vegetables rich in vitamins and minerals. (a) Vitamin A. (b) Ascorbic acid. *Source:* I, Luc Viatour/Wikimedia Commons/CC BY-SA 2.5. (c) Vitamin E. (d) Iron complex. *Source:* Lucian Milasan/Adobe Stock. (e) Zinc complex. *Source:* nila/Adobe Stock. (f) Calcium Complex. *Source:* JAVIER/ Adobe Stock.

Table 14.4 Genetic engineering strategies to enhance the production of vitamins and minerals.

Nutrients	Plant material	Target genes	References
Vitamin A	*Zea mays*	*Zmpsy1, PacrtI, PcrtW, Gllycb, PacrtB*	Aluru et al. (2008), Zhu et al. (2009), Naqvi et al. (2009)
	Triticum aestivum	*Zmpsy1, PacrtI*	Cong et al. (2009)
	Solanum tuberosum	*EuCrtB, EuCrtI, EuCrtY, BoOr, AtZEP, PacrtB,*	Diretto et al. (2007), Lopez et al. (2008), Römer et al. (2002), Ducreux et al. (2005)
	Oryza sativa	*Nppsy1, EucrtI, Zmppsy1*	Ye et al. (2000), Paine et al. (2005)
Vitamin C	*Z. mays*	*Osdhar*	Naqvi et al. (2009)
	Solanum lycopersicum	*Acggp*	Bulley et al. (2012)
	S. tuberosum	*StVTC2A*	Bulley et al. (2012)
Folic acid	*O. sativa*	*Atgtpchi, Atadcs*	Storozhenko et al. (2007)
	S. lycopersicum	*Atgch, Atadcs1*	Storozhenko et al. (2007)
Iron (Fe)	*O. sativa*	*Gm ferritin* and *Af phytase*	Johnson et al. (2011), Wirth et al. (2009), Lee et al. (2011)
	Z. mays	*Gm ferritin, Af phytase, Osnas1,* Activation tagging of *Osnas2*	Drakakaki et al. (2005)
Zinc (Zn)	*O. sativa*	*Gm ferritin, Af phytase, Osnas1,* Activation tagging of *Osnas2*	Lee et al. (2011), Johnson et al. (2011), Wirth et al. (2009)
Calcium (Ca)	*Daucus carota*	*scax1*	Park et al. (2004)
	Lactuca sativa	*scax1*	Park et al. (2009)
	S. tuberosum	*scax1, cax2b* chimeric	Park et al. (2005), Kim et al. (2006)

type (Welsch et al. 2010). Another reported strategy is the increase of the cauliflower *Or* gene in tubers, which favored better carotenoid storage by stimulating the formation of chromoplasts, which led to potatoes with orange flesh with 10-fold more than the normal amount of β-carotene (Lopez et al. 2008). Golden Potato development consisted of the expression of three *Erwinia* plant genes coding for the enzymes lycopene beta-cyclase (CrtY), phytoene synthase (CrtB), and phytoene desaturase (CrtI) in potatoes, which lead to the formation of β-carotenoids instead of α-carotene, generating an aggregation of 47 μg/g DW of β-carotene (Diretto et al. 2007).

Folic acid, too known as folate or vitamin B9, is the source of tetrahydrofolate, which take part in a very main role in DNA synthesis and other metabolic reactions. Folate is a tripartite molecule consisting of pterin, p-aminobenzoate (PABA), and glutamate residues (Wolak et al. 2017). Currently, the increase of folate in plants has been carried out by modifying the pterin and PABA pathways; however, changing only one pathway can damage the other pathway and not achieve the desired increase (Gorelova et al. 2017).

Based on this information, several articles have been carried out in rice plants where crosses became made in transgenic plants expressing genes from both pathways separately, the GCH1 gene that potentiates the cytosolic branch (pterin) and the ADCS1 gene that potentiates the PABA branch. This resulted in a 100-fold increase in folate levels compared to transgenic plants with only one pathway altered (Storozhenko et al. 2007).

Some studies have evaluated methods to enrich the folate content of tomato, rice, corn, lettuce, and bean plants. One of the strategies used is the overexpression of the gene of the enzyme dihydrofolate synthetase (DHFS), which in corn increased twofold the folate content. Also, the enzymes involved in the first step in the cytosol (CTPCHI) have been overexpressed, increasing 2-fold more the folate content in tomatoes, 8.5-fold more in lettuce, and 3-fold more in beans (Díaz de la Garza et al. 2007; Storozhenko et al. 2007; Naqvi et al. 2009; Nunes et al. 2009; Blancquaert et al. 2013; Ramírez Rivera et al. 2016).

Ascorbic acid is an enzyme cofactor and a potent antioxidant (Bartholomew 2002). In plants, several routes can generate ascorbate, and it can also be recycled when it is oxidized through an additional course using glutathione as a reducing agent. Because of this, ascorbic acid content can be improved by increasing synthesis or enhancing the rate at which it can be recycled.

In lettuce and potato plants, an increase in ascorbic acid has been reach by overexpression of the L-gulono γ-lactone oxidase (GLOase) gene generating a sixfold rise in lettuce and a twofold increase in potato vitamin C. An increase in vitamin C has also been seen in maize expressing the rice *dhar* gene involved in the ascorbate recycling pathway in a sixfold increase compared to normal conditions (Naqvi et al. 2009).

14.6.2 Tools to Increase the Production of Minerals in Plants and Crops

Besides being a structural element of the cell wall, calcium in plants fulfills several functions, acting as a second messenger through the calmodulin protein, which is responsible for the interpretation and emission of cellular signals (Taneja and Upadhyay 2021). Both tubers and lettuce leaf crops are fine sources of calcium; however, some vegetables with high content of oxalate and phytate may hinder calcium absorption (Jeong and Guerinot 2008; Upadhyay 2021b).

To enlarge calcium content in plants, metabolic engineering strategies have been implemented where calcium transporters such as CAX1 from *Arabidopsis* have been expressed in carrots and potatoes, increasing this mineral threefold (Park et al. 2005; Connolly 2008).

Zinc is a mineral required as an enzymatic cofactor for several enzymes and plays a very important role in binding transcription factors to DNA. Because zinc is a more soluble mineral than iron and is more easily absorbed from the soil, research has focused on genetic engineering strategies that facilitate transport and absorption, with cereals being the main target of study because they are poor sources of zinc (Palmgren et al. 2008). The increase of NAS/NAAT that are involved in the synthesis of phytosiderophores and the transporters Osysl15 and Osirt1 in rice plants can increase the content of zinc as well as iron because many phytosiderophores as transporters can act for both minerals (Zheng et al. 2010; Lee et al. 2011).

Iron (Fe) is an essential mineral for plants because it performs various principal roles, such as chlorophyll synthesis, maintenance of chloroplast structure, and enzyme activity.

However, iron uptake depends on soil conditions since only the ferrous form (Fe^{2+}) is soluble and bioavailable, while the ferric state (Fe^{3+}) is sequestered by soil particles forming insoluble (Gómez-Galera et al. 2012; Shumayla and Upadhyay 2022). Therefore, one of the strategies to increase iron concentrations is to increase the expression of enzymes such as ferritin, favoring the storage of iron in bioavailable form; another strategy is the expression of NAAT/NAS enzymes. In particular, transgenic rice plants expressing the NAS Osnas1, Osnas2, or Osnas3 genes accumulated up to 19 μg/g of iron in the endosperm, and the expression of NAAT/NAS enzymes in the endosperm is another strategy (Zheng et al. 2010; Johnson et al. 2011).

14.7 Metabolic Engineering to Enhance the Production of Polyunsaturated Fatty Acids

Higher plants are one of the main sources of polyunsaturated fatty acid precursors (PUFA) such as linoleic acid (LA) and α-linolenic acid (ALA), which will be unable to synthesize in mammalian tissues and are therefore required as an essential part of the daily diet. In addition, some seeds and grains are rich in PUFAs. Its main biological activities are antioxidant and anti-inflammatory (Figure 14.5).

14.7.1 Biosynthetic Pathway of Polyunsaturated Fatty Acids

The production site is in the plastids where a sequence of reactions occur in which acetyl-CoA residues are incorporated, resulting in chains of 16 or 18 carbons in length, with acetyl-CoA carboxylase and fatty acid synthase being the enzymes involved.

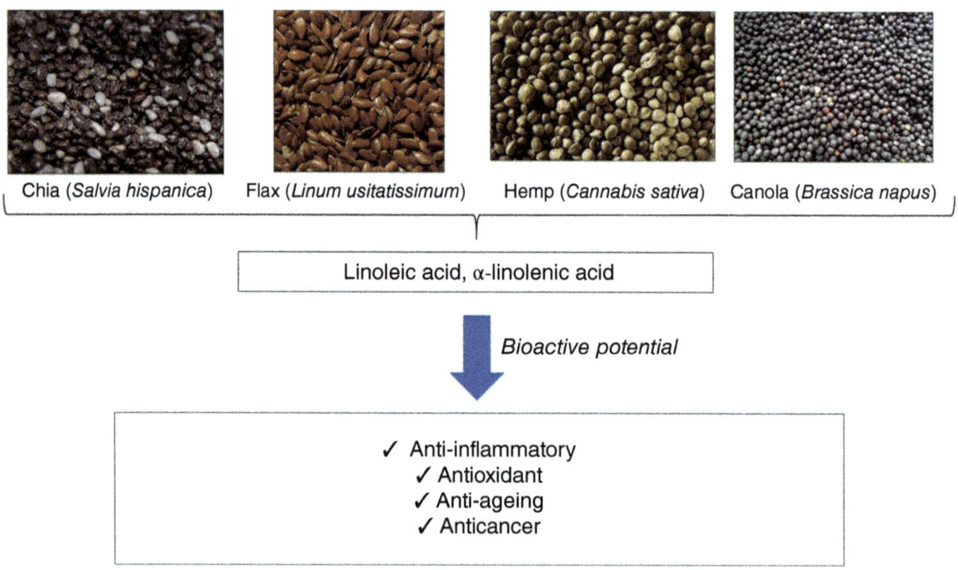

Figure 14.5 Seeds and grains rich in PUFAs and their biological activities.

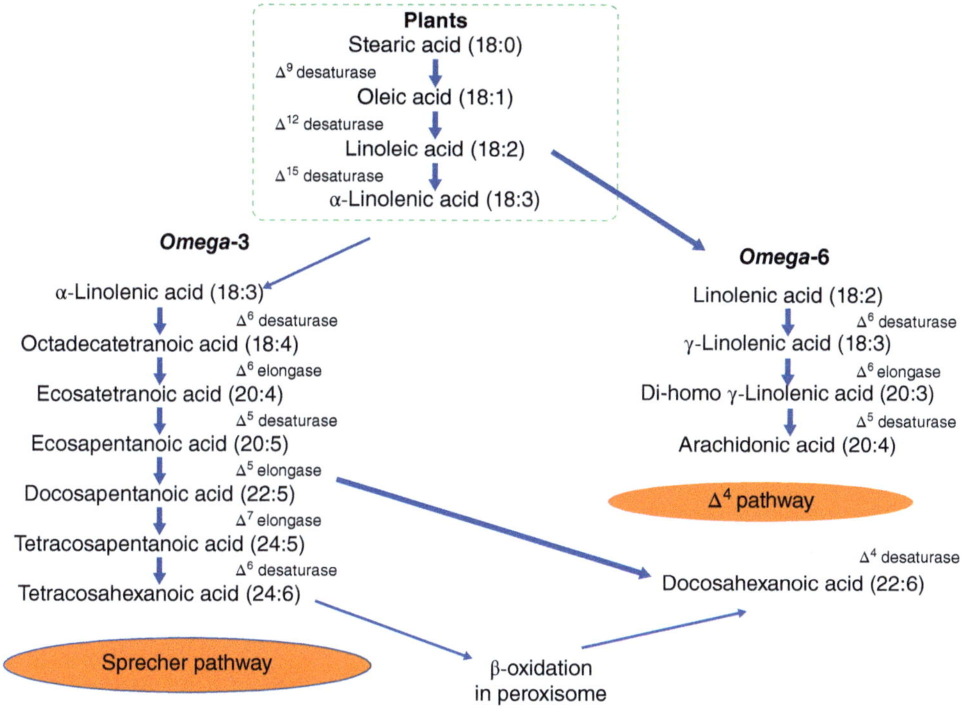

Figure 14.6 Biosynthesis of polyunsaturated fatty acids. *Source:* Adapted from Warude et al. (2006).

The PUFA synthesis pathway occurs mostly aerobically in eukaryotic organisms, starting with Δ^6 desaturations of both 18:2 n-6 and 18:3 n-3, generating γ-linolenic acid (GLA; 18:3 $\Delta^{6,9,12}$) and octadecatetraenoic acid (OTA; 18:4 $\Delta^{6,9,12,15}$), respectively. The second step leads to specific C2 elongation from 6 to 20:3 $\Delta^{8,11,14}$ and 20:4 ^{8,11,14,17} and another Δ^5-desaturation to produce arachidonic acid (AA) and eicosapentaenoic acid (EPA). There are two pathways by which DHA can be made (Figure 14.6); the first is a linear pathway where the elongation of C2 from EPA to C22: $\Delta^{5,7,10,13,16,19}$, which is desaturated by Δ^4-desaturase and specific for DHA production, takes place. The other pathway is also known as the "Sprecher" pathway, which is independent of Δ^4 desaturases, where two consecutive C2 elongation cycles are involved for the production of 24: $\Delta^{5, 7, 10, 10, 13, 16, 19}$ come after by a Δ^6 desaturation and a C2 shortening cycle by ß-oxidation in the peroxisome to produce DHA.

14.7.2 Tools to Increase the Production of Polyunsaturated Fatty Acids in Plants and Crops

Early studies focused on plants for polyunsaturated fatty acid production began with the expression of genes coding for desaturases in tobacco plants expressing the gene for Δ^6 desaturase from borage (Sayanova et al. 1997). They were also able to modify *A. thaliana* plants for the production of ARA and EPA by introducing three bacterial genes that code for the enzymes Δ^9 elongase, Δ^8 desaturase, and Δ^5 desaturase, which were isolated from *Isochrysis galbana*, *Euglena gracilis,* and *Mortierella alpine*, respectively (Qi et al. 2004).

Based on this research and the development of new genetic engineering strategies, a new approach was opened to produce unusual PUFA in several plants that lacked them. Several investigations focused on the production of GLA in *Brassica juncea* L., soybean, and safflower (*Carthamus tinctorius* L.) plants with the insertion of the gene coding for the enzyme Δ^6 desaturase from other organisms such as *Pythium irregulare*, *Borago officinalis*, *Saprolegnia diclina*, and *Mortierella alpina*.

In camelina plants (*Camelina sativa* L.), it was modified by insertion of seven genes involved in fatty acid synthesis such as Δ^6 desaturase isolated from *Ostreococcus tauri*, the Δ^6 elongase isolated from *Physcomitrella patens*, the Δ^5 desaturase isolated from *Thraustochytrium* sp, the Δ^{12} desaturase isolated from *Phytophthora sojae*, the ω3-desaturase isolated from *Phytophthora infestans*, the Δ^5 elongase from *Ostreococcus tauri* and finally the Δ^4 desaturase isolated from *Emiliania huxleyi* (Ruiz-Lopez et al. 2014). In this way, they obtained transgenic plants that synthesize oil that can accumulate fatty acids and oils with PUFA ω-3 content (EPA, DHA) in proportions similar to fish oils.

In general terms, we can mention that oilseed crops are of the greatest interest in metabolic engineering since they can naturally gather elevated amounts of fatty acids, so the various techniques focus on the modification of the PUFA biosynthetic pathway and pathways intricate in the synthesis of other value-added fatty acids (Powell 2007; Haslam et al. 2013; Maheshwari and Kovalchuk 2014).

The first research focused on producing different types of PUFA, such as LA and ALA, in barley plants (*Hordeum vulgare* L.). Likewise, it has been possible to obtain novel PUFA by expressing the Δ^6 desaturase gene obtained from the fungus *Tamnidium elegans*, getting barley plants that produced GLA and SDA in mature seeds with 0. 141% and 0.294% of the total amount of fatty acids synthesized, respectively. In contrast to wheat (*Triticum aestivum* L.) plants that were similarly grown, GLA and SDA contents varied up to 1.40% and 1.53%, respectively (Wang et al. 2007; Mihálik et al. 2014; Mihálik et al. 2015).

14.8 Metabolic Engineering to Enhance the Production of Bioactive Peptides

Bioactive peptides are also becoming very important for controlling various diseases since biotechnology is intended to improve plants that favor a greater production of bioactive peptides, which are produced mainly during enzymatic proteolysis (Daliri et al. 2017).

14.8.1 Tools to Increase the Production of Bioactive Peptides in Plants and Crops

Regarding the production of bioactive peptides using genetic engineering tools, we have the study carried out in rice plants where they were able to express a fusion protein where the storage protein glutelin is found and a bioactive peptide with antihypertensive activity obtained from ovoquinine RPLKPW (Yang et al. 2006). Likewise, a protein composed of 10 or 18 tandem repeats of novokinin, with a KDEL retention signal in the endoplasmic reticulum at the C-terminal end, has been expressed in rice plants using the glutelin promoter as a signal peptide.

In soybean plants, research has also been carried out to obtain a greater amount of bioactive peptides, where novokinin (RPLKPW) has been introduced in sequences of the α-subunit of β-conglycinin by directed mutagenesis since it has been demonstrated that RPLKPW incorporated in the alpha-subunit has antihypertensive activity, being the first sample of a genetically modified food protein with the physiological activity of a bioactive peptide (Matoba et al. 2001).

A 12-aa peptide derived from milk casein with angiotensin I-converting enzyme (ACE) inhibitory action has also been expressed in tobacco and tomato plants by viral vector-based expression using the tobacco mosaic virus (TMV) vector (Karaki et al. 1990), obtaining CP-IECA protein yields of 100 μg/g of tobacco leaf tissue and 10 μg/g of tomato fruit, which were considered enough for *in vivo* evaluation by oral administration (Hamamoto et al. 1993).

Developing strategies that favor a higher expression and accumulation of bioactive peptides with specific activities are still necessary. Current strategies focus on achieving high levels of expression of bioactive peptides in plant seeds, being of great importance to conduct research in the search for better specific promoters as well as the development of plant transformation protocols could be of great importance in the production of transgenic plants with high content of bioactive peptides (Stoger et al. 2005; Lau and Sun 2009).

Other strategies, such as RNA interference (RNAi), reduce the expression of endogenous proteins and thus enlarge the accumulation of recombinant proteins. Another alternative is the development of chimeric proteins, which will allow the production of multiple bioactive peptides in a single transformation event, ease the production of functional foods using the biomass of genetically modified plants (Nakashita et al. 2001; Quesada-Vargas et al. 2005).

14.9 Conclusions

Metabolic engineering is undoubtedly a very useful tool to increase the biosynthesis of secondary metabolites and other nutraceuticals. Thanks to the elucidation and understanding of the biosynthesis routes, improvement alternatives can be proposed to increase the functional and nutraceutical content of foods. Metabolic engineering is of vital importance for the development and maintenance of food systems in the world, mainly in developing countries. The development of new drugs derived from phytochemicals must have a comprehensive approach, including metabolic engineering, which helps to meet the great demand for dietary supplements and medicines globally.

References

Agati, G., Brunetti, C., Fini, A. et al. (2020). Are flavonoids effective antioxidants in plants? Twenty years of our investigation. *Antioxidants* 9: 1–18.

Alagoz, Y., Gurkok, T., Zhang, B. et al. (2016). Manipulating the biosynthesis of bioactive compound alkaloids for next-generation metabolic engineering in opium poppy using CRISPR-Cas 9 genome editing technology. *Scientific Reports* 6: 30910.

Aluru, M., Xu, Y., Guo, R. et al. (2008). Generation of transgenic maize with enhanced provitamin A content. *Journal of Experimental Botany* 59: 3551–3562.

Aniszewski, T. (2015a). Chapter 1 – definition, typology, and occurrence of alkaloids. In: *Alkaloids*, 2ee (ed. T. Aniszewski), 1–97. Boston: Elsevier.

Aniszewski, T. (2015b). Chapter 2 – alkaloid chemistry. In: *Alkaloids*, 2ee (ed. T. Aniszewski), 99–193. Boston: Elsevier.

Araki, N., Kusumi, K., Masamoto, K. et al. (2000). Temperature-sensitive *Arabidopsis* mutant defective in 1-deoxy-D-xylulose 5-phosphate synthase within the plastid non-mevalonate pathway of isoprenoid biosynthesis. *Physiologia Plantarum* 108: 19–24.

Ashour, M., Wink, M., and Gershenzon, J. (2010). Biochemistry of terpenoids: monoterpenes, sesquiterpenes and diterpenes. In: *Biochemistry of Plant Secondary Metabolism*. Oxford: Wiley-Blackwell.

Azzi, A. (2022). Oxidative stress: what is it? Can it be measured? Where is it located? Can it be good or bad? Can it be prevented? Can it be cured? *Antioxidants* 11: 1431.

Bai, C., Twyman, R.M., Farré, G. et al. (2011). A golden era – pro-vitamin A enhancement in diverse crops. *In Vitro Cellular & Developmental Biology – Plant* 47: 205–221.

Bai, F., Li, S., Yang, C. et al. (2019). Overexpression of the AbSAUR1 gene enhanced biomass production and alkaloid yield in *Atropa belladonna*. *Industrial Crops and Products* 140: 111705.

Bartholomew, M. (2002). James Lind's treatise of the scurvy (1753). *Postgraduate Medical Journal* 78: 695–696.

Battilana, J., Emanuelli, F., Gambino, G. et al. (2011). Functional effect of grapevine 1-deoxy-D-xylulose 5-phosphate synthase substitution K284N on Muscat flavour formation. *Journal of Experimental Botany* 62: 5497–5508.

Blancquaert, D., Storozhenko, S., Van Daele, J. et al. (2013). Enhancing pterin and para-aminobenzoate content is not sufficient to successfully biofortify potato tubers and *Arabidopsis thaliana* plants with folate. *Journal of Experimental Botany* 64: 3899–3909.

Bulley, S., Wright, M., Rommens, C. et al. (2012). Enhancing ascorbate in fruits and tubers through over-expression of the L-galactose pathway gene GDP-L-galactose phosphorylase. *Plant Biotechnology Journal* 10: 390–397.

Čermák, T., Baltes, N.J., Radim, Č. et al. (2015). High-frequency, precise modification of the tomato genome. *Genome Biology* 16: 232.

Chang, L., Wu, S., and Tian, L. (2019). Effective genome editing and identification of a regiospecific gallic acid 4-O-glycosyltransferase in pomegranate (*Punica granatum* L.). *Horticulture Research* 6: 123.

Chen, F., Tholl, D., Bohlmann, J. et al. (2011). The family of terpene synthases in plants: a mid-size family of genes for specialized metabolism that is highly diversified throughout the kingdom. *Plant Journal* 66: 212–229.

Cong, L., Wang, C., Chen, L. et al. (2009). Expression of phytoene synthase1 and carotene desaturase crtI genes result in an increase in the total carotenoids content in transgenic elite wheat (*Triticum aestivum* L.). *Journal of Agricultural and Food Chemistry* 57: 8652–8660.

Connolly, E.L. (2008). Raising the bar for biofortification: enhanced levels of bioavailable calcium in carrots. *Trends in Biotechnology* 26: 401–403.

Criollo-Mendoza, M.S., Ramos-Payán, R., Contreras-Angulo, L.A. et al. (2022). Cytotoxic activity of polyphenol extracts from three oregano species: *Hedeoma patens*, *Lippia graveolens* and *Lippia palmeri*, and antiproliferative potential of *Lippia graveolens* against two types of breast cancer cell lines (MDA-MB-231 and MCF-7). *Molecules* 27: 5240.

Cushnie, T.P.T., Cushnie, B., and Lamb, A.J. (2014). Alkaloids: an overview of their antibacterial, antibiotic-enhancing and antivirulence activities. *International Journal of Antimicrobial Agents* 44: 377–386.

D'ambrosio, C., Stigliani, A.L., and Giorio, G. (2018). CRISPR/Cas9 editing of carotenoid genes in tomato. *Transgenic Research* 27: 367–378.

Daley, S.-K. and Cordell, G.A. (2021). Alkaloids in contemporary drug discovery to meet global disease needs. *Molecules* 26: 3800.

Daliri, E.B., Oh, D.H., and Lee, B.H. (2017). Bioactive peptides. *Foods* 6: 32.

Danilo, B., Perrot, L., Botton, E. et al. (2018). The DFR locus: a smart landing pad for targeted transgene insertion in tomato. *PLoS One* 13: 1–14.

Das, G., Gonçalves, S., Heredia, B.J. et al. (2022). Cardiovascular protective effect of cinnamon and its major bioactive constituents: an update. *Journal of Functional Foods* 97: 105045.

Deng, L., Wang, H., Sun, C. et al. (2018). Efficient generation of pink-fruited tomatoes using CRISPR/Cas9 system. *Journal of Genetics and Genomics* 45: 51–54.

Díaz De La Garza, R.I., Gregory, J.F., and Hanson, A.D. (2007). Folate biofortification of tomato fruit. *PNAS* 104: 4218–4222.

Diretto, G., Al-Babili, S., Tavazza, R. et al. (2007). Metabolic engineering of potato carotenoid content through tuber-specific overexpression of a bacterial mini-pathway. *PLoS One* 2: e350.

Dixit, S., Shukla, A., Singh, V., and Upadhyay, S.K. (2021) Engineering of plant metabolic pathway for nutritional improvement. *Wiley*. https://doi.org/10.1002/9781119672425.ch20.

Dong, C., Qu, G., Guo, J. et al. (2022). Rational design of geranylgeranyl diphosphate synthase enhances carotenoid production and improves photosynthetic efficiency in *Nicotiana tabacum*. *Science Bulletin* 67: 315–327.

Drakakaki, G., Marcel, S., Glahn, R.P. et al. (2005). Endosperm-specific co-expression of recombinant soybean ferritin and *Aspergillus phytase* in maize results in significant increases in the levels of bioavailable iron. *Plant Molecular Biology* 59: 869–880.

Ducreux, L.J., Morris, W.L., Hedley, P.E. et al. (2005). Metabolic engineering of high carotenoid potato tubers containing enhanced levels of beta-carotene and lutein. *Journal of Experimental Botany* 56: 81–89.

Dudley, Q.M., Jo, S., Serna, G.D.A. et al. (2022). Reconstitution of monoterpene indole alkaloid biosynthesis in genome engineered *Nicotiana benthamiana*. *Communications Biology* 5: 949.

Enfissi, E.M.A., Fraser, P.D., Lois, L.-M. et al. (2005). Metabolic engineering of the mevalonate and non-mevalonate isopentenyl diphosphate-forming pathways for the production of health-promoting isoprenoids in tomato. *Plant Biotechnology Journal* 3: 17–27.

Engprasert, S., Taura, F., and Shoyama, Y. (2005). Molecular cloning, expression and characterization of recombinant 1-deoxy-D-xylulose-5-phosphate reductoisomerase from *Coleus forskohlii* Briq. *Plant Science* 169: 287–294.

Farré, G., Sanahuja, G., Naqvi, S. et al. (2010). Travel advice on the road to carotenoids in plants. *Plant Science* 179: 28–48.

Feng, S., Song, W., Fu, R. et al. (2018). Application of the CRISPR/Cas9 system in *Dioscorea zingiberensis*. *Plant Cell, Tissue and Organ Culture* 135: 133–141.

Ferraz, A.C.A., De Oliveira Júnior, R.G., Picot, L. et al. (2019). Pre-clinical investigations of β-carboline alkaloids as antidepressant agents: a systematic review. *Fitoterapia* 137: 104196.

Fiaz, S., Ahmar, S., Saeed, S. et al. (2021). Evolution and application of genome editing techniques for achieving food and nutritional security. *International Journal of Molecular Sciences* 22: 5585.

Fornalé, S., Lopez, E., Salazar-Henao, J.E. et al. (2013). AtMYB7, a new player in the regulation of UV-sunscreens in *Arabidopsis thaliana*. *Plant and Cell Physiology Advance* 55: 507–516.

Foyer, C.H., Kyndt, T., and Hancock, R.D. (2020). Vitamin C in plants: novel concepts, new perspectives, and outstanding issues. *Antioxidant & Redox Signalling* 32: 463–485.

Ghasemzadeh, A. and Ghasemzadeh, N. (2011). Flavonoids and phenolic acids: role and biochemical activity in plants and humans. *Journal of Medicinal Plant Research* 5: 6697–6703.

Gómez-Galera, S., Sudhakar, D., Pelacho, A.M. et al. (2012). Constitutive expression of a barley Fe phytosiderophore transporter increases alkaline soil tolerance and results in iron partitioning between vegetative and storage tissues under stress. *Plant Physiology and Biochemistry* 53: 46–53.

Gong, Y., Liao, Z., Chen, M. et al. (2005). Molecular cloning and characterization of a 1-deoxy-D-xylulose 5-phosphate reductoisomerase gene from *Ginkgo biloba*. *DNA Sequence* 16: 111–120.

González, M.N., Massa, G.A., Andersson, M. et al. (2020). Reduced enzymatic browning in potato tubers by specific editing of a polyphenol oxidase gene via ribonucleoprotein complexes delivery of the CRISPR/Cas9 system. *Frontiers in Plant Science* 10: 1–12.

Gorelova, V., Ambach, L., Rébeillé, F. et al. (2017). Folates in plants: research advances and progress in crop biofortification. *Frontiers in Chemistry* 5: 21.

Grotewold, E. (2006). *The Science of Flavonoids*. New York: Springer.

Gutensohn, M., Nagegowda, D.A., and Dudareva, N. (2013). Involvement of compartmentalization in monoterpene and sesquiterpene biosynthesis in plants. In: *Isoprenoid Synthesis in Plants and Microorganisms: New Concepts and Experimental Approaches* (ed. T.J. Bach and M. Rohmer), 155–169. New York: Springer.

Gutiérrez-Grijalva, E.P., Antunes-Ricardo, M., Acosta-Estrada, B.A. et al. (2019). Cellular antioxidant activity and *in vitro* inhibition of α-glucosidase, α-amylase and pancreatic lipase of oregano polyphenols under simulated gastrointestinal digestion. *Food Research International* 116: 676–686.

Hamamoto, H., Sugiyama, Y., Nakagawa, N. et al. (1993). A new tobacco mosaic virus vector and its use for the systemic production of angiotensin-I-converting enzyme inhibitor in transgenic tobacco and tomato. *Nature Biotechnology* 11: 930–932.

Hanafy, M.S., Matter, M.A., Asker, M.S. et al. (2016). Production of indole alkaloids in hairy root cultures of *Catharanthus roseus* L. and their antimicrobial activity. *South African Journal of Botany* 105: 9–18.

Hasebe, F., Yuba, H., Hashimoto, T. et al. (2021). CRISPR/Cas9-mediated disruption of the pyrrolidine ketide synthase gene reduces the accumulation of tropane alkaloids in *Atropa belladonna* hairy roots. *Bioscience, Biotechnology, and Biochemistry* 85: 2404–2409.

Haslam, R.P., Ruiz-Lopez, N., Eastmond, P. et al. (2013). The modification of plant oil composition via metabolic engineering – better nutrition by design. *Plant Biotechnology Journal* 11: 157–168.

He, L., Liu, H., Cheng, C. et al. (2021). RNA sequencing reveals transcriptomic changes in tobacco (*Nicotiana tabacum*) following NtCPS2 knockdown. *BMC Genomics* 22: 1–14.

Heinrich, M., Mah, J., and Amirkia, V. (2021). Alkaloids used as medicines: structural phytochemistry meets biodiversity – an update and forward look. *Molecules* 26: 1836.

Hichri, I., Heppel, S.C., Pillet, J. et al. (2010). The basic helix-loop-helix transcription factor MYC1 is involved in the regulation of the flavonoid biosynthesis pathway in grapevine. *Molecular Plant* 3: 509–523.

Huang, Y., Xie, F.J., Cao, X. et al. (2021). Research progress in biosynthesis and regulation of plant terpenoids. *Biotechnology and Biotechnological Equipment* 35: 1800–1809.

Iaffaldano, B., Zhang, Y., and Cornish, K. (2016). CRISPR/Cas9 genome editing of rubber producing dandelion *Taraxacum kok-saghyz* using *Agrobacterium rhizogenes* without selection. *Industrial Crops and Products* 89: 356–362.

Iwashina, T. (2000). The structure and distribution of the flavonoids in plants. *Journal of Plant Research* 113: 287–299.

Jan, R., Asaf, S., Numan, M. et al. (2021). Plant secondary metabolite biosynthesis and transcriptional regulation in response to biotic and abiotic stress conditions. *Agronomy* 11: 1–31.

Jeong, J. and Guerinot, M.L. (2008). Biofortified and bioavailable: the gold standard for plant-based diets. *PNAS* 105: 1777–1778.

Jiang, W., Tang, M., Yang, L. et al. (2022). Analgesic alkaloids derived from traditional Chinese medicine in pain management. *Frontiers in Pharmacology* 13: 851508.

Jin, H., Cominelli, E., Bailey, P. et al. (2000). Transcriptional repression by AtMYB4 controls production of UV-protecting sunscreens in *Arabidopsis*. *The EMBO Journal* 19: 6150–6161.

Johnson, A.A., Kyriacou, B., Callahan, D.L. et al. (2011). Constitutive overexpression of the OsNAS gene family reveals single-gene strategies for effective iron-and zinc-biofortification of rice endosperm. *PLoS One* 6: e24476.

Jung, Y.J., Ju, H., Jong, L. et al. (2019). CRISPR/Cas9 – targeted mutagenesis of F3′H, DFR and LDOX, genes related to anthocyanin biosynthesis in black rice (*Oryza sativa* L.). *Plant Biotechnology Reports* 13: 521–531.

Karaki, H., Doi, K., Sugano, S. et al. (1990). Antihypertensive effect of tryptic hydrolysate of milk casein in spontaneously hypertensive rats. *Comparative Biochemistry and Physiology Part C: Comparative Pharmacology* 96: 367–371.

Karkute, S.G., Singh, A.K., Gupta, O.P. et al. (2017). CRISPR/Cas9 mediated genome engineering for improvement of horticultural crops. *Frontiers in Plant Science* 8: 1635.

Kim, C.K., Han, J.S., Lee, H.S. et al. (2006). Expression of an *Arabidopsis* CAX2 variant in potato tubers increases calcium levels with no accumulation of manganese. *Plant Cell Reports* 25: 1226–1232.

Kim, Y.J., Lee, O.R., Oh, J.Y. et al. (2014). Functional analysis of 3-hydroxy-3-methylglutaryl coenzyme a reductase encoding genes in triterpene saponin-producing ginseng. *Plant Physiology* 165: 373–387.

Klimek, M., Tomasz, C., Levi, O. et al. (2018). Efficient CRISPR/Cas9-based genome editing in carrot cells. *Plant Cell Reports* 37: 575–586.

Koes, R.E., Quattrocchio, F., and Mol, J.N.M. (1994). The flavonoid biosynthetic pathway in plants: function and evolution. *BioEssays* 16: 123–132.

Kumar, S. and Pandey, A.K. (2013). Chemistry and biological activities of flavonoids: an overview. *The Scientific World Journal* 2013: 16.

Kumar, D., Yadav, A., Ahmad, R. et al. (2022). CRISPR-based genome editing for nutrient enrichment in crops: a promising approach toward global food security. *Frontiers in Genetics* 13: 932859.

Lange, B.M. and Croteau, R. (1999). Isoprenoid biosynthesis via a mevalonate-independent pathway in plants. *Archives of Biochemistry and Biophysics* 365: 170–174.

Lau, O.S. and Sun, S.S.M. (2009). Plant seeds as bioreactors for recombinant protein production. *Biotechnology Advances* 27: 1015–1022.

Lee, S., Persson, D.P., Hansen, T.H. et al. (2011). Bio-available zinc in rice seeds is increased by activation tagging of nicotianamine synthase. *Plant Biotechnology Journal* 9: 865–873.

Leyva-López, N., Gutierrez-Grijalva, E.P., Ambriz-Perez, D.L. et al. (2016). Flavonoids as cytokine modulators: a possible therapy for inflammation-related diseases. *International Journal of Molecular Sciences* 17: 921.

Li, B., Cui, G., Shen, G. et al. (2017). Targeted mutagenesis in the medicinal plant *Salvia miltiorrhiza*. *Scientific Reports* 7: 1–9.

Li, X., Cao, L., Jiao, B. et al. (2022). The bHLH transcription factor AcB2 regulates anthocyanin biosynthesis in onion (*Allium cepa* L.). *Horticulture Research* 9: 1–11.

Lichman, B.R. (2021). The scaffold-forming steps of plant alkaloid biosynthesis. *Natural Product Reports* 38: 103–129.

Limera, C., Sabbadini, S., Sweet, J.B. et al. (2017). New biotechnological tools for the genetic improvement of major woody fruit species. *Frontiers in Plant Science* 8: 1–16.

Lin, S.X., Curtis, M.A., and Sperry, J. (2020). Pyridine alkaloids with activity in the central nervous system. *Bioorganic & Medicinal Chemistry* 28: 115820.

Lopez, A.B., Van Eck, J., Conlin, B.J. et al. (2008). Effect of the cauliflower or transgene on carotenoid accumulation and chromoplast formation in transgenic potato tubers. *Journal of Experimental Botany* 59: 213–223.

Maheshwari, P. and Kovalchuk, I. (2014). Genetic engineering of oilseed crops. *Biocatalysis and Agricultural Biotechnology* 3: 31–37.

Maioli, A., Gianoglio, S., Moglia, A. et al. (2020). Simultaneous CRISPR/Cas9 editing of three PPO genes reduces fruit flesh browning in *Solanum melongena* L. *Frontiers in Plant Science* 11: 607161.

Marchiosi, R., Dos Santos, W.D., Constantin, R.P. et al. (2020). Biosynthesis and metabolic actions of simple phenolic acids in plants. *Phytochemistry Reviews* 19: 865–906.

Masyita, A., Mustika, S.R., Astuti, A.D. et al. (2022). Terpenes and terpenoids as main bioactive compounds of essential oils, their roles in human health and potential application as natural food preservatives. *Food Chemistry: X* 13: 100217.

Matoba, N., Doyama, N., Yamada, Y. et al. (2001). Design and production of genetically modified soybean protein with anti-hypertensive activity by incorporating potent analogue of ovokinin (2–7). *FEBS Letters* 497: 50–54.

Mihálik, D., Gubišová, M., Klempová, T. et al. (2014). Transgenic barley producing essential polyunsaturated fatty acids. *Biologia Plantarum* 58: 348–354.

Mihálik, D., Klčová, L., Ondreičková, K. et al. (2015). Biosynthesis of essential polyunsaturated fatty acids in wheat triggered by expression of artificial gene. *International Journal of Molecular Sciences* 16: 30046–30060.

Mohammad, S., Šamec, D., Tomczyk, M. et al. (2020). Flavonoid biosynthetic pathways in plants: versatile targets for metabolic engineering. *Biotechnology Advances* 38: 107316.

Mohanta, T.K., Bashir, T., Hashem, A. et al. (2017). Genome editing tools in plants. *Genes* 8: 1–24.

Mondal, A., Gandhi, A., Fimognari, C. et al. (2019). Alkaloids for cancer prevention and therapy: current progress and future perspectives. *European Journal of Pharmacology* 858: 172472.

Mora-Vásquez, S., Wells-Abascal, G.G., Espinosa-Leal, C. et al. (2022). Application of metabolic engineering to enhance the content of alkaloids in medicinal plants. *Metabolic Engineering Communications* 14: e00194.

Muñoz-Bertomeu, J., Arrillaga, I., Ros, R. et al. (2006). Up-regulation of 1-Deoxy-D-xylulose-5-phosphate synthase enhances production of essential oils in transgenic spike lavender. *Plant Physiology* 142: 890–900.

Nagegowda, D.A. (2010). Plant volatile terpenoid metabolism: biosynthetic genes, transcriptional regulation and subcellular compartmentation. *FEBS Letters* 584: 2965–2973.

Naik, B.J., Vasamsetti, B.M.K., Kim, S.-C. et al. (2022). Genome editing in medicinal plants using CRISPR/Cas9 tool. In: *Phytochemical Genomics: Plant Metabolomics and Medicinal Plant Genomics* (ed. M.K. Swamy and A. Kumar). Singapore: Springer.

Nakashita, H., Arai, Y., Shikanai, T. et al. (2001). Introduction of bacterial metabolism into higher plants by polycistronic transgene expression. *Bioscience, Biotechnology, and Biochemistry* 65: 1688–1691.

Naqvi, S., Zhu, C., Farre, G. et al. (2009). Transgenic multivitamin corn through biofortification of endosperm with three vitamins representing three distinct metabolic pathways. *PNAS* 106: 7762–7767.

Ngaki, M.N., Louie, G.V., Philippe, R.N. et al. (2012). fatty-acid binding to stereospecific catalysis. *Nature* 485: 530–533.

Ninkuu, V., Zhang, L., Yan, J. et al. (2021). Biochemistry of terpenes and recent advances in plant protection. *International Journal of Molecular Sciences* 22: 5710.

Nishihara, M., Higuchi, A., Watanabe, A. et al. (2018). Application of the CRISPR/Cas9 system for modification of flower color in *Torenia fournieri*. *BMC Plant Biology* 18: 1–9.

Nitarska, D., Boehm, R., Debener, T. et al. (2021). First genome edited poinsettias: targeted mutagenesis of flavonoid 3′-hydroxylase using CRISPR/Cas9 results in a colour shift. *Plant Cell, Tissue and Organ Culture* 147: 49–60.

Nunes, A.C., Kalkmann, D.C., and Aragao, F.J.L. (2009). Folate biofortification of lettuce by expression of a codon optimized chicken GTP cyclohydrolase I gene. *Transgenic Research* 18: 661–667.

Oldfield, E. and Lin, F.Y. (2012). Terpene biosynthesis: modularity rules. *Angewandte Chemie International Edition* 51: 1124–1137.

Paine, J.A., Shipton, C.A., Chaggar, S. et al. (2005). Improving the nutritional value of golden rice through increased pro-vitamin A content. *Nature Biotechnology* 23: 482–487.

Palmgren, M.G., Clemens, S., Williams, L.E. et al. (2008). Zinc biofortification of cereals: problems and solutions. *Trends in Plant Science* 13: 464–473.

Pandita, D. and Pandita, A. (2022). Omics technology for the promotion of nutraceuticals and functional foods. *Frontiers in Physiology* 13: 817247.

Park, S., Kim, C.-K., Pike, L.M. et al. (2004). Increased calcium in carrots by expression of an *Arabidopsis* $H+/Ca\ 2+$ transporter. *Molecular Breeding* 14: 275–282.

Park, S., Kang, T.S., Kim, C.K. et al. (2005). Genetic manipulation for enhancing calcium content in potato tuber. *Journal of Agricultural and Food Chemistry* 53: 5598–5603.

Park, S., Elless, M.P., Park, J. et al. (2009). Sensory analysis of calcium-biofortified lettuce. *Plant Biotechnology Journal* 7: 106–117.

Parks, H.M., Cinelli, M.A., Bedewitz, M.A. et al. (2023). Redirecting tropane alkaloid metabolism reveals pyrrolidine alkaloid diversity in *Atropa belladonna*. *New Phytologist* 237: 1810–1825.

Pérez-Massot, E., Banakar, R., Gómez-Galera, S. et al. (2013). The contribution of transgenic plants to better health through improved nutrition: opportunities and constraints. *Genes & Nutrition* 8: 29–41.

Powell, K. (2007). Functional foods from biotech – an unappetizing prospect? *Nature Biotechnology* 25: 525–531.

Qi, B., Fraser, T., Mugford, S. et al. (2004). Production of very long chain polyunsaturated omega-3 and omega-6 fatty acids in plants. *Nature Biotechnology* 22: 739–745.

Qu, D., Yan, F., Meng, R. et al. (2016). Identification of MicroRNAs and their targets associated with fruit-bagging and subsequent sunlight re-exposure in the "Granny Smith" apple exocarp using high-throughput sequencing. *Frontiers in Plant Science* 7: 1–14.

Quesada-Vargas, T., Ruiz, O.N., and Daniell, H. (2005). Characterization of heterologous multigene operons in transgenic chloroplasts: transcription, processing, and translation. *Plant Physiology* 138: 1746–1762.

Ramírez Rivera, N.G., García-Salinas, C., Aragão, F.J. et al. (2016). Metabolic engineering of folate and its precursors in Mexican common bean (*Phaseolus vulgaris* L.). *Plant Biotechnology Journal* 14: 2021–2032.

Römer, S., Lübeck, J., Kauder, F. et al. (2002). Genetic engineering of a zeaxanthin-rich potato by antisense inactivation and co-suppression of carotenoid epoxidation. *Metabolic Engineering* 4: 263–272.

Ruiz-Lopez, N., Haslam, R.P., Napier, J.A. et al. (2014). Successful high-level accumulation of fish oil omega-3 long-chain polyunsaturated fatty acids in a transgenic oilseed crop. *Plant Journal* 77: 198–208.

Sadre, R., Anthony, T.M., Grabar, J.M. et al. (2022). Metabolomics-guided discovery of cytochrome P450s involved in pseudotropine-dependent biosynthesis of modified tropane alkaloids. *Nature Communications* 13: 3832.

Saito, K., Yonekura-Sakakibara, K., Nakabayashi, R. et al. (2013). Plant physiology and biochemistry the flavonoid biosynthetic pathway in *Arabidopsis*: structural and genetic diversity. *Plant Physiology and Biochemistry* 72, 1–14.

Sayanova, O., Smith, M.A., Lapinskas, P. et al. (1997). Expression of a borage desaturase cDNA containing an N-terminal cytochrome b5 domain results in the accumulation of high levels of delta6-desaturated fatty acids in transgenic tobacco. *PNAS* 94: 4211–4216.

Schijlen, E., Jonker, H., Molthoff, J. et al. (2006). Pathway engineering for healthy phytochemicals leading to the production of novel flavonoids in tomato fruit. *Plant Biotechnology Journal* 4: 433–444.

Schweizer, F., Colinas, M., Pollier, J. et al. (2018). An engineered combinatorial module of transcription factors boosts production of monoterpenoid indole alkaloids in *Catharanthus roseus*. *Metabolic Engineering* 48: 150–162.

Shumayla and Upadhyay, S.K. (2022). Mechanism of iron uptake and homeostasis in plants. *Elsevier*. https://doi.org/10.1016/B978-0-323-85790-1.00015-4.

Singh, A., Saraf, S., Dasgupta, I. et al. (2016). Identification and validation of a virus-inducible ta-siRNA-generating TAS4 locus in tomato. *Journal of Biosciences* 41: 109–118.

Singh, S.K., Patra, B., Paul, P. et al. (2020). Revisiting the ORCA gene cluster that regulates terpenoid indole alkaloid biosynthesis in *Catharanthus roseus*. *Plant Science* 293: 110408.

Stoger, E., Ma, J.K., Fischer, R. et al. (2005). Sowing the seeds of success: pharmaceutical proteins from plants. *Current Opinion in Biotechnology* 16: 167–173.

Storozhenko, S., De Brouwer, V., Volckaert, M. et al. (2007). Folate fortification of rice by metabolic engineering. *Nature Biotechnology* 25: 1277–1279.

Stracke, R., Jahns, O., Keck, M. et al. (2010). Analysis of production of flavonol glycosides-dependent flavonol glycoside accumulation in *Arabidopsis thaliana* plants reveals glycoside accumulation. *New Phytologist* 188: 985–1000.

Sudheeran, P.K., Ovadia, R., Galsarker, O. et al. (2020). Glycosylated flavonoids: fruit's concealed antifungal arsenal. *New Phytologist* 225: 1788–1798.

Sunitha, S. and Rock, C.D. (2020). CRISPR/Cas9-mediated targeted mutagenesis of TAS4 and MYBA7 loci in grapevine rootstock 101-14. *Transgenic Research* 29: 355–367.

Sushmita, Kaur, G., Upadhyay, S.K., and Verma, P.C. (2021). An overview of genome-engineering methods. *Wiley*. https://doi.org/10.1002/9781119672425.ch1.

Suzuki, M., Kamide, Y., Nagata, N. et al. (2004). Loss of function of 3-hydroxy-3-methylglutaryl coenzyme A reductase 1 (HMG1) in *Arabidopsis* leads to dwarfing, early senescence and male sterility, and reduced sterol levels. *Plant Journal* 37: 750–761.

Swinnen, G., De Meyer, M., Pollier, J. et al. (2022). The basic helix–loop–helix transcription factors MYC1 and MYC2 have a dual role in the regulation of constitutive and stress-inducible specialized metabolism in tomato. *New Phytologist* 236: 911–928.

Taneja, M. and Upadhyay, S.K. (2021). An introduction to the calcium transport elements in plants. *Elsevier*. https://doi.org/10.1016/B978-0-12-821792-4.00019-9.s.

Tang, H.V., Berryman, D.L., Mendoza, J. et al. (2022). Dedicated farnesyl diphosphate synthases circumvent isoprenoid-derived growth-defense tradeoffs in *Zea mays*. *Plant Journal* 112: 207–220.

Tu, M., Fang, J., Zhao, R. et al. (2022). CRISPR/Cas9-mediated mutagenesis of VvbZIP36 promotes anthocyanin accumulation in grapevine (*Vitis vinifera*). *Horticulture Research* 9: 1–14.

United Nations (2017). World population prospects: the 2017 revision. https://desapublications.un.org/publications/world-population-prospects-2017-revision (accessed 23 August 2023).

United Nations (2023). The sustainable development goals report 2023: special edition. https://unstats.un.org/sdgs/report/2023/The-Sustainable-Development-Goals-Report-2023.pdf (accessed 23 August 2023).

Upadhyay, S.K. (2021a). Genome engineering for crop improvement. *Wiley*. https://doi.org/10.1002/9781119672425.

Upadhyay, S.K. (2021b). Calcium transport elements in plants. *Elsevier*. https://doi.org/10.1016/C2019-0-04172-6.

Upadhyay, S.K. and Singh, S.P. (2021). Bioprospecting of plant biodiversity for industrial molecules. *Wiley*. https://doi.org/10.1002/9781119718017.

Upadhyay, S.K. and Singh, S.P. (2023). Plants as bioreactors for industrial molecules. *Wiley*. https://doi.org/10.1002/9781119875116.

Waltz, E. (2016). Gene-edited CRISPR mushroom escapes US regulation. *Nature* 532: 293.

Wang, D., Li, M., Wei, D. et al. (2007). Identification and functional characterization of the delta 6-fatty acid desaturase gene from *Thamnidium elegans*. *Journal of Eukaryotic Microbiology* 54: 110–117.

Wang, D., Song, Y., Chen, Y. et al. (2013). Metabolic pools of phenolic acids in *Salvia miltiorrhiza* are enhanced by co-expression of *Antirrhinum majus* Delila and Roseal transcription factors. *Biochemical Engineering Journal* 74: 115–120.

Wang, Y., Yang, Q., Zhu, Y. et al. (2022). MrTPS3 and MrTPS20 are responsible for β-caryophyllene and α-pinene production, respectively, in red bayberry (*Morella rubra*). *Frontiers in Plant Science* 12: 1–13.

Warude, D., Joshi, K., and Harsulkar, A. (2006). Polyunsaturated fatty acids: biotechnology. *Critical Reviews in Biotechnology* 26: 83–93.

Watanabe, K., Kobayashi, A., Endo, M. et al. (2017). CRISPR/Cas9-mediated mutagenesis of the dihydroflavonol- 4-reductase-B (DFR-B) locus in the Japanese morning glory ipomoea. *Scientific Reports* 7: 1–9.

Wei, H., Xu, C., Movahedi, A. et al. (2019). Characterization and function of 3-hydroxy-3-methylglutaryl-CoA reductase in *Populus trichocarpa*: overexpression of PtHMGR enhances terpenoids in transgenic poplar. *Frontiers in Plant Science* 10: 1–18.

Wei, C., Li, M., Cao, X. et al. (2022). Linalool synthesis related PpTPS1 and PpTPS3 are activated by transcription factor PpERF61 whose expression is associated with DNA methylation during peach fruit ripening. *Plant Science* 317: 111200.

Welch, R.M. and Graham, R.D. (2005). Agriculture: the real nexus for enhancing bioavailable micronutrients in food crops. *Journal of Trace Elements in Medicine and Biology* 18: 299–307.

Welsch, R., Arango, J., Bär, C. et al. (2010). Provitamin A accumulation in cassava (*Manihot esculenta*) roots driven by a single nucleotide polymorphism in a phytoene synthase gene. *The Plant Cell* 22: 3348–3356.

Wen, D., Wu, L., Wang, M. et al. (2022). CRISPR/Cas9-mediated targeted mutagenesis of FtMYB45 promotes flavonoid biosynthesis in *Tartary buckwheat* (*Fagopyrum tataricum*). *Frontiers in Plant Science* 13: 879390.

Widhalm, J.R. and Dudareva, N. (2015). A familiar ring to it: biosynthesis of plant benzoic acids. *Molecular Plant* 8: 83–97.

Wirth, J., Poletti, S., Aeschlimann, B. et al. (2009). Rice endosperm iron biofortification by targeted and synergistic action of nicotianamine synthase and ferritin. *Plant Biotechnology Journal* 7: 631–644.

Wolak, N., Zawrotniak, M., Gogol, M. et al. (2017). Vitamins B1, B2, B3 and B9 – occurrence, biosynthesis pathways and functions in human nutrition. *Mini Reviews in Medicinal Chemistry* 17: 1075–1111.

Xiang, L.-L., Liu, X.-F., Li, X. et al. (2015). A novel bHLH transcription factor involved in regulating anthocyanin biosynthesis in Chrysanthemums (*Chrysanthemum morifolium* Ramat.). *PLoS One* 10: 1–17.

Xiao, H. and Zhong, J.J. (2016). Production of useful terpenoids by higher-fungus cell factory and synthetic biology approaches. *Trends in Biotechnology* 34: 242–255.

Yang, L., Tada, Y., Yamamoto, M.P. et al. (2006). A transgenic rice seed accumulating an anti-hypertensive peptide reduces the blood pressure of spontaneously hypertensive rats. *FEBS Letters* 580: 3315–3320.

Yang, F., Cai, J., Yang, Y. et al. (2013). Overexpression of microRNA828 reduces anthocyanin accumulation in *Arabidopsis*. *Plant Cell, Tissue and Organ Culture* 115: 159–167.

Ye, X., Al-Babili, S., Klöti, A. et al. (2000). Engineering the provitamin A (beta-carotene) biosynthetic pathway into (carotenoid-free) rice endosperm. *Science* 287: 303–305.

Yu, J., Tu, L., Subburaj, S. et al. (2021). Simultaneous targeting of duplicated genes in petunia protoplasts for flower color modification via CRISPR – Cas9 ribonucleoproteins. *Plant Cell Reports* 40: 1037–1045.

Yuan, G., Guan, Y., Yi, H. et al. (2021). Antibacterial activity and mechanism of plant flavonoids to gram-positive bacteria predicted from their lipophilicities. *Scientific Reports* 11: 10471.

Zeb, A. (2020). Concept, mechanism, and applications of phenolic antioxidants in foods. *Journal of Food Biochemistry* 44: e13394.

Zeng, L., Zhang, Q., Jiang, C. et al. (2021). Development of *Atropa belladonna* L. plants with high-yield hyoscyamine and without its derivatives using the CRISPR/Cas9 system. *International Journal of Molecular Sciences* 22: 1731.

Zeng, J., Liu, M., Zeng, L. et al. (2022). Engineering the production of medicinal tropane alkaloids through enhancement of tropinone and littorine biosynthesis in root cultures of *Atropa belladonna*. *Industrial Crops and Products* 189: 115778.

Zhang, Y., Yan, Y.P., and Wang, Z.Z. (2010). The *Arabidopsis* PAP1 transcription factor plays an important role in the enrichment of phenolic acids in *Salvia miltiorrhiza*. *Journal of Agricultural and Food Chemistry* 58: 12168–12175.

Zhang, S., Ma, P., Yang, D. et al. (2013). Cloning and characterization of a putative R2R3 MYB transcriptional repressor of the rosmarinic acid biosynthetic pathway from *Salvia miltiorrhiza*. *PLoS One* 8: 1–12.

Zhang, X., He, Y., Li, L. et al. (2021). Involvement of the R2R3-MYB transcription factor MYB21 and its homologs in regulating flavonol accumulation in *Arabidopsis* stamen. *Journal of Experimental Botany* 72: 4319–4332.

Zhang, J., Zhou, Q., Zhang, D. et al. (2022a). The agronomic traits, alkaloids analysis, FT-IR and 2DCOS-IR spectroscopy identification of the low-nicotine-content nontransgenic tobacco edited by CRISPR-Cas9. *Molecules* 27: 3817.

Zhang, Q., Liang, M., Zeng, J. et al. (2022b). Engineering tropane alkaloid production and glyphosate resistance by overexpressing AbCaM1 and G2-EPSPS in *Atropa belladonna*. *Metabolic Engineering* 72: 237–246.

Zhao, T., Li, S., Wang, J. et al. (2020). Engineering tropane alkaloid production based on metabolic characterization of ornithine decarboxylase in *Atropa belladonna*. *ACS Synthetic Biology* 9: 437–448.

Zheng, L., Cheng, Z., Ai, C. et al. (2010). Nicotianamine, a novel enhancer of rice iron bioavailability to humans. *PLoS One* 5: e10190.

Zhi, J., Liu, X., Li, D. et al. (2020). CRISPR/Cas9 – mediated SlAN2 mutants reveal various regulatory models of anthocyanin biosynthesis in tomato plants. *Plant Cell Reports* 39: 799–809.

Zhou, Z., Tan, H., Li, Q. et al. (2018). Phytochemistry CRISPR/Cas9-mediated efficient targeted mutagenesis of RAS in *Salvia miltiorrhiza*. *Phytochemistry* 148: 63–70.

Zhu, C., Naqvi, S., Capell, T. et al. (2009). Metabolic engineering of ketocarotenoid biosynthesis in higher plants. *Archives of Biochemistry and Biophysics* 483: 182–190.

Zhu, H., Xia, R., Zhao, B. et al. (2012). Unique expression, processing regulation, and regulatory network of peach (*Prunus persica*) miRNAs. *BMC Plant Biology* 12: 149.

Zhu, C., Sanahuja, G., Yuan, D. et al. (2013). Biofortification of plants with altered antioxidant content and composition: genetic engineering strategies. *Plant Biotechnology Journal* 11: 129–141.

15

Exploration of Genome Editing Tools for microRNA Engineering in Plants

Hengyi Xu

Departments of Molecular Biosciences and Oncology, Institute of Cellular and Molecular Biology, University of Texas at Austin, Austin, TX, USA

15.1 Introduction

Comparing to the prokaryote and the lower clade of the eukaryote's organisms, the transcriptome of mammalian and plant are more complex in the biotype, function mode, and regulation of the RNA. In mammalian cells, it is believed that less than 2% of the genome code for protein (Kosinska-Selbi et al. 2020). Similarly, a large portion of the transcripts in plant cells (80–90%) do not produce protein thereby are only of specialized function or regulatory role as non-coding RNAs (Anand and Pandi 2021). Except for the well-known rRNA and tRNA, ncRNA by definition can be grouped into long noncoding RNA (lncRNAs) and small non-coding RNA (sncRNAs) by size at 200 nt (Perkel 2013) as the cut-off. Via the dynamic folded RNA secondary structure, lncRNAs exhibited a variety function mode including but not limited to scaffold, decoy, signal and guide (Wang and Chang 2011; Taneja et al. 2021) via its ability to interact with chromatin, RNA or protein (Statello et al. 2021). For most lncRNA to carry out their biological functions, proteins play extensive roles in the lncRNA. Each lncRNAs works with the distinct sets of accessory proteins in helping them to fold into the correct structure conformations. The lncRNA bind to its evolved proteins partners to form and function as ribonucleoprotein (RNP) complex e.g. HOTAIR-PRC2 or TERC-telomerase (Tsai et al. 2010; Zhang et al. 2011a; Upadhyay 2021a). Unlike lncRNA that each one evolved a specialized set of proteins to carry out the function, a subset of sncRNA, sized from 21 to 35 nt, are called small RNA (sRNA) exhibit a regulatory role via a uniformed process called RNA silencing. In plant, based on the origin, processing procedure and interaction mode to the downstream effector proteins, the sRNA can be further categorized into two main families: microRNA (miRNA), small interfering RNA (siRNA) (Axtell 2013). Recent studies showed that another small family of the sRNA is derived from tRNA (tRNA-derived small RNA or tsRNAs) (Alves and Nogueira 2021). Based on the gene repression mode, sRNA mediates the regulation of gene expression through two RNA silencing pathways: transcriptional gene silencing (TGS) and posttranscriptional gene silencing (PTGS). In either case, the sRNA was used as the key to

Applications of Genome Engineering in Plants, First Edition. Edited by Santosh Kumar Upadhyay.
© 2024 John Wiley & Sons Ltd. Published 2024 by John Wiley & Sons Ltd.

recognize the target gene by the sequence complementarity. In plant, for example, siRNAs and its associated specific downstream effector proteins could modulate the DNA methylation/chromatin status, the change of which alters the gene accessibility and therefore inhibit the targeted gene transcription (Blevins et al. 2015; Singh et al. 2019). A similar TGS process in mammalian cell was mediated by PIWI-interacting RNA or piRNA instead (Sienski et al. 2012). However, the plant miRNA and its associated specific downstream effector proteins mainly repress gene expression via PTGS pathways. The RNA silencing is essential in almost all aspects of plant life cycle for its proper growth and development (Meins et al. 2005; Zhu et al. 2011). Meanwhile, the arsenals of the sRNA and RNA silencing machineries play a crucial role in coping with the abiotic stress, for example, heat shock or drought (Guan et al. 2013; Ferdous et al. 2015), as well as biotic stress from the invasion by the other organism such as plant virus and microbes (Zhang et al. 2006; Zhang et al. 2011b). Therefore, miRNA in plant is of great research interests in previous years and manipulating miRNA with proper genome editing tools allow researchers to acquire the wanted trait of plants, especially the ones for agriculture. This review will mainly focus on the pathway of RNA silencing mediated by miRNA and miRNA engineering with the recent advancement of genome editing tools.

15.2 The Biogenesis of the miRNA and RNA Silencing in Plant

Understanding the miRNA biogenesis and regulation could provide insights in designing the correct strategies in engineering the MIRs. The studies of the miRNA biogenesis and RNA silencing pathways were extensively carried out by using the plant model organism *Arabidodpsis thaliana,* resulting in the paradigm and reference of the related themes for other plant organisms (Figure 15.1).

The miRNA biogenesis consists of several events in both nucleus and cytoplasm. In nucleus, the miRNA was transcribed by Pol II from MIR gene locus in the plant genome as a long noncoding RNA called pri-miRNA. Like other Pol II transcripts, the pri-miRNA undergoes the conventional RNA posttranscriptional modification: 5′ cap, 3′ poly-A tail process and when necessary intron splicing (Cai et al. 2004; Xie et al. 2005; Rogers and Chen 2013). The pri-miRNA fold into a structure of the stem-loop with 5′ and 3′ single-stranded RNA arms. Although the length of stem-loop varies from 60 nt to longer than 500 nt, the folded hairpin structure is evolutionary conserved that can be recognized and cleaved by the Dicer-like (DCLs) proteins which contain the activity of RNase III endonucleases. The copy number of the DCLs genes are various among plant species (Liu et al. 2009). In *A. thaliana,* there are four DCL genes and the miRNA production is predominantly mediated by DCL1. The double-stranded RNA binding protein Hyponastic Leaves 1 (HYL1) and zinc-finger domain containing protein Serrate (SE) facilitate the miRNA production alongside with DCL1 (Fang and Spector 2007; Dong et al. 2008). Other DCLs may also catalyze the biogenesis of some specific miRNAs as the AtDCL4 produces miR822 and miR839 (Rajagopalan et al. 2006) while SlDCL2 is involved in the 22 nt miR6026 production in tomato for the virus resistance (Wang et al. 2018). The DCL protein "sit" at the stem of the hair-pin structure. The miRNA can be cleaved by two different sequential orders either from lower stem to loop or from loop to lower stem

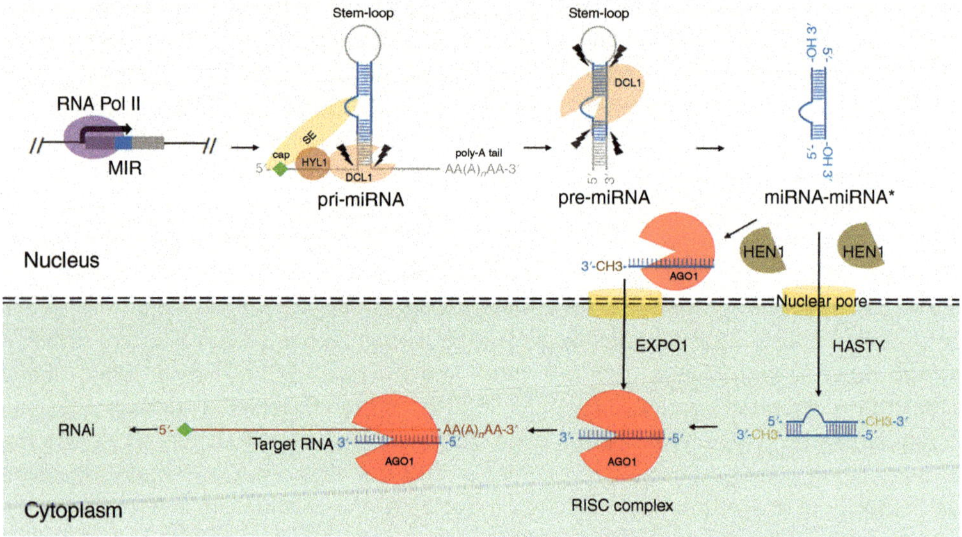

Figure 15.1 The biogenesis of miRNA and RISC complex-mediated RNAi. The MIR genes are transcribed by the RNA Pol II from the genomic loci. The nascent transcript of each MIR gene is a long noncoding RNA called primary-miRNA (pri-miRNA) that folds into a conventional structure of hair-pin stem-loop flanked with 5′ and 3′ arms. After 5′ cap and 3′ poly-A processing, the correctly folded pri-miRNA is then recognized by DCL1. Faciliated by HYL1 and SE, the DCL1 cut at two stem-arm regions and generate an intermediate stem-loop precursor-miRNA (pre-miRNA). The pre-miRNA is then further processed by DCL1, leaving only the miRNA-miRNA* double strand RNA duplex. The 3′ end of the miRNA duplex are methylated by HEN1 on both strands. Majority of the methylated miRNA duplexes are transported by HASTY to the cytoplasm, where one strand is selected and loaded into the Agonaute protein 1 (AGO1) to form the RISC complex. However, a subset of the miRNA undergoes the RISC complex formation in nucleus and the complex is transported by EXPO1 to the cytoplasm. The target RNA harboring the sequence to be reverse complementary to the miRNA can be recognized by the RISC, which will consequently trigger the RNAi by destabilizing the target RNA or repressing the translation efficiency for protein-coding genes.

(Addo-Quaye et al. 2009; Bologna et al. 2009), with precursor miRNA (pre-miRNA) as an intermediate, ending up with a double-stranded nascent miRNA/miRNA* duplex structure. Pri-miRNA or pre-miRNA with branched terminal loop could "confuse" the DCL protein "sitting and facing" to the wrong direction, ending up the cleavage at the wrong site to cause the abortive processing (Zhu et al. 2013). The miRNA/miRNA* generated by the functional DCL processing is typically ended with 2 nt overhang at both 3′ sides and each top and bottom RNA strand are with 5′ phosphate and 3′ hydroxyls (2′ and 3′—OH). Hua Enhancer1 (HEN1), which is a sRNA specialized methyltransferase, deposit the methyl group on the 2′—OH only when the 2 nt overhang and both 2′—OH and 3′—OH existed on the miRNA duplex (Yang et al. 2006). Hasty (HST), a homologous protein of the human Expotin 5 for miRNA exportation, exports the methylated miRNA/miRNA* duplexes from the nucleus to the cytoplasm, where the guide strand miRNA is selectively loaded in to the effector protein, in this case mostly AGO1 (Vaucheret et al. 2004), to form the RISC complex. Recent evidence showed that a minority of the RISC complex assembly is in the nucleus and exported to the cytoplasm by EXPO1 (Bologna et al. 2018). Like other RNPs with distinct role of RNA and protein, miRNA-RISC complex mediates

RNA silencing by the following described mode: the miRNA guide the AGO to the target gene transcript of either mRNA or non-coding RNAs by base-pairing; AGO protein cleaves the RNA molecule with the endonuclease activity (Pratt and MacRae 2009) or inhibits the translation if the target is mRNA (Iwakawa and Tomari 2013). The cleavage of RISC on some lncRNA could be a substrate of RDR6 and the downstream accessory proteins resulting in the generation phase small interfering RNAs (phasiRNAs), that could be loaded to AGOs for target gene silencing (Yoshikawa et al. 2005; Axtell et al. 2006). Comparing to the animal miRNA-mediated silencing, the plant miRNA requires a more stringent base-paring to the target, with almost perfect paring in the 5′ region (<1 mismatch) and <4 mismatches at the 3′ region (Schwab et al. 2005; Axtell and Meyers 2018). The functional analysis of the slicer-deficient AGO suggest that the cleavage based gene silencing is indispensable for the plant growth, while the translation inhibition is a widespread mechanism for gene silencing (Carbonell et al. 2012; Arribas-Hernández et al. 2016).

15.3 MIRs as a Family in Plant

Plant MIR genes, like their counterparts for protein coding, can experience the gene duplication, inversion, mutation, and amplification during the genome evolution (Baldrich et al. 2018). A group of miRNAs that derive from a common ancestor could be grouped as a miRNA family, that also could be conserved among different plant species (Arteaga-Vázquez et al. 2006). Family that shares the high levels of similarity and shared target genes are typically categorized as superfamily. Normally, the mature miRNA sequence has higher similarity in a family while the primary sequence and secondary structure of the stem-loop structure could vary, which in many cases can result in different efficiency of the biogenesis of miRNA (Zhu et al. 2013) and miRNA abundance (Jasinski et al. 2010; Bologna et al. 2013). The similarity of mature miRNA sequence determines similar physiological functions shared by the miRNA family members. Hence, the potential functional redundancy among miRNA family member needs to be aware when aiming to manipulate them.

The cluster of MIRs in a family exist in plant (Xia et al. 2015). In rice, 18 members of miR2118 are clustered at two genomic loci (Johnson et al. 2009) while two clusters of miR482 were found in the genome of litchi (Ma et al. 2018). The cluster of MIRs provide a convenience in manipulation because they could be treated as one unit, for example, by a genome editing tools to delete a large fragment in order to knock out all MIRs in the cluster.

Sometimes, the unique mutation in one family member may evolve the miRNA with a specialized localization and function. For example, *Arabidopsis* miR164a/b/c have function redundancy in inflorescence development, while their expression patterns vary and showed different influences in regulating the target gene CUC1 and CUC2 (Sieber et al. 2007). Therefore, the knowledge of MIR family member and target prediction seems to be essential for the successful MIR engineer. For the miRNA of research interest, the previously published research articles and public miRNA archived database such as miRBase could be a good resources for these knowledges. For miRNA with unknown target or target without known miRNA, several prediction tools were developed including miRClassify (Zou et al. 2014) and miRDB (Chen and Wang 2020) and PmiREN specific for plant miRNA (Guo et al. 2020).

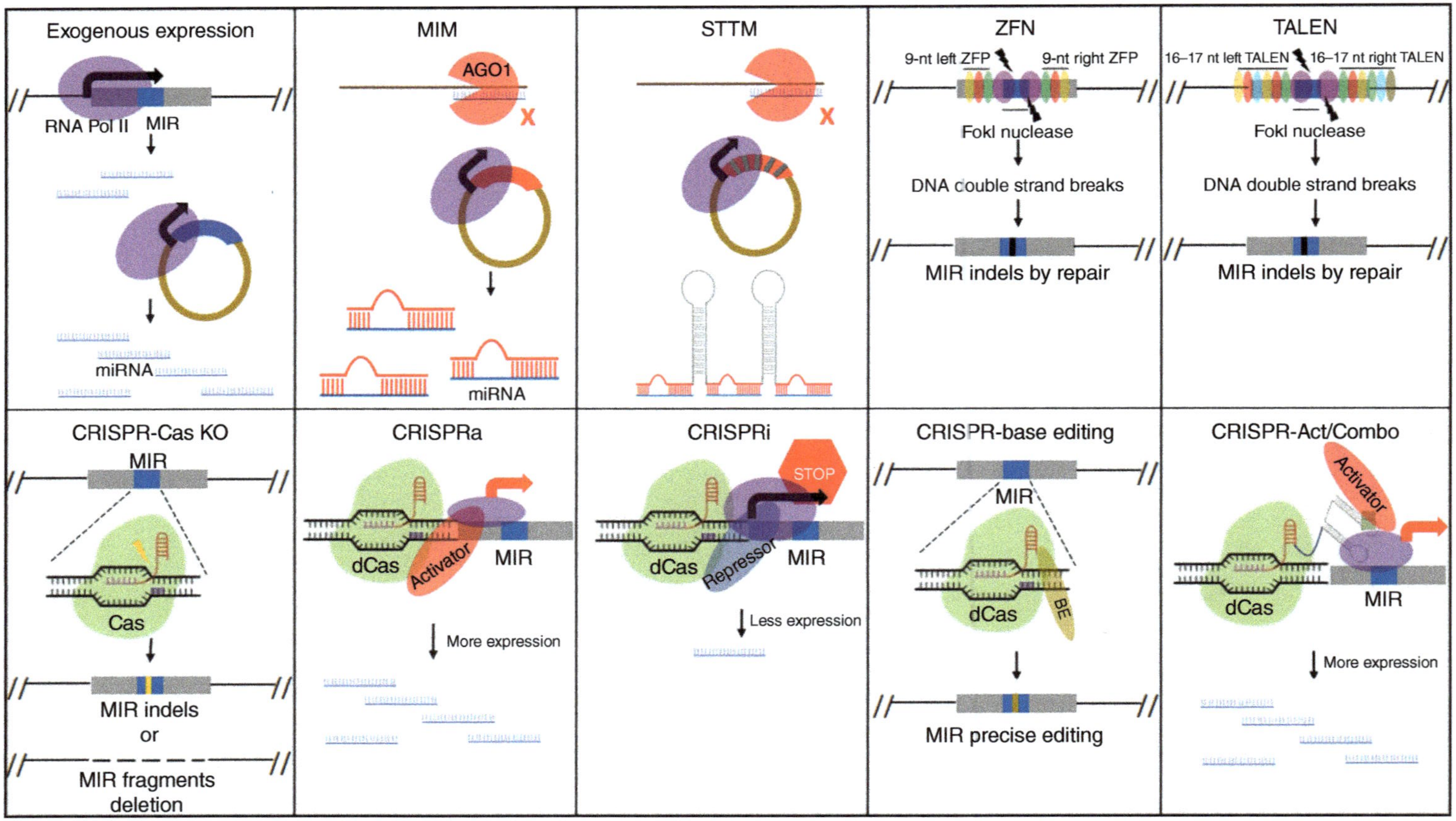

Figure 15.2 The miRNA engineer tool kits in plant. The miRNA abundance can be increased by the exogenous expression of the miRNA genes driven by the native promoter or a promoter that overexpress the miRNAs. To repress the miRNA function, miRNA target mimicry (MIM) can be realized by exogenous expressing a construct that is basically reverse complementary sequence of the miRNA (red in MIM). The miRNA mimicry molecule sponge the miRNA from the target RNA or RISC complex resulting in a decreased RNAi. The STTM method is evolved based on MIM where a construct containing multiple copies of the mimicry sequences was concatenated by some linker sequences, which are more efficient to sequester the miRNA from the RNAi process. At the genomic level, the miRNA genes can be edited by the sequence specific nuclease including the ZFN and TALEN as well as CRISPR-Cas system. The sequence recognition by ZFN is mediated by ZFP proteins with the fusion of the FokI nuclease. The dimerized FokI nuclease generate the double strand

15.4 The miRNA Engineering Methods in Plant

A couple methods so far has been proposed and developed to engineer the function of miRNA. Since miRNA negatively regulate the targeted mRNA, the further downregulation of the target mRNA gene could be achieved by the overexpression of the MIR genes with either a strong or specific promoter (Betti et al. 2021). On the other hand, alleviation of the miRNA repression effect to the targeted mRNA can be realized by the RNAi on the miRNA or engineering the production of the long non-coding RNA that share high-level identity of the mRNAs. These lncRNA serve as a "sponge" that deplete the endogenous miRNA, therefore release the mRNA target from the control of the miRNA-RISC complex-mediated regulation. Such methods include miRNA artificial target mimicry (MIM) and short tandem target mimicry (STTM) (Todesco et al. 2010; Tang et al. 2012).

Unlike changing the abundance of the miRNA and its targeting effect after the MIR transcription, a couple methods are available to disrupt MIR gene function at the genomic level. Before the birth of the CRISPR/Cas and its application in plant genome editing, two technologies widely used for targeted gene editing were ZNF proteins and transcription activator-like effector nucleases (TALENs). Both ZNFs and TALENs contains endonuclease and sequence determining elements to achieve the specific gene disruption (Sushmita et al. 2021). The principle behind both methods is the endonuclease guided by the sequence determining elements can generate the DNA double-stranded breaks at the target loci and the mistakes on DNA after the DNA repair by chance will leave indels or mutations, leading to the disruption of the gene function. However, these two methods are not widely used in targeting the miRNA genes in plant. Until recently only very few studies manipulate miRNA genes in plant by the TALEN methods (Bi et al. 2020). Similarly, gene editing by CRISPR-Cas system is another sequence specific nuclease (SSN) method, that relies on the errors left by the DNA repair machinery at the targeted gene loci (Upadhyay 2021b). Unlike the ZNF proteins and TALENs recognition of DNA target by the protein engineer, a single stranded spacer guide RNA (gRNA) is employed by the CRISPR to guide the Cas nuclease to the targeted loci (Figure 15.2).

Figure 15.2 (Continued) DNA breaks (DSBs). A similar process can be achieved by TALEN with a sequence recognition by a longer pattern. The cleaved dsDNA strand can be repaired and by chance the repair can leave an indel mistake. The idea to introduce mutation to the target gene is same for the CRISPR-Cas system. Unlike the ZFN and TALEN, the sequence recognition motif was realized by an engineered guide RNA (gRNA) that can be base-pair to the target region. The PAM motif adjacent to the gRNA-DNA region (purple box in CRISPR-Cas KO) is required for the sequence recognition and successful DNA cleavage. A couple CRISPR derivative methods with success on the protein-coding gene manipulation can be potentially used in plant miRNA. For example, the activator or repressor protein can be fused to the nuclease-dead Cas as CRISPRa and CRISPRi systems. By carefully selecting the target sequence around the gene promoter region, increased or decreased gene expression can be achieved. The precise base-editing can also be realized by fusing the base editor (BE), which are proteins from nucleotide deaminase family including adenosine base-editor (ABE) and cytosine base-editor (CBE). Last, the gRNA can be further engineered with aptamer RNA motif that could be bound by the RBPs. By fusing these RBPs to the effector proteins such as activator protein, the activator can be recruited proximal to the target site. By including multiple copies of the RNA aptamers, a much higher activation than the CRISPRa can be achieved by having cluster of the activator proteins close by the target site. Furthermore, the combination of the engineer in both gRNA and dCas protein was established in plant called CRISPR-Combo system, which can simultaneously activate the target gene and base-edit it precisely.

In all methods for gene editing, construct design, delivery of the constructs and screening for the plant transformants are the three main steps to achieve the cell lines with the wanted mutated genes. As ZNFs and TALENs rely on DNA-protein interaction for gene targeting, the option of the sequence targeting designs are limited and making those constructions are time and labor consuming. On the other hand, Cas protein can be easily reconstructed with nuclear localization and gRNA sequence can be optimized to better target the sequence with web-based tools. In practice, the design and the realization of the construct is much easier for CRISPR-Cas than the other tools. Moreover, more Cas variants and systems have been developed and employed with advantages including higher specificity and the ability to target broad gene range. Therefore, the CRISPR-Cas system is the most popular gene editing system due to its accuracy, simplicity and efficiency (Alok et al. 2021). miRNA overexpression, MIM or STTM are also considered as transgene-based methods. Considering these miRNA engineering methods cannot acquire a genetic stable lineage and usually can be silenced after a few generations, genome editing by endonuclease is more preferable and of great interests and significance for crop breeding. Taking together of all merits of the CRISPR-Cas and its advantages comparing to the other available SSN methods, this review below will then mainly focus on the recent advancement of CRISPR-Cas system application in plant miRNA engineering. However, the ideas of the construct design and strategies could be a reference for other engineering tools for miRNA purpose.

15.5 The PAM of CRISPR/Cas and Strategy in Construct Design for miRNA Knock-Out

Unlike the ZNF and TALEN, CRISPR mediated gene editing can be highly sequence specific with the guidance by a spacer guide RNA (gRNA). The classic CRISPR/Cas used for gene editing is the SpCRISPR/Cas9 which is isolated from *Streptococcus pyogenes*. SpCRISPR/Cas9 requires the NGG protospacer adjacent motif (PAM) sequence at the DNA target site. In the case of gene editing, the gRNA can be engineered with the sequence of the target gene containing the PAM sequence. The endonuclease of CRISPR-associated protein (Cas9)'s subsequently introduces the DNA double strand breaks at the gene of interests. In animal and plant, such breaks then can be repaired by the non-homologous end joining (NHEJ), which is an error prone DNA repair system or the homologous recombination (HR), which is a high fidelity DNA repair pathway but could ends up with a large fragments deletion during genome editing (Kapusi et al. 2017; Cai et al. 2018). Depending on the choice of the DNA repair modes, the common DNA repair mistakes are the indels, while some other studies showed large fragments deletion and replacement can be achieved in cases (Xue and Greene 2021). The mistakes generated after DNA DSB repair end up with the disruption of the gene function.

Typically, the miRNA:miRNA* duplex region is the target for CRISPR/Cas design, which can abolish the miRNA biogenesis and function. Since the duplex is short, not all MIR genes can be edited by the SpCRISPR/Cas9. To overcome the target limitation, multiple Cas9 variants were developed with a changed PAM pattern, which provide more flexibility in sequence recognition. For example, the PAM of the *Staphylococcus aureus* Cas9 (SaCas9) is NNGRRT, while the *Neisseria meningitidis* (NmCas9) and *Campylobacter jejuni* (CjCas9) recognize NNNNGATT and NNNNACAC respectively (Hou et al. 2013; Friedland et al. 2015; Kim et al. 2017). Another potential solution to solve the target range limitation is by protein engineering on SpCas9 or other Cas variants. For example, SpCas9 PAM engineered variants showed robust gene editing potentials with altered PAM recognition (Kleinstiver et al. 2015).

Besides the Cas9 for gene editing, Cas12a is another extensively developed RNA-guided CRISPR/Cas system that can broaden the genome targeting range. Acidaminococcus sp. (AsCas12a) and Lachnospiraceae bacterium (LbCas12a) recognized T-rich PAM (Zetsche et al. 2015), that are recently used in plant gene editing (Jia et al. 2019). Like SpCas9, effort in protein engineering on LbCas12a (Tóth et al. 2020) and Moraxella bovoculi (Mb2Cas12a) (Zetsche et al. 2020) further expand the PAM selections. The third effective CRISPR/Cas used in plant genome editing is the Cas12b. Cas12b from *Bacillus hisashii* (BhCas12b) require the protein engineering to achieve higher cleavage activity (Strecker et al. 2019). Comparing to the other two Cas systems, the engineered BhCas12b from showed higher sequence specificity and among other Cas12b variants can be successfully applied in genome editing in plant (Ming et al. 2020).

When the genomic sequence of miRNA:miRNA* duplex are not inscribed with PAMs matching to any identified Cas or variants, the choice of the target sites should be considered for the one with the PAM on the pri-miRNA with the closest distance to the duplex region, which can potentially achieve the same effect in disrupting the miRNA biogenesis (Yang and McJunkin 2020). On the other side, broaden the MIR gene targeting sites could be another strategy. For example, many MIR genes contain introns, where the splicing sites can be considered for targeting to disrupt the generation of the pri-miRNA. If the MIR gene is not embedded in a mRNA gene exon or intron and has independent genomic features for transcription, the MIR gene's promoter region can be targeted to solely disrupt the MIR gene's nascent transcription. Moreover, because the pri-miRNA is usually a lncRNA, the effort to disrupt sites important for the pri-miRNA to form the correct secondary structure plus the sites crucial for the recognition and cleavage by Dicer protein could be viable. But these proposed methods require a more sophisticated and systematic experimental design on each miRNA via *in vitro* RNA biochemistry and RNA secondary structure prediction tools.

For the PAM-less sequence targeting, recent studies and efforts identified that SpRY and SpG Cas9 recognize NRN/NYN PAM sequence (Walton et al. 2020; Liang et al. 2022). This PAM pattern by SpRY is basically considered as a near PAM-less Cas system. Therefore, the development of SpRY tool box breaks the PAM restriction for gene editing. A couple examples of SpRY application in plant genome editing were reported, suggesting its potential in plant miRNA genes editing (Ren et al. 2021; Xu et al. 2021).

15.6 Evolving CRISPR/Cas Tools, Strategies, and Their Potential Uses in MIR Regulation

As mentioned above, the functional CRISPR/Cas is a RNP that requires the coordination between the guide RNA and the endonuclease protein for gene editing. Therefore, the flexibility of the application by using the CRISPR/Cas can be realized from two directions: RNA and protein.

From the protein side, "dead" Cas (dCas) protein with mutated endonuclease active site can be implemented to recognized the target sequence without cleaving the DNA. The dCas can then be a platform to re-engineer the CRISPR-Cas system for different purposes by fusing with other effector proteins. In plant, these developed effectors so far include modules for gene activation, repression, epigenetic reprogramming, and precise base editing (Gallego-Bartolomé et al. 2018; Zhang et al. 2019; Pan et al. 2021). The toolkit of dCas applied in plant genome editing by so far include dCas9, dCas12a, and engineered variants dzCas9, dzCas9-NG, dSpRY, and dAaCas12b (Ming et al. 2020; Moradpour and Abdulah 2020). Because MIR genes are mainly transcribed by RNA polymerase II, indicating the regulation of their transcription can be via a similar manner to the protein-coding genes. Therefore, the transcriptional activity of MIR genes can be regulated by using the developed CRISPR activation (CRISPRa) and CRISPR interference (CRISPRi) systems or their derivatives. Up- or downregulation of the MIR gene expression have been realized in plant to change the AtmiR319 and AtmiR159A/B expression level (Lowder et al. 2018). For the other dCas derivatives, no applications to up- or downregulate MIR genes are reported in plant yet. On the other side, fusing the base editing motif to the dCas9 allows the precise base editing in genes while an epigenomic modulator such as methyltransferase fused to dCas9 enable the epigenome/chromatin status modulation. These tools were successfully applied to the coding gene targeting in multiple plant species including the plant model Arabidospsis as well as the crops of tomato, rice, maize, and *Brassica napus* (Shimatani et al. 2017; Zong et al. 2017; Kang et al. 2018; Ghoshal et al. 2020). However, these CRISPR-Cas tools remain to be applied in the miRNA genes. With growing numbers of successes in plant manipulation, it will be exciting to apply all these tools in fine-tuning MIR genes expression in future studies.

From the side on the guide RNAs, multiplex targeting can be realized by including more than one guide RNA in the construct. The strategy of co-expressing multiple targeting guide RNA by CRISPR was well summarized and reviewed previously (McCarty et al. 2020). The multiplex CRISPR/Cas targeting has at least two obvious significances in the context of the miRNA knock-out purpose. First, multiple gRNAs can be designed to target onto one MIR gene by targeting to the different places of the stem-loop pri-miRNA, which could potentially end up with a large fragment deletion or results in a complete disruption of the miRNA biogenesis. This can be especially useful for the MIR with PAM-less sequence in the duplex region. Next, as mentioned above, several miRNAs function as a super family that harboring sequence with high similarity or regulating onto the same target. Therefore, by carefully design guide RNA to different members in a family, multiplex mutagenesis allows the dissection of the specific roles of each miRNA and help understanding the biological significance of the whole family to the host. Successful examples include the identification of rice OsmiR2118 MIRs' functions in anther wall development by a multiplex gRNA based 1.6 kb fragment deletion in the 14 OsmiR2118 gene cluster (Araki et al. 2020). Another example is the resistance to the *Phytophthora infestans* caused late blight disease in tomato by simultaneously knocking-out both miR482b and miR482c (Hong et al. 2021). Besides, the function of five MIR172 genes with either specific roles or redundant roles in plant development was deciphered by independent knockout each MIR172 genes using a strategy of two gRNA-based MIR deletion (Lian et al. 2021). A similar independent knockout strategy was applied in tomato to understand the distinct roles of each MIR164 in tomato development where the STTM-based method suggested pluripotent function of MIR164 family in different aspects and failed to dissect each MIR164's roles (Gupta et al. 2021).

Beyond fusing an effector to the Cas protein, the effector protein can be recruited by the gRNA when an RNA aptamer sequence was properly linked to the gRNA sequences. By using this idea, the CRISPR-Act 2.0 (Lowder et al. 2018; Malzahn et al. 2019) or 3.0 (Pan et al. 2021) was developed and showed a four to sixfold activation efficiency than the conventional CRISPRa system. Recent developed CRISPR-Combo system can realize gene editing (base editing) and gene activation simultaneously with shortened plant life cycle and reduced efforts in plant screening (Pan et al. 2022). With the robust evolution of the new CRISPR tools and its flexible usage with the long noncoding RNA and protein engineer, more tools can be applied on the miRNA to achieve the trait of interests that are of both biological and agricultural significance.

15.7 Conclusion and Future Perspectives

Comparing to the mRNA, manipulating noncoding RNA is more challenging and requires more sophisticated considerations. Usually, the reasonable design to disrupt the noncoding RNAs requires the prerequisite knowledges in their biogenesis and functions. Among noncoding RNAs, the miRNA due to its short length calls for the engineering tools with a much higher resolution and sensitivity to pinpoint to a very small region for the MIR gene engineering. Given the significant roles of miRNAs involved in almost all aspects of the plant life events and in the adaptation of plant to the environmental changes, it is indispensable to develop efficient tools in manipulating miRNAs in the evolutionary conserved lineages to enhance our understanding of these miRNAs' function and their evolutionary history. Versatile miRNA engineers tools were developed in the past decades at both transcriptional and posttranscriptional levels. Because miRNA manipulation could potentially improve the crop's productivity, quality, and tolerance to abiotic or biotic stresses, the genome editing is preferable for the agriculturally important plants since the wanted traits can be stably inherited down to the next generations. Comparing to the predecessors ZFN and TALEN, the toolkits of CRISPR/Cas and their successes in plant genome editing proves it as a promising and suitable one in plant miRNA manipulation due to its resolution, efficiency, and feasibility. Many derivative system of the CRISPR-Cas based on the engineered dCas-effector or dCas-RNA-aptamer-effector allows fine-tuning the expression of the genes including the MIRs. Due to its great flexibility and feasibility, without any doubt, the CRISPR-based genome engineer tools will be dominant in the near future for the gene manipulation and gene engineering. It is also expected that the experiment design and tips on the miRNA manipulation by CRISPR could be extend to the use for the other small noncoding RNAs such as siRNA or tsRNAs.

The construct design strategy, delivery methods, the CRISPR-system optimization, and the new CRISPR tools development are and will remain the momentum with the extensive interests in the near future studies. However, the idea of fusing an effector protein to dCas can be borrowed to the ZNF and TALENs. Recently, such a success included a fusion of a DRM domain for DNA methylation regulation to the zinc finger protein could improve the bacteria blight disease resistance in Cassava by changing the gene expression at the epigenome level (Veley et al. 2023). On the other side, the FokI nuclease used in ZFN and TALEN can be fused to a CRISPR-dCas system and enable the gRNA directed gene targeting for the dimerized FokI to carry out the DNA cleavage (Tsai et al. 2014). The FokI-dCas reduces the off-target effects by 10,000-fold (Guilinger et al. 2014; Tsai et al. 2014), while the application in *Arabidopsis* has a large trade-off in the efficiency (Paul and Qi 2016). Because the accuracy and efficiency of the genetic engineer tools is of great significance to miRNA engineering, future studies of developing a better genetic engineer tools by exploiting the full spectrum of the ZNF and TALEN in combination with the CRISPR-Cas or the CRISPR-dCas derived strategy could be an interesting direction.

New types of the Cas protein, such as Cas13a RNA dependent RNA targeting tools could be potentially applied to the genes editing in plant (Paul and Qi 2016). Similar to the MIM and STTM action mode, CRISPR/Cas system that recognize RNA sequences for RNA editing is promising due to it does not change the genomic sequence. Another promising direction is the genome editing by *in vitro* assembled CRISPR RNP. The RNP editing has been proved efficient in plant (Jiang et al. 2021). This should be viable in either RNA or DNA CRISPR-Cas system and could be very useful especially when miRNA sequence is AT-rich. Basically, in the sequence used for base-pairing to the miRNA target, modification such as LNA nucleotide can be incorporated in the synthetic guide RNA at the base-paring nucleotide to increase the Tm. Or incorporating a 2-M Methyl group at the 3′ end of the guide RNA can be proposed to increase the gRNA stability. Thus, proper selections and designs of the modified positions onto the gRNA in combination with a pool of the nucleotide modifications could potentially increase the target specificity and efficiency plus the flexibility in genome editing by CRISPR toolkit.

In conclusion, we could expect to have more successes in MIR gene regulations in future by continuing the previous experience in MIR manipulation and transplanting the recent achievements on how those developed genetic engineering tools were applied to target to the protein coding genes. Expanding the genome engineer for MIR genes purpose will accelerate the plant biology research and address the urgent needs in agriculture. Further enhancing these advancements in crop breeding could help cope with the environmental change and will ultimately promote the food security especially in facing to the explosion of the population growth in this planet.

References

Addo-Quaye, C., Snyder, J.A., Park, Y.B. et al. (2009). Sliced microRNA targets and precise loop-first processing of MIR319 hairpins revealed by analysis of the physcomitrella patens degradome. *RNA* 15: 2112–2121.

Alok, A., Chauhan, H., Upadhyay, S.K. et al. (2021). Compendium of plant-specific CRISPR vectors and their technical advantages. *Life* 11: 1021. https://doi.org/10.3390/life11101021.

Alves, C.S. and Nogueira, F.T.S. (2021). Plant small RNA world growing bigger: tRNA-derived fragments, longstanding players in regulatory processes. *Frontiers in Molecular Biosciences* 8: 638911.

Anand, A. and Pandi, G. (2021). Noncoding RNA: an insight into chloroplast and mitochondrial gene expressions. *Life (Basel)* 11: 49.

Araki, S., Le, N.T., Koizumi, K. et al. (2020). miR2118-dependent U-rich phasiRNA production in rice anther wall development. *Nature Communications* 11: 3115.

Arribas-Hernández, L., Kielpinski, L.J., and Brodersen, P. (2016). mRNA decay of most *Arabidopsis* miRNA targets requires slicer activity of AGO1. *Plant Physiology* 171: 2620–2632.

Arteaga-Vázquez, M., Caballero-Pérez, J., and Vielle-Calzada, J.P. (2006). A family of microRNAs present in plants and animals. *Plant Cell* 18: 3355–3369.

Axtell, M.J. (2013). Classification and comparison of small RNAs from plants. *Annual Review of Plant Biology* 64: 137–159.

Axtell, M.J. and Meyers, B.C. (2018). Revisiting criteria for plant MicroRNA annotation in the era of big data. *Plant Cell* 30: 272–284.

Axtell, M.J., Jan, C., Rajagopalan, R., and Bartel, D.P. (2006). A two-hit trigger for siRNA biogenesis in plants. *Cell* 127: 565–577.

Baldrich, P., Beric, A., and Meyers, B.C. (2018). Despacito: the slow evolutionary changes in plant microRNAs. *Current Opinion in Plant Biology* 42: 16–22.

Betti, F., Ladera-Carmona, M.J., Weits, D.A. et al. (2021). Exogenous miRNAs induce post-transcriptional gene silencing in plants. *Nature Plants* 7: 1379–1388.

Bi, H., Fei, Q., Li, R. et al. (2020). Disruption of miRNA sequences by TALENs and CRISPR/Cas9 induces varied lengths of miRNA production. *Plant Biotechnology Journal* 18: 1526–1536.

Blevins, T., Podicheti, R., Mishra, V. et al. (2015). Identification of Pol IV and RDR2-dependent precursors of 24 nt siRNAs guiding de novo DNA methylation in *Arabidopsis*. *eLife* 4: e09591.

Bologna, N.G., Mateos, J.L., Bresso, E.G., and Palatnik, J.F. (2009). A loop-to-base processing mechanism underlies the biogenesis of plant microRNAs miR319 and miR159. *The EMBO Journal* 28: 3646–3656.

Bologna, N.G., Schapire, A.L., Zhai, J. et al. (2013). Multiple RNA recognition patterns during microRNA biogenesis in plants. *Genome Research* 23: 1675–1689.

Bologna, N.G., Iselin, R., Abriata, L.A. et al. (2018). Nucleo-cytosolic shuttling of ARGONAUTE1 prompts a revised model of the plant MicroRNA pathway. *Molecular Cell* 69: 709–719. e705.

Cai, X., Hagedorn, C.H., and Cullen, B.R. (2004). Human microRNAs are processed from capped, polyadenylated transcripts that can also function as mRNAs. *RNA* 10: 1957–1966.

Cai, Y., Chen, L., Sun, S. et al. (2018). CRISPR/Cas9-mediated deletion of large genomic fragments in soybean. *International Journal of Molecular Sciences* 19: 3835.

Carbonell, A., Fahlgren, N., Garcia-Ruiz, H. et al. (2012). Functional analysis of three *Arabidopsis* ARGONAUTES using slicer-defective mutants. *Plant Cell* 24: 3613–3629.

Chen, Y. and Wang, X. (2020). miRDB: an online database for prediction of functional microRNA targets. *Nucleic Acids Research* 48: D127–d131.

Dong, Z., Han, M.H., and Fedoroff, N. (2008). The RNA-binding proteins HYL1 and SE promote accurate in vitro processing of pri-miRNA by DCL1. *Proceedings of the National Academy of Sciences of the United States of America* 105: 9970–9975.

Fang, Y. and Spector, D.L. (2007). Identification of nuclear dicing bodies containing proteins for microRNA biogenesis in living *Arabidopsis* plants. *Current Biology* 17: 818–823.

Ferdous, J., Hussain, S.S., and Shi, B.J. (2015). Role of microRNAs in plant drought tolerance. *Plant Biotechnology Journal* 13: 293–305.

Friedland, A.E., Baral, R., Singhal, P. et al. (2015). Characterization of *Staphylococcus aureus* Cas9: a smaller Cas9 for all-in-one adeno-associated virus delivery and paired nickase applications. *Genome Biology* 16: 257.

Gallego-Bartolomé, J., Gardiner, J., Liu, W. et al. (2018). Targeted DNA demethylation of the *Arabidopsis* genome using the human TET1 catalytic domain. *Proceedings of the National Academy of Sciences of the United States of America* 115: E2125–e2134.

Ghoshal, B., Vong, B., Picard, C.L. et al. (2020). A viral guide RNA delivery system for CRISPR-based transcriptional activation and heritable targeted DNA demethylation in *Arabidopsis thaliana*. *PLoS Genetics* 16: e1008983.

Guan, Q., Lu, X., Zeng, H. et al. (2013). Heat stress induction of miR398 triggers a regulatory loop that is critical for thermotolerance in *Arabidopsis*. *The Plant Journal* 74: 840–851.

Guilinger, J.P., Thompson, D.B., and Liu, D.R. (2014). Fusion of catalytically inactive Cas9 to FokI nuclease improves the specificity of genome modification. *Nature Biotechnology* 32: 577–582.

Guo, Z., Kuang, Z., Wang, Y. et al. (2020). PmiREN: a comprehensive encyclopedia of plant miRNAs. *Nucleic Acids Research* 48: D1114–d1121.

Gupta, S.K., Vishwakarma, A., Kenea, H.D. et al. (2021). CRISPR/Cas9 mutants of tomato MICRORNA164 genes uncover their functional specialization in development. *Plant Physiology* 187: 1636–1652.

Hong, Y., Meng, J., He, X. et al. (2021). Editing miR482b and miR482c simultaneously by CRISPR/Cas9 enhanced tomato resistance to *Phytophthora infestans*. *Phytopathology* 111: 1008–1016.

Hou, Z., Zhang, Y., Propson, N.E. et al. (2013). Efficient genome engineering in human pluripotent stem cells using Cas9 from *Neisseria meningitidis*. *Proceedings of the National Academy of Sciences of the United States of America* 110: 15644–15649.

Iwakawa, H.O. and Tomari, Y. (2013). Molecular insights into microRNA-mediated translational repression in plants. *Molecular Cell* 52: 591–601.

Jasinski, S., Vialette-Guiraud, A.C., and Scutt, C.P. (2010). The evolutionary-developmental analysis of plant microRNAs. *Philosophical Transactions of the Royal Society of London. Series B, Biological Sciences* 365: 469–476.

Jia, H., Orbović, V., and Wang, N. (2019). CRISPR-LbCas12a-mediated modification of citrus. *Plant Biotechnology Journal* 17: 1928–1937.

Jiang, W., Bush, J., and Sheen, J. (2021). A versatile and efficient plant protoplast platform for genome editing by Cas9 RNPs. *Frontiers in Genome Editing* 3: 719190.

Johnson, C., Kasprzewska, A., Tennessen, K. et al. (2009). Clusters and superclusters of phased small RNAs in the developing inflorescence of rice. *Genome Research* 19: 1429–1440.

Kang, B.C., Yun, J.Y., Kim, S.T. et al. (2018). Precision genome engineering through adenine base editing in plants. *Nature Plants* 4: 427–431.

Kapusi, E., Corcuera-Gómez, M., Melnik, S., and Stoger, E. (2017). Heritable genomic fragment deletions and small indels in the putative ENGase gene induced by CRISPR/Cas9 in barley. *Frontiers in Plant Science* 8: 540.

Kim, E., Koo, T., Park, S.W. et al. (2017). In vivo genome editing with a small Cas9 orthologue derived from *Campylobacter jejuni*. *Nature Communications* 8: 14500.

Kleinstiver, B.P., Prew, M.S., Tsai, S.Q. et al. (2015). Engineered CRISPR-Cas9 nucleases with altered PAM specificities. *Nature* 523: 481–485.

Kosinska-Selbi, B., Mielczarek, M., and Szyda, J. (2020). Review: long non-coding RNA in livestock. *Animal* 14: 2003–2013.

Lian, H., Wang, L., Ma, N. et al. (2021). Redundant and specific roles of individual MIR172 genes in plant development. *PLoS Biology* 19: e3001044.

Liang, F., Zhang, Y., Li, L. et al. (2022). SpG and SpRY variants expand the CRISPR toolbox for genome editing in zebrafish. *Nature Communications* 13: 3421.

Liu, Q., Feng, Y., and Zhu, Z. (2009). Dicer-like (DCL) proteins in plants. *Functional & Integrative Genomics* 9: 277–286.

Lowder, L.G., Zhou, J., Zhang, Y. et al. (2018). Robust transcriptional activation in plants using multiplexed CRISPR-Act2.0 and mTALE-Act systems. *Molecular Plant* 11: 245–256.

Ma, W., Chen, C., Liu, Y. et al. (2018). Coupling of microRNA-directed phased small interfering RNA generation from long noncoding genes with alternative splicing and alternative polyadenylation in small RNA-mediated gene silencing. *The New Phytologist* 217: 1535–1550.

Malzahn, A., Zhang, Y., and Qi, Y. (2019). CRISPR-Act2.0: an improved multiplexed system for plant transcriptional activation. *Methods in Molecular Biology* 1917: 83–93.

McCarty, N.S., Graham, A.E., Studená, L., and Ledesma-Amaro, R. (2020). Multiplexed CRISPR technologies for gene editing and transcriptional regulation. *Nature Communications* 11: 1281.

Meins, F. Jr., Si-Ammour, A., and Blevins, T. (2005). RNA silencing systems and their relevance to plant development. *Annual Review of Cell and Developmental Biology* 21: 297–318.

Ming, M., Ren, Q., Pan, C. et al. (2020). CRISPR-Cas12b enables efficient plant genome engineering. *Nature Plants* 6: 202–208.

Moradpour, M. and Abdulah, S.N.A. (2020). CRISPR/dCas9 platforms in plants: strategies and applications beyond genome editing. *Plant Biotechnology Journal* 18: 32–44.

Pan, C., Wu, X., Markel, K. et al. (2021). CRISPR-Act3.0 for highly efficient multiplexed gene activation in plants. *Nature Plants* 7: 942–953.

Pan, C., Li, G., Malzahn, A.A. et al. (2022). Boosting plant genome editing with a versatile CRISPR-combo system. *Nature Plants* 8: 513–525.

Paul, J.W. 3rd and Qi, Y. (2016). CRISPR/Cas9 for plant genome editing: accomplishments, problems and prospects. *Plant Cell Reports* 35: 1417–1427.

Perkel, J.M. (2013). Visiting "noncodarnia". *Biotechniques* 54 (301): 303–304.

Pratt, A.J. and MacRae, I.J. (2009). The RNA-induced silencing complex: a versatile gene-silencing machine. *The Journal of Biological Chemistry* 284: 17897–17901.

Rajagopalan, R., Vaucheret, H., Trejo, J., and Bartel, D.P. (2006). A diverse and evolutionarily fluid set of microRNAs in *Arabidopsis thaliana*. *Genes & Development* 20: 3407–3425.

Ren, Q., Sretenovic, S., Liu, S. et al. (2021). PAM-less plant genome editing using a CRISPR-SpRY toolbox. *Nature Plants* 7: 25–33.

Rogers, K. and Chen, X. (2013). Biogenesis, turnover, and mode of action of plant microRNAs. *Plant Cell* 25: 2383–2399.

Schwab, R., Palatnik, J.F., Riester, M. et al. (2005). Specific effects of microRNAs on the plant transcriptome. *Developmental Cell* 8: 517–527.

Shimatani, Z., Kashojiya, S., Takayama, M. et al. (2017). Targeted base editing in rice and tomato using a CRISPR-Cas9 cytidine deaminase fusion. *Nature Biotechnology* 35: 441–443.

Sieber, P., Wellmer, F., Gheyselinck, J. et al. (2007). Redundancy and specialization among plant microRNAs: role of the MIR164 family in developmental robustness. *Development* 134: 1051–1060.

Sienski, G., Dönertas, D., and Brennecke, J. (2012). Transcriptional silencing of transposons by Piwi and maelstrom and its impact on chromatin state and gene expression. *Cell* 151: 964–980.

Singh, J., Mishra, V., Wang, F. et al. (2019). Reaction Mechanisms of Pol IV, RDR2, and DCL3 Drive RNA Channeling in the siRNA-Directed DNA Methylation Pathway. *Molecular Cell* 75: 576–589.e575.

Statello, L., Guo, C.J., Chen, L.L., and Huarte, M. (2021). Gene regulation by long non-coding RNAs and its biological functions. *Nature Reviews. Molecular Cell Biology* 22: 96–118.

Strecker, J., Jones, S., Koopal, B. et al. (2019). Engineering of CRISPR-Cas12b for human genome editing. *Nature Communications* 10: 212.

Sushmita, Kaur, G., Upadhyay, S.K., and Verma, P.C. (2021). An overview of genome-engineering methods. *Wiley*. https://doi.org/10.1002/9781119672425.ch1.

Taneja, M., Shumayla, and Upadhyay, S.K. (2021). An overview of long non-coding RNA in plants. *Elsevier*. https://doi.org/10.1016/B978-0-12-821452-7.00001-5.

Tang, G., Yan, J., Gu, Y. et al. (2012). Construction of short tandem target mimic (STTM) to block the functions of plant and animal microRNAs. *Methods* 58: 118–125.

Todesco, M., Rubio-Somoza, I., Paz-Ares, J., and Weigel, D. (2010). A collection of target mimics for comprehensive analysis of microRNA function in *Arabidopsis thaliana*. *PLoS Genetics* 6: e1001031.

Tóth, E., Varga, É., Kulcsár, P.I. et al. (2020). Improved LbCas12a variants with altered PAM specificities further broaden the genome targeting range of Cas12a nucleases. *Nucleic Acids Research* 48: 3722–3733.

Tsai, M.C., Manor, O., Wan, Y. et al. (2010). Long noncoding RNA as modular scaffold of histone modification complexes. *Science* 329: 689–693.

Tsai, S.Q., Wyvekens, N., Khayter, C. et al. (2014). Dimeric CRISPR RNA-guided foki nucleases for highly specific genome editing. *Nature Biotechnology* 32: 569–576.

Upadhyay, S.K. (2021a). Long non-coding RNAs in plants: roles in development and stress. *Elsevier*. https://doi.org/10.1016/C2019-0-03200-1.

Upadhyay, S.K. (2021b). Genome engineering for crop improvement. *Wiley*. https://doi.org/10.1002/9781119672425.

Vaucheret, H., Vazquez, F., Crété, P., and Bartel, D.P. (2004). The action of ARGONAUTE1 in the miRNA pathway and its regulation by the miRNA pathway are crucial for plant development. *Genes & Development* 18: 1187–1197.

Veley, K.M., Elliott, K., Jensen, G. et al. (2023). Improving cassava bacterial blight resistance by editing the epigenome. *Nature Communications* 14: 85.

Walton, R.T., Christie, K.A., Whittaker, M.N., and Kleinstiver, B.P. (2020). Unconstrained genome targeting with near-PAMless engineered CRISPR-Cas9 variants. *Science* 368: 290–296.

Wang, K.C. and Chang, H.Y. (2011). Molecular mechanisms of long noncoding RNAs. *Molecular Cell* 43: 904–914.

Wang, Z., Hardcastle, T.J., Canto Pastor, A. et al. (2018). A novel DCL2-dependent miRNA pathway in tomato affects susceptibility to RNA viruses. *Genes & Development* 32: 1155–1160.

Xia, R., Ye, S., Liu, Z. et al. (2015). Novel and recently evolved MicroRNA clusters regulate expansive F-BOX gene networks through phased small interfering RNAs in wild diploid strawberry. *Plant Physiology* 169: 594–610.

Xie, Z., Allen, E., Fahlgren, N. et al. (2005). Expression of *Arabidopsis* MIRNA genes. *Plant Physiology* 138: 2145–2154.

Xu, Z., Kuang, Y., Ren, B. et al. (2021). SpRY greatly expands the genome editing scope in rice with highly flexible PAM recognition. *Genome Biology* 22: 6.

Xue, C. and Greene, E.C. (2021). DNA repair pathway choices in CRISPR-Cas9-mediated genome editing. *Trends in Genetics: TIG* 37: 639–656.

Yang, B. and McJunkin, K. (2020). CRISPR screening strategies for microRNA target identification. *The FEBS Journal* 287: 2914–2922.

Yang, Z., Ebright, Y.W., Yu, B., and Chen, X. (2006). HEN1 recognizes 21–24 nt small RNA duplexes and deposits a methyl group onto the 2′ OH of the 3′ terminal nucleotide. *Nucleic Acids Research* 34: 667–675.

Yoshikawa, M., Peragine, A., Park, M.Y., and Poethig, R.S. (2005). A pathway for the biogenesis of trans-acting siRNAs in *Arabidopsis*. *Genes & Development* 19: 2164–2175.

Zetsche, B., Gootenberg, J.S., Abudayyeh, O.O. et al. (2015). Cpf1 is a single RNA-guided endonuclease of a class 2 CRISPR-Cas system. *Cell* 163: 759–771.

Zetsche, B., Abudayyeh, O.O., Gootenberg, J.S. et al. (2020). A survey of genome editing activity for 16 Cas12a orthologs. *The Keio Journal of Medicine* 69: 59–65.

Zhang, X., Yuan, Y.R., Pei, Y. et al. (2006). Cucumber mosaic virus-encoded 2b suppressor inhibits *Arabidopsis* argonaute1 cleavage activity to counter plant defense. *Genes & Development* 20: 3255–3268.

Zhang, Q., Kim, N.K., and Feigon, J. (2011a). Architecture of human telomerase RNA. *Proceedings of the National Academy of Sciences of the United States of America* 108: 20325–20332.

Zhang, W., Gao, S., Zhou, X. et al. (2011b). Bacteria-responsive microRNAs regulate plant innate immunity by modulating plant hormone networks. *Plant Molecular Biology* 75: 93–105.

Zhang, R., Liu, J., Chai, Z. et al. (2019). Generation of herbicide tolerance traits and a new selectable marker in wheat using base editing. *Nature Plants* 5: 480–485.

Zhu, H., Hu, F., Wang, R. et al. (2011). *Arabidopsis* argonaute10 specifically sequesters miR166/165 to regulate shoot apical meristem development. *Cell* 145: 242–256.

Zhu, H., Zhou, Y., Castillo-González, C. et al. (2013). Bidirectional processing of pri-miRNAs with branched terminal loops by *Arabidopsis* dicer-like1. *Nature Structural & Molecular Biology* 20: 1106–1115.

Zong, Y., Wang, Y., Li, C. et al. (2017). Precise base editing in rice, wheat and maize with a Cas9-cytidine deaminase fusion. *Nature Biotechnology* 35: 438–440.

Zou, Q., Mao, Y., Hu, L. et al. (2014). miRClassify: an advanced web server for miRNA family classification and annotation. *Computers in Biology and Medicine* 45: 157–160.

16

Application of Genome Editing in Pulses

Nikhil Malhotra

ICAR-National Bureau of Plant Genetic Resources Regional Station, Shimla, Himachal Pradesh, India

16.1 Introduction

Pulses, often referred to as grain legumes, are a key staple crop that are grown all over the world and provide a nutrient-rich diet for both animals and people. It is a food source that doesn't even require industrial or post-harvest processing. In terms of contributing to global food production, pulses are second only to cereals as one of the most significant crops in the agricultural system. In sustainable farming practices, they contribute significantly to enhancing soil health by engaging in symbiotic nitrogen fixation and releasing valuable organic matter into the soil. Notwithstanding the ecological importance and health benefits, low crop yields have an impact on pulse production. Adoption of novel approaches like genomics-assisted selection (also known as marker-assisted selection or genomic selection) and precision breeding (genome editing) will help with the challenging task of increasing yield potential and reliability. Huge strides have been made in the structural analysis and sequencing of pulse genomes over the past 10 years. Draft genome and transcriptome assemblages have been created for many crop species (Bauchet et al. 2019). By using these fundamental resources, molecular markers are being discovered and used to select complex traits and create improved cultivars of grain legume crops (Varshney et al. 2019). In this chapter, we emphasize on application of genome editing technologies as a selective breeding system for genetic enhancement of pulses.

Genome editing is the process of generating precise, targeted alterations to the genome of an organism using synthetic nucleases and cellular DNA repair processes. Over 30 years ago, it was discovered that a meganuclease called *I-SceI* could be used to introduce precise double-stranded breaks in chromosomes. This finding sparked the development of genome editing techniques (Rouet et al. 1994). However, the use of meganucleases for genome editing was limited by the short rate of target sites in the majority of genes. The development of programmable zinc finger nucleases (ZFNs;) and transcription activator-like effector nucleases (TALENs;) provided a significant boost to efforts to create effective genome editing tools (Bibikova et al. 2002; Li et al. 2011; Zhang et al. 2011; Sushmita et al. 2021).

Applications of Genome Engineering in Plants, First Edition. Edited by Santosh Kumar Upadhyay.
© 2024 John Wiley & Sons Ltd. Published 2024 by John Wiley & Sons Ltd.

Nevertheless, the key development in this ground-breaking method was the effective use of the RNA-guided Cas9 nuclease from the type II bacterial clustered regularly interspaced short palindromic repeats (CRISPR) adaptive immune system for genome editing in eukaryotic cells (Cong et al. 2013; Mali et al. 2013; Alok et al. 2018, 2021; Upadhyay 2021). As a consequence, the recent past has seen several innovations aimed at enhancing various agronomic traits and nutritional attributes of pulses.

16.2 Genome Editing for Crop Improvement in Pulses

With the new developments in knowledge, the genome editing methods have been substantially expanded. The possible uses of these systems encompass exploring gene expression, gene silencing and gene overexpression. Researchers have successfully employed ZFNs, TALENs and CRISPR techniques in the context of grain legume agriculture, presenting an opportunity to develop high-quality, environmentally friendly crops such as pulses. Notably, genome editing has been successfully implemented in several grain legume crops like chickpea, cowpea and soybean, with established standardized transformation protocols outlined in Table 16.1. Following, we go over the current status and prospects of employing various genome editing tools for improvement of numerous traits in pulses.

16.2.1 Chickpea (*Cicer arietinum*)

Gene editing methods can be utilized to solve production problems with the chickpea (*C. arietinum* L.), a crop that is significant commercially throughout the world. In a work by Badhan et al. (2021), the genes *4-coumarate ligase* (*4CL*) and *Reveille 7* (*RVE7*) that are linked to drought tolerance were targeted for CRISPR/Cas9 editing in chickpea protoplast. The knockout of the *RVE7* gene demonstrated high-efficiency editing in vivo. These findings showed that genes linked to drought tolerance in chickpea can be edited using CRISPR/Cas9 DNA-free gene editing by using protoplast. This is the only study to date that has employed CRISPR/Cas9 gene editing in chickpea.

Table 16.1 Examples of genome editing in pulses.

Crop species	Target genes	References
Chickpea	*4-coumarate ligase* (*4CL*) and *Reveille 7* (*RVE7*)	Badhan et al. (2021)
Cowpea	*Symbiosis receptor-like kinase* and *VuSPO11-1*	Ji et al. (2019), Juranić et al. (2020), Che et al. (2021)
Soybean	*Chalcone-flavanone isomerase, DDM1 orthologs, dicer-like3a, double-stranded RNA-binding2a/2b, fatty acid desaturase 2, glucosyltransferase, glutamine synthase, MET1 orthologs, phytoene desaturase* and *squamosa promoter-binding protein-like*	Jacobs et al. (2015), Li et al. (2015), Michno et al. (2015), Sun et al. (2015), Du et al. (2016), Curtin et al. (2018), Al Amin et al. (2019), Bao et al. (2019), Bao et al. (2020)

16.2.2 Cowpea (*Vigna unguiculata*)

Cowpea (*V. unguiculata* L. Walp.), a type of grain legume crop, is abundant in nutrients and have numerous health advantages. It can symbiotically fix nitrogen (*SNF*) effectively, can withstand low rainfall and needs less fertilizers. It became one of the most significant pulses in the world because of these agronomically significant characteristics (Ji et al. 2019; Che et al. 2021). These factors are driving greater interest in cowpea genome editing techniques. Ji et al. (2019) showed CRISPR/Cas9-mediated genome editing in non-inheritable hairy roots of cowpea by targeting *SNF* genes. They noticed that in the mutants with both alleles disrupted, nodule development was totally prevented. Then, Juranić et al. (2020) discovered three cowpea meiosis genes, viz. *OSD1* (encodes *Ophiostoma scytalone* dehydratase protein, which promotes meiotic progression), *REC8* (encodes meiotic recombination protein) and *SPO11-1* (encodes SPO11 protein, which is the initiator of meiotic double-strand breaks). They discovered that the *SPO11-1* gene had biallelic mutations in exons 1 and 3, which caused problems in meiosis and resulted in full male and female sterility in the T0 plants. Following them, Che et al. (2021) recently used CRISPR/Cas9 to knock down the cowpea meiosis gene *VuSPO11-1* and found alterations at the target.

16.2.3 Soybean (*Glycine max*)

Soybean (*G. max* L. Merr.) has emerged as a key candidate for genetic modification using the CRISPR/Cas9 technology. This is primarily due to its significance as a crop with high oil and protein contents, making it an essential agricultural commodity (Bao et al. 2020). The palaeotetraploid that became the soybean is a diploid species. Given that its genome is heavily replicated, knowing how genes work using traditional genetic methods will be extremely difficult. The limited success of Agrobacterium-mediated transformation in soybean, which can be influenced by tissue type or cultivar, poses a significant challenge. To address this issue, researchers have widely employed *Agrobacterium rhizogenes*-mediated hairy root transformation for rapid evaluation of gRNA efficiency before proceeding to whole plant transformation. This method is preferred due to its shorter timeframe for generating transgenic hairy roots (Li et al. 2019). Notably, in 2015, several scientific studies reported successful soybean genome editing using the CRISPR/Cas9 technique (Cai et al. 2015; Jacobs et al. 2015; Li et al. 2015; Sun et al. 2015; Bhalothia et al. 2020).

Following these first breakthroughs, the genome editing method has continued to be improved by the soybean scientific community. In a study conducted by Di et al. in (2019), the effectiveness of 11 different GmU6 promoters in driving gRNA expression was evaluated in hairy roots of soybean. The findings revealed that the GmU6-8 and GmU6-10 promoters exhibited higher activity, resulting in a notable enhancement of editing competence (20.3% and 20.6%, respectively). In contrast, the remaining nine GmU6 promoters displayed lower activity ranging from 2.8% to 17%. Several reports demonstrated that the simultaneous use of dual gRNAs to target adjacent loci on the same chromosome in soybean led to significant genomic deletions, ranging from ~1 kb to over 4.5 kb. This extensive deletion strategy ensures complete knockout of the target genes and also offers opportunities for exploring the roles of regulatory elements or non-coding genes, contributing to advancements in various research domains (Cai et al. 2018; Li et al. 2018; Do et al. 2019).

Of late, Zheng et al. (2020) examined egg cell-specific promoters and discovered that one of them might result in stable transgenic soybean lines with good genome editing effectiveness (26.8%). More importantly, attempts have been undertaken to improve/manipulate agronomic features like flowering period, kernel oil and storage proteins and plant architecture as well as understanding how genes function. Two homologous genes, *GmPPD1* and *GmPPD2*, which encode orthologs of the *Arabidopsis PEAPOD*, were edited by Kanazashi et al. (2018) using a single gRNA. Double mutants that were produced had severely aberrant leaf and pod characteristics. Cai et al. (2018) conducted a study in soybeans using CRISPR/Cas9-assisted mutagenesis, where they observed a delay in flowering in soybean *FLOWERING LOCUS T2a* (*GmFT2a*) homozygous mutants. They further investigated the effects by combining the *ft2a* mutant with the *GmFT5a* mutation, resulting in the creation of *ft2aft5a* double mutants. These double mutants exhibited a later blooming time and produced more pods and kernels compared to the wild-type plants under short-day conditions (Cai et al. 2020). In a separate study, Bao et al. (2019) utilized a CRISPR/Cas9-based multiplex genome editing approach to modify four *GmSPL9* genes. The resulting soybean mutants displayed variations in node number on the main stem and branch number. CRISPR/Cas9 technology has also been employed to improve various kernel-related traits in soybeans, such as seed oil profile (Do et al. 2019), reduction of undesirable beany flavour in soybean seed products (Wang et al. 2020), modification of isoflavone content and enhancement of resistance to soybean mosaic virus (Zhang et al. 2020). Furthermore, Li et al. (2019) focused on generating mutants of genes responsible for seed storage protein synthesis, which can be beneficial for breeding soybeans with desirable food traits. Notably, a comprehensive soybean CRISPR library encompassing over 100 target genes has been successfully established. Also, a library of mutant soybean lines was created utilizing upgraded techniques (Bai et al. 2020). These efforts to compile mutants of every soybean gene will prove crucial for soybean genetics and biotechnology.

16.2.4 Non-Edited Grain Legumes

16.2.4.1 Common Bean (*Phaseolus vulgaris*)

The most consumed grain legume in the world is the common bean (*P. vulgaris* L.), which contains a variety of nutrients, such as proteins, minerals and vitamins (Cichy et al. 2009). The complete genomic sequence of the common bean has been accessible since 2014. Researchers also discovered genes linked to increased foliage and kernel mass in common bean (Schmutz et al. 2014). Despite the paucity of effective studies on genetic transformation in common bean, numerous indirect and direct gene transfer efforts have been made (Nadeem et al. 2021). One of those experiments involved transferring the *Rep-TrAP-REn* and *BC1* genes to common bean using the gene gun technique in order to create transgenic lines which exhibit resistance to the golden mosaic geminivirus (BgMV-BR) (Arago et al. 1998). *A. rhizogenes* was used by Xue et al. (2017) to transfer the *PvPOX1* gene from a Fusarium wilt-resistant cultivar to a Fusarium wilt-susceptible cultivar. Besides, different techniques were used to enhance the regeneration proficiency of crop, similar to other pulses (Nadeem et al. 2021). In light of this, Barraza et al. (2015) reported that the alteration of *PvTRX1h* gene, the ortholog of a plant histone lysine methyltransferase involved in biosynthesis of plant hormones, could aid in resolving problems with obstinate regeneration in the crop as it controls somatic embryogenesis in callus derived from common bean. Moreover, scientists have worked to increase the resistance of common bean to various biotic (such as anthracnose, microbial blight, rust and white mold) and abiotic stresses. Moreover, common bean possesses higher qualitative and quantitative characteristics alongside nutritional features that are important breeding considerations (Assefa et al. 2019). Genetic research on common beans can utilize genome editing technologies to develop stress-resistant crops and achieve breeding objectives, even in the absence of extensive utilization of whole genome data, CRISPR/Cas, or other genome editing techniques in common bean studies.

16.2.4.2 Dry Pea (*Pisum sativum*)

Dry pea (*P. sativum* L.), a significant pulse crop in the globe, is a source of fibre, dietary proteins, minerals and nutrients (Bastianelli et al. 1998). Because of its symbiotic nitrogen-fixing capacity, it is an important crop for the advancements of systems that increase soil productivity (Mabrouk et al. 2018). Moreover, Mendel originally utilized the crop as a model plant to develop the principles of genetics (Ellis et al. 2011). Its genome is ~4.45 Gb in size. In 2019, a reference genome that sheds light on the evolution of the crop genome was released (Kreplak et al. 2019). Finding the genes for crucial traits and creating genomic tools for crop development are both made possible by the genomics approach. Although there has been substantial success until now, crop development and enhancing agricultural quality and productivity should carry on in order to fed the expanding world population.

Abiotic stress factors like cold, drought, heat and salinity, as well as bacteria, fungus, parasitic weeds and viruses, affect growth and yield of dry pea. The *α-amylase inhibitor* gene from common bean and *cry1Ac* gene from *Bacillus thuringiensis* were successfully transferred into the crop for insect tolerance using the Agrobacterium-mediated transformation approach (Negawo et al. 2013). Another attempt at transforming antifungal genes against Fusarium wilt into dry pea failed to provide sustained expression after three years of field testing (Kahlon et al. 2018). In a recent study, zeatin was added to the selection medium utilizing Agrobacterium-mediated transformation where prolonged infection times increased the efficiency and frequency of seed-to-seed regeneration (Aftabi et al. 2017). The utilization of genome editing techniques in crops is largely influenced by the presence of effective transformation and regeneration applications, as well as the accessibility of genome sequence information (Cardi et al. 2017). Until now, no experiments have been carried out using novel genome editing methods in this particular crop due to its restricted capacity for regeneration (Pandey et al. 2021). The creation of novel genome editing techniques may open up new possibilities in dry pea breeding to increase production and develop new varieties with high nutritional value.

16.2.4.3 Faba Bean (*Vicia faba*)

Since faba bean (*V. faba* L.) grow better in colder climates than its counterparts, it can be used in sustainable agricultural practises (Temesgen et al. 2015). Also, faba bean has beneficial systems that increase soil fertility, just like other grain legumes. Breeding faba bean, a significant source of protein, fibre and other nutrients, keeps up the demand for future food and feed (Khazaei et al. 2021). However, the scarcity of publicly available genome sequencing data for faba beans may be attributed to the challenges associated with assembling their vast genome (Cooper et al. 2017). Transformation and regeneration efficiency are typically difficult for faba beans, despite the availability of results from Agrobacterium-mediated transformation with low transformation competence (Cardi et al. 2017; Maalouf et al. 2019). To establish stable transgenic lines, Böttinger et al. (2001) employed in vitro growth of internode stem sections harboring Agrobacterium, while Hanafy et al. (2005) removed Agrobacterium-penetrated embryo axes. Additionally, Hanafy et al. (2013) utilized the same transformation technique to transfer the *PR10a* gene from potato to faba beans, resulting in the development of abiotic stress-tolerant lines. Various biotic and abiotic stress factors, such as heat, pests, viruses and weeds, contribute to reduced faba bean yields (Cernay et al. 2015; Maalouf et al. 2019). To confer disease resistance in plants, TALEN technology, another genome editing approach, has been employed. Römer et al. (2009) successfully altered TAL effector binding sites in sugar transporter gene promoters to enhance resistance, while Li et al. (2012) and Xiao (2012) demonstrated the use of TALENs to enhance resistance against pathogens in rice.

Furthermore, by destroying the virus-related genome or altering the crop genome for resistance, CRISPR/Cas9 is a useful alternative strategy for gaining plant virus resistance (Lenka et al. 2020). Moreover, vicine and convicine, which have harmful effects on humans due to a change in the enzyme glucose-6-phosphate dehydrogenase, were identified as pyrimidine glycosides in the cotyledons of the faba bean (Luzzatto and Arese 2018; Khazaei et al. 2021). With the aid of novel faba bean genome editing technologies, those chemical levels can be managed.

16.2.4.4 Mung Bean (*Vigna radiata*)

Warm climate-adapted mung bean (*V. radiata* L. Wilczek) is a grain legume in the Fabaceae family that grows quickly. Compared to other species of pulses, mung bean provides more essential nutrients in the kernels (Keatinge et al. 2011). Moreover, mung bean plants can retain aerial nitrogen, thereby improving soil productiveness (Maalouf et al. 2019). The genomic investigation and molecular breeding of the crop have been made possible by the entire genome sequence of *V. radiata* var. *sublobata*, which was released earlier (Kang et al. 2014). The lack of mung bean germplasm and incompatibility with crop wild relatives, however, have an impact on mung bean breeding (Aasim et al. 2019).

Furthermore, susceptibility to various diseases and environmental factors lower the yield of mung bean. Thus, advancements in transformation technologies aid scientists in creating lines that can withstand biotic and abiotic stresses. Although the genetic transformation efficiency of mung bean was low, numerous investigations using Agrobacterium-mediated transformation have been done for various agronomic traits, including biotic (Saini et al. 2007) and abiotic stress tolerance (Yadav et al. 2012; Sahoo et al. 2016). According to Tripathi et al. (2021), the choice of the explants, transformation vectors and selective agents are the key elements in those findings that significantly influence the accomplishment of the transformation competence. Although there haven't been any research on the use of innovative genome editing techniques in mung bean, the accessibility of whole genome sequences and advancements in ex vivo revival and transformation technologies may make it possible to use more advanced genome editing techniques to produce crops with desired characteristics.

16.2.4.5 Lentil (*Lens culinaris*)

Lentil (*L. culinaris* Medik.) is a diploid plant belonging to the Fabaceae family, known for its ability to self-pollinate having high content of various micro- and macronutrients. It can help prevent malnutrition and micronutrient deficiencies in impoverished nations and in those whose diets don't rely much on pricey animal protein (Kumar et al. 2015). In addition, lentils enrich the soil with nitrogen, which in turn increases its quality (Kumar et al. 2021). An extensive draft whole genome sequence of lentil, which can be accessed at https://knowpulse.usask.ca/lentil-genome, has been developed through bulk sequencing, gene annotation and prediction of the assembled 2.6 Gb genome (Bett et al. 2014; Bett et al. 2016). These available draft genome sequences offer valuable assistance in identifying specific candidate genes responsible for various traits such as abiotic and biotic stress responses, herbicide resistance and nutrient value, among others. Lentil breeding can be made better with the aid of genomic tools and technologies. Some genetic transformation studies are available for this important grain legume. The introduction of the *dehydration-responsive element binding gene* (*DREB1A*), known for its role in plant reactions to abiotic stresses, into lentil plants using Agrobacterium-mediated transformation demonstrated successful transfer. This process yielded transgenic lentil plants exhibiting enhanced resilience to drought and salinity conditions, as reported by Khatib et al. (2011). Another crucial issue is in vitro regeneration after transformation. According to a study, decapitated embryos were more effective at producing shoots than other tissues (Sarker et al. 2003). These endeavours gave modern genome editing research some useful tools. In lentil, numerous candidate genes for agronomic traits as well as biotic and abiotic stresses have been identified (Kumar et al. 2021). Yet, till date, no novel genome editing studies employing ZFNs, TALENs or CRISPR/Cas9 technologies have been conducted on potential genes in lentil. The function of those candidate genes can be easily and affordably determined using genome editing technologies, giving cultivars the desired qualities like agronomic attributes and stress tolerance (Bhowmik et al. 2021).

16.3 Conclusion and Future Prospects

Crop breeding and functional genomics have advanced quickly due to genome editing research, particularly the CRISPR system that won the Nobel Prize. The accuracy and effectiveness of altering genes by adding or removing genetic material has risen as a result of recent developments in genome editing research. Moreover, genome editing tools have been utilized to make it easier to manipulate one or more genetic loci in many plants. Researchers looking to make better use of genome editing tools can learn a lot from the recently sequenced variable pulse genomes. Thus, effective genome editing requires an in-depth knowledge of the genomic sequences of grain legumes and their characteristics. With the elucidation of genes that regulate certain properties in pulses like flavour, size, biotic and abiotic stresses, new grain legume varieties may be created harbouring agronomically superior features along with imperative nutritional traits (Figure 16.1).

Despite the presence of a few promising endeavours, the primary challenges appear to lie in the limited effectiveness of transformation and the persistent issues surrounding regeneration in these pulses. These factors could potentially explain the lack of new initiatives in

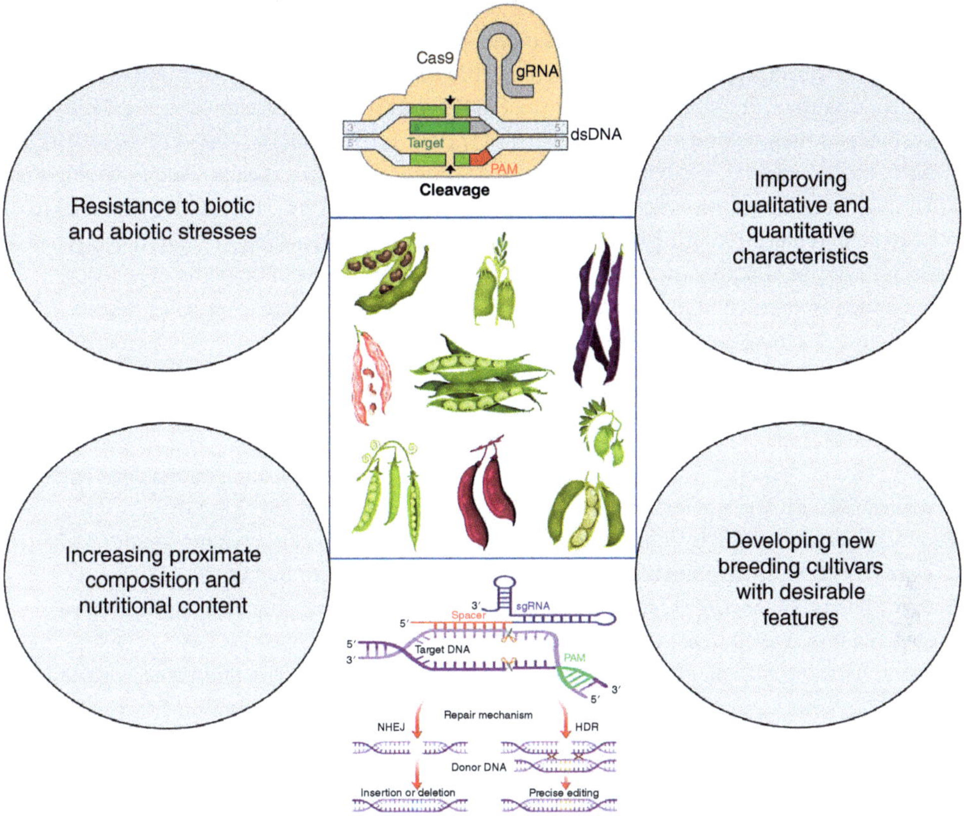

Figure 16.1 Approaches for enhancing genetic gains in pulses by genome editing.

the field of genome editing. For a big portion of the world to be fed adequately, new methods to increase transformation and regeneration efficiency are essential in order to achieve the significant improvement that current genome editing methods promise in grain legumes with solemn potential. Last but not least, all non-governmental organizations (NGOs), business sectors and healthcare authorities need to work together in a robust manner to ensure that genome editing is used in a legal and effective way to produce new pulse cultivars to end hidden hunger.

References

Aasim, M., Barpete, S., Usta, A., and Sevinc, C. (2019). An insight into *Agrobacterium tumefaciens*-mediated genetic transformation studies in mungbean (*Vigna radiata* L. Wilczek). *Journal of Global Innovations in Agricultural and Social Sciences* 7 (2): 47–51.

Aftabi, M., Teressa Negawo, A., and Hassan, F. (2017). Improved protocol for agrobacterium-mediated transformation of pea (*Pisum sativum*). *Molecular Biology* 7: 1–6.

Al Amin, N., Ahmad, N., Wu, N. et al. (2019). CRISPR-Cas9 mediated targeted disruption of FAD2-2 microsomal omega-6 desaturase in soybean (*Glycine max* L). *BMC Biotechnology* 19: 9.

Alok, A., Kumar, J., and Upadhyay, S.K. (2018). Engineering in hairy roots using CRISPR/Cas9-mediated editing. In: *Hairy Roots* (ed. V. Srivastava, S. Mehrotra, and S. Mishra). Singapore: Springer. https://doi.org/10.1007/978-981-13-2562-5_15.

Alok, A., Chauhan, H., Upadhyay, S.K. et al. (2021). Compendium of plant-specific CRISPR vectors and their technical advantages. *Life* 11: 1021. https://doi.org/10.3390/life11101021.

Aragão, F.J.L., Ribeiro, S.G., Barros, L.M.G. et al. (1998). Transgenic beans (*Phaseolus vulgaris* L.) engineered to express viral antisense RNAs show delayed and attenuated symptoms to bean golden mosaic geminivirus. *Molecular Breeding* 4 (6): 491–499.

Assefa, T., Assibi Mahama, A., Brown, A.V. et al. (2019). A review of breeding objectives, genomic resources, and marker-assisted methods in common bean (*Phaseolus vulgaris* L.). *Molecular Breeding* 39: 20.

Badhan, S., Ball, A.S., and Mantri, N. (2021). First report of CRISPR/Cas9 mediated DNA-free editing of *4CL* and *RVE7* genes in chickpea protoplasts. *International Journal of Molecular Sciences* 22: 396.

Bai, M., Yuan, J., Kuang, H. et al. (2020). Generation of a multiplex mutagenesis population via pooled CRISPR-Cas9 in soya bean. *Plant Biotechnology Journal* 18: 721–731.

Bao, A., Chen, H., Chen, L. et al. (2019). CRISPR/Cas9-mediated targeted mutagenesis of *GmSPL9* genes alters plant architecture in soybean. *BMC Plant Biology* 19: 131.

Bao, A., Zhang, C., Huang, Y. et al. (2020). Genome editing technology and application in soybean improvement. *Oil Crop Science* 5: 31–40.

Barraza, A., Cabrera-Ponce, J.L., Gamboa-Becerra, R. et al. (2015). The *Phaseolus vulgaris PvTRX1h* gene regulates plant hormone biosynthesis in embryogenic callus from common bean. *Frontiers in Plant Science* 6: 577.

Bastianelli, D., Grosjean, F., Peyronnet, C. et al. (1998). Feeding value of pea (*Pisum sativum* L.) 1. Chemical composition of different categories of pea. *Animal Science* 67 (3): 609–619.

Bauchet, G.J., Bett, K.E., Cameron, C.T. et al. (2019). The future of legume genetic data resources: challenges, opportunities, and priorities. *Legume Science* 1: e16.

Bett, K., Ramsay, L., Sharpe, A., et al. (2014). Lentil genome sequencing: establishing a comprehensive platform for molecular breeding. *Proceedings of International Food Legumes Research Conference (IFLRC-VI) and ICCLG-VII*. Crop Development Center, SK, Saskatoon, p. 19.

Bett, K., Ramsay, L., Chan, C., et al. (2016). Lentil 1.0 and beyond. *PAG XXIV: Plant and Animal Genomics Conference*, 8–13 January 2016, San Diego, California, USA

Bhalothia, P., Yajnik, K., Alok, A., and Upadhyay, S.K. (2020). The current progress of CRISPR/Cas9 development in plants. *Elsevier*. https://doi.org/10.1016/B978-0-12-818140-9.00011-8.

Bhowmik, P., Konkin, D., Polowick, P. et al. (2021). CRISPR/Cas9 gene editing in legume crops: opportunities and challenges. *Legume Science* 3 (3): https://doi.org/10.1002/leg3.96.

Bibikova, M., Golic, M., Golic, K.G., and Carroll, D. (2002). Targeted chromosomal cleavage and mutagenesis in *Drosophila* using zinc-finger nucleases. *Genetics* 161: 1169–1175.

Böttinger, P., Steinmetz, A., Schieder, O., and Pickardt, T. (2001). Agrobacterium-mediated transformation of *Vicia faba*. *Molecular Breeding* 8: 243–254.

Cai, Y., Chen, L., Liu, X. et al. (2015). CRISPR/Cas9-mediated genome editing in soybean hairy roots. *PLoS One* 10: e0136064.

Cai, Y., Chen, L., Sun, S. et al. (2018). CRISPR/Cas9-mediated deletion of large genomic fragments in soybean. *International Journal of Molecular Sciences* 19: 3835.

Cai, Y., Wang, L., Chen, L. et al. (2020). Mutagenesis of *GmFT2a* and *GmFT5a* mediated by CRISPR/Cas9 contributes for expanding the regional adaptability of soybean. *Plant Biotechnology Journal* 18: 298–309.

Cardi, T., D'Agostino, N., and Tripodi, P. (2017). Genetic transformation and genomic resources for next-generation precise genome engineering in vegetable crops. *Frontiers in Plant Science* 8: 241.

Cernay, C., Ben-Ari, T., Pelzer, E. et al. (2015). Estimating variability in grain legume yields across Europe and the Americas. *Scientific Reports* 5: 11171.

Che, P., Chang, S., Simon, M.K. et al. (2021). Developing a rapid and highly efficient cowpea regeneration, transformation and genome editing system using embryonic axis explants. *The Plant Journal* https://doi.org/10.1111/tpj.15202.

Cichy, K.A., Caldas, G.V., Snapp, S.S., and Blair, M.W. (2009). QTL analysis of seed iron, zinc, and phosphorus levels in an andean bean population. *Crop Science* 49: 1742–1750.

Cong, L., Ran, F.A., Cox, D. et al. (2013). Multiplex genome engineering using CRISPR/Cas systems. *Science* 339: 819–823.

Cooper, J.W., Wilson, M.H., Derks, M.F.L. et al. (2017). Enhancing faba bean (*Vicia faba* L.) genome resources. *Journal of Experimental Botany* 68 (8): 1941–1953.

Curtin, S.J., Xiong, Y., Michno, J.M. et al. (2018). CRISPR/Cas9 and TALENs generate heritable mutations for genes involved in small RNA processing of *Glycine max* and *Medicago truncatula*. *Plant Biotechnology Journal* 16: 1125–1137.

Di, Y.-H., Sun, X.-J., Hu, Z. et al. (2019). Enhancing the CRISPR/Cas9 system based on multiple GmU6 promoters in soybean. *Biochemical and Biophysical Research Communications* 519: 819–823.

Do, P.T., Nguyen, C.X., Bui, H.T. et al. (2019). Demonstration of highly efficient dual gRNA CRISPR/Cas9 editing of the homeologous *GmFAD2–1A* and *GmFAD2–1B* genes to yield a high oleic, low linoleic and α-linolenic acid phenotype in soybean. *BMC Plant Biology* 19: 311.

Du, H., Zeng, X., Zhao, M. et al. (2016). Efficient targeted mutagenesis in soybean by TALENs and CRISPR/Cas9. *Journal of Biotechnology* 217: 90–97.

Ellis, T.H.N., Hofer, J.M.I., Timmerman-Vaughan, G.M. et al. (2011). Mendel, 150 years on. *Trends in Plant Science* 16: 590–596.

Hanafy, M., Pickardt, T., Kiesecker, H., and Jacobsen, H.-J. (2005). Agrobacterium-mediated transformation of faba bean (*Vicia faba* L.) using embryo axes. *Euphytica* 142: 227–236.

Hanafy, M.S., El-Banna, A., Schumacher, H.M. et al. (2013). Enhanced tolerance to drought and salt stresses in transgenic faba bean (*Vicia faba* L.) plants by heterologous expression of the *PR10a* gene from potato. *Plant Cell Reports* 32: 663–674.

Jacobs, T.B., LaFayette, P.R., Schmitz, R.J., and Parrott, W.A. (2015). Targeted genome modifications in soybean with CRISPR/Cas9. *BMC Biotechnology* 15: 1–10.

Ji, J., Zhang, C., Sun, Z. et al. (2019). Genome editing in cowpea *Vigna unguiculata* using CRISPR-Cas9. *International Journal of Molecular Sciences* 20: 2471.

Juranić, M., Nagahatenna, D.S.K., Salinas-Gamboa, R. et al. (2020). A detached leaf assay for testing transient gene expression and gene editing in cowpea (*Vigna unguiculata* [L.] walp.). *Plant Methods* 16: 88.

Kahlon, J.G., Jacobsen, H.-J., Chatterton, S. et al. (2018). Lack of efficacy of transgenic pea (*Pisum sativum* L.) stably expressing antifungal genes against fusarium spp. in three years of confined field trials. *GM Crops & Food* 9: 90–108.

Kanazashi, Y., Hirose, A., Takahashi, I. et al. (2018). Simultaneous site-directed mutagenesis of duplicated loci in soybean using a single guide RNA. *Plant Cell Reports* 37: 553–563.

Kang, Y.J., Kim, S.K., Kim, M.Y. et al. (2014). Genome sequence of mungbean and insights into evolution within *Vigna* species. *Nature Communications* 5 (1): 1–9.

Keatinge, J.D.H., Easdown, W.J., Yang, R.Y. et al. (2011). Overcoming chronic malnutrition in a future warming world: the key importance of mungbean and vegetable soybean. *Euphytica* 180: 129–141.

Khatib, F., Makris, A., Yamaguchi-Shinozaki, K. et al. (2011). Expression of the DREB1A gene in lentil (*Lens culinaris* Medik. subsp. *culinaris*) transformed with the agrobacterium system. *Crop & Pasture Science* 62 (6): 488–495.

Khazaei, H., O'Sullivan, D.M., Stoddard, F.L. et al. (2021). Recent advances in faba bean genetic and genomic tools for crop improvement. *Legume Science* 3 (3): e75.

Kreplak, J., Madoui, M.-A., Cápal, P. et al. (2019). A reference genome for pea provides insight into legume genome evolution. *Nature Genetics* 51 (9): 1411–1422.

Kumar, S., Rajendran, K., Kumar, J. et al. (2015). Current knowledge in lentil genomics and its application for crop improvement. *Frontiers in Plant Science* 6: 78.

Kumar, J., Sen Gupta, D., Baum, M. et al. (2021). Genomics-assisted lentil breeding: current status and future strategies. *Legume Science* 3 (3): e71.

Lenka, B., Satapathy, S.N., and Senapati, M. (2020). Engineering plant virus resistance: gene silencing to genome editing. *International Journal of Current Microbiology and Applied Sciences* 9 (10): 3086–3096.

Li, T., Huang, S., Zhao, X. et al. (2011). Modularly assembled designer TAL effector nucleases for targeted gene knockout and gene replacement in eukaryotes. *Nucleic Acids Research* 39: 6315–6325.

Li, T., Liu, B., Spalding, M.H. et al. (2012). High-efficiency TALEN-based gene editing produces disease-resistant rice. *Nature Biotechnology* 30: 390–392.

Li, Z., Liu, Z.-B., Xing, A. et al. (2015). Cas9-guide RNA directed genome editing in soybean. *Plant Physiology* 169: 960–970.

Li, C., Chen, C., Chen, H. et al. (2018). Verification of DNA motifs in *Arabidopsis* using CRISPR/Cas9-mediated mutagenesis. *Plant Biotechnology Journal* 16: 1446–1451.

Li, C., Nguyen, V., Liu, J. et al. (2019). Mutagenesis of seed storage protein genes in soybean using CRISPR/Cas9. *BMC Research Notes* 12: 176.

Luzzatto, L. and Arese, P. (2018). Favism and glucose-6-phosphate dehydrogenase deficiency. *The New England Journal of Medicine* 378: 60–71.

Maalouf, F., Hu, J., O'Sullivan, D.M. et al. (2019). Breeding and genomics status in faba bean (*Vicia faba*). *Plant Breeding* 138 (4): 465–473.

Mabrouk, Y., Hemissi, I., Salem, I.B. et al. (2018). Potential of rhizobia in improving nitrogen fixation and yields of legumes. *Symbiosis* 33: 107.

Mali, P., Yang, L., Esvelt, K.M. et al. (2013). RNA-guided human genome engineering via Cas9. *Science* 339: 823–826.

Michno, J., Wang, X., Liu, J. et al. (2015). CRISPR/Cas mutagenesis of soybean and *Medicago truncatula* using a new web-tool and a modified Cas9 enzyme. *GM Crops & Food* 6: 243–252.

Nadeem, M.A., Yeken, M.Z., Shahid, M.Q. et al. (2021). Common bean as a potential crop for future food security: an overview of past, current and future contributions in genomics, transcriptomics, transgenics and proteomics. *Biotechnology and Biotechnological Equipment* 35 (1): 759–787.

Negawo, A.T., Aftabi, M., Jacobsen, H.-J. et al. (2013). Insect resistant transgenic pea expressing *cry1Ac* gene product from *Bacillus thuringiensis*. *Biological Control* 67: 293–300.

Pandey, A.K., Rubiales, D., Wang, Y. et al. (2021). Omics resources and omics-enabled approaches for achieving high productivity and improved quality in pea (*Pisum sativum* L.). *Theoretical and Applied Genetics* 134: 755–776.

Römer, P., Recht, S., and Lahaye, T. (2009). A single plant resistance gene promoter engineered to recognize multiple TAL effectors from disparate pathogens. *Proceedings of the National Academy of Sciences United States of America* 106: 20526–20531.

Rouet, P., Smih, F., and Jasin, M. (1994). Introduction of double-strand breaks into the genome of mouse cells by expression of a rare-cutting endonuclease. *Molecular and Cellular Biology* 14: 8096–8106.

Sahoo, D.P., Kumar, S., Mishra, S. et al. (2016). Enhanced salinity tolerance in transgenic mungbean overexpressing *Arabidopsis* antiporter (*NHX1*) gene. *Molecular Breeding* https://doi.org/10.1007/s11032-016-0564-x.

Saini, R., Singh, R.P., and Jaiwal, P.K. (2007). *Agrobacterium tumefaciens*-mediated transfer of *Phaseolus vulgaris* α-amylase Inhibitor-1 gene into mungbean *Vigna radiata* (L.) Wilczek using bar as selectable marker. *Plant Cell Reports* 26 (2): 187–198.

Sarker, R.H., Biswas, A., Mustafa, B.M. et al. (2003). Agrobacterium-mediated transformation of lentil (*Lens culinaris* Medik.). *Plant Tissue Culture* 13 (1): 1–12.

Schmutz, J., McClean, P.E., Mamidi, S. et al. (2014). A reference genome for common bean and genome-wide analysis of dual domestications. *Nature Genetics* 46: 707–713.

Sun, X., Hu, Z., Chen, R. et al. (2015). Targeted mutagenesis in soybean using the CRISPR-Cas9 system. *Scientific Reports* 5: 1–10.

Sushmita, K.G., Upadhyay, S.K., and Verma, P.C. (2021). An Overview of Genome-Engineering Methods. In: *Genome Engineering for Crop Improvement* (ed. S.K. Upadhyay). Wiley. https://doi.org/10.1002/9781119672425.ch1.

Temesgen, T., Keneni, G., Sefera, T., and Jarso, M. (2015). Yield stability and relationships among stability parameters in faba bean (*Vicia faba* L.) genotypes. *Crop Journal* 3 (3): 258–268.

Tripathi, A., Debnath, S., and Shukla, S. (2021). Review on biotechnological advances in *Vigna radiata* and their future prospects. *Journal of Crop Science and Biotechnology* 24: 245–258.

Upadhyay, S.K. (2021). *Genome Engineering for Crop Improvement*. Wiley. https://doi.org/10.1002/9781119672425.

Varshney, R.K., Pandey, M.K., Bohra, A. et al. (2019). Toward the sequence-based breeding in legumes in the post-genome sequencing era. *Theoretical and Applied Genetics* 132: 797–816.

Wang, J., Kuang, H., Zhang, Z. et al. (2020). Generation of seed lipoxygenase-free soybean using CRISPR-Cas9. *Crop Journal* 8: 432–439.

Xiao, S. (2012). Protecting crops from pathogens: novel approaches to an old problem. *Gene Technology* https://doi.org/10.4172/2329-6682.1000e103.

Xue, R., Wu, X., Wang, Y. et al. (2017). Hairy root transgene expression analysis of a secretory peroxidase (*PvPOX1*) from common bean infected by fusarium wilt. *Plant Science* 260: 1–7.

Yadav, S.K., Katikala, S., Yellisetty, V. et al. (2012). Optimization of agrobacterium mediated genetic transformation of cotyledonary node explants of *Vigna radiata*. *SpringerPlus* 1 (1): 1–8.

Zhang, F., Cong, L., Lodato, S. et al. (2011). Efficient construction of sequence-specific TAL effectors for modulating mammalian transcription. *Nature Biotechnology* 29: 149–153.
Zhang, P., Du, H., Wang, J. et al. (2020). Multiplex CRISPR/Cas9-mediated metabolic engineering increases soya bean isoflavone content and resistance to soyabean mosaic virus. *Plant Biotechnology Journal* 18: 1384–1395.
Zheng, N., Li, T., Dittman, J.D. et al. (2020). CRISPR/Cas9-based gene editing using egg cell-specific promoters in *Arabidopsis* and soybean. *Frontiers in Plant Science* 11: 800.

17

Genome Editing for Microbial Pathogens Resistance in Crops

Mudasir Ahmad Bhat[#], Saima Jan[#], Sumreen Amin Shah, and Arif Tasleem Jan

School of Biosciences and Biotechnology, Baba Ghulam Shah Badshah University, Rajouri, Jammu and Kashmir, India

17.1 Introduction

The development of resistant crops is a crucial aspect of agriculture, especially in light of the current and projected impacts of climate change that cause alterations in precipitation patterns, temperature regimes and atmospheric carbon dioxide levels (Anderegg et al. 2020; Hassani et al. 2020; Raymond et al. 2020; Van Houtan et al. 2021; Zandalinas et al. 2021). The impact of climatic change on the quality and productivity of food increases concern over food security highlighting the urgent need to have crops better adapted to these changing conditions (Ritchie and Roser 2019; Chen et al. 2019a). Increasing demand for food for the growing population projects a requirement to have an adequate supply of different food items that projects a need to have a several-fold increase in their productivity worldwide. For these reasons, efforts are being made to increase the productivity of a wide range of crop plants using all possible methods. Relying on the conventional approaches of breeding to produce a cultivar having desired traits usually takes more than 10 years of cultivation with many generations of crosses and backcrosses in order to accumulate desired QTLs for the target trait (Nagamine and Ezura 2022). Keeping in view the time constraint in responding to the needs of the growing population, there appears an urgent need to have rapid but reliable techniques that can facilitate accurate, precise, effective, efficient and targeted alterations at genomic loci by inserting, deleting, or substituting the DNA bases in the host genome (Chen et al. 2019b; Gao 2021).

Technological advances in biological research have led to innovations happening at a rapid pace over the past few decades. Of the different emerged approaches, the versatility of editing a genome with greater accuracy and in a precise manner has resulted in spellbinding progress in terms of engineering genomes for the desirable traits (Bhat et al. 2020, 2021). Compared to other contemporary technologies adopted in editing a genome, the use of

[#] Authors contributed equally to the writing of different sections of the chapter

Applications of Genome Engineering in Plants, First Edition. Edited by Santosh Kumar Upadhyay.
© 2024 John Wiley & Sons Ltd. Published 2024 by John Wiley & Sons Ltd.

different classes of sequence-specific nucleases such as ZFNs (zinc finger nucleases), TALENs (transcription activator-like effector nucleases) and most advanced CRISPR/Cas (clustered regularly interspaced palindromic repeats associated nucleases) show tremendous progress in terms of their applicability for use in genome editing (Sushmita et al. 2021; Upadhyay 2021). Target genome editing at specific sequences with ease and in a time-saving manner not only increases the specificity and efficiency of the process but also helps in minimizing the off-target effects. The ZFNs involving introduction of the target DSBs and stimulating the cellular DNA repair mechanisms were extensively used for the targeted genome alterations in numerous plant species such as Arabidopsis (Lloyd et al. 2005; Osakabe et al. 2010), tobacco and maize (Shukla et al. 2009; Townsend et al. 2009), etc. The TALENs work in the same manner as ZNFs and were initially identified in plant pathogenic bacteria (*Xanthomonas*) (Boch et al. 2009; Christian et al. 2010). The genome editing techniques ZFN and TALEN were superseded by the recently emerged CRISPR/Cas system in achieving editing of the genomes more precisely and in an accurate manner. The present study focuses on key concepts of CRISPR/Cas system, a detailed description of CRISPR/Cas genome editing technique in creating cultivars resistant to climate change, its advantages and applications to fulfill the needs of the growing population as part of the 'Zero Hunger' sustainable goal set by the United Nations.

17.2 Effects of Climate Change on Crop Productivity

The consequences of climate change are expected to have a direct impact on global losses in yield and crop adaptability. The climatic change-induced abiotic stress largely affects different crop species that significantly impacts its productivity. The main factors that restrict the growth and as such performance under changing climatic conditions are increased CO_2 concentration, extremes of temperature, drought, salinity, etc. (Van der Kooi et al. 2016). The reduction in transpiration was seen with increased atmospheric CO_2 concentration and caused a rise in temperature of 0.42 ± 0.02 K (Malhi et al. 2021). It was reported by Zhu et al. (2018) that a decline in iron, protein and zinc content in rice plants was observed in response to rising CO_2 concentrations. A study on climate change shows that there will be a decline in crop yields by up to 15% by 2050, with the greatest reduction in sub-Saharan Africa, South Asia and the Middle East (Aryal et al. 2020). In another study, it was found that increasing changes in temperature and precipitation patterns could reduce crop yields and increase the frequency of crop failures in many regions of the world (Talari et al. 2022). It was found that climate change could lead to a significant increase in the incidence of mycotoxins, harmful substances produced by certain types of fungi that can contaminate crops which pose a significant health risk to humans and livestock (Misiou and Koutsoumanis 2022).

Though heat stress adversely affects plant growth, physiology and development while drought stress causes a decrease in production yield (Fahad et al. 2017). Drought stress affects turgor and water potential which interfere with normal plant processes (Nawaz et al. 2015). It was demonstrated that drought stress could reduce the nutritional quality of crops, particularly in terms of protein content and it was also found that it could reduce the protein content of wheat by up to 10% (Zahra et al. 2021; Iqbal et al. 2022; Zhao et al. 2022).

Several plant species like *Tagetes erecta* (Asrar and Elhindi 2011), *Oryza sativa* (Chen et al. 2011), *Triticum aestivum* (Arzanesh et al. 2011), *Jatropha curcas* (Sapeta et al. 2013) and *Cuminum cyminum* (Soleymani and Shahrajabian 2018) have been examined for growth reduction owing to drought stress. Under drought stress, a decrease in the starch content of wheat grains has been noted (Noori and Taliman 2022). In a study, it was observed that drought stress could reduce the yield of maize, soybean and wheat by 19%, 11% and 5% respectively (Daryanto et al. 2016). The process of fertilization and anthesis in rice is severely disrupted by drought stress. The harvest index drops to 60% because of the water shortage, which also lowers the amount of grain set (Garrity et al. 1994). The reproductive stage of plants is altered by water deficit and higher temperatures. In a study, the flower development of cereals was found adversely affected by water stress (Winkel et al. 1997). Due to a lack of water during the meiotic stage, the yield of wheat and rice was reduced by 35–75% (Qamar and Javeed 2022). In cereals, an increase in temperature (above 30 °C) causes flowers to become sterile during their developmental stages and reduces the crop yield (Rabara et al. 2021).

Climate change and desertification were probably going to reduce the agricultural yield, especially in the regions having dry climatic conditions. With decreased productivity, Asian and African continents would have the largest populations at risk from accelerated desertification. At present, studies have reported that crops grown in higher-altitude regions are producing more, while those in lower latitudes are starting to see a reduction in yield (Iizumi et al. 2018; Shukla et al. 2019). Plants under salt stress experience decreased germination, photosynthesis, hydration, oxidative stress and productivity (Choi et al. 2010; Xu et al. 2011; Shekari et al. 2017). In another study, it was demonstrated that salinity stress could reduce the yield of crops such as wheat, rice and maize by up to 50% (Choudhary et al. 2022). Rahneshan et al. (2018) tested the effects of NaCl (50–150 mM) on two Pistachio (*Pistacia vera*) rootstocks and discovered that moderate to high salt concentrations have a detrimental effect on the shoot and root length of the plant. It is predicted that extreme weather changes will disrupt the world's food supply as already becoming evident and surely will get worse in times to come if not addressed in a time-bound manner.

17.3 CRISPR/Cas-Mediated Genome Editing in Plants

The use of CRISPR/Cas in editing genome was demonstrated on observing type II CRISPR/Cas developed by the bacterium *Streptococcus pyogenes* as the most flexible and efficient genome editing tool capability of altering the DNA as per the target specificity (Jinek et al. 2012). The use of CRISPR/Cas in genome editing requires two RNAs to function in the natural system: a crRNA that specifies the target and a tracrRNA (trans-activating crRNA) that stabilizes the entire structure and allow Cas9 to cleave the target sequence as directed by the crRNA (Mojica et al. 2009). Following this, a concept of developing a chimeric sgRNA (single guide RNA) by combining features of crRNA and tracrRNA emerged that works well with the Cas9 in cleaving the DNA having a short-conserved photo spacer adjacent motif (PAM) sequence such as 5′-NGG-3′ downstream of the target sequence; thereby changes the concept from three-component system into two-component system (Jinek et al. 2012; Bhalothia et al. 2020). Compared to ZFNs and TALENs that requires designing of proteins, CRISPR/Cas enables editing of a wide variety of genomes (both plants and animals) following recognition of DNA sequences in a relatively simple, adaptable and cost-effective manner. Unlike ZFNs and TALENs requiring multiple gRNAs for editing each target sequence, CRISPR/Cas enables multitasking easier with a sgRNA that works in combination with the Cas9 for editing multiple positions. Additionally, CRISPR constructs are smaller and easy to transport, inherit ability to detect methylated DNA thus allowing it to target locations that ZNFs and TALENs are unable to reach. CRISPR/Cas9 constructs are introduced into plants either through delivering Cas9 protein and *in vitro* produced guide RNA as pre-assembled Ribonucleoproteins (RNPs) to the cells for transient gene expression or through use of binary vectors which carries the Cas9 gene and a gRNA cassette on two distinct plasmids or both of them as a single entity on a single plasmid. In the previous method, it is possible to employ mRNA that also expresses Cas9 and gRNA together into the cells. Since the RNPs are broken down inside the cells, the previous method is preferred as the technique is considered as transgene free with minimum off-target effects (Waltz 2012). The cell transformation method in the previous technique includes use of cell-penetrating peptides, mesoporous silica nanoparticles (MSN), polyethylene glycol (PEG) and liposome mediated transformation, electroporation and microinjection (Farooq et al. 2018), while in the latter the adapted delivery modules include microinjection, electroporation, heat shock and Agrobacterium-mediated transformation. Though, in the latter method transgenes can be removed via genetic segregation, people still considers plants edited with this method as transgenic due to the stable integration of transgenes.

17.3.1 CRISPR/cpf1

CRISPR/Cpf1 is a gene editing tool that uses a different type of enzymes, i.e. Cas12a in place of the more widely used Cas9. It is a smaller protein than Cas9 and has different characteristics that make it attractive for use in gene editing. It was originally used to modify a mammalian cell and was recently redesignated as a member of class II type V nuclease (Mahfouz 2017). Although Cpf1 endonuclease lacks the HNH domain as observed in Cas9, it does have a RuvC-type nuclease domain. In place of the blunt end points found in Cas9, Cpf1 generates DSBs with sticky or staggered ends. By using HDR-directed-repair, it permits gene deletion, insertion or substitution (Zetsche et al. 2015; Mehta et al. 2020; Alok et al. 2021). Contrary to Cas9 capable of cleaving DNA only once, Cpf1 cleaves DNA at distinct locations from the PAM. In contrast to the Cas9 utilizing a G-rich PAM, Cpf1 uses a T-rich PAM and shorter crRNA. Cpf1 possess higher specificity than Cas9 as of which is used to increase the number of edits within a plant genome (Stella et al. 2017; Tang et al. 2017b).

17.3.2 CRISPRi

CRISPRi or CRISPR interference variant makes use of dCas9 (catalytically inactive Cas9), in interfering with the ability of gene to function (Qi et al. 2013). In both prokaryotic organisms and complex eukaryotic ones, CRISPRi performs the role of a disruptor involved in the sequence-specific inhibition of transcription (Huang et al. 2016). It was originally developed to interfere with expression of silent genes, but it is currently employed in achieving the epigenetic alterations and transcriptional activation by exploiting the functionality of gene through methylation or chromatin modifications in different stress responses. Its nature is highly precise and specialized, with some slight off-target consequences (Dominguez et al. 2016). It was observed that the amount of expression pertaining to different genes are lowered by 99% when dCas9 and kruppel-associated box (KARB) are used in combination in human cells (Gilbert et al. 2013). Therefore, this method can be effectively used to identify stress-responsive genes in plants and drive their transcription through activation or repression process (Piatek et al. 2015). The transcriptional activation was boosted by combining the dCas9 with the transcriptional activator (Gilbert et al. 2013). The effectiveness of certain target genes can be controlled by CRISPRi, and its impact can be increased (Lo and Qi 2017).

17.4 CRISPR-Based Engineering of Crop Plants

17.4.1 Gene Disruption via Indel in Coding Sequences

The CRISPR-Cas9 mediated gene disruption is prominent that get benefit from the cell's own NHEJ DNA-repair system. In gene disruption, insertion or deletion of nucleotides (InDel) takes place at the sgRNA-guided site which progresses with a frame shift mutation and interfering with the synthesis process. The approach employs disruption of the disease susceptible gene that cause modification in the interaction pattern of pathogens with the plants (reducing fitness of pathogen on the plants); thereby imparts disease resistance to the plants (Zaidi et al. 2018). Several crops like rice and wheat have been modified with this method to introduce desirable features of resistance to different pathogens (Chen et al. 2019b). One outstanding example is provided by CRISPR/Cas9 mediated induction of changes in the eukaryotic initiation factor 4E (eIF4E) towards developing resistant lines of Arabidopsis and cucumber (*Cucumis sativus*) for protection against multiple RNA viruses (Chandrasekaran et al. 2016; Pyott et al. 2016).

17.4.2 Gene Disruption via Indel in Promoter Regions

The promoter associated with a gene can be altered by inserting InDels using a similar approach as used for the coding region of the gene. The sensitivity of plant to pathogens is disrupted on editing promoters using CRISPR/Cas technique in one of two ways: Disruption of the promoter sequence to stop gene expression or destruction of an effector-binding site in order to stop pathogen binding to the promoter. By using the approach of disrupting the effector binding site, the promoter region of the sugar transporter gene OsSWEET14 in rice was altered, it served the purpose of preventing the binding of effector for a bacterial blight pathogen and as such attribute it with blight resistance (Li et al. 2012; Blanvillain-Baufume et al. 2017; Oliva et al. 2019; Xu et al. 2019). CRISPR enlightens the path to change the gene regulation in addition to editing the promoter via direct modification of the upstream open reading frame (ORF) and as such modify the cis-regulatory elements (Si et al. 2020).

17.4.3 Gene Deletion via Multiplex sgRNAs

The introduction of multiple DSBs at specific targets with in a genome using sgRNAs was achieved using CRISPR system. For instance, sgRNAs binding to start and stop codon of the target genes would result in DSBs at those loci. The fragments of DNA having the target gene are subsequently removed as a result of these DSBs, followed by DSB repair by the cellular NHEJ machinery. This strategy has been utilized to delete both individual genes and huge chromosomal segments since sgRNAs can be generated for any genomic area having PAM trinucleotide sequence on to the right side (Zhou et al. 2014; Cai et al. 2018). Rational designing of the special Cas, Sp-Cas9-NG with a softer stringency than standard NGG PAMs can detect NG PAMs and facilitates even this more (Nishimasu et al. 2018). Talking about engineered pathogen resistance, the gene clusters are very much helpful. In susceptible gene clusters, which are regions of the chromosome where multiple susceptibility related genes are present, any deletion of the chromosomal fragments is anticipated in providing long-lasting resistance toward the target pathogen (Ashfield et al. 2012).

17.4.4 Gene Insertion via Homology-Directed Repair

The aforesaid approaches of CRISPR have been applied to susceptible gene alteration to produce disease resistance in plants. All plant proteins, produced by susceptible genes are vital and typically serve many functions; therefore, altering these proteins have a negative impact on the health or production of plants. At certain instances, the use of resistance (R) genes against pathogens is made towards understanding the interplay of the plant-pathogen interaction. After sgRNA-directed DSB brought by Cas9, CRISPR-mediated insertion at a specific target do employs a different pathway that make use of the cellular HDR mechanism instead of NHEJ. The R gene can be inserted in such conditions using the CRISPR toolkit. In this approach, sgRNA Cas9 contains R gene and sequences homologous to the ends of DSBs. By following the HDR, the R gene is inserted using this cassette as a guide in between the two DSB sites (Van Vu et al. 2019). Despite the possibility, Liu et al. (2019) found that the efficacy of HDR in plants is extremely less, making it difficult to apply gene insertion in plants even though new methods to enhance are currently being developed (Butt et al. 2017). The study of important susceptible gene activities as well as the development of disease resistance can be accomplished by CRISPR-mediated gene insertion. For its authentication, Wang et al. (2017c) inserted the glutathione-S-transferase locus with green fluorescent protein (GFP) fused in frame into rice. An analysis of the susceptible gene features such as localization and spatiotemporal regulation of the gene expression are the potential uses of this technology. To examine the expression and distribution of host proteins involved in infection, they can be directly tagged with GFP in the genome and any alteration to gene structure can change expression and location of the protein.

17.5 CRISPR/Cas in Imparting Tolerance to Biotic Factors

To address the growing problems posed by the climate change in view of the expanding global population, different projects were taken that encourages adoption and promotion of sustainable agricultural approaches. Of them, imparting resistance to crops against different stresses is considered as a promising strategy toward improving growth and overall productivity of the crops. The following sections deals with development of resistance to biotic factors that exerts great effect on the crop productivity.

17.5.1 CRISPR in Developing Resistance to Viruses

Significant efforts have been made in recent years to combat viruses using defense systems derived from pathogens. With the rapid advancement of CRISPR/Cas9-mediated gene editing, it has become possible to look into efficient ways to build viral resistance. Targeting host components involved in virus replication is one way to increase virus resistance (Borrelli et al. 2018) or directly targeted the viral genome to bring an end to virus replication (Ali et al. 2015). The genomes of Geminivirus have been used for the majority of viral resistance tests mediated by CRISPR/Cas (Kang et al. 2006; Baltes et al. 2015; Ji et al. 2015). As the viruses are responsible for considerable agricultural loss across the globe, *N. benthamiana* and Arabidopsis were examined for their ability to resist the pathogenicity of *Beet Sever Curly Top Virus* (BSCTV), *Geminiviruses* and *Bean Yellow Dwarf Virus* (BeYDV) (Duan et al. 2016; Meng et al. 2017). Targeting both coding and non-coding regions in the BSCTV genome, Ji et al. (2015) examined forty-three possible sgRNA/Cas9 target sites. Despite the variable degrees of difference to which each sgRNA/Cas9 was constructed, it reduced viral accumulation in the inoculated leaves of *N. benthamiana* and Arabidopsis and as such improved resistance to virus infection. Similarly, Baltes and co-workers (2015) observed that targeting Rep-binding sites in the genome of BeYDV with different sgRNAs, it is possible to reduce the targeted viral burden in *N. benthaminia* by up to 87%. Since spCas9 from *Streptococcus pyogenes* exclusively detects dsDNA, it was found often challenging to protection against RNA viruses. To this, the discovery and characterization of nucleases that can bind and cut RNA, such as *Francisella novicida* FnCas9 or *Leptotrichia wadei* LwaCAS13 was reported. In an initial study, sgRNA specific to CMV and TMV were used to demonstrate mechanism of resistance to different viruses with RNA genome in *N. benthamiana* and Arabidopsis (Kang et al. 2006) with transgenic plants showing up to 40–80% reduction in the TMV and CMV infection as compared to control plants (Table 17.1).

17.5.2 CRISPR in Developing Resistance to Fungal Pathogens

Numerous methods were developed to increase fungal resistance in plant based on the information pertaining to molecular pathways implicated in the interaction of pathogens with plants. At present, several viable CRISPR/Cas9 editing targets have been reported with potential in imparting resistance to fungi. It was reported that plants such as tomato, grapevine and bread wheat have MLO (encoding membrane localized protein) loci that is often targeted by Cas9 endonuclease (Malnoy et al. 2016; Nekrasov et al. 2017). In the homozygous loss of function mutant condition, MLO known as one of the susceptible gene was found imparting resistance to powdery mildew among tomato, barley and Arabidopsis (Piffanella et al. 2004; Consonni et al. 2006; Bai et al. 2008). CRISPR/Cas9 mediated regulation of OsERF922 and OsSEC3A in rice imparts resistance to blast disease caused by *M. oryza* without interfering with the plant's normal growth (Wang et al. 2016). CRISPR/Cas mediated targeting of PMR4 gene ortholog SSIPMR4 imparts resistance to Powdery

Table 17.1 CRISPR/Cas- based crop improvement with resistance to bacterial, viral and fungal diseases.

Crop	Disease resistance	Targeted gene	CRISPR/Cas system	Transformation method	Result	Reference
Watermelon	Fungus	*Clpsk1*	CRISPR/Cas9	*Agrobacterium*	Imparts resistance to *Fusarium oxysporum* f. sp. niveum	Zhang et al. (2020a)
Arabidopsis	Potyvirus	*eIF(iso)4E*	CRISPR/Cas9	Floral dip	Imparts resistance to *Turnip Mosaic Virus* (TuMV)	Pyott et al. (2016)
	Bacterial	*AtIAN9*	CRISPR/Cas9	*Agrobacterium*	Imparts resistance to *Pseudomonas syringae*	Wang et al. (2019b)
	Fungus	*At2OGO* (DMR6-like)	CRISPR/Cas9	*Agrobacterium*	Imparts resistance to *Fusarium graminearum*	Low et al. (2020)
	Fungus	*AtErf019*	CRISPR/Cas9	*Agrobacterium*	Imparts resistance to *Phytophthora parasitica*	Lu et al. (2020)
Rice	Bacteria	*Os8n3*	CRISPR/Cas9	*Agrobacterium*	Imparts resistance to *Xanthomonas oryzae* pv. *Oryzae*	Kim et al. (2019)
	Viral	*eIF4G*	CRISPR/Cas9	*Agrobacterium*	Imparts resistance to *Rice Tungro Spherical Virus*	Macovei et al. (2018)
	Bacteria	*OsSWEET14*	CRISPR/Cas9	*Agrobacterium*	Imparts resistance to *Xanthomonas oryzae* pv. *Oryzae*	Duy et al. (2021)
	Fungus	*OsERF922*	CRISPR/Cas9	*Agrobacterium*	Imparts resistance to *Magnaporthe oryzae*	Zhang et al. (2018c)
Apple	Bacteria	*MpDIPM-1/2/4*	CRISPR/Cas9 RNP	*Agrobacterium*	Imparts resistance to *Erwinia amylovora*	Malnoy et al. (2016)
Tomato	Bacteria	*SlJAZ2*	CRISPR/Cas9	*Agrobacterium*	Imparts resistance to *Pseudomonas syringae* pv. *Tomato (pto) DC3000*	Ortigosa et al. (2019)
	Fungus	*SiMlo*	CRISPR/Cas9	*Agrobacterium*	Imparts resistance to Powdery mildew	Nekrasov et al. (2017)

	Bacteria	*SIDMR6*	CRISPR/Cas9	*Agrobacterium*	Imparts resistance to *Pseudomonas syringae, Xanthomonas* spp. *Phytophthora capsici,*	Paula de Toledo Thomazella et al. (2016)
	Fungus	*SIPMR4*	CRISPR/Cas9	*Agrobacterium*	Imparts resistance to *Oidium neolycopersici*	Santillán Martínez et al. (2020)
Barley	Fungus	*HvMorc1*	CRISPR/Cas9	*Agrobacterium*	Imparts resistance to *Blumeria graminis* (Bgh), *Fusarium graminearum*	Kumar et al. (2018)
Citrus	Bacteria	*CsLOBT*	CRISPR/Cas9	*Agrobacterium*	Imparts resistance to *Xanthomonas citri*. subsp. *citri*	Peng et al. (2017)
Orange	Bacteria	*CsWRKY22*	CRISPR/Cas9	*Agrobacterium*	Imparts resistance to *Xanthomonas citri*. subsp. *citri*	Wang et al. (2019a)
Grape	Fungus	*VvMLO-7*	CRISPR/Cas9 RNP	*Agrobacterium*	Imparts resistance to *Erysiphe necator*	Malnoy et al. (2016)
	Fungus	*VvWYRKY52*	CRISPR/Cas9	*Agrobacterium*	Imparts resistance to *Botrytis cinerea*	Wang et al. (2018)
Cucumber	Virus	*Cseif4e*	CRISPR/Cas9	*Agrobacterium*	Imparts resistance to *Cucumber Vein Yellowing Virus, Zucchini Yellow Mosaic Virus, Papaya Ring-spot Mosaic Virus-W*	Chandrasekaran et al. (2016)
Cotton	Fungus	*14-3-3*	CRISPR/Cas9	*Agrobacterium*	Imparts resistance to *Verticillium dahliae*	Zhang et al. (2018c)

(*Continued*)

Table 17.1 (Continued)

Crop	Disease resistance	Targeted gene	CRISPR/Cas system	Transformation method	Result	Reference
Wheat	Fungus	*Mlo*	CRISPR/Cas9	Gene gun and protoplast	Imparts resistance to Powdery mildew	Wang et al. (2014)
	Fungus	*TaEDR1*	CRISPR/Cas9	Gene gun and protoplast	Imparts resistance to *Blumeria graminis*	Zhang et al. (2017)
	Fungus	*TaNFXL1*	CRISPR/Cas9	Gene gun and protoplast	Imparts resistance to *Fusarium graminearum*	Brauer et al. (2020)
Duncan grapefruit	Bacteria	*Cslob1*	CRISPR/Cas9 Promoter	*Agrobacterium*	Imparts resistance to *Xanthomonas citri*	Jia et al. (2016a)
Cassava	Viral	*MenCBP-1/2*	CRISPR/Cas9	*Agrobacterium*	Imparts resistance to *Cassava Brown Streak Virus* (CBSV)	Gomez et al. (2019)
N. benthamiana	Viral	*GFP1, GFP2, HC-Pro, CP*	CRISPR/Cas9	*Agrobacterium*	Imparts resistance to *Turnip Mosaic Virus* (TuMV)	Aman et al. (2018)
Oilseed rape	Fungus	*BnWRKY70*	CRISPR/Cas9	*Agrobacterium*	Imparts resistance to *Sclerotinia sclerotiorum*	Sun et al. (2018)
	Fungus	*BnCRT1a*	CRISPR/Cas9	*Agrobacterium*	Imparts resistance to *Verticillium longisporum*	Pröbsting et al. (2020)
Cacao	Fungus	*TcNPR3*	CRISPR/Cas9	*Agrobacterium*	Imparts resistance to *Phytophthora tropicalis*	Fister et al. (2018)
Soybean	Viral	*GmF3H1/2, FNSII-1*	CRISPR/Cas9	*Agrobacterium*	Imparts resistance to *Soybean Mosaic Virus* (SMV)	Zhang et al. (2020b)

mildew causing agent *Oidium neolycopersici* (Imam et al. 2016; Martinez et al. 2020). It is reported that CRISPR/Cas mediated alteration of the mitogen-activated protein kinase 3 (SIMAPK3) in tomato attribute it with resistance to *B. cinerea* (Zhang et al. 2018b). The enhanced level of resistance to *P. palmivora* was achieved by using the CRISPR/Cas9 approach to produce a mutant papaya plant expressing a functional Ppa1EPIC8 (cysteine protease inhibitor) gene (Gumtow et al. 2018). CRISPR/Cas mediated targeting of the FcWRKY40 gene in Fusarium crown-rot susceptible wheat was found associated with alteration in the expression of genes involved in defence responses, leading to enhanced resistance to fungus (Pavase et al. 2021).

17.5.3 CRISPR in Developing Resistance to Different Bacteria

There are fewer studies reported on the application of CRISPR/Cas towards imparting resistance to bacterial infections than that on viral and fungal resistance. The proteobacteria *Xanthomonas oryzae* pv. Oryza employs type III transcription activator factors (TALEs) in stimulating the gene expression that imparts resistance to host. Transfer of OsSWEET13 (a PthXo2 effector protein for *X. oryzae* pv. Oryzae) from indica rice IR24 to japonica rice increases resistance to bacterial blight (Zhou et al. 2015). In the bacterial speck disease on tomato, *Pseudomonas syringae* pv. Tomato mediates production of coronatine (COR) that promotes bacterial invasion through stomatal opening. In Arabidopsis, the COR co-receptor AtJAZ2 seems essential for the stomatal response to COR (Gimenez-Ibanez et al. 2017). The shortened version of JAZ2 (JAZ21jas) lacking the essential C-terminal Jas domain, prevents stomatal opening in response to COR. With the help of CRISPR/Cas9, the tomato homolog of AtJAZ2 (SIJAZ2) was modified in such a way that it prevented COR-induced stomatal opening and as such impart resistance to biotrophic bacteria. CRISPR offers a resistance option to most significant bacterial disease, Citrus canker (Tigosa et al. 2019). One of the most harmful citrus disease, brought on by the bacterium *Xanthomonas citri* subsp. citri, reduces yields in nations where citrus is grown. Recent research from two investigations showed that citrus trees might be made resistant to bacterial canker using CRISPR Cas9. Jia et al. (2016a) modified the PthA4 effector binding site in the promoter region of CsLOB1 gene that imparts canker resistant among Duncan grape fruit. The mutant lines showed a reduction in the symptoms of canker disease without any alteration in other phenotypic characteristics. It was also reported that no additional LOB family genes appeared to have any a possible off-target changes after PCR sequencing (Jia et al. 2016b). Peng et al. (2017) observed a strong correlation between CsLOB1 promoter activity and susceptibility of Wanjincheng orange (*Citrus sinensis*) to CBC disease. There was a rise in CBC resistance on deleting the EBEthA4 sequence from CsLOB1 alleles (Peng et al. 2017). The enterobacterium *Erwinia amylovora* is a causative organism for the fire blight disease in apples and other members of the Rosaceae family. The DspE interacting proteins of *E. amylovora* on interacting with DIPM genes including DIPM 1, 2, 3 and 4 impart resistance to fire blight disease (Malnoy et al. 2016; Pinto et al. 2017). It has been reported that preassembled sgRNA/Cas9 ribonucleoproteins (RNPs) directly into plant protoplast offers a number of advantages such as quick targeting effectiveness, improved target performance with significant off-target activity (Malnoy et al. 2016).

17.6 CRISPR/Cas in Abiotic Stress Tolerance in Crops

To address the growing problems posed by the climate change in view of the expanding global population, different projects were taken that encourages adoption and promotion of sustainable agricultural approaches. Of them, imparting resistance to crops against different stresses is considered as a promising strategy towards improving growth and overall productivity of the crops. In the current day scenario of advancement in genetics and breeding procedures, structural as well as functional components of crop genomes are better elucidated that increases the likelihood of possible improvements in their genomes on applying the modern genomic editing approaches (Yamamoto et al. 2018; Salava et al. 2021). To combat the negative impact of abiotic stresses on crops, it is being formulated that there is an urgent need to develop new crop varieties or enhance the productivity of existing ones using both conventional selection or breeding techniques or boost the breeding system via adoption of different platforms of genome editing technologies. As CRISPR/Cas system is one of these subtypes that has undergone a great deal of research in achieving precise alteration in the genetic makeup of different crops, the following sections deals with alteration and analysis of the gene function as part of enhancement in the crop productivity.

17.6.1 CRISPR/Cas in Temperature Stress Tolerance

Agriculture has been severely impacted by increase in heat and cold pressures. Cold stress has an impact on grain yield, biomass build-up and leaf photosynthesis (Jiang et al. 2020). In response to cold stress, proline-rich proteins (PRPs) play important role in regulating different physiological and biochemical functioning; thereby to overall growth of the plants. On removing the OsPRP1encoding proline-rich protein from rice using CRISPR/Cas approach, it was found to have more sensitiveness to the cold (Nawaz et al. 2019). In another study, removal of the R2R3-type MYB gene, OsMYB30 that shows greater response to cold by adopting CRISPR/Cas, the mutant was upholding more resistance to cold than wild-type rice (Zeng et al. 2020). On exposing *Arabidopsis thaliana* to low temperature, a significant enhancement in the expression of CBF (CCAAT-binding factor) genes was observed. CRISPR/Cas mediated generation of single, double and triple CBF gene mutants (represented as CBF1, CBF2 & CBF3) in *Arabidopsis thaliana* has confirmed their contribution to cold acclimatization (Jia et al. 2016b). Of them, the triple mutant was found exhibiting a highly sensitive phenotype to freezing temperatures compared with the cbf1-cbf3 double mutants and cbf3 mutants. The transcription of COR genes is encouraged by the CBF proteins, which increases the resistance of plants to freezing (Zhao et al. 2016). SlCBF1, the only member of the CBF gene family in tomato, also exhibit high inducibility to cold (Li et al. 2018). Salicyclic acid and H_2O_2 induced enhancement in the expression of SlCBF1 was found attributing cold tolerance to tomato (Wang et al. 2017b). CRISPR/Cas mediated generation of the SlCBF1 mutants on exposure to cold stress exacerbate membrane damage; thereby shows electrolyte leakage and higher accumulation of malondialdehyde (MDA) as compared to wild type (Li et al. 2018). In a similar manner, the cis-regulatory elements constituting the promoter region of OsAnn5 in rice was found having numerous similar sequence elements. Of the, two unique sequences denoted as MYB recognition site and dehydration responsive element (DRE) are implicated to have an important role in the CBF-mediated cold response in rice. In OsAnn5, elements controlling plant hormone production and response to light were found organized between start codon and the region represented around 2082 bp; thereby suggesting regulation of the expression of OsAnn5 by a series of transcription factors. CRISPR/Cas mediated deletion of the OsAnn5 was found increasing relative electrical conductivity (REC) and decreased survival as compared with the wild type rice; thereby attributing the function of cold stress tolerance with greater survival rate for growth of rice under cold stress conditions (Shen et al. 2017; Boyidi et al. 2021).

Excess of heat exerts detrimental effect on varied biological processes, being it be turgor, photosynthetic rate, nutrient acquisition, cellular metabolism, chlorophyl concentration and exchange of gases at the leaf surface, which progresses with decrease in the productivity of crops (Lisar et al. 2012). In rice, fructokinase-like protein 2 encoded by heat-stress sensitive albino 1 (OsHSA1) gene was found having a prominent role in the chloroplast production and heat stress defence. CRISPR/Cas mediated knockdown of the OsHSA1 gene results in the development of sensitive mutants than the wild type rice plants. A significant decrease in the cellular antioxidant enzyme (catalase, ascorbate peroxidase etc.) activity was observed on creating the knockdown of calcium dependent protein kinase-28 (Slcpk28) in tomato using CRISPR/Cas approach, which progresses with increased accumulation of reactive oxygen species (ROS) and an enhanced protein oxidation level (Hu et al. 2021). A class of steroid plant hormones known as Brassinosteroid can increase a plant's ability to with stand a variety of abiotic stresses (Fariduddin et al. 2014). In tomato, the Brassinosteroid response (BR) is significantly regulated by the Brassinazole Resistant-1(SlBZR1) gene. In response to enhancement in the ROS, the effect of damage of heat stress on bzr1 increases that progresses through loss in the tolerance attributed by Brassinosteroids (Yin et al. 2018). Considering the critical nature of antioxidants activated in response to oxidative stress, CRISPR/Cas based removal of the OsNAC006 attributed it with the function of improving heat tolerance via, regulation in the production of antioxidants and subsequent enhancement in the photosynthetic activity (Wang et al. 2020a).

A large number of studies have suggested the possibility of modifications in the genes such as phytochrome (PHY, referred to as thermosensor) attributing plants with the property to tolerate variation in temperature (Legris et al. 2019). In tomato and Arabidopsis, alteration of the PHY gene was found having significant effect on heat stress tolerance (Abdellatif et al. 2022). High initial rate respiration in pollen may be produced by the depletion of endogenous respiratory substrates, subsequent ageing and a decline in mitochondrial activity brought on by high temperatures, which can impair fruit set and cause the pollen to abort (Sita et al. 2017). Using modern approaches such as CRISPR/Cas, it was reported that parthenocarpic phenotype in tomato is associated with the SlAGAMOUS-like-6 (SlAGL6) gene encoding the MADS-box. With significant increase in the yield, the AGL6 mutant lines of tomato produce seedless fruits resembling wild-type seeded fruits in terms of weight and shape with facultative parthenocarpy without any signs of pleiotropic effects (Klap et al. 2017). Compared to wild-type tomato seedlings with compound leaves, the CRISPR/Cas developed mutant lines harbour simple leaves with parthenocarpic fruits developed as a result of fruit initiation and development happening prior to fertilization (Ueta et al. 2017). As gibberellin signalling is negatively regulated by DELLA, tomatoes with SlDELLA loss-of-function mutations display the parthenocarpic phenotype

(Shinozaki et al. 2018). The Target-AID system's base-edition generated SlDELLA mutant displayed strong gibberellin sensitivity and a parthenocarpic phenotype; thereby suggesting possibility of enhancement in achieving parthenocarpy in tomatoes following genome-editing approaches (Kashojiya et al. 2022).

17.6.2 Drought Stress Responses

Drought – an important environmental variable, affects plant growth and drastically reduce agricultural productivity. In tomato, SlLBD40 (member of the Lateral Organ Boundaries Doman (LBD) family) gene was found highly expressed in root and fruits of the plant. In a study, knockout of SlLBD40 gene using CRISPR/Cas resulted in reducing water loss compared with the wild type plants (Liu et al. 2020a). The leaves of the knockdown mutant were found to withstand the wilting even after 24 hours of watering, while the wildtype plants succumb to wilting (Chen et al. 2021a). In Arabidopsis, knockdown of the transcription factor AITR by CRISPR/Cas method were found to improve their resistance to salt and drought, with a simultaneous decrease in their susceptibility to abiotic stressors like ABA (Chen et al. 2021b). CRISPR/Cas based mutation in the open stomata-2 (OST-2) gene encoding plasma membrane embedded H^+-ATPase with a finite role in controlling the stomatal response, reduces loss of water via, a high check on the activity of stomatal closure; thereby ability of Arabidopsis to with stand drought (Osakabe et al. 2016; Yamauchi et al. 2016). The histone acetyltransferase (HAT) from Arabidopsis improves chromatin relaxation and encourages gene expression. AREB1 (ABA-responsive element binding protein) is considered to have an active role as a positive regulator in the drought stress response in plants. In Arabidopsis, histone acetyltransferase (HAT) improves chromatin relaxation and encourages gene expression. In a study, HAT coupled with the dCas9 was found boosting promoter activity of AREB1; thereby improving survival of plants to drought stress (Roca Paixão et al. 2019). By employing the CRISPR/Cas approach, translocation of GOS2 promoter to a locus associated in controlling ARGOS8 transcript levels, the mutant lines were found increasing maize grain production with no significant decline for growth under drought stress conditions (Shi et al. 2017). In rice, OsERA1 functions as a negative regulator of drought stress responses. CRISPR/Cas mediated generation of the mutant lines on growing under drought stress conditions were found to have lower relative stomatal conductance rates along with enhanced sensitivity to ABA (Ogata et al. 2020). CRISPR/Cas mediated editing of the SRL1 and SRL2 genes governing the leaf shape were found exhibiting rolled leaf phenotype with higher amount of ABA and a lower level of MDA for growth under drought stress conditions, confirming the condition of being more tolerant to drought than their wild-counterparts (Liao et al. 2019).

Addition to rice, several crops including wheat (Nazir et al. 2022), soybean (Wang et al. 2021), maize (Zhang et al. 2018a) and tomato (Tran et al. 2021), have been benefited from CRISPR/Cas9 genome editing. As large number of physiological and biochemical processes are disrupted by drought, roles of different genes and phytohormone signalling pathways was elucidated and were found to have an active role in drought stress responses (Amnan et al. 2022). As ABA regulates the relative water content of the plants and their response to drought stress, main focus was put on the genes implicated in ABA signalling as part of developing drought stress tolerance lines in different crops. In this procedure, the function of OsABA8ox2 encoding ABA-8′ hydroxylase was identified in the drought tolerance lines of rice (Zhang et al. 2020c). CRISPR/Cas-mediated development of the OsABA8ox2 deletion lines in rice was found to have a higher amount of ABA in root, attributing it with the drought tolerance properties. Contrarily, rice with over-expressed OsABA8ox2 gene displays shows a significant reduction in root growth and as such hypersensitivity to drought stress (Zhang et al. 2020c). Alteration of the ERA1 encoding a subunit component of Farnesyltransferase in Japonica rice (Nipponbare variety) led to mutant lines that showed improved ABA sensitivity and as such increases the probability of its use in generating drought tolerant variety of different crops (Ogata et al. 2020). Yin et al. (2017) reported a significant reduction (~8 fold) in the stomatal density of CRISPR/Cas edited rice plants for OsEPFL9 (Epidermal Patterning Factor like-9). Considering their potential to withstand drought stress, lowering of stomatal density to ~50% in barley and wheat led to significant in carbon sequestration and conductance and improvement in efficiency of water usage (Hughes et al. 2017; Caine et al. 2019). In grapevine, CRISPR/Cas mediated editing of VvEPFL9-1 caused about 60% decline in stomatal density, with impaired carbon sequestration with respect to wild type (Clemens et al. 2022). Similarly, in tomatoes CRISPR/Cas mediated editing of SlMAPK3 caused severe damage to cell membrane and enhancement in the wilting signs on exposure to drought stress (Wang et al. 2017b). Contrary to gene knockout (KO) approach, the CRISPR/Cas gene editing technique also opens up the possibility to construct knockins of any set of desired genes. In this approach, Shimatani et al. (2017) inserted a desired promoter for the ARGOS8 gene and observed an enhancement in the grain productivity for growth under water stress conditions. The complexity of abiotic stress tolerance, which requires manipulation of multiple genes to change the trait of interest, accounts for the paucity in research findings on targeting different genes for abiotic stress tolerance. Together, these findings provide foundation for setting up CRISPR/Cas system in alteration of the plant genome and enables us to learn more about the function of genes involved in abiotic stress response.

17.6.3 Salinity Stress Responses

The CRISPR/Cas technique is used in improving the crop survival under changing environmental conditions. High salt concentration endangers plant growth and development via, development of osmotic stress, nutritional imbalance, ionic toxicity and oxidative stress (Shrivastava and Kumar 2015). Inside the plants, the salt concentration are kept in check by controlling biochemical, physiological and molecular mechanisms that modify ionic tolerance, tissue-specific tolerance and osmotic balance. Many TFs serve as beneficial salt-stress-responsive factors (Hrmova and Hussain 2021). A common strategy adopted for elucidating information of the molecular pathways is achieved by understanding the role of TFs in salt-response pathways following CRISPR mediated targeting of TFs. Following knockout of salt stress associated genes including TFs, CRISPR/Cas mediated editing of plants either increases susceptibility of plants to stress or imparts tolerance for growth under salt stress conditions. Based on the KO effect, positive regulators are thought to include genetic variables that promote stress tolerance. CRISPR/Cas mediated knockdown of the OsRR22 gene results in developing cultivars of rice with enhanced resistance to salt stress (Zhang et al. 2019). Similarly, targeting of the OsNAC041 in rice results in developing cultivars that uphold a taller phenotype compared to their wild relatives (Bo et al. 2019). RAV encoding a TF of AP2/ERF family in involved in adapting plants to salt stress (Xie et al. 2019; Faraji et al. 2020). Knowing induction in the expression of OsRAV2 gene in response to salt stress, Duan et al. (2016) ascertained the function of GT-1 region in the promoter region of OsRAV2 gene in conferring resistance to salt stress. Similarly, Liu et al. (2020b) observed salt hypersensitive phenotype for the GT-2 mutation in the rice cultivars having OsGT-2 deletion. By employing the CRISPR/Cas approach, a large series of kinases, phosphatases or TFs were studied in plants with respect to their susceptibility to salt stress. Examples of kinase and phosphatase genes that have been studied include FLN2 and BBS1 in rice (Zeng et al. 2020), MIR528 in rice (Zhou et al. 2015), ITPK1 in barley (Vlcko and Ohnoutkova 2020) and SAUR41 in Arabidopsis. The plants edited for these genes show reduced salt-responsive phenotypes (Qiu et al. 2020).

CRISPR/Cas mediated knockdown of the RR22 (a type-B response regulator) having role in the cytokine signalling attributed with enhancement in tolerant to salt in rice (Zhang et al. 2019). Development of the transgene free lines using T1 segregation, the homozygous lines present possibility of an opportunity for trait introgression in related crop varieties. CRISPR/Cas mediated editing of the SPL10 (members of the SPL family) in rice results in the development of lines with salt tolerance phenotype (Kausar and Komatsu 2022). In MTU1010 cultivar of indica rice, base-edited mutation of the zinc finger TF, DST results in the agronomic enhancement of performance under drought and salinity stress conditions (Saradadevi et al. 2021). Examples of CRISPR mediated editing of the negative regulators includes HyPRP1 (a cell wall embedded structural protein) mutation in tomato and C/VIF1 (proteinaceous inhibitor of fructosidases) mutation in rice (Wang et al. 2022). Generally, plants use transporter proteins for movement of ions across short (cell to cell) and long (across different tissues) distances (Gill et al. 2021). It is reported that plants use different ion transporters to maintain the proper Na^+/K^+ balance and as such withstand the salt stress. To keep a check on the operating mechanisms, it is imperative to study different members of the Na^+ transporter family (ENA, NHX, HKT, ATPases and other cation exchangers) along with KT/HAK/KUP family, VGICs and KCO constituting members of the K^+ transporter family (Shelake et al. 2022). CRISPR mediated editing of the HKT1 in maize following a 5 bp deletion or truncation of a 34 bp fragment helped in establishing a positive regulatory role of HKT1 (Zhang et al. 2018a). Similarly, CRISPR mediated editing of the ion transporters NHX5 (Khan et al. 2020) and CLCg (Raza et al. 2022) among soybean and maize enhanced their tolerance to salt on having alteration at these loci (Wang et al. 2020b).

17.6.4 Metal Stress Tolerance

The build-up of metal ions in plants disturbs the ionic balance that progresses with build-up of heavy metals that negatively impact the productivity of the plants. Inside the plants, heavy metals exert their effect through enhancement in the oxidative stress followed by damage of the cellular structures (Yadav 2010). To minimize the negative effects following build-up of the heavy metals, plants have developed a series of detoxifying strategies including effective neutralization of metal ions in the protoplast, regulated removal of harmful ions from roots and appropriation or transfer to distant organs (Sruthi et al. 2017).

To increase tolerance to heavy metal exposure, different genes control these pathways (Mahmud et al. 2019). CRISPR/Cas technology was employed in several research to lessen mineral toxicity. Nieves-Cordones et al. (2017) created low cesium-containing rice plants by employing the CRISPR/Cas9 system to inactivate the K^+ transporter OsHAK1. OsNramp5 and OsARM1 were knocked out in rice, and the results demonstrated increased arsenic tolerance (Wang et al. 2017a) and minimal cadmium accumulation (Tang et al. 2017a). It has been suggested that the loss-of-function mutants of OXP1 and α-glutamylcyclotransferase exhibit heavy metal and xenobiotic detoxification due to increased glutathione (GSH) accumulation because, for instance, the loss-of-function mutant of α-glutamylcyclotransferase displayed defensive characteristics against heavy metal toxicity (Singh and Laxmi 2015). Therefore, creating plant mutants using the CRISPR-Cas9 system would be helpful to combat the heavy metal stress in plants. Baeg et al. (2021) recently developed oxp1/CRISPR mutant Arabidopsis plants that shown resistance to Cd suggests that mutant plants are better at heavy metal detoxification than WT Col0 plants. As a result, this study demonstrated how to use the gene-editing technique to confer resistance to xenobiotics and heavy metals in plants by indel mutations (Baeg et al. 2021). With the help of the transporters OsNramp1, OsNramp5 and OsCd1, rice roots take Cd from the soil. Xylem loading is adversely regulated by OsHMA3's participation in Cd sequestration within the root vacuole, and OsLCT1 is implicated in Cd transport to the grains (Chen et al. 2019a). Cd in the grain crop has been somewhat reduced by genome editing techniques that manipulate the expression of these transporter genes. Low Cd levels were produced in rice by OsNramp5 and OsLCT1 gene mutants created using CRISPR-Cas9. (Tang et al. 2017a). Similar to how OsARM1 controls transporter genes related to as in rice. The basal and upper nodes of the vascular bundle's phloem express it. The OSARM1 is improved by CRISPR knockout whereas its overexpression makes people more sensitive to As. (Wang et al. 2017a). The main mechanism for Cs^+ absorption and translocation in rice is the Cs^+-permeable OsHAK1 transporter. The CRISPR-Cas technology was utilised to create transgenic plants lacking the function of OsHAK1 in order to reduce the radioactive caesium (Cs) uptake by rice plants in Fukusima soil contaminated with 137 Cs^+. The radioactive caesium contents and 137 Cs^+ levels in the roots were noticeably lower in the OsHAK1 knockout plants (Nieves-Cordones et al. 2017). Another use of the crisper-cas-based editing in rice is to understand the role of a potential target, OsPRX2, for enhanced tolerance to potassium deprivation. In a K^+ limiting situation, OsPRX2 is known to lessen ROS generation. It was discovered that OsPRX2 knockdown results in major deficiencies in leaves phenotypic and the stomata opening under the K^+-deficiency tolerance, whereas OsPRX2 overexpression causes the stomatal closing and K^+ deficiency tolerance to rise.

17.7 Conclusion

Genome engineering has emerged as a powerful tool for the development of climate-resilient crop plants. Zinc finger nucleases (ZNFs), transcription activator-like effector nucleases (TALENS) and clustered regularly interspaced short Palindromic Repeats (CRISPR) associated protein 9 (Cas9) are some of the most commonly used genome editing tools for the development of such crops. These tools have proven to be effective in introducing precise and targeted changes in the genome, thus leading to improved abiotic stress tolerance in plants. For example, ZNFs have been used to modify drought tolerance genes in crops, while TALENS and CRISPR/Cas9 have been used to edit genes that regulate salt and temperature stress responses. The development of climate resilient crops will have a significant impact on global food security, particularly in regions affected by the changing climate patterns. However, as with any new technology, there are ethical and regulatory concerns that must be addressed. Despite, these challenges, the future of genome editing technique in the development of climate-resilient crop plants looks promising, and we can expect to see continued advances in this field in the coming years.

Author Contributions

All authors contributed equally in generating the draft of the different sections and in the repeated editing of the contents present in the finalized version of the manuscript. All authors have read and agreed to the publication of the final version of the manuscript.

Funding

The work was supported by J&K Science Technology and Innovation Council (J&K ST&IC), India grant (JK ST&IC/SRE/996-998).

Acknowledgements

Authors would like to thank their colleagues, whose suggestions helped in improving the contents of the manuscript.

Conflicts of Interest

The authors declare no conflict of interest.

References

Abdellatif, I.M., Yuan, S., Na, R. et al. (2022). Functional characterization of tomato phytochrome A and B1B2 mutants in response to heat stress. *International Journal of Molecular Sciences* 23 (3): 1681.

Ali, Z., Abulfaraj, A., Idris, A. et al. (2015). CRISPR/Cas9-mediated viral interference in plants. *Genome Biology* 16: 1–11.

Alok, A., Chauhan, H., Upadhyay, S.K. et al. (2021). Compendium of plant-specific CRISPR vectors and their technical advantages. *Life* 11: 1021. https://doi.org/10.3390/life11101021.

Aman, R., Ali, Z., Butt, H. et al. (2018). RNA virus interference via CRISPR/Cas13a system in plants. *Genome Biology* 19 (1): 1–9.

Amnan, M.A.M., Aizat, W.M., Khaidizar, F.D., and Tan, B.C. (2022). Drought stress induces morpho-physiological and proteome changes of *Pandanus amaryllifolius*. *Plants* 11 (2): 221.

Anderegg, W.R., Trugman, A.T., Badgley, G. et al. (2020). Climate-driven risks to the climate mitigation potential of forests. *Science* 368 (6497): eaaz7005.

Aryal, J.P., Sapkota, T.B., Khurana, R. et al. (2020). Climate change and agriculture in South Asia: adaptation options in smallholder production systems. *Environment, Development and Sustainability* 22 (6): 5045–5075.

Arzanesh, M.H., Alikhani, H.A., Khavazi, K. et al. (2011). Wheat (*Triticum aestivum* L.) growth enhancement by Azospirillum sp. under drought stress. *World Journal of Microbiology and Biotechnology* 27: 197–205.

Ashfield, T., Egan, A.N., Pfeil, B.E. et al. (2012). Evolution of a complex disease resistance gene cluster in diploid phaseolus and tetraploid Glycine. *Plant Physiology* 159 (1): 336–354.

Asrar, A.W.A. and Elhindi, K.M. (2011). Alleviation of drought stress of marigold (*Tagetes erecta*) plants by using arbuscular mycorrhizal fungi. *Saudi Journal of Biological Sciences* 18 (1): 93–98.

Baeg, G.J., Kim, S.H., Choi, D.M. et al. (2021). CRISPR/Cas9-mediated mutation of 5-oxoprolinase gene confers resistance to sulfonamide compounds in Arabidopsis. *Plant Biotechnology Reports* 15: 753–764.

Bai, Y., Pavan, S., Zheng, Z. et al. (2008). Naturally occurring broad-spectrum powdery mildew resistance in a Central American tomato accession is caused by loss of Mlo function. *Molecular Plant-Microbe Interactions* 21 (1): 30–39.

Baltes, N.J., Hummel, A.W., Konecna, E. et al. (2015). Conferring resistance to geminiviruses with the CRISPR–Cas prokaryotic immune system. *Nature Plants* 1 (10): 1–4.

Bhalothia, P., Yajnik, K., Alok, A., and Upadhayay, S.K. (2020). The current progress of CRISPR/Cas9 development in plants. In: *Genome Engineering via CRISPR-Cas9 Systems* (ed. V. Singh and P.K. Dhar), 123–129. Academic Press. https://doi.org/10.1016/B978-0-12-818140-9.00011-8.

Bhat, M.A., Bhat, M.A., Kumar, V. et al. (2020). The era of editing plant genomes using CRISPR/Cas: a critical appraisal. *Journal of Biotechnology* 324: 34–60.

Bhat, M.A., Mir, R.A., Kumar, V. et al. (2021). Mechanistic insights of CRISPR/Cas-mediated genome editing towards enhancing abiotic stress tolerance in plants. *Physiologia Plantarum* 172 (2): 1255–1268.

Blanvillain-Baufumé, S., Reschke, M., Solé, M. et al. (2017). Targeted promoter editing for rice resistance to *Xanthomonas oryzae* pv. oryzae reveals differential activities for SWEET 14-inducing TAL effectors. *Plant Biotechnology Journal* 15 (3): 306–317.

Bo, W., Zhaohui, Z., Huanhuan, Z.H.A.N.G. et al. (2019). Targeted mutagenesis of NAC transcription factor gene, OsNAC041, leading to salt sensitivity in rice. *Rice Science* 26 (2): 98–108.

Boch, J., Scholze, H., Schornack, S. et al. (2009). Breaking the code of DNA binding specificity of TAL-type III effectors. *Science* 326 (5959): 1509–1512.

Borrelli, V.M., Brambilla, V., Rogowsky, P. et al. (2018). The enhancement of plant disease resistance using CRISPR/Cas9 technology. *Frontiers in Plant Science* 9: 1245.

Boyidi, P., Trishla, V.S., Botta, H.K. et al. (2021). Heterologous expression of rice annexin OsANN5 potentiates abiotic stress tolerance in transgenic tobacco through ROS amelioration. *Plant Stress* 2: 100022.

Brauer, E.K., Balcerzak, M., Rocheleau, H. et al. (2020). Genome editing of a deoxynivalenol-induced transcription factor confers resistance to *Fusarium graminearum* in wheat. *Molecular Plant-Microbe Interactions* 33 (3): 553–560.

Butt, H., Eid, A., Ali, Z. et al. (2017). Efficient CRISPR/Cas9-mediated genome editing using a chimeric single-guide RNA molecule. *Frontiers in Plant Science* 8: 1441.

Cai, Y., Chen, L., Sun, S. et al. (2018). CRISPR/Cas9-mediated deletion of large genomic fragments in soybean. *International Journal of Molecular Sciences* 19 (12): 3835.

Caine, R.S., Yin, X., Sloan, J. et al. (2019). Rice with reduced stomatal density conserves water and has improved drought tolerance under future climate conditions. *New Phytologist* 221 (1): 371–384.

Chandrasekaran, J., Brumin, M., Wolf, D. et al. (2016). Development of broad virus resistance in non-transgenic cucumber using CRISPR/Cas9 technology. *Molecular Plant Pathology* 17 (7): 1140–1153.

Chen, W., Yao, X., Cai, K., and Chen, J. (2011). Silicon alleviates drought stress of rice plants by improving plant water status, photosynthesis and mineral nutrient absorption. *Biological Trace Element Research* 142: 67–76.

Chen, J., Zou, W., Meng, L. et al. (2019a). Advances in the uptake and transport mechanisms and QTLs mapping of cadmium in rice. *International Journal of Molecular Sciences* 20 (14): 3417.

Chen, K., Wang, Y., Zhang, R. et al. (2019b). CRISPR/Cas genome editing and precision plant breeding in agriculture. *Annual Review of Plant Biology* 70: 667–697.

Chen, M., Zhu, X., Liu, X. et al. (2021a). Knockout of auxin response factor SlARF4 improves tomato resistance to water deficit. *International Journal of Molecular Sciences* 22 (7): 3347.

Chen, S., Zhang, N., Zhou, G. et al. (2021b). Knockout of the entire family of AITR genes in *Arabidopsis* leads to enhanced drought and salinity tolerance without fitness costs. *BMC Plant Biology* 21: 1–15.

Choi, T.S., Kang, E.J., Kim, J.H., and Kim, K.Y. (2010). Effect of salinity on growth and nutrient uptake of *Ulva pertusa* (*Chlorophyta*) from an eelgrass bed. *Algae* 25 (1): 17–26.

Choudhary, M., Chandra, P., Dixit, B. et al. (2022). Plant growth-promoting microbes: role and prospective in amelioration of salt stress. *Communications in Soil Science and Plant Analysis* 53 (13): 1692–1711.

Christian, M., Cermak, T., Doyle, E.L. et al. (2010). Targeting DNA double-strand breaks with TAL effector nucleases. *Genetics* 186 (2): 757–761.

Clemens, M., Faralli, M., Lagreze, J. et al. (2022). VvEPFL9-1 knock-out via CRISPR/Cas9 reduces stomatal density in grapevine. *Frontiers in Plant Science* 13: 878001.

Consonni, C., Humphry, M.E., Hartmann, H.A. et al. (2006). Conserved requirement for a plant host cell protein in powdery mildew pathogenesis. *Nature Genetics* 38 (6): 716–720.

Daryanto, S., Wang, L., and Jacinthe, P.A. (2016). Global synthesis of drought effects on maize and wheat production. *PLoS One* 11 (5): e0156362.

Dominguez, A.A., Lim, W.A., and Qi, L.S. (2016). Beyond editing: repurposing CRISPR–Cas9 for precision genome regulation and interrogation. *Nature Reviews Molecular Cell Biology* 17 (1): 5–15.

Duan, Y.B., Li, J., Qin, R.Y. et al. (2016). Identification of a regulatory element responsible for salt induction of rice OsRAV2 through ex situ and in situ promoter analysis. *Plant Molecular Biology* 90: 49–62.

Duy, P.N., Lan, D.T., Pham Thu, H. et al. (2021). Improved bacterial leaf blight disease resistance in the major elite vietnamese rice cultivar TBR225 via editing of the OsSWEET14 promoter. *PLoS One* 16 (9): e0255470.

Fahad, S., Bajwa, A.A., Nazir, U. et al. (2017). Crop production under drought and heat stress: plant responses and management options. *Frontiers in Plant Science* 8: 1147.

Faraji, S., Filiz, E., Kazemitabar, S.K. et al. (2020). The AP2/ERF gene family in *Triticum durum*: genome-wide identification and expression analysis under drought and salinity stresses. *Genes* 11 (12): 1464.

Fariduddin, Q., Yusuf, M., Ahmad, I., and Ahmad, A. (2014). Brassinosteroids and their role in response of plants to abiotic stresses. *Biologia Plantarum* 58 (1): 9–17.

Farooq, R., Hussain, K., Nazir, S. et al. (2018). CRISPR/Cas9; A robust technology for producing genetically engineered plants. *Cellular and Molecular Biology* 64 (14): 31–38.

Fister, A.S., Landherr, L., Maximova, S.N., and Guiltinan, M.J. (2018). Transient expression of CRISPR/Cas9 machinery targeting TcNPR3 enhances defense response in *Theobroma cacao*. *Frontiers in Plant Science* 9: 268.

Gao, C. (2021). Genome engineering for crop improvement and future agriculture. *Cell* 184 (6): 1621–1635.

Garrity, D.P. and O'Toole, J.C. (1994). Screening rice for drought resistance at the reproductive phase. *Field Crops Research* 39 (2–3): 99–110.

Gilbert, L.A., Larson, M.H., Morsut, L. et al. (2013). CRISPR-mediated modular RNA-guided regulation of transcription in eukaryotes. *Cell* 154 (2): 442–451.

Gill, R.A., Ahmar, S., Ali, B. et al. (2021). The role of membrane transporters in plant growth and development, and abiotic stress tolerance. *International Journal of Molecular Sciences* 22 (23): 12792.

Gimenez-Ibanez, S., Boter, M., Ortigosa, A. et al. (2017). JAZ 2 controls stomata dynamics during bacterial invasion. *New Phytologist* 213 (3): 1378–1392.

Gomez, M.A., Lin, Z.D., Moll, T. et al. (2019). Simultaneous CRISPR/Cas9-mediated editing of cassava eIF 4E isoforms nCBP-1 and nCBP-2 reduces cassava brown streak disease symptom severity and incidence. *Plant Biotechnology Journal* 17 (2): 421–434.

Gumtow, R., Wu, D., Uchida, J., and Tian, M. (2018). A *Phytophthora palmivora* extracellular cystatin-like protease inhibitor targets papain to contribute to virulence on papaya. *Molecular Plant-Microbe Interactions* 31 (3): 363–373.

Hassani, A., Azapagic, A., and Shokri, N. (2020). Predicting long-term dynamics of soil salinity and sodicity on a global scale. *Proceedings of the National Academy of Sciences* 117 (52): 33017–33027.

Hrmova, M. and Hussain, S.S. (2021). Plant transcription factors involved in drought and associated stresses. *International Journal of Molecular Sciences* 22 (11): 5662.

Hu, Z., Li, J., Ding, S. et al. (2021). The protein kinase CPK28 phosphorylates ascorbate peroxidase and enhances thermotolerance in tomato. *Plant Physiology* 186 (2): 1302–1317.

Huang, C.H., Shen, C.R., Li, H. et al. (2016). CRISPR interference (CRISPRi) for gene regulation and succinate production in cyanobacterium *S. elongatus* PCC 7942. *Microbial Cell Factories* 15 (1): 1–11.

Hughes, J., Hepworth, C., Dutton, C. et al. (2017). Reducing stomatal density in barley improves drought tolerance without impacting on yield. *Plant Physiology* 174 (2): 776–787.

Iizumi, T., Shiogama, H., Imada, Y. et al. (2018). Crop production losses associated with anthropogenic climate change for 1981–2010 compared with preindustrial levels. *International Journal of Climatology* 38 (14): 5405–5417.

Imam, J., Singh, P.K., and Shukla, P. (2016). Plant microbe interactions in post genomic era: perspectives and applications. *Frontiers in Microbiology* 7: 1488.

Iqbal, N., Sehar, Z., Fatma, M. et al. (2022). Nitric oxide and abscisic acid mediate heat stress tolerance through regulation of osmolytes and antioxidants to protect photosynthesis and growth in wheat plants. *Antioxidants* 11 (2): 372.

Ji, X., Zhang, H., Zhang, Y. et al. (2015). Establishing a CRISPR–Cas-like immune system conferring DNA virus resistance in plants. *Nature Plants* 1 (10): 1–4.

Jia, H., Orbovic, V., Jones, J.B., and Wang, N. (2016a). Modification of the PthA4 effector binding elements in Type I Cs LOB 1 promoter using Cas9/sg RNA to produce transgenic Duncan grapefruit alleviating XccΔpthA4: dCs LOB 1.3 infection. *Plant Biotechnology Journal* 14 (5): 1291–1301.

Jia, Y., Ding, Y., Shi, Y. et al. (2016b). The cbfs triple mutants reveal the essential functions of CBF s in cold acclimation and allow the definition of CBF regulons in *Arabidopsis*. *New Phytologist* 212 (2): 345–353.

Jiang, Y.Y., Chai, Y.P., Lu, M.H. et al. (2020). Prime editing efficiently generates W542L and S621I double mutations in two ALS genes in maize. *Genome Biology* 21: 1–10.

Jinek, M., Chylinski, K., Fonfara, I. et al. (2012). A programmable dual-RNA–guided DNA endonuclease in adaptive bacterial immunity. *Science* 337 (6096): 816–821.

Kang, S.W., Rane, N.S., Kim, S.J. et al. (2006). Substrate-specific translocational attenuation during ER stress defines a pre-emptive quality control pathway. *Cell* 127 (5): 999–1013.

Kashojiya, S., Lu, Y., Takayama, M. et al. (2022). Modification of tomato breeding traits and plant hormone signaling by Target-AID, the genome-editing system inducing efficient nucleotide substitution. *Horticulture Research* 9: uhab004.

Kausar, R. and Komatsu, S. (2022). Proteomic approaches to uncover salt stress response mechanisms in crops. *International Journal of Molecular Sciences* 24 (1): 518.

Khan, I., Mohamed, S., Regnault, T. et al. (2020). Constitutive contribution by the rice OsHKT1; 4 Na+ transporter to xylem sap desalinization and low Na+ accumulation in young leaves under low as high external Na+ conditions. *Frontiers in Plant Science* 11: 1130.

Kim, Y.A., Moon, H., and Park, C.J. (2019). CRISPR/Cas9-targeted mutagenesis of Os8N3 in rice to confer resistance to *Xanthomonas oryzae* pv. oryzae. *Rice* 12 (1): 1–13.

Klap, C., Yeshayahou, E., Bolger, A.M. et al. (2017). Tomato facultative parthenocarpy results from Sl AGAMOUS-LIKE 6 loss of function. *Plant Biotechnology Journal* 15 (5): 634–647.

Kumar, N., Galli, M., Ordon, J. et al. (2018). Further analysis of barley MORC 1 using a highly efficient RNA-guided Cas9 gene-editing system. *Plant Biotechnology Journal* 16 (11): 1892–1903.

Legris, M., Ince, Y.Ç., and Fankhauser, C. (2019). Molecular mechanisms underlying phytochrome-controlled morphogenesis in plants. *Nature Communications* 10 (1): 5219.

Li, T., Liu, B., Spalding, M.H. et al. (2012). High-efficiency TALEN-based gene editing produces disease-resistant rice. *Nature Biotechnology* 30 (5): 390–392.

Li, R., Zhang, L., Wang, L. et al. (2018). Reduction of tomato-plant chilling tolerance by CRISPR–Cas9-mediated SlCBF1 mutagenesis. *Journal of Agricultural and Food Chemistry* 66 (34): 9042–9051.

Liao, S., Qin, X., Luo, L. et al. (2019). CRISPR/Cas9-induced mutagenesis of semi-rolled leaf1, 2 confers curled leaf phenotype and drought tolerance by influencing protein expression patterns and ROS scavenging in rice (*Oryza sativa* L.). *Agronomy* 9 (11): 728.

Lisar, S.Y., Motafakkerazad, R., Hossain, M.M., and Rahman, I.M. (2012). Causes, effects and responses. *Water Stress* 25 (1): 33.

Liu, M., Rehman, S., Tang, X. et al. (2019). Methodologies for improving HDR efficiency. *Frontiers in Genetics* 9: 691.

Liu, L., Zhang, J., Xu, J. et al. (2020a). CRISPR/Cas9 targeted mutagenesis of SlLBD40, a lateral organ boundaries domain transcription factor, enhances drought tolerance in tomato. *Plant Science* 301: 110683.

Liu, X., Wu, D., Shan, T. et al. (2020b). The trihelix transcription factor OsGTγ-2 is involved adaption to salt stress in rice. *Plant Molecular Biology* 103: 545–560.

Lloyd, A., Plaisier, C.L., Carroll, D., and Drews, G.N. (2005). Targeted mutagenesis using zinc-finger nucleases in *Arabidopsis*. *Proceedings of the National Academy of Sciences* 102 (6): 2232–2237.

Lo, A. and Qi, L. (2017). Genetic and epigenetic control of gene expression by CRISPR–Cas systems. *F1000Research* 6: F1000.

Low, Y.C., Lawton, M.A., and Di, R. (2020). Validation of barley 2OGO gene as a functional orthologue of *Arabidopsis* DMR6 gene in fusarium head blight susceptibility. *Scientific Reports* 10 (1): 1–13.

Lu, W., Deng, F., Jia, J. et al. (2020). The *Arabidopsis thaliana* gene AtERF019 negatively regulates plant resistance to *Phytophthora parasitica* by suppressing PAMP-triggered immunity. *Molecular Plant Pathology* 21 (9): 1179–1193.

Macovei, A., Sevilla, N.R., Cantos, C. et al. (2018). Novel alleles of rice eIF4G generated by CRISPR/Cas9-targeted mutagenesis confer resistance to Rice tungro spherical virus. *Plant Biotechnology Journal* 16 (11): 1918–1927.

Mahfouz, M.M. (2017). Genome editing: the efficient tool CRISPR–Cpf1. *Nature Plants* 3 (3): 1–2.

Mahmud, J.A., Bhuyan, M.H.M.B., Anee, T.I. et al. (2019). Reactive oxygen species metabolism and antioxidant defense in plants under metal/metalloid stress. In: *Plant Abiotic Stress Tolerance* (ed. M. Hasanuzzaman, K. Hakeem, K. Nahar, and H. Alharby), 221–257. Cham, Switzerland: Springer.

Malhi, G.S., Kaur, M., and Kaushik, P. (2021). Impact of climate change on agriculture and its mitigation strategies: a review. *Sustainability* 13 (3): 1318.

Malnoy, M., Viola, R., Jung, M.H. et al. (2016). DNA-free genetically edited grapevine and apple protoplast using CRISPR/Cas9 ribonucleoproteins. *Frontiers in Plant Science* 7: 1904.

Martínez, M.I., Bracuto, V., Koseoglou, E. et al. (2020). CRISPR/Cas9-targeted mutagenesis of the tomato susceptibility gene PMR4 for resistance against powdery mildew. *BMC Plant Biology* 20 (1): 1–13.

Mehta, S., Lal, S.K., Sahu, K.P. et al. (2020). CRISPR/Cas9-edited rice: a new frontier for sustainable agriculture. *New Frontiers in Stress Management for Durable Agriculture* 427–458.

Meng, Y., Hou, Y., Wang, H. et al. (2017). Targeted mutagenesis by CRISPR/Cas9 system in the model legume *Medicago truncatula*. *Plant Cell Reports* 36: 371–374.

Misiou, O. and Koutsoumanis, K. (2022). Climate change and its implications for food safety and spoilage. *Trends in Food Science & Technology* 126: 142–152.

Mojica, F.J., Díez-Villaseñor, C., García-Martínez, J., and Almendros, C. (2009). Short motif sequences determine the targets of the prokaryotic CRISPR defence system. *Microbiology* 155 (3): 733–740.

Nagamine, A. and Ezura, H. (2022). Genome editing for improving crop nutrition. *Frontiers in Genome Editing* 4: 850104.

Nawaz, F., Ahmad, R., Ashraf, M.Y. et al. (2015). Effect of selenium foliar spray on physiological and biochemical processes and chemical constituents of wheat under drought stress. *Ecotoxicology and Environmental Safety* 113: 191–200.

Nawaz, G., Han, Y., Usman, B. et al. (2019). Knockout of OsPRP1, a gene encoding proline-rich protein, confers enhanced cold sensitivity in rice (*Oryza sativa* L.) at the seedling stage. *3 Biotech* 9: 1–18.

Nazir, R., Mandal, S., Mitra, S. et al. (2022). Clustered regularly interspaced short palindromic repeats (CRISPR)/CRISPR-associated genome-editing toolkit to enhance salt stress tolerance in rice and wheat. *Physiologia Plantarum* 174 (2): e13642.

Nekrasov, V., Wang, C., Win, J. et al. (2017). Rapid generation of a transgene-free powdery mildew resistant tomato by genome deletion. *Scientific Reports* 7 (1): 1–6.

Nieves-Cordones, M., Mohamed, S., Tanoi, K. et al. (2017). Production of low-Cs+ rice plants by inactivation of the K+ transporter Os HAK 1 with the CRISPR-Cas system. *The Plant Journal* 92 (1): 43–56.

Nishimasu, H., Shi, X., Ishiguro, S. et al. (2018). Engineered CRISPR-Cas9 nuclease with expanded targeting space. *Science* 361 (6408): 1259–1262.

Noori, M.S. and Taliman, N.A. (2022). Effect of drought stress and chemical fertilizers on wheat productivity and grain and quality. *Plant Production and Genetics* 3 (1): 43–56.

Ogata, T., Ishizaki, T., Fujita, M., and Fujita, Y. (2020). CRISPR/Cas9-targeted mutagenesis of OsERA1 confers enhanced responses to abscisic acid and drought stress and increased primary root growth under nonstressed conditions in rice. *PLoS One* 15 (12): e0243376.

Oliva, R., Ji, C., Atienza-Grande, G. et al. (2019). Broad-spectrum resistance to bacterial blight in rice using genome editing. *Nature Biotechnology* 37 (11): 1344–1350.

Ortigosa, A., Gimenez-Ibanez, S., Leonhardt, N., and Solano, R. (2019). Design of a bacterial speck resistant tomato by CRISPR/Cas9-mediated editing of Sl JAZ 2. *Plant Biotechnology Journal* 17 (3): 665–673.

Osakabe, K., Osakabe, Y., and Toki, S. (2010). Site-directed mutagenesis in *Arabidopsis* using custom-designed zinc finger nucleases. *Proceedings of the National Academy of Sciences* 107 (26): 12034–12039.

Osakabe, Y., Watanabe, T., Sugano, S.S. et al. (2016). Optimization of CRISPR/Cas9 genome editing to modify abiotic stress responses in plants. *Scientific Reports* 6 (1): 26685.

Pavese, V., Moglia, A., Corredoira, E. et al. (2021). First report of CRISPR/Cas9 gene editing in *Castanea sativa* mill. *Frontiers in Plant Science* 12: 728516.

Peng, A., Chen, S., Lei, T. et al. (2017). Engineering canker-resistant plants through CRISPR/Cas9-targeted editing of the susceptibility gene Cs LOB 1 promoter in citrus. *Plant Biotechnology Journal* 15 (12): 1509–1519.

Piatek, A., Ali, Z., Baazim, H. et al. (2015). RNA-guided transcriptional regulation in planta via synthetic dC as9-based transcription factors. *Plant Biotechnology Journal* 13 (4): 578–589.

Piffanelli, P., Ramsay, L., Waugh, R. et al. (2004). A barley cultivation-associated polymorphism conveys resistance to powdery mildew. *Nature* 430 (7002): 887–891.

Pinto, L., Yaseen, T., Caputo, L. et al. (2017). Application of passive refrigeration and gaseous ozone to reduce postharvest losses on red chicory. *VI International Conference Postharvest Unlimited* 1256: 419–426.

Pröbsting, M., Schenke, D., Hossain, R. et al. (2020). Loss of function of CRT1a (calreticulin) reduces plant susceptibility to *Verticillium longisporum* in both *Arabidopsis thaliana* and oilseed rape (*Brassica napus*). *Plant Biotechnology Journal* 18 (11): 2328–2344.

Pyott, D.E., Sheehan, E., and Molnar, A. (2016). Engineering of CRISPR/Cas9-mediated potyvirus resistance in transgene-free *Arabidopsis* plants. *Molecular Plant Pathology* 17 (8): 1276–1288.

Qamar, R. and Javeed, H.M.R. (2022). Rice physiology under changing climate. In: *Modern Techniques of Rice Crop Production* (ed. N. Sarwar, Atique-ur-Rehman, S. Ahmad, and M. Hasanuzzaman), 165–186. Springer Singapore: Singapore.

Qi, L.S., Larson, M.H., Gilbert, L.A. et al. (2013). Repurposing CRISPR as an RNA-guided platform for sequence-specific control of gene expression. *Cell* 152 (5): 1173–1183.

Qiu, T., Qi, M., Ding, X. et al. (2020). The SAUR41 subfamily of SMALL AUXIN UP RNA genes is abscisic acid inducible to modulate cell expansion and salt tolerance in *Arabidopsis thaliana* seedlings. *Annals of Botany* 125 (5): 805–819.

Rabara, R.C., Msanne, J., Basu, S. et al. (2021). Coping with inclement weather conditions due to high temperature and water deficit in rice: an insight from genetic and biochemical perspectives. *Physiologia Plantarum* 172 (2): 487–504.

Rahneshan, Z., Nasibi, F., and Moghadam, A.A. (2018). Effects of salinity stress on some growth, physiological, biochemical parameters and nutrients in two pistachio (*Pistacia vera* L.) rootstocks. *Journal of Plant Interactions* 13 (1): 73–82.

Raymond, C., Matthews, T., and Horton, R.M. (2020). The emergence of heat and humidity too severe for human tolerance. *Science Advances* 6 (19): eaaw1838.

Raza, M.A., Yasin, H.S., Gul, H. et al. (2022). Maize/soybean strip intercropping produces higher crop yields and saves water under semi-arid conditions. *Frontiers in Plant Science* 13: 1006720.

Ritchie, H. and Roser, M. (2019). Land Use. Published online at Our World InData.org (accessed 22 November 2020).

Roca Paixão, J.F., Gillet, F.X., Ribeiro, T.P. et al. (2019). Improved drought stress tolerance in *Arabidopsis* by CRISPR/dCas9 fusion with a histone acetylTransferase. *Scientific Reports* 9 (1): 8080.

Salava, H., Thula, S., Mohan, V. et al. (2021). Application of genome editing in tomato breeding: mechanisms, advances, and prospects. *International Journal of Molecular Sciences* 22 (2): 682.

Santillán Martínez, M.I., Bracuto, V., Koseoglou, E. et al. (2020). CRISPR/Cas9-targeted mutagenesis of the tomato susceptibility gene PMR4 for resistance against powdery mildew. *BMC Plant Biology* 20 (1): 1–13.

Sapeta, H., Costa, J.M., Lourenco, T. et al. (2013). Drought stress response in *Jatropha curcas*: growth and physiology. *Environmental and Experimental Botany* 85: 76–84.

Saradadevi, G.P., Das, D., Mangrauthia, S.K. et al. (2021). Genetic, epigenetic, genomic and microbial approaches to enhance salt tolerance of plants: a comprehensive review. *Biology* 10 (12): 1255.

Shekari, F., Abbasi, A., and Mustafavi, S.H. (2017). Effect of silicon and selenium on enzymatic changes and productivity of dill in saline condition. *Journal of the Saudi Society of Agricultural Sciences* 16 (4): 367–374.

Shelake, R.M., Kadam, U.S., Kumar, R. et al. (2022). Engineering drought and salinity tolerance traits in crops through CRISPR-mediated genome editing: targets, tools, challenges, and perspectives. *Plant Communications* 3: 100417.

Shen, C., Que, Z., Xia, Y. et al. (2017). Knock out of the annexin gene OsAnn3 via CRISPR/Cas9-mediated genome editing decreased cold tolerance in rice. *Journal of Plant Biology* 60: 539–547.

Shi, J., Gao, H., Wang, H. et al. (2017). ARGOS 8 variants generated by CRISPR-Cas9 improve maize grain yield under field drought stress conditions. *Plant Biotechnology Journal* 15 (2): 207–216.

Shimatani, Z., Kashojiya, S., Takayama, M. et al. (2017). Targeted base editing in rice and tomato using a CRISPR-Cas9 cytidine deaminase fusion. *Nature Biotechnology* 35 (5): 441–443.

Shinozaki, Y., Ezura, K., Hu, J. et al. (2018). Identification and functional study of a mild allele of SlDELLA gene conferring the potential for improved yield in tomato. *Scientific Reports* 8 (1): 1–15.

Shrivastava, P. and Kumar, R. (2015). Soil salinity: a serious environmental issue and plant growth promoting bacteria as one of the tools for its alleviation. *Saudi Journal of Biological Sciences* 22 (2): 123–131.

Shukla, V.K., Doyon, Y., Miller, J.C. et al. (2009). Precise genome modification in the crop species *Zea mays* using zinc-finger nucleases. *Nature* 459 (7245): 437–441.

Shukla, P. R., Skea, J., Calvo Buendia, E. et al. (2019). IPCC, 2019: Climate Change and Land: An IPCC Special Report on Climate Change, Desertification, Land Degradation, Sustainable Land Management, Food Security, and Greenhouse Gas Fluxes in Terrestrial Ecosystems.

Si, X., Zhang, H., Wang, Y. et al. (2020). Manipulating gene translation in plants by CRISPR–Cas9-mediated genome editing of upstream open reading frames. *Nature Protocols* 15 (2): 338–363.

Singh, D. and Laxmi, A. (2015). Transcriptional regulation of drought response: a tortuous network of transcriptional factors. *Frontiers in Plant Science* 6: 895.

Sita, K., Sehgal, A., Kumar, J. et al. (2017). Identification of high-temperature tolerant lentil (*Lens culinaris* Medik.) genotypes through leaf and pollen traits. *Frontiers in Plant Science* 8: 744.

Soleymani, A. and Shahrajabian, M.H. (2018). Changes in germination and seedling growth of different cultivars of cumin to drought stress. *Cercetări Agronomice în Moldova* 51: 91–100.

Sruthi, P., Shackira, A.M., and Puthur, J.T. (2017). Heavy metal detoxification mechanisms in halophytes: an overview. *Wetlands Ecology and Management* 25: 129–148.

Stella, S., Alcón, P., and Montoya, G. (2017). Structure of the Cpf1 endonuclease R-loop complex after target DNA cleavage. *Nature* 546 (7659): 559–563.

Sun, Q., Lin, L., Liu, D. et al. (2018). CRISPR/Cas9-mediated multiplex genome editing of the BnWRKY11 and BnWRKY70 genes in *Brassica napus* L. *International Journal of Molecular Sciences* 19 (9): 2716.

Sushmita, K.G., Upadhyay, S.K., and Verma, P.C. (2021). An overview of genome-engineering methods. In: *Genome Engineering for Crop Improvement* (ed. S.K. Upadhyay), 1–21. Wiley.

Talari, G., Cummins, E., McNamara, C., and O'Brien, J. (2022). State of the art review of big data and web-based decision support systems (DSS) for food safety risk assessment with respect to climate change. *Trends in Food Science & Technology* 126: 192–204.

Tang, L., Mao, B., Li, Y. et al. (2017a). Knockout of OsNramp5 using the CRISPR/Cas9 system produces low Cd-accumulating indica rice without compromising yield. *Scientific Reports* 7 (1): 14438.

Tang, X., Lowder, L.G., Zhang, T. et al. (2017b). A CRISPR–Cpf1 system for efficient genome editing and transcriptional repression in plants. *Nature plants* 3 (3): 1–5.

Tigosa Urbieta, A., Giménez-Ibáñez, S., Leonhardt, N., and Solano, R. (2019). Design of a bacterial speck resistant tomato by CRISPR/Cas9-mediated editing of SlJAZ2. *Plant Biotechnology Journal* 17: 665–673.

de Toledo, P., Thomazella, D., Brail, Q. et al. (2016). CRISPR-Cas9 mediated mutagenesis of a DMR6 ortholog in tomato confers broad-spectrum disease resistance. *BioRxiv* 064824. https://doi.org/10.1101/064824.

Townsend, J.A., Wright, D.A., Winfrey, R.J. et al. (2009). High-frequency modification of plant genes using engineered zinc-finger nucleases. *Nature* 459 (7245): 442–445.

Tran, M.T., Doan, D.T.H., Kim, J. et al. (2021). CRISPR/Cas9-based precise excision of SlHyPRP1 domain (s) to obtain salt stress-tolerant tomato. *Plant Cell Reports* 40: 999–1011.

Ueta, R., Abe, C., Watanabe, T. et al. (2017). Rapid breeding of parthenocarpic tomato plants using CRISPR/Cas9. *Scientific Reports* 7 (1): 507.

Upadhyay, S.K. (2021). *Genome Engineering for Crop Improvement*. Wiley. https://doi.org/10.1002/9781119672425.

Van der Kooi, C.J., Reich, M., Löw, M. et al. (2016). Growth and yield stimulation under elevated CO2 and drought: a meta-analysis on crops. *Environmental and Experimental Botany* 122: 150–157.

Van Houtan, K.S., Tanaka, K.R., Gagné, T.O., and Becker, S.L. (2021). The geographic disparity of historical greenhouse emissions and projected climate change. *Science Advances* 7 (29): eabe4342.

Van Vu, T., Sung, Y.W., Kim, J. et al. (2019). Challenges and perspectives in homology-directed gene targeting in monocot plants. *Rice* 12 (1): 1–29.

Vlcko, T. and Ohnoutkova, L. (2020). Allelic variants of CRISPR/Cas9 induced mutation in an inositol trisphosphate 5/6 kinase gene manifest different phenotypes in barley. *Plants* 9 (2): 195.

Waltz, E. (2012). Tiptoeing around transgenics: new techniques for manipulating plant genomes are yielding plants touted as nontransgenic. Will that relieve regulatory burden? Emily Waltz investigates. *Nature Biotechnology* 30 (3): 215–218.

Wang, Y., Cheng, X., Shan, Q. et al. (2014). Simultaneous editing of three homoeoalleles in hexaploid bread wheat confers heritable resistance to powdery mildew. *Nature Biotechnology* 32 (9): 947–951.

Wang, F., Wang, C., Liu, P. et al. (2016). Enhanced rice blast resistance by CRISPR/Cas9-targeted mutagenesis of the ERF transcription factor gene OsERF922. *PLoS One* 11 (4): e0154027.

Wang, F.Z., Chen, M.X., Yu, L.J. et al. (2017a). OsARM1, an R2R3 MYB transcription factor, is involved in regulation of the response to arsenic stress in rice. *Frontiers in Plant Science* 8: 1868.

Wang, L., Chen, L., Li, R. et al. (2017b). Reduced drought tolerance by CRISPR/Cas9-mediated SlMAPK3 mutagenesis in tomato plants. *Journal of Agricultural and Food Chemistry* 65 (39): 8674–8682.

Wang, M., Lu, Y., Botella, J.R. et al. (2017c). Gene targeting by homology-directed repair in rice using a geminivirus-based CRISPR/Cas9 system. *Molecular Plant* 10 (7): 1007–1010.

Wang, X., Tu, M., Wang, D. et al. (2018). CRISPR/Cas9-mediated efficient targeted mutagenesis in grape in the first generation. *Plant Biotechnology Journal* 16 (4): 844–855.

Wang, L., Chen, S., Peng, A. et al. (2019a). CRISPR/Cas9-mediated editing of CsWRKY22 reduces susceptibility to *Xanthomonas citri* subsp. citri in Wanjincheng orange (*Citrus sinensis* (L.) Osbeck). *Plant Biotechnology Reports* 13: 501–510.

Wang, Y., Li, Y., Rosas-Diaz, T. et al. (2019b). The IMMUNE-ASSOCIATED NUCLEOTIDE-BINDING 9 protein is a regulator of basal immunity in *Arabidopsis thaliana*. *Molecular Plant-Microbe Interactions* 32 (1): 65–75.

Wang, B., Zhong, Z., Wang, X. et al. (2020a). Knockout of the OsNAC006 transcription factor causes drought and heat sensitivity in rice. *International Journal of Molecular Sciences* 21 (7): 2288.

Wang, W., Chen, S., Cheng, Y. et al. (2020b). Cell wall/vacuolar inhibitor of fructosidase 1 regulates ABA response and salt tolerance in *Arabidopsis*. *Plant Signaling & Behavior* 15 (4): 1744293.

Wang, T., Xun, H., Wang, W. et al. (2021). Mutation of GmAITR genes by CRISPR/Cas9 genome editing results in enhanced salinity stress tolerance in soybean. *Frontiers in Plant Science* 2752.

Wang, R., You, X., Zhang, C. et al. (2022). An ORFeome of rice E3 ubiquitin ligases for global analysis of the ubiquitination interactome. *Genome Biology* 23 (1): 1–21.

Winkel, T., Renno, J.F., and Payne, W.A. (1997). Effect of the timing of water deficit on growth, phenology and yield of pearl millet (*Pennisetum glaucum* (L.) R. Br.) grown in Sahelian conditions. *Journal of Experimental Botany* 48 (5): 1001–1009.

Xie, Z., Nolan, T.M., Jiang, H., and Yin, Y. (2019). AP2/ERF transcription factor regulatory networks in hormone and abiotic stress responses in *Arabidopsis*. *Frontiers in Plant Science* 10: 228.

Xu, S., Hu, B., He, Z. et al. (2011). Enhancement of salinity tolerance during rice seed germination by presoaking with hemoglobin. *International Journal of Molecular Sciences* 12 (4): 2488–2501.

Xu, Z., Xu, X., Gong, Q. et al. (2019). Engineering broad-spectrum bacterial blight resistance by simultaneously disrupting variable TALE-binding elements of multiple susceptibility genes in rice. *Molecular Plant* 12 (11): 1434–1446.

Yadav, S.K. (2010). Heavy metals toxicity in plants: an overview on the role of glutathione and phytochelatins in heavy metal stress tolerance of plants. *South African Journal of Botany* 76 (2): 167–179.

Yamamoto, T., Kashojiya, S., Kamimura, S. et al. (2018). Application and development of genome editing technologies to the solanaceae plants. *Plant Physiology and Biochemistry* 131: 37–46.

Yamauchi, S., Takemiya, A., Sakamoto, T. et al. (2016). The plasma membrane H+-ATPase AHA1 plays a major role in stomatal opening in response to blue light. *Plant Physiology* 171 (4): 2731–2743.

Yin, X., Biswal, A.K., Dionora, J. et al. (2017). CRISPR-Cas9 and CRISPR-Cpf1 mediated targeting of a stomatal developmental gene EPFL9 in rice. *Plant Cell Reports* 36: 745–757.

Yin, Y., Qin, K., Song, X. et al. (2018). BZR1 transcription factor regulates heat stress tolerance through FERONIA receptor-like kinase-mediated reactive oxygen species signaling in tomato. *Plant and Cell Physiology* 59 (11): 2239–2254.

Zahra, N., Wahid, A., Hafeez, M.B. et al. (2021). Grain development in wheat under combined heat and drought stress: Plant responses and management. *Environmental and Experimental Botany* 188: 104517.

Zaidi, S.S.E.A., Mukhtar, M.S., and Mansoor, S. (2018). Genome editing: targeting susceptibility genes for plant disease resistance. *Trends in Biotechnology* 36 (9): 898–906.

Zandalinas, S.I., Fritschi, F.B., and Mittler, R. (2021). Global warming, climate change, and environmental pollution: recipe for a multifactorial stress combination disaster. *Trends in Plant Science* 26 (6): 588–599.

Zeng, Y., Wen, J., Zhao, W. et al. (2020). Rational improvement of rice yield and cold tolerance by editing the three genes OsPIN5b, GS3, and OsMYB30 with the CRISPR–Cas9 system. *Frontiers in Plant Science* 10: 1663.

Zetsche, B., Gootenberg, J.S., Abudayyeh, O.O. et al. (2015). Cpf1 is a single RNA-guided endonuclease of a class 2 CRISPR-Cas system. *Cell* 163 (3): 759–771.

Zhang, Y., Bai, Y., Wu, G. et al. (2017). Simultaneous modification of three homoeologs of Ta EDR 1 by genome editing enhances powdery mildew resistance in wheat. *The Plant Journal* 91 (4): 714–724.

Zhang, M., Cao, Y., Wang, Z. et al. (2018a). A retrotransposon in an HKT1 family sodium transporter causes variation of leaf Na+ exclusion and salt tolerance in maize. *New Phytologist* 217 (3): 1161–1176.

Zhang, S., Wang, L., Zhao, R. et al. (2018b). Knockout of SlMAPK3 reduced disease resistance to *Botrytis cinerea* in tomato plants. *Journal of Agricultural and Food Chemistry* 66 (34): 8949–8956.

Zhang, Z., Ge, X., Luo, X. et al. (2018c). Simultaneous editing of two copies of Gh14-3-3d confers enhanced transgene-clean plant defense against *Verticillium dahliae* in allotetraploid upland cotton. *Frontiers in Plant Science* 9: 842.

Zhang, A., Liu, Y., Wang, F. et al. (2019). Enhanced rice salinity tolerance via CRISPR/Cas9-targeted mutagenesis of the OsRR22 gene. *Molecular Breeding* 39: 1–10.

Zhang, M., Liu, Q., Yang, X. et al. (2020a). CRISPR/Cas9-mediated mutagenesis of Clpsk1 in watermelon to confer resistance to *Fusarium oxysporum* f. sp. niveum. *Plant Cell Reports* 39: 589–595.

Zhang, P., Du, H., Wang, J. et al. (2020b). Multiplex CRISPR/Cas9-mediated metabolic engineering increases Soybean isoflavone content and resistance to *Soybean mosaic virus*. *Plant Biotechnology Journal* 18 (6): 1384–1395.

Zhang, Y., Wang, X., Luo, Y. et al. (2020c). OsABA8ox2, an ABA catabolic gene, suppresses root elongation of rice seedlings and contributes to drought response. *The Crop Journal* 8 (3): 480–491.

Zhao, C., Zhang, Z., Xie, S. et al. (2016). Mutational evidence for the critical role of CBF transcription factors in cold acclimation in Arabidopsis. *Plant Physiology* 171 (4): 2744–2759.

Zhao, K., Tao, Y., Liu, M. et al. (2022). Does temporary heat stress or low temperature stress similarly affect yield, starch, and protein of winter wheat grain during grain filling? *Journal of Cereal Science* 103: 103408.

Zhou, H., Liu, B., Weeks, D.P. et al. (2014). Large chromosomal deletions and heritable small genetic changes induced by CRISPR/Cas9 in rice. *Nucleic Acids Research* 42 (17): 10903–10914.

Zhou, J., Peng, Z., Long, J. et al. (2015). Gene targeting by the TAL effector PthXo2 reveals cryptic resistance gene for bacterial blight of rice. *The Plant Journal* 82 (4): 632–643.

Zhu, C., Kobayashi, K., Loladze, I. et al. (2018). Carbon dioxide (CO2) levels this century will alter the protein, micronutrients, and vitamin content of rice grains with potential health consequences for the poorest rice-dependent countries. *Science Advances* 4 (5): eaaq1012.

18

Genome Editing for Raising Crops for Arid Lands: A Perspective of Increasing Stress Tolerance

Pooja Jangir, Purva Khandelwal, and Praveen Soni

Department of Botany, University of Rajasthan, Jaipur, Rajasthan, India

Abbreviations

ABA	Abscisic acid
APX	Ascorbate peroxidase
AREB1	Abscisic acid (ABA)-responsive element binding protein 1
ARF	Auxin response factor
BZR1	Brassinazole resistant 1
CAT	Catalase
CDPK	Calcium-dependent protein kinase
CRISPR/Cas9	Clustered regularly interspaced short palindromic repeats/CRISPR associated protein 9
CRISPR/Cpf1	CRISPR from *Prevotella* and *Francisella* 1
DPA4	Development related PcG target in the APEX4
DST	Drought and salt tolerance
EL	Electrolyte leakage
ERA1	Enhanced response to ABA1
FLN2	Fructokinase-like protein 2
GA	Gibberellic acid
GE	Genome editing
GID1	Gibberellin-insensitive dwarf 1
H_2O_2	Hydrogen peroxide
HSF	Heat shock transcription factor
HSP	Heat shock protein
KO	Knockout
MAPK	Mitogen-activated protein kinase
MBF1	Multiprotein bridging factor 1
MDA	Malondialdehyde

Applications of Genome Engineering in Plants, First Edition. Edited by Santosh Kumar Upadhyay.
© 2024 John Wiley & Sons Ltd. Published 2024 by John Wiley & Sons Ltd.

MNs	Meganucleases
NPR1	Non-expresser of pathogenesis-related gene 1
NtAITRs	*Nicotiana tabacum* ABA-induced transcription repressors
O_2^-	Superoxide anion
POD	Peroxidase
PSII	Photosystem II
RBOH1	Respiratory burst oxidase homolog 1
ROS	Reactive oxygen species
RUBISCO	Ribulose bisphosphate carboxylase-oxygenase
RWC	Relative water content
SAPK2	Osmotic stress/ABA–activated protein kinase 2
SlAGL6	*Solanum lycopersicum* AGAMOUS-LIKE 6
SRL1	Semi-rolled leaf 1
SRL2	Semi-rolled leaf 2
SOD	Superoxide dismutase
SOD7	Suppressor of da1-1
SSNs	Sequence-specific nucleases
TALENs	Transcription activator-like effector nucleases
TF	Transcription factor
ZFNs	Zinc finger nucleases

18.1 Introduction

Several abiotic stresses including salt, water scarcity, severe temperatures, flood, radiation, and metal toxicity exert detrimental impacts on crop productivity and decrease average output up to 50% (Vaughan et al. 2018). Among them, drought causes a severe danger to food supply, especially in arid or semi-arid regions of the world, which make up one-third of the land area of Earth and are home to some of the world's poorest human populations (Malagnoux et al. 2008). They face inadequate and/or irregular rainfall patterns. Here, high temperature is another dominant environmental constraint to agriculture. Rainfall is a major source of water supply for arid land agriculture. Rain-fed agricultural land produces 60% of global food (FAO 2008). According to an estimate from the Food and Agricultural Organization (FAO), the population of the globe will reach 10 billion in 2050, a rise of almost 30% when compared to data from 2020 (FAO 2017, 2022). This fact just highlights how much more food will be required in the future years. There will be a need to increase global food production by 60–100%. This need will have to be met in the scenario of the challenges of frequent drought conditions and rising global temperature due to climate change. A rise in global temperature is a big threat to agricultural output as it not only badly affects crop production but also deteriorates the quality of harvested products. With each 1 °C increase in the world mean temperature, the yields of major crops including maize, wheat, rice, and soybeans decline by 7.4%, 6.0%, 6.2%, and 3.1%, respectively (Parthasarathi et al. 2022). A 1.5 °C increase will result in longer warm seasons, rising heat waves, and shorter cold seasons. With a 2 °C increase, heat extremes would occur more frequently and exceed thresholds of tolerability for agriculture (IPCC 2021). In addition to the aforementioned restrictions, dwindling availability of water resources and agricultural land is also imposing significant challenges to food production. Between 1983 and 2009, drought reduced the world's production of major grain crops including soybean, maize, rice, and wheat by over 75%. (Kim et al. 2019). Thus, these stresses have resulted in a huge reduction in major crop yields over time. So, it is crucial to develop stress-tolerant crops. Numerous techniques have been used to increase the tolerance of crops such as conventional breeding techniques, agronomic techniques like mutagenesis (gamma rays and EMS), application of plant growth-promoting microorganisms (PGPM), and genetic engineering, etc. (Hasanuzzaman et al. 2013; Jangir et al. 2021; Seleiman et al. 2021). Though conventional methods have historically been important for the improvement of available germplasm, these methods can't meet the increasing demand for food in the modern period due to limitations like a need for a lengthy selection process for the germplasm, which is arduous and slow. In the case of genetic engineering, genetically modified crops face strict regulations and ethical issues. Over the last few years, genome editing (GE) techniques have transformed the process of precise improvement of crops with respect to their yields and qualities. It has been achieved through the development of effective GE tools namely transcription activator-like effector nucleases (TALENs), meganucleases (MNs), zinc-finger nucleases (ZFNs), clustered regularly interspaced short palindromic repeats-CRISPR associated protein 9 (CRISPR/Cas9) and CRISPR from *Prevotella* and *Francisella 1* (CRISPR/Cpf1) (Sushmita et al. 2021; Upadhyay 2021). These have made it possible for plant scientists to alter target genes in various crops. With the use of these tools, crop improvement programs have gained momentum and now new varieties can be produced more effectively and quickly. These varieties are developed in view of specific environmental constrain or regional consumer demands. GE technology is also helping in deciphering multiple biological pathways and understanding the function of genetic and epigenetic regulators in different plant species. It has the potential to significantly improve sustainable food production, benefiting both people and the environment.

18.2 Genome Editing Toolbox

In GE technology, sequence-specific nucleases (SSNs) (or site-directed nucleases, SDNs) are utilized to carry out highly efficient and precise targeted mutations in plant genomes. This technology is very useful, especially for those crop plants that have complex genomes which are challenging to improve using traditional breeding techniques (Feng et al. 2013; Georges and Ray 2017). GE entails the capacity to make precise DNA alterations, like deletion, insertion, or replacement of DNA by sequence-targeted recombination, which has accelerated the speed, simplicity, and reproducibility of producing local DNA alterations. GE technology also helps in gene characterization in addition to crop improvement. The synthetically created SSNs such as MNs, ZFNs, CRISPR/Cas, and TALENs have been widely used for producing high-efficiency targeted mutagenesis in a number of crops and model plants (Razzaq et al. 2019; Xu et al. 2019). The SSNs make precise double-strand breaks (DSBs) at the intended places in genome and use the cell's own machinery to repair the DSBs by non-homologous end-joining (NHEJ) or homologous recombination (HR). Without using a homologous template, the NHEJ pathway ligates broken ends to repair DSBs, this process frequently outcomes in deletions or and single nucleotide polymorphisms (SNPs) at the breakpoint which leads to modifications like frameshift or nonsense mutations. Fundamentally, the NHEJ mechanism generates random mutations which makes it error-prone. Whereas, by substituting the DNA sequence with a template at the break site, HR permits gene replacement, which makes precise change at desired genome site (Voytas and Gao 2014). By using homologous recombination, Paszkowski et al. (1988) invented the use of GE for integrating genes into the tobacco genome. A transcription factor (TF) and a promoter must interact for sequence-specific DNA binding. The aforementioned nucleases can target specific sequences in order to create DSBs for GE guided by either RNA-DNA base-pairing (CRISPR/Cas9 and CRISPR/Cpf1) or protein-DNA contact (MNs, ZFNs, and TALENs).

MN also known as homing nuclease is the first nuclease among GE tools which was initially found in the genomes of microbes, organelles and unicellular eukaryotes (Silva et al. 2011). MN is a two-domain system consisting of a DNA binding domain and a DNA cleavage domain. MNs cleave both strands of DNA in a site-specific way by identifying DNA sequence components with lengths between 12 and 40 base pairs. DSB at the recognition site is repaired by HR. They can be delivered using viral vectors (Dunn and Pinkert 2014). However, because of few restrictions, their use in genome engineering is less common. For example, firstly the enzyme's catalytic activity is compromised by the overlap between its DNA binding and cleavage domains. Secondly, they lack the modular architecture of DNA-binding domains found in ZFNs and TALENs, and thirdly they can occasionally exhibit sequence degeneration which raises the probability of off-target (an unintended target) binding (Argast et al. 1998; Stoddard 2011). Comparatively, their performance in terms of precision, efficiency, ease of design, cost-effectiveness, and specificity level is less than those of other tools. Additionally, they are not able to make multiplex editing.

ZFNs are the first precise GE tool. Like MNs, they consist of a nuclease domain (FokI endonuclease) and a specific zinc finger DNA-binding domain that binds to a 3-bp DNA target. Longer nucleotide sequences are frequently recognized by many ZFNs, enhancing specificity and enabling customization to a target of interest (Kim et al.1996). Their target site length varies from 24 to 36–40 base pairs. They are the first generation of chimerically created nuclease-based GE tools (Gaj et al. 2013). They were applied successfully for GE in *Arabidopsis* for the first time in 2005, and then in tobacco in the same year (Lloyd et al. 2005). Relatively, ZFNs perform better than MNs but poorer than TALEN and CRISPR/Cas9 in terms of precision, efficiency, ease of design, cost-effectiveness, and specificity level. Multiplex editing can be done with ZFNs, although it is challenging.

TALENs appeared as a replacement for ZFNs with enormous potential for precise GE in 2011 (Christian et al. 2010; Chen and Gao 2013). Its characteristics are almost similar to ZFNs. Like them, TALENs are also chimeric nucleases created by joining 13–28 transcriptional activator-like effector (TALE) repeats (which constitute the DNA binding domain) and FokI endonuclease (DNA cleavage domain) (Boch et al. 2009). TALENs are derived from the bacterium *Xanthomonas* (Boch and Bonas 2010; Bogdanove et al. 2010). In contrast to ZFNs, each TALE repeat only targets one nucleotide, permitting a more flexible target design and a greater range of target locations. This GE tool has been used in case of different plants like *Arabidopsis*, flax, sugarcane, rice, potato, maize, tomato, barley, *Nicotiana*, wheat, rapeseed, and soybean (Martinez-Fortun et al. 2017; Ran et al. 2017). It was applied for crop improvement for the first time in *Oriza sativa* by interrupting

OsSWEET14, a bacterial blight susceptibility gene (Li et al. 2012). It also has drawbacks. Its delivery is difficult because the cDNA encoding TALEN is about 3 kb in size.

ZFNs and TALENs were the two most common GE techniques until 2013. An important innovation in GE techniques relied on RNA-directed nucleases. CRISPR/Cas9 and CRISPR/Cpf1 are RNA-guided powerful nucleases. Based on proteins and associated RNAs (both are core components), the CRISPR/Cas9 editing system is of three types (type I, II, and III) (Makarova and Koonin 2015). The type II system (from *Streptococcus pyogenes*) is the most prevalent one. In 2013, CRISPR/Cas9 was utilized for the first time for plant gene editing (Zhang et al. 2017). On account of its simplicity, flexibility, high efficiency, and ease of multiplexing, it is the most productive GE tool for creating transgene-free GE plants (Langner et al. 2018).

Subsequently, in order to improve the competence and accuracy of editing, base editing (BE) technique has been developed (Tabassum et al. 2021). BE utilizes CRISPR components to make point mutations without causing DSBs to rely on HDR (Anzalone et al. 2020). Adenine base editors and cytosine base editors are the two different types of base editors. Recently the invention of prime editing technology has further enhanced the efficiency of the GE system (Molla et al. 2021).

Nonetheless, for the successful implementation of genome editing for crop improvement there are a couple of obstacles that must be overcome. Off-target mutations are a serious cause for concern in this area (Mao et al. 2019). GE tools particularly CRISPR/Cas9, frequently detect an inaccurate fit and cut nucleotide sequences without specificity. The efficient application of GE in plants is seriously jeopardized by these off-target alterations, even though they are predicted to be separated by rigorous breeding and selection procedures during crop improvement programs. Researchers have developed various software, such as CRISPR-GE and CRISPR-P that can help in precisely creating the guide RNA in order to address the issue of off-target mutations (Lei et al. 2014; Xie et al. 2017; Alok et al. 2021). Another problem is that different regulatory bodies have strong disputes about how to regulate and commercialize crops that have had their genomes altered (EFSA 2012; Pauwels et al. 2014). It is necessary to establish a set of rules for genome-edited crops which are universally recognized. The main issue with their regulation is the conundrum of whether such crops should be taken into account under the regulations for genetically modified organisms (GMOs). In this regard, in 2012 the United States Department of Agriculture (USDA) approved the farming and commercialization of CRISPR/Cas9 edited mushrooms and waxy maize by escaping them from GMO regulation (Waltz 2012, 2016; Turnbull et al. 2021). On the other hand, The European Court of Justice (ECJ) has brought genome-edited crops under the GMO legislation which might impede their commercialization in Europe or have an impact on agricultural cooperation with European nations (Urnov et al. 2018).

18.3 Plants' Responses to Drought and Heat

Drought stress occurs when a plant's water needs are not met which injures the plant (Blum 2005). The manner in which plants respond to drought stress depends on the plant species, habitat as well as timing, and severity of the drought. In order to cope with drought stress, plants employ different mechanisms which have been grouped into four main categories: drought avoidance, drought escape, drought recovery, and drought tolerance (Fang and Xiong 2015). A plant using a drought avoidance mechanism can retain normal physiological activities under moderate to severe drought stress, by changing morphological structures or growth rates to decrease transpiration or increase water intake to maintain high water levels inside it (Levitt 1980). Reduced water loss through stomatal closure, leaf rolling, and increased wax buildup on the surface of the leaf are a few examples of these changes (Zhang et al. 2005). Enhanced water intake by modified roots can also avoid drought (Tardieu 2013). To escape dry spells, the drought escape strategy entails modulation of the maturation rate, developmental flexibility, and hasty physiological development (Turner 1979). This permits plants to finish their life span before the start of drought period. Drought recovery refers to the ability of some plants to recover from extreme drought stress by resuming growth and productivity (Luo 2010). Finally, in certain conditions, plants employ a drought tolerance strategy that involves retaining plant fitness and economic output in spite of poor water potential, by regulating numerous genes and a number of metabolic pathways to decrease and/or fix the injury brought by drought stress (Turner 1979; Blum 2000). Hence, even under extreme drought stress, a plant is able to maintain a certain degree of physiological activity. Together, drought avoidance and tolerance mechanisms are considered to as drought resistance, which is the capacity of a plant to survive, develop and reproduce successfully in situations of a restricted and erratic water supply or under intermittent water deficit conditions.

Drought is caused by various factors; high temperature is one of them. It has a significant impact on plant growth and development, especially at the reproductive stage which ultimately affects the yield of the plants (Jiang et al. 2015). It causes alterations at physiological, morphological, biochemical, and molecular levels. Exposure of plants to lethal temperatures results in swollen chloroplasts with abnormal stacking of grana which eventually causes the disintegration of their membranes (Daniell et al. 1969; Hasanuzzaman et al. 2013). Ion leakage from heated leaf cells and change in the photosystem's energy distribution accompany such structural changes. It also causes a reduction in the photosynthetic efficiency due to reduced activity of PSII and RUBISCO enzyme and loss of photosynthetic pigments. The formation of reactive oxygen species (ROS) (singlet oxygen, hydrogen peroxide, superoxide radical, and hydroxyl radical) in heat-exposed plants results in oxidative stress (Tyagi et al. 2019; Zhao et al. 2020). All of these alterations lead to lower crop yields and quality due to reduced germination rates, plant emergence, and poor seedling vigor (Hasanuzzaman et al. 2013). Plants respond to such stress by maintaining membrane stability, producing antioxidants for scavenging ROS, accumulating and adjusting compatible solutes, and triggering protein kinases-mediated cascades resulting into transcriptional induction and chaperone activation (Wahid et al. 2007). Plant heat responses at the molecular level are controlled by a signaling cascade of heat shock proteins (HSPs) and heat shock transcription factors (HSFs). HSPs serve as chaperons for denatured proteins, and their expression is controlled by HSFs, which is crucial to ensure thermotolerance in plants (Zhao et al. 2020). Basal and acquired thermotolerance are the two categories of heat tolerance. Basal thermotolerance denotes to the capacity to endure high temperatures, while acquired thermotolerance refers to the capacity to withstand fatal high temperatures after an acclimation time (Chen et al. 2022). All these molecularly controlled events allow plants to survive and grow under heat stress.

18.4 Increasing Drought Tolerance in Plants Through Genome Editing

Due to the polygenic characteristic of drought tolerance, a variety of genetic variables have been investigated to know their functions in drought response or to develop varieties that are resistant to drought. When genes encoding positive regulators of drought signaling are overexpressed, it leads to an improvement in plant tolerance to drought (Fang and Xiong 2015). In contrast, the expression of negative regulators worsens conditions in plants. A fundamental understanding of biochemical, physiological, and gene regulatory networks is necessary to design crop plants that are highly resistant to the effects of drought stress. For this, genes related to drought avoidance mechanism like leaf rolling, stomatal closure, water conservation and cuticle wax accumulation have been targeted using GE tools. Moreover, a variety of other tactics have also been investigated, namely transcription factor (TF) editing, manipulation of phytohormone signaling and metabolic pathway, and modulating the expression of different genes via miRNA editing, epigenetic modification, and promoter editing as mentioned in Table 18.1. The next sections provide an overview of the application of GE technology for the improvement of drought tolerance by targeting genes of the aforementioned categories.

18.4.1 Transcription Factors

Many genes, particularly those encoding TFs that function as molecular switches for gene expression, help plants in enduring adverse circumstances of drought (Joshi et al. 2016). These genes continue to be promising genomic candidates for widespread crop breeding. Auxin response factors (ARFs) are the TFs by which auxin affects the transcriptional control of target genes. Loss of function of *SlARF4* in *Solanum lycopersicum* using CRISPR/Cas9 enhanced drought tolerance via improving chlorophyll content, RWC, soluble sugars content, ABA content, root development and density, and decreased stomatal conductance (Bouzroud et al. 2020). In another example, knocking out a zinc finger TF gene *drought and salt tolerance* (*OsDST*) using CRISPR/Cas9 enhanced drought tolerance in mutants (Santosh Kumar et al. 2020). The increase in leaf water retention, improved leaf architecture (reduced stomatal density and broader width of leaves), and stomatal conductance provided drought resistance in mutants. In *Nicotiana tabacum,* ABA-induced transcription repressors (NtAITRs) are novel regulators of drought tolerance. They negatively regulate drought response by changing the expression of ABA signaling genes. *ntaitrs* mutant tobacco developed through CRISPR/Cas9 technology showed improved survival rate and ABA sensitivity (Li et al. 2022). Similarly, targeted mutations in *DPA4* (*development-related PcG target in APEX4*) and *SOD7* (*suppressor of da1-1*) (both are transcription repressors engaged in the regulation of seed size) in *Arabidopsis*, increased seed size in *dpa4 sod7* double mutants (Chen et al. 2019). Interestingly, stacking this double mutation with *aitr2 aitr5 aitr6* (*aitr256*) triple mutation also improved drought tolerance (improving recovery rate and reduced ABA sensitivity) in the *dpa4 sod7 aitr256* quintuple mutants. Thus, this finding shows that CRISPR/Cas9 editing is successful in integrating traits related to seed size and abiotic stress tolerance. A basic leucine zipper (bZIP) TF

Table 18.1 Improvement of drought tolerance in plants through genome editing.

S. no.	Plant	Variety	GE tool	Target gene	Classification of gene	Function	Mutation type	Mutant's phenotype with respect to drought stress	Observed modifications	Reference
1	Tomato	Micro-Tom	CRISPR/Cas9	*SlLBD40*	TF	Lateral organ development	KO (insertion)	Tolerant	Improved water-holding ability of plant and decreased wilting symptoms	Liu et al. (2020b)
		M82	CRISPR/Cas9	*GID1*	Receptor	Gibberellin signaling	KO (insertion)	Tolerant	Improved leaf water content and decreased whole plant transpiration and suppressed xylem vessel proliferation and expansion	Illouz-Eliaz et al. (2020)
		Micro-Tom	CRISPR/Cas9	*SlARF4*	TF	Plant responses to environmental stresses	KO (deletion)	Tolerant	Improved root development and root density, chlorophyll content, RWC, soluble sugars content, ABA content and decreased stomatal conductance	Bouzroud et al. (2020)
		Ailsa Craig	CRISPR/Cas9	*SlNPR1*	Receptor	Plant defense response to pathogens	KO (insertion, deletion)	Sensitive	Higher stomatal aperture, EL, MDA and H_2O_2 levels, and reduced antioxidant enzymes activities, downregulated expression of drought-related genes *SlGST, SlDHN* and *SlDREB*	Li et al. (2019)

		Ailsa Craig	CRISPR/Cas9	*SlMAPK3*	Signaling	Involved in plant drought response	KO (insertion, deletion)	Sensitive	Severe wilting symptom, high H_2O_2 content and EL level and lower APX, CAT, POD, and SOD activities	Wang et al. (2017)
2	Rice	Nipponbare	CRISPR/Cas9	*OsmiR535*	Regulator	Involved in stress response	KO (deletion)	Tolerant	Improved survival rate of plants, shoot length, number of lateral roots, primary root length, ABA insensitivity, and decreased leaf senescence	Yue et al. (2020)
		Restorer rice lines (GXU16, GXU20 and GXU28)	CRISPR/Cas9	*SRL1*, *SRL2*	Metabolism	Regulation of leaf rolling on the adaxial side	KO (deletion)	Tolerant	Improved survival rate, ABA content, SOD and CAT activities, panicle number and grain filling, and decreased MDA content, reduced stomatal conductance, number of stomata, and transpiration rate	Liao et al. (2019)
		Nipponbare	CRISPR/Cas9	*OsERA1*	Signaling	ABA signaling and drought response	KO (insertion)	Tolerant	Increased sensitivity to ABA and decreased stomatal conductance	Ogata et al. (2020)
		Sasanishiki	CRISPR/Cas9	*OsGS3*, *OsDEP1*	Signaling	Regulation of yield-related traits	KO	Tolerant	Dwarf phenotype	Cui et al. (2020)
		Japonica	CRISPR/Cas9	*OsSAPK2*	Signaling	ABA signaling	Deletion	Sensitive	Decreased ABA signaling and stomatal conductance	Lou et al. (2017)

(*Continued*)

Table 18.1 (Continued)

S. no.	Plant	Variety	GE tool	Target gene	Classification of gene	Function	Mutation type	Mutant's phenotype with respect to drought stress	Observed modifications	Reference
		MTU1010	CRISPR/Cas9	*DST*	TF	Drought and salt response	KO (deletion, insertion)	Tolerant	Improved leaf traits with reduced stomatal density; broader width of leaves and enhanced leaf water retention	Santosh Kumar et al. (2020)
		IR-96	CRISPR/Cas9	*OsPYL9*	Receptor	ABA receptors	KO (deletion, insertion)	Tolerant	Improved survival rate, plant height, grain yield (1000-grain weight, grain number per panicle, grain width, grain length, yield per plant), chlorophyll content, ABA accumulation, leaf cuticular wax, CAT and SOD activities, and decreased stomatal conductance and transpiration rate	Usman et al. (2020)
3	Wheat	Giza168	CRISPR/Cas9	*TaSal1* (5 homologous genes)	Metabolism	Hydrolase enzyme for phosphoric monoester bond	KO (deletion, insertion)	Tolerant	Improved seed germination, seedling growth, leaf rolling, size of bulliform cells, and stomatal closure	Abdallah et al. (2022)

4	Tobacco	K326	CRISPR/Cas9	*NtAITR*	TF	ABA-induced transcription repressor	KO (deletion, insertion)	Tolerant	Improved survival rate of plants and ABA sensitivity	Li et al. (2022)
5	Maize	–	CRISPR/Cas9	*ARGOS8*	Signaling	Negative regulator of ethylene responses	Replacement of native promotor with GOS2 promoter	Tolerant	Higher grain yield	Shi et al. (2017)
		–	CRISPR/Cas9	*abh2*	Metabolism	Drought response	KO (deletion)	Tolerant	Improved survival rate, less water loss, and more ABA content	Liu et al. (2020a)
6	Rapeseed	Westar	CRISPR/Cas9	*BnaA6.RGA*	Signaling	Involved in GA signaling	KO (deletion, insertion)	Tolerant	Improved survival rate of plants and RWC, decreased ion leakage, MDA content, and water loss by closing stomata	Wu et al. (2020)
7	*Arabidopsis*	Col-0 ecotype	CRISPR/Cas9	*MIR169a*	Regulatory	Involved in adaptation to multiple stresses	KO (deletion)	Tolerant	Improve survival rate	Zhao et al. (2016)
		Col-0	CRISPR/Cas9	*DPA4* and *SOD7*	TF	Seed size regulators	KO (deletion, insertion)	Tolerant	Improved seed size and recovery rate, and reduced ABA sensitivity	Chen et al. (2019)
		Col-0	CRISPR/ $dCas9^{HAT}$	*AREB1*	TF	ABA signaling-mediated drought tolerance	Epigenome editing	Tolerant	Improved chlorophyll content and survival rate	Roca Paixão et al. (2019)

(*Continued*)

Table 18.1 (Continued)

S. no.	Plant	Variety	GE tool	Target gene	Classification of gene	Function	Mutation type	Mutant's phenotype with respect to drought stress	Observed modifications	Reference
8	Soybean	Tianlong no. 1	CRISPR/Cas9	*GmHdz4*	TF	Involved in drought and salt tolerance regulation	KO (deletion, insertion)	Tolerant	Improved root-shoot ratio, dry weight of root and shoot, root system architecture (total root length, superficial area and root tip number), osmolytes accumulation (soluble sugar, proline), antioxidant enzymes (CAT, SOD, POD) decreased H_2O_2, O^{2-}, and MDA content	Zhong et al. (2022)
		Huachun 6	CRISPR/Cas9	*RS2*, *RS3*, *RS4* and *STS*	Metabolism	Anti-nutritive factors	KO (deletion, insertion)	Tolerant	Reduced raffinose and stachyose contents in seeds	Lin et al. (2023)

known as abscisic acid (ABA)-responsive element binding protein 1 (AREB1) positively regulates drought tolerance via ABA signaling pathway. Overexpression of *AREB1* via epigenome editing using CRISPR/dCas9^{HAT} enhanced drought tolerance of *Arabidopsis* plants. Generated plants exhibited improved chlorophyll content, survival rate and faster stomatal aperture (Roca Paixao et al. 2019). Therefore, CRISPR/dCas9 mediated epigenome editing of drought-responsive genes is also a promising approach for enhancing drought tolerance. The homeodomain-leucine zipper (HD-Zip) TFs are crucial in secondary metabolism, plant development, and abiotic stress responses. Soybean *gmhdz4* mutant showed improved root architecture (root tip number, total root length, and root superficial area), root-shoot ratio, dry weight of root and shoot, osmolytes accumulation (soluble sugar and proline), antioxidant enzymes (CAT, SOD, and POD), and decreased H_2O_2, O^{2-}, and malondialdehyde (MDA) content (Zhong et al. 2022). The tomato TF *SlLBD40* gene, which encodes a protein that belongs to lateral organ boundaries domain (LBD) subfamily II, shows higher expression in fruits and roots. Compared to wild-type tomato plants, the average water loss rate in *Sllbd40* knockout (KO) mutants created using CRISPR/Cas9 was substantially lower, which resulted in improved water-holding ability and decreased wilting symptoms (Liu et al. 2020b). Thus, these findings highlight the broad application of TF targeting for improving plants' ability to withstand drought stress.

18.4.2 Phytohormone Signaling

For responding to drought stress plants have evolved complex signaling pathways. One of the plants' initial responses to drought stress is the production of ABA, which serves as the first line of defense against it. Plants then react to drought employing ABA-independent or dependent signaling cascades (Budak et al. 2015). In rice, OsERA1 (enhanced response to ABA 1) negatively regulates drought stress responses. *Osera1* mutants were made using CRISPR/Cas9 and they had lower seedlings stomatal conductance rates and higher sensitivity to ABA than those of the wild-type (Ogata et al. 2020). Thus, targeting *OsERA1*, drought tolerance was increased in rice seedlings. In rice, SnRK2 family enzyme OsSAPK2 (osmotic stress/ABA-activated protein kinase 2) mediates ABA signaling. Its mutants created by CRISPR/Cas9 showed ABA insensitivity and susceptibility to oxidative and drought stresses (Lou et al. 2017). This study implies that *OsSAPK2* is essential for rice to withstand drought. Abscisic acid 8′-hydroxylase encoding gene *ABH2* participates in the negative regulation of drought tolerance in maize (Liu et al. 2020a). Disruption of this gene via CRISPR/Cas9 approach enhanced drought tolerance by improving water conservation and accumulation of ABA content.

The mitogen-activated protein kinase (MAPK) cascade is among the most widely used signaling channels for translating environmental cues into intracellular reactions (Wang et al. 2010). Protein kinases modify substrate proteins by adding phosphate groups which affect their activities. The MAPK3 enzyme in tomatoes was functionally characterized by using CRISPR/Cas9-based method. *SlMAPK3* KO produced a sensitive phenotype in tomato (Wang et al. 2017). KO lines exhibited severe wilting symptoms with high H_2O_2 content and electrolyte leakage (EL) and lower APX, CAT, POD, and SOD activities. *Brassica napus* genome contains 4 *BnaRGA* genes engaged in DELLA protein synthesis that are negative regulators of gibberellic acid (GA) signaling. CRISPR/Cas9 technique was applied to generate KO mutant of *BnaA6.RGA,* which exhibited enhanced survival rate and RWC, and decreased EL, MDA content, and water loss (Wu et al. 2020).

Editing of phytohormone receptor genes also offers crop improvement to combat recurrent drought spells. For example, in tomato *Gibberellin-Insensitive Dwarf1* (*GID1a* encoding a GA receptor) mutant produced by CRISPR/Cas9 displayed low GA activity and high drought tolerance (Illouz-Eliaz et al. 2020). *gid1a* mutant plants exhibited a reduction in transpiration and maintained higher leaf water content. In addition to this, they also showed suppressed xylem vessel proliferation and expansion. It restricted hydration conductance under drought. In another example, *OsPYL9* (*pyrabactin resistant-like 9*) gene encoding ABA receptor was targeted using CRISPR/Cas9 in rice (Usman et al. 2020). Its mutants had higher plant height, grain yield (in terms of 1000-grain weight, grain number per panicle, grain width, grain length, and yield per plant) ABA content, leaf cuticular wax, chlorophyll content, survival rate, CAT and SOD activities, and lower stomata number, stomatal conductance and transpiration rate than that of wild type. NPR1 (non-expresser of pathogenesis-related gene 1), a receptor of salicylic acid, has a substantial role in controlling plant defense mechanisms (Wu et al. 2012). CRISPR/Cas9-mediated KO of this gene in tomato revealed it also positively regulates drought response as *slnpr1* mutants showed drought-sensitive phenotype (Li et al. 2019). Poor drought tolerance in *slnpr1* mutants was conveyed by higher stomatal aperture, EL, MDA, and H_2O_2 levels, and lower antioxidant enzymes activities, downregulated expression of *SlGST, SlDHN*, and *SlDREB* drought-related genes.

18.4.3 Morphology and Drought Avoidance

Leaf controls transpiration, a crucial factor in drought tolerance. Thus, modifying leaf morphological traits is a critical strategy to design drought-tolerant genotypes for sufficient agricultural production in drought-prone areas. Increasing the number of bulliform cells promotes adaxial leaf rolling which is crucial for reducing water loss during drought. A rolled leaf phenotype was obtained in rice through CRISPR/Cas9-mediated KO of *Semi-Rolled Leaf 1* and *2* (*OsSRL1* and *OsSRL2*) (Liao et al. 2019). The mutants exhibited curled leaves, fewer stomata number, stomatal conductance, transpiration rate, and MDA content under drought. In addition, better CAT and SOD activities, survival rate and yield performance were also observed. Recently, mutant plants were produced in wheat by applying multiplex sgRNA-CRISPR/Cas9 system by targeting five homologs of *TaSal1* gene (Abdallah et al. 2022). These mutant plants were drought resistant because they possessed bigger bulliform cells, which caused rolling leaf phenotypes under drought. In addition, growth traits like improved seed germination and seedling growth were observed in mutants under drought condition.

18.4.4 MicroRNAs

MicroRNAs (miRNA; small non-coding RNAs) upon binding with ARGONAUTE proteins make miRNA-induced silencing complex, that can identify target genes and block them at the post-transcriptional level thus they play a regulatory role in plant development (Jones-Rhoades et al. 2006). Using a new combinatory dual-sgRNA/SpCas9 vector, simultaneous gene deletion and replacement in *Arabidopsis* led to the generation of mutants of *mir169a* with improved drought tolerance phenotype (Zhao et al. 2016). *OsmiR535* belongs to a highly conserved miRNA family (miR156/miR529/miR535) in plants. *OsmiR535* is engaged in panicle and grain length determination, modulation of plant growth and regulation of cold stress response by suppression of expression of *OsSPL7/12/16* (Sun et al. 2019). *OsmiR535* negatively regulates the drought stress response in rice. Mutants of *OsmiR535* produced by CRISPR/Cas9 showed improved resistance to osmotic, salt, and drought stresses in rice (Yue et al. 2020). Additionally, *osmiR535* mutants displayed higher shoot length, number of lateral roots, primary root length, ABA insensitivity and improved survival rate, and decreased leaf senescence under drought conditions (Yue et al. 2020).

18.4.5 Nutrient and Yield Traits

Improvement of nutrient quality and agricultural production is crucial in a world where the climate is changing drastically. This has been achieved using the GE technique. Stachyose, raffinose, and oligosaccharides of raffinose family are thought to be anti-nutritive components in soybean seeds because they are indigestible. Using multiplex CRISPR/Cas9 system, three raffinose synthase genes (*RS2, RS3* and *RS4*) and one stachyose synthase gene (*STS*) were mutated. *rs2/rs3/rs4/sts* quadruple mutants had improved drought tolerance and reduced raffinose and stachyose contents in seed (Lin et al. 2023).

Maize *ARGOS8* interacts with ethylene response factors and negatively regulates ethylene response (Shi et al. 2017). Its overexpression enhances plant growth by promoting cell division. CRISPR/Cas9 tool was used to increase its expression. It was achieved by the replacement of the native *ARGOS8* promoter with *GOS2* promoter harboring drought-responsive *cis*-elements. *GOS2 promoter-ARGOS8* GE lines showed better drought resistance and higher grain yield. Insertion of *GOS2* promoter in 5′UTR (untranslated region) of the *ARGOS8* gene also provided the same result. The role of G protein encoding genes in yield traits and stress tolerance was validated by CRISPR/Cas9 approach in rice (Cui et al. 2020). KO of G protein encoding genes *gs3* and *dep1* enabled rice plants to produce more desirable agronomic traits such as increased panicle number and grain size under drought condition. Thus, all of these examples show that GE is helping in increasing agricultural productivity in drought-prone areas.

18.5 Increasing Heat Tolerance in Plants Through Genome Editing

GE technology is efficient in enhancing heat tolerance in plants. Although the number of such reports is limited today (Table 18.2), it is likely to increase in near future. GE is also aiding in understanding the functions of heat-responsive genes. For example, CRISPR/Cas9 mediated gene editing of various TFs and other signaling proteins has provided a better understanding of heat stress responses in tomato (Hu et al. 2021). Multiprotein bridging factor 1 (MBF1) is a transcriptional co-factor that helps in mediating the

Table 18.2 Increasing heat tolerance in plants through genome editing.

S. no.	Plant	Variety	GE tool	Target gene	Classification of gene	Function	Mutation type	Mutant's phenotype with respect to heat stress	Observed modification	Reference
1	Tomato	Condine Red	CRISPR/Cas9	*SlCPK28*	Protein kinase	Ca^{2+} sensor	KO (Deletion)	Sensitive	Decrease in shoot height and efficiency of PSII, increase in EL, MDA content, H_2O_2 content, accumulation of ROS, and increase in the level of protein carbonylation	Hu et al. (2021)
		Ailsa Craig	CRISPR/Cas9	*SlMAPK3*	Protein kinase	Negative regulator of thermotolerance	KO (deletion)	Tolerant	Less wilting and membrane damage, decrease in MDA content, ion leakage and ROS contents, and increase in levels of antioxidant enzymes, elevated expressions of *HSFs* and *HSPs*	Yu et al. (2019)
		Condine Red	CRISPR/Cas9	*SlBZR1*	TF	Regulator of brassinosteroid response	Null mutation	Sensitive	More severe wilting, decrease in quantum efficiency of PSII, Increase in the EL and MDA content	Yin et al. (2018)
		M82	CRISPR/Cas9	*SlAGL6*	Agamous-like MADS-box 6	Negative regulator of facultative parthenocarpy	KO (Deletion)	Tolerant	Higher seedless fruit yield and TSS content even under heat stress	Klap et al. (2017)
2	Rice	Nipponbare	CRISPR/Cas9	*OsHSA1*	Fructokinase-like protein 2	Regulator of chloroplast development	KO (Deletion)	Sensitive	Severe albino phenotype, delayed biogenesis of chlorophyll, shorter plants, reduced number of primary branches and effective tillers, shorter panicles, reduction in seed-setting rate, grain number per panicle, and 1000-grain weight of seeds	Qiu et al. (2018)
		Nipponbare	CRISPR/Cas9	*OsNAC006*	TF	Regulates stress responses	KO (deletion)	Sensitive	Increase in O_2^-, H_2O_2 and MDA levels, decrease in chlorophyll content and activities of antioxidant enzymes	Wang et al. (2020)
3	Wheat	CB037	CRISPR/Cas9	*TaMBF1c*	Transcription coactivator	Positive regulator of thermotolerance	KO (deletion)	Sensitive	Decrease in fresh weight and plant height	Tian et al. (2022)
4	Maize	Chang7-2	CRISPR/Cas9	*ZmCDPK7*	Protein kinase	ABA signaling and regulator of heat tolerance	KO (deletion)	Sensitive	Decrease in chlorophyll content, photosynthetic rate, APX and CAT activities, higher H_2O_2 and MDA content	Zhao et al. (2021)

interaction between TFs and TATA box binding factors. CRISPR/Cas9-mediated mutation of *TaMBF1c* gene increased heat sensitivity in wheat plants (Tian et al. 2022). Interestingly, TaMBF1c regulates heat-responsive genes including HSPs at the translational level also. Calcium-dependent protein kinases (CDPKs) are the calcium ion sensors that are found to regulate thermotolerance in tomato. CRISPR/Cas9 mediated KO mutant *slcpk28* displayed heat sensitivity and reduced growth as compared to the wild-type plants when subjected to temperatures of 45 °C suggesting its positive role in thermotolerance (Hu et al. 2021). Similarly, ZmCDPK7 acts as a positive regulator of thermotolerance in maize plants (Zhao et al. 2021). It interacts with a small HSP (sHSP17.4) and RBOHB (respiratory burst oxidase homolog) to regulate their activity. Tomato MAP kinase *SlMAPK3* was found to be the negative regulator of thermotolerance as CRISPR/Cas9 mediated mutant exhibited a higher survival rate and elevated expression of *HSFs* and *HSPs* suggesting a correlation between MAPK signaling pathway and expression of heat-responsive genes (Yu et al. 2019). HSFs control the expression of heat stress-induced genes including *HSPs*. BZR1 (Brassinazole resistant 1), a modulator of brassinosteroid response, affects thermotolerance in tomato by regulating ROS signaling through RBOH1 (Yin et al. 2018). bzr1 mutants become insensitive to 24-epibrassinolide and show decreased growth and reduced ability to tolerate heat (Yin et al. 2018). Heat stress severely affects sexual reproduction especially the microsporogenesis process resulting in yield loss. It also severely hinders fertilization-dependent fruit set. To overcome this problem, parthenocarpic fruits (fertilization-independent seedless fruits) have been produced in tomato. It was achieved by CRISPR/Cas9-mediated KO of *SlAGL6* (*AGAMOUS-LIKE 6*) gene (Klap et al. 2017). The weight and shape of parthenocarpic fruits were found to be similar to those of normally developed fruits. *FLN2* encoding chloroplasts localized fructokinase-like protein 2 was found to be essential for chloroplast formation and its protection from heat stress (Qiu et al. 2018). *oshsa1* mutant generated through CRISPR/Cas9, displayed delayed chloroplast biogenesis and severe albinism under heat stress at an early stage of leaf development.

18.6 Conclusion and Future Perspective

GE has revolutionized crop improvement process due to its effectiveness, ease, and specificity, and it also offers a new platform in the areas of functional genomics. Among the available GE tools, CRISPR/Cas9 is being widely used in modern crop improvement programs as it is accurate, rapid, simple, and offers multiplex editing. Due to genetic redundancy, mutations that result in a loss of function are typically ineffective in polyploid species. However, CRISPR/Cas9 system allows the use of multiple gRNAs and the introduction of mutations in various alleles. Being cutting-edge and precise methods for producing point mutations in plants, novel GE technologies such as base editing and prime editing provide an open platform for developing stress-resistant plants without running into problems with DNA DSBs and HDR. However, to exploit GE to reform global agriculture, a number of technical and legal obstacles still need to be overcome. As biotechnology develops, breakthroughs in the genome-editing system will overcome existing hurdles.

Author Contributions

PJ wrote the first draft of the chapter. PK also collected the literature and wrote the heat-related part of the chapter. PS conceptualized, edited and proofread the chapter. All authors approved the chapter.

Conflicts of Interest

The authors assert that they have no competing interests and personal relationships which could have affected the work of this chapter.

Acknowledgment

PS would like to acknowledge the Ministry of human resource development, Department of higher education, Government of India for the RUSA 2.0 programme (Thematic Project III) for financial support. PJ and PK acknowledge Department of Botany, University of Rajasthan for research facilities. PJ also acknowledges University Grants Commission, New Delhi for senior research fellowship.

References

Abdallah, N.A., Elsharawy, H., Abulela, H.A. et al. (2022). Multiplex CRISPR/Cas9-mediated genome editing to address drought tolerance in wheat. *GM Crops & Food* 1–17: https://doi.org/10.1080/21645698.2022.2120313.

Alok, A., Chauhan, H., Upadhyay, S.K. et al. (2021). Compendium of plant-specific CRISPR vectors and their technical advantages. *Life* 11: 1021. https://doi.org/10.3390/life11101021.

Anzalone, A.V., Koblan, L.W., and Liu, D.R. (2020). Genome editing with CRISPR-Cas nucleases, base editors, transposases and prime editors. *Nature Biotechnology* 38 (7): 824–844. https://doi.org/10.1038/s41587-020-0561-9.

Argast, G.M., Stephens, K.M., and Emond, M.J. (1998). Monnat Jr RJ. I-PpoI and I-CreI homing site sequence degeneracy determined by random mutagenesis and sequential in vitro enrichment. *Journal of Molecular Biology* 280 (3): 345–353. https://doi.org/10.1006/jmbi.1998.1886.

Blum, A. (2000). Towards standard assays of drought resistance in crop plants. Production in water-Limited environments. In: *Molecular Approaches for the Genetic Improvement of Cereals for Stable Production in Water-limited Environments* (ed. J.-M. Ribaut), 29–35. CIMMYT.

Blum, A. (2005). Drought resistance, water-use efficiency, and yield potential – are they compatible, dissonant, or mutually exclusive? *Australian Journal of Agricultural Research* 56 (11): 1159. https://doi.org/10.1071/ar05069.

Boch, J. and Bonas, U. (2010). *Xanthomonas* AvrBs3 family-type III effectors: discovery and function. *Annual Review of Phytopathology* 48 (1): 419–436. https://doi.org/10.1146/annurev-phyto-080508-081936.

Boch, J., Scholze, H., Schornack, S. et al. (2009). Breaking the code of DNA binding specificity of TAL-type III effectors. *Science* 326 (5959): 1509–1512. https://doi.org/10.1126/science.1178811.

Bogdanove, A.J., Schornack, S., and Lahaye, T. (2010). TAL effectors: finding plant genes for disease and defense. *Current Opinion in Plant Biology* 13 (4): 394–401. https://doi.org/10.1016/j.pbi.2010.04.010.

Bouzroud, S., Gasparini, K., Hu, G. et al. (2020). Down regulation and loss of auxin response factor 4 function using CRISPR/Cas9 alters plant growth, stomatal function and improves tomato tolerance to salinity and osmotic stress. *Genes* 11 (3): 272. https://doi.org/10.3390/genes11030272.

Budak, H., Hussain, B., Khan, Z. et al. (2015). From genetics to functional genomics: improvement in drought signaling and tolerance in wheat. *Frontiers in Plant Science* 6: https://doi.org/10.3389/fpls.2015.01012.

Chen, K. and Gao, C. (2013). TALENs: customizable molecular DNA scissors for genome engineering of plants. *Yi Chuan Xue Bao [Journal of Genetics and Genomics]* 40 (6): 271–279. https://doi.org/10.1016/j.jgg.2013.03.009.

Chen, Z., Galli, M., and Gallavotti, A. (2022). Mechanisms of temperature-regulated growth and thermotolerance in crop species. *Current Opinion in Plant Biology* 65: 102134. https://doi.org/10.1016/j.pbi.2021.102134.

Chen, S., Zhang, N., Zhang, Q. et al. (2019). Genome editing to integrate seed size and abiotic stress tolerance traits in *Arabidopsis* reveals a role for DPA4 and SOD7 in the regulation of inflorescence architecture. *International Journal of Molecular Sciences* 20 (11): 2695. https://doi.org/10.3390/ijms20112695.

Christian, M., Cermak, T., Doyle, E.L. et al. (2010). Targeting DNA double-strand breaks with TAL effector nucleases. *Genetics* 186 (2): 757–761. https://doi.org/10.1534/genetics.110.120717.

Cui, Y., Jiang, N., Xu, Z., and Xu, Q. (2020). Heterotrimeric G protein are involved in the regulation of multiple agronomic traits and stress tolerance in rice. *BMC Plant Biology* 20 (1): 90. https://doi.org/10.1186/s12870-020-2289-6.

Daniell, J.W., Chappell, W.E., and Couch, H.B. (1969). Effect of sublethal and lethal temperature on plant cells. *Plant Physiology* 44 (12): 1684–1689. https://doi.org/10.1104/pp.44.12.1684.

Dunn, D.A. and Pinkert, C.A. (2014). Gene editing. In: *Transgenic Animal Technology* (ed. C.A. Pinkert), 229–248. Elsevier.

EFSA (2012). Scientific opinion addressing the safety assessment of plants developed using Zinc Finger Nuclease 3 and other Site-Directed Nucleases with similar function: plants developed by ZFN-3. *EFSA Journal* 10 (10): 2943. https://doi.org/10.2903/j.efsa.2012.2943.

Fang, Y. and Xiong, L. (2015). General mechanisms of drought response and their application in drought resistance improvement in plants. *Cellular and Molecular Life Sciences: CMLS* 72 (4): 673–689. https://doi.org/10.1007/s00018-014-1767-0.

FAO. (2017). The Future of Food and Agriculture—Trends and Challenges. Rome. Available online: https://www.fao.org/3/i6583e/i6583e.pdf (accessed on 21 March 2023).

FAO (2022). *Food and Agriculture Organization of the United Nations, International Fund for Agricultural Development, United Nations Children's Fund United Nations World Food*

Programme, World Health Organization. The State of Food Security and Nutrition in the World 2022. Repurposing food and agricultural policies to make healthy diets more affordable. Rome: FAO https://doi.org/10.4060/cc0639en.

Feng, Z., Zhang, B., Ding, W. et al. (2013). Efficient genome editing in plants using a CRISPR/Cas system. *Cell Research* 23 (10): 1229–1232. https://doi.org/10.1038/cr.2013.114.

Food and Agriculture Organization of the United Nations. Food Outlook: Global Market Analysis (2008). *Food and Agriculture Organization of the United Nations. Food Outlook: Global Market Analysis*. FAO.

Gaj, T., Gersbach, C.A., and Barbas, C.F. 3rd. (2013). ZFN, TALEN, and CRISPR/Cas-based methods for genome engineering. *Trends in Biotechnology* 31 (7): 397–405. https://doi.org/10.1016/j.tibtech.2013.04.004.

Georges, F. and Ray, H. (2017). Genome editing of crops: a renewed opportunity for food security. *GM Crops & Food* 8 (1): 1–12. https://doi.org/10.1080/21645698.2016.1270489.

Hasanuzzaman, M., Nahar, K., Alam, M.M. et al. (2013). Physiological, biochemical, and molecular mechanisms of heat stress tolerance in plants. *International Journal of Molecular Sciences* 14 (5): 9643–9684. https://doi.org/10.3390/ijms14059643.

Hu, Z., Li, J., Ding, S. et al. (2021). The protein kinase CPK28 phosphorylates ascorbate peroxidase and enhances thermotolerance in tomato. *Plant Physiology* 186 (2): 1302–1317. https://doi.org/10.1093/plphys/kiab120.

Illouz-Eliaz, N., Nissan, I., Nir, I. et al. (2020). Mutations in the tomato gibberellin receptors suppress xylem proliferation and reduce water loss under water-deficit conditions. *Journal of Experimental Botany* 71 (12): 3603–3612. https://doi.org/10.1093/jxb/eraa137.

IPCC (2021). Summary for policymakers. In: *Climate change 2021: the physical science basis*. Contribution of Working Group I to the sixth assessment report of the Intergovernmental Panel on Climate Change (ed. V. Masson-Delmotte, P. Zhai, A. Pirani, et al.). Cambridge University Press.

Jangir, P., Shekhawat, P.K., Bishnoi, A. et al. (2021). Role of *Serendipita indica* in enhancing drought tolerance in crops. *Physiological and Molecular Plant Pathology* 116 (101691): 101691. https://doi.org/10.1016/j.pmpp.2021.101691.

Jiang, Y., Lahlali, R., Karunakaran, C. et al. (2015). Seed set, pollen morphology and pollen surface composition response to heat stress in field pea: heat stress on reproductive development in pea. *Plant, Cell & Environment* 38 (11): 2387–2397. https://doi.org/10.1111/pce.12589.

Jones-Rhoades, M.W., Bartel, D.P., and Bartel, B. (2006). MicroRNAS and their regulatory roles in plants. *Annual Review of Plant Biology* 57 (1): 19–53. https://doi.org/10.1146/annurev.arplant.57.032905.105218.

Joshi, R., Wani, S.H., Singh, B. et al. (2016). Transcription factors and plants response to drought stress: current understanding and future directions. *Frontiers in Plant Science* 7: 1029. https://doi.org/10.3389/fpls.2016.01029.

Kim, Y.G., Cha, J., and Chandrasegaran, S. (1996). Hybrid restriction enzymes: zinc finger fusions to Fok I cleavage domain. *Proceedings of the National Academy of Sciences of the United States of America* 93 (3): 1156–1160. https://doi.org/10.1073/pnas.93.3.1156.

Kim, W., Iizumi, T., and Nishimori, M. (2019). Global patterns of crop production losses associated with droughts from 1983 to 2009. *Journal of Applied Meteorology and Climatology* 58 (6): 1233–1244. https://doi.org/10.1175/jamc-d-18-0174.1.

Klap, C., Yeshayahou, E., Bolger, A.M. et al. (2017). Tomato facultative parthenocarpy results from Sl AGAMOUS-LIKE 6 loss of function. *Plant Biotechnology Journal* 15 (5): 634–647. https://doi.org/10.1111/pbi.12662.

Langner, T., Kamoun, S., and Belhaj, K. (2018). CRISPR crops: plant genome editing toward disease resistance. *Annual Review of Phytopathology* 56 (1): 479–512. https://doi.org/10.1146/annurev-phyto-080417-050158.

Lei, Y., Lu, L., Liu, H.-Y. et al. (2014). CRISPR-P: a web tool for synthetic single-guide RNA design of CRISPR-system in plants. *Molecular Plant* 7 (9): 1494–1496. https://doi.org/10.1093/mp/ssu044.

Levitt, J. (1980). *Responses of Plants to Environmental Stresses. Volume II. Water, Radiation, Salt, and Other Stresses*. New York: Academic Press.

Li, T., Liu, B., Spalding, M.H. et al. (2012). High-efficiency TALEN-based gene editing produces disease-resistant rice. *Nature Biotechnology* 30 (5): 390–392. https://doi.org/10.1038/nbt.2199.

Li, R., Liu, C., Zhao, R. et al. (2019). CRISPR/Cas9-Mediated SlNPR1 mutagenesis reduces tomato plant drought tolerance. *BMC Plant Biology* 19 (1): 38. https://doi.org/10.1186/s12870-018-1627-4.

Li, G., Ma, Y., Wang, X. et al. (2022). CRISPR/Cas9 gene editing of NtAITRs, a family of transcription repressor genes, leads to enhanced drought tolerance in tobacco. *International Journal of Molecular Sciences* 23 (23): 15268. https://doi.org/10.3390/ijms232315268.

Liao, S., Qin, X., Luo, L. et al. (2019). CRISPR/Cas9-induced Mutagenesis of semi-rolled Leaf1,2 confers curled leaf phenotype and drought tolerance by influencing protein expression patterns and ROS scavenging in rice (*Oryza sativa* L.). *Agronomy (Basel, Switzerland)* 9 (11): 728. https://doi.org/10.3390/agronomy9110728.

Lin, W., Kuang, H., Bai, M. et al. (2023). Multiplex genome editing targeting soybean with ultra-low anti-nutritive oligosaccharides. *The Crop Journal* https://doi.org/10.1016/j.cj.2023.01.001.

Liu, L., Zhang, J., Xu, J. et al. (2020a). CRISPR/Cas9 targeted mutagenesis of SlLBD40, a lateral organ boundaries domain transcription factor, enhances drought tolerance in tomato. *Plant Science: An International Journal of Experimental Plant Biology* 301 (110683): 110683. https://doi.org/10.1016/j.plantsci.2020.110683.

Liu, S., Li, C., Wang, H. et al. (2020b). Mapping regulatory variants controlling gene expression in drought response and tolerance in maize. *Genome Biology* 21 (1): 163. https://doi.org/10.1186/s13059-020-02069-1.

Lloyd, A., Plaisier, C.L., Carroll, D., and Drews, G.N. (2005). Targeted mutagenesis using zinc-finger nucleases in *Arabidopsis*. *Proceedings of the National Academy of Sciences of the United States of America* 102 (6): 2232–2237. https://doi.org/10.1073/pnas.0409339102.

Lou, D., Wang, H., Liang, G., and Yu, D. (2017). OsSAPK2 confers abscisic acid sensitivity and tolerance to drought stress in rice. *Frontiers in Plant Science* 8: 993. https://doi.org/10.3389/fpls.2017.00993.

Luo, L.J. (2010). Breeding for water-saving and drought-resistance rice (WDR) in China. *Journal of Experimental Botany* 61 (13): 3509–3517. https://doi.org/10.1093/jxb/erq185.

Makarova, K.S. and Koonin, E.V. (2015). Annotation and classification of CRISPR-Cas systems. *Methods in Molecular Biology* 1311: 47–75. https://doi.org/10.1007/978-1-4939-2687-9_4.

Malagnoux, M., Sène, E.H., and Atzmon, N. (2008). Forests, trees and water in arid lands: a delicate balance. *Unasylva-Fao* 229: 24–29.

Mao, Y., Botella, J.R., Liu, Y., and Zhu, J.-K. (2019). Gene editing in plants: progress and challenges. *National Science Review* 6 (3): 421–437. https://doi.org/10.1093/nsr/nwz005.

Martínez-Fortún, J., Phillips, D.W., and Jones, H.D. (2017). Potential impact of genome editing in world agriculture. *Emerging Topics in Life Sciences* 1 (2): 117–133. https://doi.org/10.1042/etls20170010.

Molla, K.A., Sretenovic, S., Bansal, K.C., and Qi, Y. (2021). Precise plant genome editing using base editors and prime editors. *Nature Plants* 7 (9): 1166–1187. https://doi.org/10.1038/s41477-021-00991-1.

Ogata, T., Ishizaki, T., Fujita, M., and Fujita, Y. (2020). CRISPR/Cas9-targeted mutagenesis of OsERA1 confers enhanced responses to abscisic acid and drought stress and increased primary root growth under nonstressed conditions in rice. *PLoS One* 15 (12): e0243376. https://doi.org/10.1371/journal.pone.0243376.

Parthasarathi, T., Firdous, S., Mariya David, E. et al. (2022). Effects of high temperature on crops. In: *Advances in Plant Defense Mechanisms* (ed. J.N. Kimatu). IntechOpen. https://doi.org/10.5772/intechopen.105945.

Paszkowski, J., Baur, M., Bogucki, A., and Potrykus, I. (1988). Gene targeting in plants. *The EMBO Journal* 7 (13): 4021–4026. https://doi.org/10.1002/j.1460-2075.1988.tb03295.x.

Pauwels, K., Podevin, N., Breyer, D. et al. (2014). Engineering nucleases for gene targeting: safety and regulatory considerations. *New Biotechnology* 31 (1): 18–27. https://doi.org/10.1016/j.nbt.2013.07.001.

Qiu, Z., Kang, S., He, L. et al. (2018). The newly identified heat-stress sensitive albino 1 gene affects chloroplast development in rice. *Plant Science: An International Journal of Experimental Plant Biology* 267: 168–179. https://doi.org/10.1016/j.plantsci.2017.11.015.

Ran, Y., Liang, Z., and Gao, C. (2017). Current and future editing reagent delivery systems for plant genome editing. *Science China. Life Sciences* 60 (5): 490–505. https://doi.org/10.1007/s11427-017-9022-1.

Razzaq, A., Saleem, F., Kanwal, M. et al. (2019). Modern trends in plant genome editing: an inclusive review of the CRISPR/Cas9 toolbox. *International Journal of Molecular Sciences* 20 (16): 4045. https://doi.org/10.3390/ijms20164045.

Roca Paixão, J.F., Gillet, F.-X., Ribeiro, T.P. et al. (2019). Improved drought stress tolerance in *Arabidopsis* by CRISPR/dCas9 fusion with a histone AcetylTransferase. *Scientific Reports* 9 (1): 8080. https://doi.org/10.1038/s41598-019-44571-y.

Santosh Kumar, V.V., Verma, R.K., Yadav, S.K. et al. (2020). CRISPR-Cas9 mediated genome editing of drought and salt tolerance (OsDST) gene in indica mega rice cultivar MTU1010. *Physiology and Molecular Biology of Plants: An International Journal of Functional Plant Biology* 26 (6): 1099–1110. https://doi.org/10.1007/s12298-020-00819-w.

Seleiman, M.F., Al-Suhaibani, N., Ali, N. et al. (2021). Drought stress impacts on plants and different approaches to alleviate its adverse effects. *Plants* 10 (2): 259. https://doi.org/10.3390/plants10020259.

Shi, J., Gao, H., Wang, H. et al. (2017). ARGOS 8 variants generated by CRISPR-Cas9 improve maize grain yield under field drought stress conditions. *Plant Biotechnology Journal* 15 (2): 207–216. https://doi.org/10.1111/pbi.12603.

Silva, G., Poirot, L., Galetto, R. et al. (2011). Meganucleases and other tools for targeted genome engineering: perspectives and challenges for gene therapy. *Current Gene Therapy* 11 (1): 11–27. https://doi.org/10.2174/156652311794520111.

Stoddard, B.L. (2011). Homing endonucleases: from microbial genetic invaders to reagents for targeted DNA modification. *Structure (London, England: 1993)* 19 (1): 7–15. https://doi.org/10.1016/j.str.2010.12.003.

Sun, M., Shen, Y., Li, H. et al. (2019). The multiple roles of OsmiR535 in modulating plant height, panicle branching and grain shape. *Plant Science: An International Journal of Experimental Plant Biology* 283: 60–69. https://doi.org/10.1016/j.plantsci.2019.02.002.

Sushmita, K.G., Upadhyay, S.K., and Verma, P.C. (2021). An overview of genome-engineering methods. In: *Genome Engineering for Crop Improvement* (ed. S.K. Upadhyay). Wiley. https://doi.org/10.1002/9781119672425.ch1.

Tabassum, J., Ahmad, S., Hussain, B. et al. (2021). Applications and potential of genome-editing systems in rice improvement: current and future perspectives. *Agronomy (Basel, Switzerland)* 11 (7): 1359. https://doi.org/10.3390/agronomy11071359.

Tardieu, F. (2013). Plant response to environmental conditions: assessing potential production, water demand, and negative effects of water deficit. *Frontiers in Physiology* 4: 17. https://doi.org/10.3389/fphys.2013.00017.

Tian, X., Qin, Z., Zhao, Y. et al. (2022). Stress granule-associated TaMBF1c confers thermotolerance through regulating specific mRNA translation in wheat (*Triticum aestivum*). *The New Phytologist* 233 (4): 1719–1731. https://doi.org/10.1111/nph.17865.

Turnbull, C., Lillemo, M., and Hvoslef-Eide, T.A.K. (2021). Global regulation of genetically modified crops amid the gene edited crop boom – a review. *Frontiers in Plant Science* 12: 630396. https://doi.org/10.3389/fpls.2021.630396.

Turner, N.C. (1979). Drought resistance and adaptation to water deficits in crop plants. *Stress Physiology in Crop Plants* 11: 343–372.

Tyagi, S., Shumayla, S., and S.P., Upadhyay, S.K. (2019). Role of superoxide dismutases (SODs) in stress tolerance in plants. In: *Molecular Approaches in Plant Biology and Environmental Challenges. Energy, Environment, and Sustainability* (ed. S. Singh, S. Upadhyay, A. Pandey, and S. Kumar). Singapore: Springer. https://doi.org/10.1007/978-981-15-0690-1_3.

Upadhyay, S.K. (2021). *Genome Engineering for Crop Improvement*. Wiely. https://doi.org/10.1002/9781119672425.

Urnov, F.D., Ronald, P.C., and Carroll, D. (2018). A call for science-based review of the European court's decision on gene-edited crops. *Nature Biotechnology* 36 (9): 800–802. https://doi.org/10.1038/nbt.4252.

Usman, B., Nawaz, G., Zhao, N. et al. (2020). Precise editing of the OsPYL9 gene by RNA-guided Cas9 nuclease confers enhanced drought tolerance and grain yield in rice (*Oryza sativa* L.) by regulating circadian rhythm and abiotic stress responsive proteins. *International Journal of Molecular Sciences* 21 (21): 7854. https://doi.org/10.3390/ijms21217854.

Vaughan, M.M., Block, A., Christensen, S.A. et al. (2018). The effects of climate change associated abiotic stresses on maize phytochemical defenses. *Phytochemistry Reviews: Proceedings of the Phytochemical Society of Europe* 17 (1): 37–49. https://doi.org/10.1007/s11101-017-9508-2.

Voytas, D.F. and Gao, C. (2014). Precision genome engineering and agriculture: opportunities and regulatory challenges. *PLoS Biology* 12 (6): e1001877. https://doi.org/10.1371/journal.pbio.1001877.

Wahid, A., Gelani, S., Ashraf, M., and Foolad, M.R. (2007). Heat tolerance in plants: an overview. *Environmental and Experimental Botany* 61 (3): 199–223. https://doi.org/10.1016/j.envexpbot.2007.05.011.

Waltz, E. (2012). Tiptoeing around transgenics: new techniques for manipulating plant genomes are yielding plants touted as nontransgenic. Will that relieve regulatory burden? Emily Waltz investigates. *Nature Biotechnology* 30 (3): 215–218. https://doi.org/10.1038/nbt.2143.

Waltz, E. (2016). Gene-edited CRISPR mushroom escapes US regulation. *Nature* 532 (7599): 293. https://doi.org/10.1038/nature.2016.19754.

Wang, P., Du, Y., Li, Y. et al. (2010). Hydrogen peroxide-mediated activation of MAP kinase 6 modulates nitric oxide biosynthesis and signal transduction in *Arabidopsis*. *The Plant Cell* 22 (9): 2981–2998. https://doi.org/10.1105/tpc.109.072959.

Wang, L., Chen, L., Li, R. et al. (2017). Reduced drought tolerance by CRISPR/Cas9-mediated SlMAPK3 Mutagenesis in tomato plants. *Journal of Agricultural and Food Chemistry* 65 (39): 8674–8682. https://doi.org/10.1021/acs.jafc.7b02745.

Wang, B., Zhong, Z., Wang, X. et al. (2020). Knockout of the OsNAC006 transcription factor causes drought and heat sensitivity in rice. *International Journal of Molecular Sciences* 21 (7): 2288. https://doi.org/10.3390/ijms21072288.

Wu, Y., Zhang, D., Chu, J.Y. et al. (2012). The *Arabidopsis* NPR1 protein is a receptor for the plant defense hormone salicylic acid. *Cell Reports* 1 (6): 639–647. https://doi.org/10.1016/j.celrep.2012.05.008.

Wu, J., Yan, G., Duan, Z. et al. (2020). Roles of the *Brassica napus* DELLA protein BnaA6.RGA, in modulating drought tolerance by interacting with the ABA signaling component BnaA10.ABF2. *Frontiers in Plant Science* 11: 577. https://doi.org/10.3389/fpls.2020.00577.

Xie, X., Ma, X., Zhu, Q. et al. (2017). CRISPR-GE: a convenient software toolkit for CRISPR-based genome editing. *Molecular Plant* 10 (9): 1246–1249. https://doi.org/10.1016/j.molp.2017.06.004.

Xu, J., Hua, K., and Lang, Z. (2019). Genome editing for horticultural crop improvement. *Horticulture Research* 6 (1): 113. https://doi.org/10.1038/s41438-019-0196-5.

Yin, Y., Qin, K., Song, X. et al. (2018). BZR1 transcription factor regulates heat stress tolerance through FERONIA receptor-like kinase-mediated reactive oxygen species signaling in tomato. *Plant & Cell Physiology* 59: 2239–2254.

Yu, W., Wang, L., Zhao, R. et al. (2019). Knockout of SlMAPK3 enhances tolerance to heat stress involving ROS homeostasis in tomato plants. *BMC Plant Biology* 19 (1): 354. https://doi.org/10.1186/s12870-019-1939-z.

Yue, E., Cao, H., and Liu, B. (2020). OsmiR535, a potential genetic editing target for drought and salinity stress tolerance in *Oryza sativa*. *Plants* 9 (10): https://doi.org/10.3390/plants9101337.

Zhang, J.-Y., Broeckling, C.D., Blancaflor, E.B. et al. (2005). Overexpression of WXP1, a putative *Medicago truncatula* AP2 domain-containing transcription factor gene, increases cuticular wax accumulation and enhances drought tolerance in transgenic alfalfa

(*Medicago sativa*): WXP1 activates wax production in alfalfa. *The Plant Journal: For Cell and Molecular Biology* 42 (5): 689–707. https://doi.org/10.1111/j.1365-313X.2005.02405.x.

Zhang, H., Zhang, J., Lang, Z. et al. (2017). Genome editing – principles and applications for functional genomics research and crop improvement. *Critical Reviews in Plant Sciences* 36 (4): 291–309. https://doi.org/10.1080/07352689.2017.1402989.

Zhao, Y., Zhang, C., Liu, W. et al. (2016). An alternative strategy for targeted gene replacement in plants using a dual-sgRNA/Cas9 design. *Scientific Reports* 6 (1): 23890. https://doi.org/10.1038/srep23890.

Zhao, J., Lu, Z., Wang, L., and Jin, B. (2020). Plant responses to heat stress: physiology, transcription, noncoding RNAs, and epigenetics. *International Journal of Molecular Sciences* 22 (1): 117. https://doi.org/10.3390/ijms22010117.

Zhao, Y., Du, H., Wang, Y. et al. (2021). The calcium-dependent protein kinase ZmCDPK7 functions in heat-stress tolerance in maize. *Journal of Integrative Plant Biology* 63 (3): 510–527. https://doi.org/10.1111/jipb.13056.

Zhong, X., Hong, W., Shu, Y. et al. (2022). CRISPR/Cas9 mediated gene-editing of GmHdz4 transcription factor enhances drought tolerance in soybean (*Glycine max* [L.] Merr.). *Frontiers in Plant Science* 13: 988505. https://doi.org/10.3389/fpls.2022.988505.

19

Genome Engineering for the Development of Climate-Resilient Crop Plants

Bhavuk Gupta[1]*, Ayush Khandelwal*[2]*, Brijesh Kumar*[3]*, and Purva Bhalothia*[4]

[1] *National Centre for Cell Science, Pune, Maharashtra, India*
[2] *South Asian University, New Delhi, India*
[3] *Patanjali Herbal Research Division, Patanjali Research Institute, Haridwar, Uttarakhand, India*
[4] *Birla Institute of Scientific Research (BISR), Jaipur, Rajasthan, India*

19.1 Introduction

The steady accumulation of greenhouse gases, particularly CO_2, has resulted in a worrisome rise in the temperature of the overall land surface and oceans, known as global warming. Climate change is the result of this process, which causes inflation in the frequency and intensity of different stress events for plants like drought, heat waves, cold snaps and flooding. Changes in land usage, soil microbiota, water balance, etc. are all results of changing climatic circumstances that are caused by temperature swings and unexpected rainfall patterns. Plants adapt to changing climatic conditions through a variety of mechanisms, including altered blooming and fruiting times, changes in floral bloom and growth detention or acceleration. Nonetheless, these methods lead in a significant decrease in plant yield. The impact of poor climatic circumstances is particularly obvious in crop plants, the majority of which have been developed for enhanced production and have lower tolerance for environmental restrictions (FAO.org. 2016). Abiotic stresses like drought and high salt content have been made worse by climate change, and it also hikes the count of pests and pathogens by affecting their terrain range and altering the interactions between pests, their natural foe and the host plant (CGIAR 2022). As a result, creating climate-resilient crops is critical for fulfilling rising food and feed demands, necessitating fresh solutions. Therefore, a dominant challenge for plant researcher, including plant breeders and plant biotechnologists, is creating climate-resilient crops that can withstand temperature variations and environmental strain, principally aggregation of these stresses.

Biotechnological breakthroughs, notably in the post-genomic age, laid the groundwork for long-term agricultural intensification as well as increased crop adaptability to climate change. To overcome the challenges of the environment, breeding approaches are complemented with modern biotechnological technologies such as genomics, proteomics and transcriptomics to generate stress-tolerant crop plants with as little negative influence on production as

Applications of Genome Engineering in Plants, First Edition. Edited by Santosh Kumar Upadhyay.
© 2024 John Wiley & Sons Ltd. Published 2024 by John Wiley & Sons Ltd.

feasible while keeping higher yields (Srivastav et al. 2018). It is prominent that plants respond to environmental signals at the physiological, biochemical and molecular levels in a complex and well-coordinated manner. For a cell to respond to such unfavourable circumstances, stress-induced disruption of normal cell function necessitates swift and extensive reprogramming at the molecular level (Woodrow et al. 2017). The foundation of this reprogramming is the stringent control of the expression of genes that counter stress, notably at the transcriptional and posttranscriptional stages. The micro RNAs (miRNAs) have become important posttranscriptional and translational regulators of gene expression that play a role in controlling how the body responds to stress, frequently by slowing the development of stressed plants (Xu et al. 2019). The accessibility of a huge plant breeding populace and germplasm collections, effective high-throughput technologies, vast data management tools and subsequent biotechnology and molecular breeding operations are all prerequisites for next-generation breeding. The scientific community will be able to quickly define one or more ideotypes that will delight the breeding need and find remarkable alleles and haplotypes to use in breeding programs. Furthermore, the latest developments in genomic knowledge, as well as the increased accessibility of information on genes and *in vitro* regeneration technologies, enable the creation & application of second-generation biotechnologies based on cisgenesis and genome editing (Cardi and Stewart 2016). Together with the tools, bioinformatics will need to continue building and maintaining integrated databases to support the analysis and presentation of multi-omics data. Decoded plant genomes now make it possible to quickly find candidate genes, molecular markers, quantitative trait loci (QTLs) and genome-wide consortium studies that may be useful in high throughput marker-assisted breeding. In current years, genomics and marker-assisted breeding have been used to uncover the mechanisms behind improved crop productivity and quality (Naqvi et al. 2022).

19.2 Effect of Climate Change on Crop Plants

The climate changes drastically when the Earth's temperature rises and becomes abiotically harsh. Environmental changes can be exceedingly harmful and put naturally occurring crop species in danger in many ways as shown in (Figure 19.1). The most common stresses in the field are heat and drought, and both have a big impact on plants. The production and yield of grains are negatively impacted by heat stress, sterility is caused by cold stress and the morphophysiology of plants is negatively impacted by drought stress (Raza et al. 2019).

Similar to how organisms adapt to their changing environment, an abiotic component that is subject to a natural change, benefits the ecosystem as a whole. However, if the alteration is artificial, or man-made, then it might result in many extreme or unfavourable changes to the creature, ultimately influencing other organisms that are dependent on it directly or indirectly and ultimately harming the ecosystem. The impact of climate change on crop plants is similar to the current situation. Everything was normal in the past and up until the pre-industrial era, but when the industrial revolution arrived, it brought a boom to the manufacturing of products, increasing both world output and population. However, as soon as the industrial revolution became apparent, everything began to suffer from its severe repercussions, including the environment, natural systems, agricultural systems and human health (Raza et al. 2019).

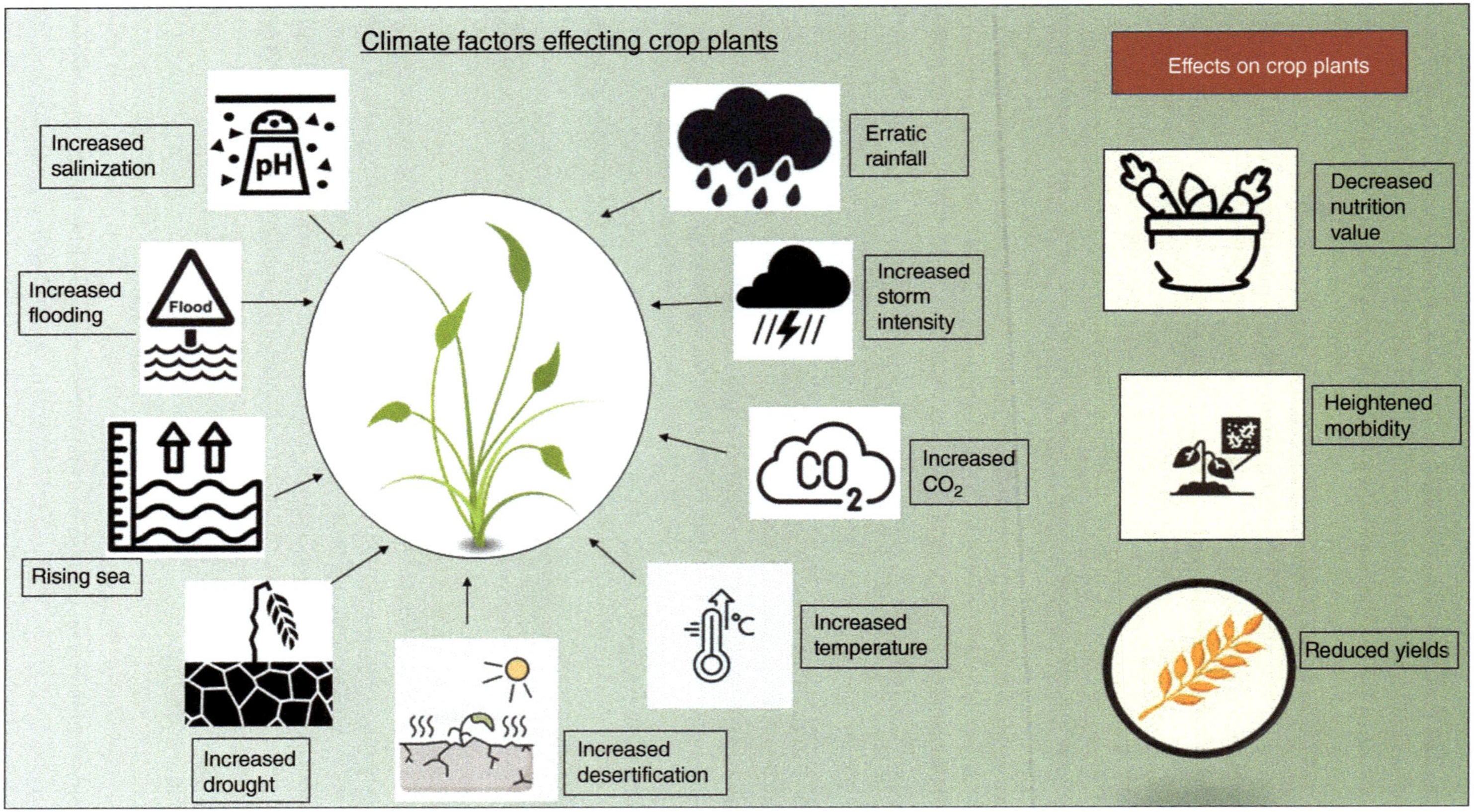

Figure 19.1 Schematic representation of effect of the different abiotic stress factors on crop plant.

Due to increased human activity including deforestation, water pollution and air pollution brought on by the industrial revolution and the subsequent rise in world population, these factors collectively constitute the primary driver of climate change, which is adversely affecting crop plants. The demand for food has increased because of the growing population. However, due to climate change and deforestation, the food supply is unable to keep up with the demand, creating a food shortage. By 2050, it is predicted that there will be 11.3 billion people on the planet, meaning that food production would need to rise by 50% to keep up with demand (DaMatta et al. 2010; Raza et al. 2019).

The primary cause of climate change can be attributed to air pollution, which is further exacerbated by a rise in the atmospheric concentration of greenhouse gases. Carbon dioxide (CO_2), which has increased to roughly 400 parts per million (ppm), is the principal greenhouse effect contributor (Raza et al. 2019). The annual increase in the earth's temperature brought on by the greenhouse effect has a direct impact on crop plants. Climate change can lower food production by decreasing the yield of major crops and also decreasing g the nutrient quality of food crops. This effect will be seen mostly in developing countries, located in the tropical regions of the earth. The negative effect of climate change on food crops has already been started (DaMatta et al. 2010).

By altering the climatic circumstances for plants and then varying the physical parameters like temperature, soil moisture and precipitation, studies have demonstrated the harmful effects of the greenhouse effect on plants. Every year, there is more volatility in these physical parameters, which causes the agricultural output to fluctuate more. Simulation models, crop growth models and regression approaches are the most common methods used to study the impact of climate change on plants (Isik and Devadoss 2006; Brouder and Volenec 2008; Blanc 2012).

The impact of climate change on crop plants has not yet been fully understood through various experiments and model studies. More research is required to analyse the effects and develop new techniques to stabilise or increase crop yield in response to changing environmental conditions and meet the demands of a growing world population.

19.2.1 Effect on Photosynthesis and CO_2 Fixation

Solar energy enables plants to fix carbon in the form of food. As the amount of CO_2 in the atmosphere has increased due to climate change and plants need CO_2 as their main source of carbon, this may appear like a good thing for crops as it could lead to an increase in crop output. Different plant types fix carbon in different ways. C3 plants fix carbon immediately through the Calvin cycle, whereas C4 plants first mix CO_2 with phosphoenolpyruvate (PEP), which is then transformed into oxaloacetate before entering the Calvin cycle. Accordingly, it has been established that an increase in CO_2 concentration causes C3 plants' photosynthetic capacity to increase by blocking the photorespiration process. However, C4 plants did not experience a similarly substantial shift (DaMatta et al. 2010).

19.2.2 Effect of Temperature

The primary variable influencing crop productivity and resulting in variance in crop yield is the change in the environmental temperature of crop plants. Photosynthesis, which fixes carbon, is temperature-dependent, and temperature changes can affect the carbon balance in plants. A temperature rise can affect the electron transport chain in chloroplasts and reduce the rate of carbon fixation in C3 plants through a process known as photorespiration. Additionally, it has been observed that higher temperatures cause the stomata to close to stop water loss. This lowers the CO_2 concentration inside the leaves and interferes with photosynthesis (DaMatta et al. 2010). The annual increase in global temperature caused by the 'greenhouse effect' leads to drought conditions and fluctuating precipitation, which have an impact on crop productivity (Deressa and Hassan 2009).

19.2.3 Effect of Change in Precipitation

Changes in precipitation may be a side effect of global warming, which also has an impact on crop plants. Evapotranspiration and water availability are both impacted by precipitation. The soil is losing water due to the temperature rise, and the pattern of rainfall is changing, which causes the crop fields to become drier. This increases the need for irrigation, which will subsequently cause the groundwater level to drop. Because of excessive rainfall in some areas during the past few years, crops in those areas have suffered flood damage. Due to this irregular pattern of precipitation, climate change has occurred (Kumar 2016).

19.2.4 Effect of Salinity

In recent decades, soil salinity has also become a major threat to food crops, as most of the food crops are sensitive to salt. This problem is mostly faced by semi-arid and arid regions of the world. Soil salinity also increases due to irrigation leading to secondary salinisation. This salinisation remain persists when there is insufficient precipitation and damages the crops (Shrivastava and Kumar 2015). Another reason for an increase in the soil salinity is the rising sea level in coastal regions that has increased the soil salinity by over 33% (Ullah et al. 2021). Accumulation of salt in the soil leads to a decrease in nutrient and water uptake further resulting in ion toxicity, osmotic stress and nutrient imbalance in crop plants. Higher salt concentrations in soil can further lead to chlorosis, and early leaf senescence and may affect photosynthetic active leaves (Zörb et al. 2019).

19.3 Genome Engineering in Crop Improvement

In the past few decades, new technology has come up showing the potential to improve crop production. So far, from the knowledge of central dogma, we know that DNA is the sole control system of the organism, and the new technology uses this DNA, and manipulate it to get the desired trait (Sedeek et al. 2019). This technology came to be known as genetic engineering. Now, genetic engineering has revolutionised the research in the improvement of crop plants. This technique uses modern DNA manipulation technology by using tools like CRISPR (clustered regularly interspaced short palindromic repeats), recombinant DNA technology and bioinformatics as shown in Figure 19.2. Due to the change in the climate conditions many food crops are affected because they could not cope with the environmental stress and the major example of the such crop was rice (Solanke

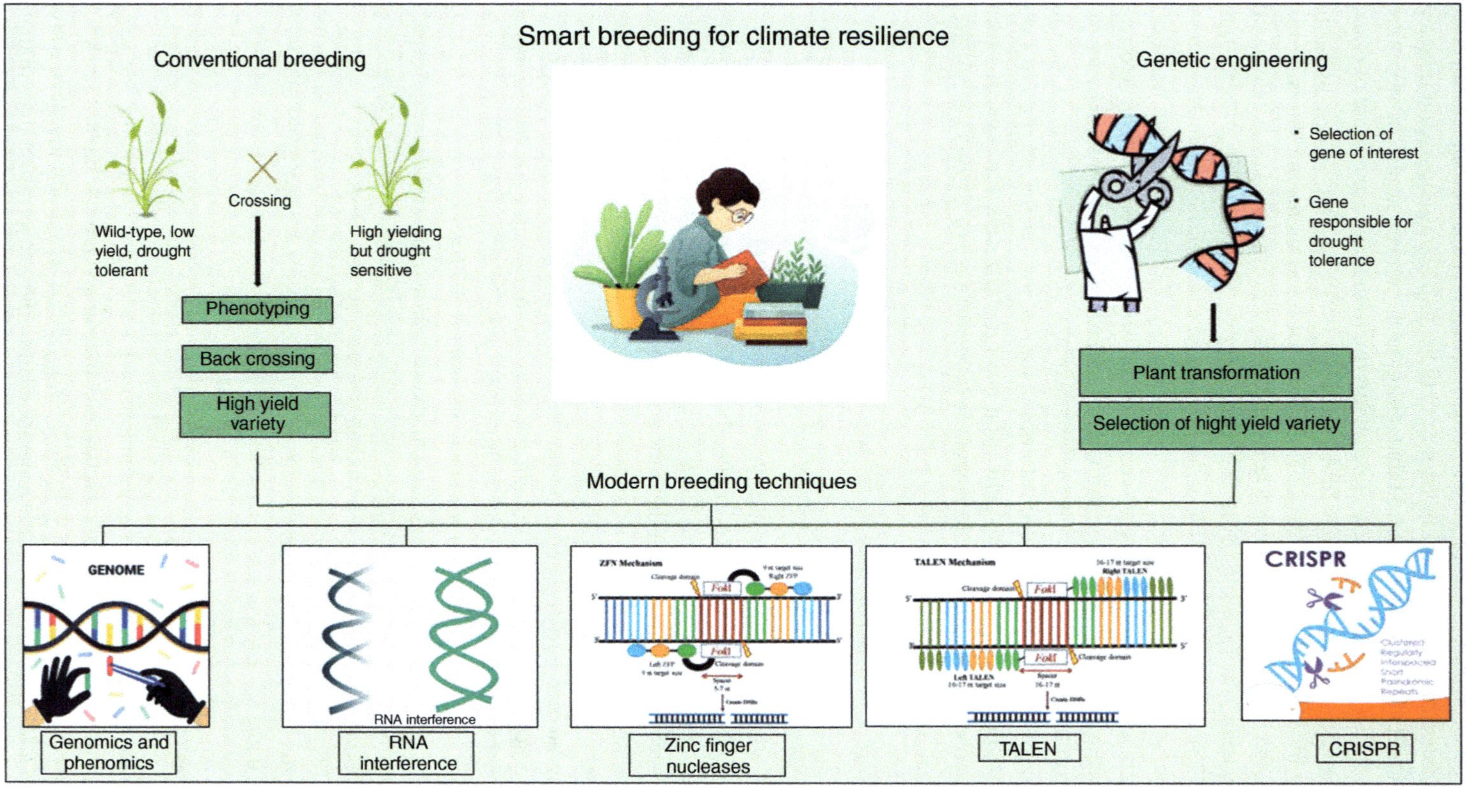

Figure 19.2 Different tools and techniques to develop genome engineered crop plants.

et al. 2021). Now with the use of genetic engineering, crops are designed to acquire new traits like increased yield, increase resistance to biotic and abiotic stress and increased crop stability (Fiaz et al. 2021). Genome editing techniques have been applied to many crops like wheat, maize, rice, potato and many more giving rise to the next generation of plant breeding by exploiting techniques like CRISPR-Cas and mutational breeding. Genetic engineering is done by editing the DNA which is further achieved using nucleases. Some major nucleases other than CRISPR which are widely used are zinc-finger nucleases (ZFNs), sequence-specific nucleases (SSNs) and transcription activator-like effector nucleases (TALENs). These help create double-strand breaks at the specific target sites and genome editing is carried out by a precise genome repair pathway (Gao 2021). Genome editing techniques may be quite complex rather to conventional breeding methods to develop new crop varieties but, it is less time consuming and offers precise efficiency with less risk of off-target effects (Rai and Singh 2021). Another advantage of genome editing is the shortened process of developing new enhanced varieties of crop plants compared to the conventional time-consuming techniques as the genes can be edited directly. Through this technique, the developed variety will be widely accessible to the whole world (Pixley et al. 2022).

19.4 Traditional and Modern Molecular Breeding for Crop Improvement

Traditional agricultural processes aim to improve plant output and nutritional composition, while contemporary modern approaches offer a diverse range of possibilities and advancements in breeding methods. These modern established strategies can be utilised to endure catastrophic plant biotic and abiotic stressors, as well as to contest rising food commodity demand (Naqvi et al. 2022). Crop development has been substantially influenced by the introduction of cutting-edge genomics techniques. Novel strategies, including genomic selection, speed breeding and high-throughput crop phenotyping (HTTP), have been demonstrated to accelerate the plant breeding system in this decade. Biotechnological treatments, such as genetic engineering techniques (gene transformation), have also been critical in the production of crops with desirable features. Other approaches, like as whole genome sequencing (WGS), genomics, gene identification, gene isolation and rapid molecular markers, are being used to improve cisgenesis, intra-genesis, mutation and polyploidy breeding (Murovec et al. 2017). It is a challenging statistical problem that is being partially solved by the application of modern machine learning (ML) techniques to correctly discover biological traits and pathways using the vastly expanding repositories of genomic and phenomics data.

19.4.1 Classical Plant Breeding

The technique of choosing reproducing plants with enticing qualities and removing those with bad characteristics is known as classical plant breeding. It is founded on a straightforward selection procedure wherein plants with superior attributes are chosen from a heterogeneous population and propagated over many generations until those features are fixed in the population. Crossing is the deliberate mating of distantly or closely related species in order to produce crop varieties with desired traits. However, these breeding strategies have restrictions, such as only mating between two plants that are sexually suitable with each other, selecting just genes that generate a distinct observable phenotype and linkage drag. Because of these factors, traditional breeding is time consuming, labour-intensive and expensive.

19.4.2 Genetic Engineering

The application of genetic engineering methods can alleviate the majority of the problems associated with breeding methodologies. No matter the source of the crop, desired genes can be transferred using genetic engineering or transgenic technology. Precision, freedom from interspecies incompatibility, lack of linkage drag and, most importantly faster, are all advantages of this technology over breeding methods. Moreover, genetic engineering makes it simpler to add genes for any relevant attribute or traits, enabling the development of crops with a variety of desirable properties. Yet, 24 years after their introduction, genetically altered crops have gained widespread acceptance, accounting for over 191.7 million hectares in 2018. For instance, freshly developed genetic varieties of drought-tolerant maize and sugarcane add to the total area of transgenic crops (ISAAA.org 2017).

Detailed knowledge of the genes engaged in the stress response is needed when engineering for abiotic stress tolerance. A massive number of candidate genes implicated in abiotic stress responses have been discovered by functional genomics and genetics and stress-receptive cellular signal transduction pathways have been mapped. Typically, the mentioned experiments manipulate the expression of genes. Using functional genomics, some of the genes encoding ion transporters, antioxidant enzymes, osmoregulatory metabolites, transcription factors (TFs), etc. have been found to modulate abiotic stress response in plants. Key studies have demonstrated enhanced tolerance to abiotic stimuli because of modulating the expression of these genes implicated in the stress condition.

19.4.3 RNA Interference

By developing knockdown transgenic lines, RNA interference has emerged as a potent tool for the inhibition of the expression of target genes and the functional investigation of genes. There are two primary methods for knocking down genes: exogenously produced siRNAs that are introduced precisely inside the cells, or by genetic engineering methods that entail the substantial integration of expression constructs that explicit antisense RNA, short hairpin RNAs, or long hairpin RNAs that are targeted specifically at a marked RNA. For Example- Rice exhibits improved tolerance to saline stress as a result of OsCKX2 downregulation utilising the RNAi method (Pradhan et al. 2015).

19.4.4 Phenomics and Genomics

For morphological traits to grow and develop from the cell to the whole plant level, multiplex interactions between genotypes (G) & a variety of environmental variables must take place. These interactions lead to phenotypes (P). Crop phenomics analyses the phenotypes that underpin important agronomic characteristics, such as yield under particular environmental variables, such as structure size, flowering period and seed size. Deep learning (DL), one of the newest ML techniques, is already being used in a variety of fields to advance automated picture recognition and machine vision. DL has been used more frequently in phenomics to extract crop phenotypes from large datasets of complicated images. To enable the connection of phenotypes with genomic variation, Deep Learning models have been employed to deliver robust and authentic feature extraction, identifying the existence of morphological attributes in both the root and the stem. To extract phenotypes from crops like wheat, maize, barley and rice, DL models have been successfully trained (Mao et al. 2019).

Genomic research can aid in the creation of climate-ready crops by characterising distinct cultivars and wild crop relatives, identifying genetic diversity and genes underpinning adaptive features and characterising functional response pathways linked to biotic and abiotic challenges. Large-scale re-sequencing across several crop species has enabled population-level genomic studies of crop germplasm viable, thanks to relatively low DNA sequencing costs and powerful bioinformatics analysis pipelines. The most popular bioinformatics tool for SNP calling, Genome Analysis Toolkit (GATK), uses modules from HaplotypeCaller and GenotypeGVCFs to provide scalable SNP calling for several samples with excellent efficiency. Researchers can describe the possible impact of an SNP at a specific locus thanks to automated variant annotation tools like ANNOVAR and SNPef, which operate the function of SNPs based on the neighbouring sequence (Cinoglani et al. 2012).

19.4.5 Role of miRNAs

It is widely accepted that biotic and abiotic stress situations affect the gene expression patterns of plants, which has a detrimental effect on their growth, development and productivity. The regulators, which primarily consist of transcription factors and miRNAs, govern the expression of this gene. A growing body of research supports the idea that altering the amounts of miRNA can change how plants react to their constantly changing surroundings, including biotic and abiotic stressors (Brant and Budak 2018). Different plant species' biotic and abiotic challenges, such as drought, salinity, nutritional deficiencies, exposure to UV radiation, metal stress, excessive CO_2 levels, constant light and oxidative stress, have been shown to affect several miRNAs. A single miRNA has also been implicated in plant development and stress responses. Yang et al. (2011), for example, investigated the functional features and expression patterns of two overexpressed miR319 transcripts in rice. Plants with overexpressed miR319a and miR319b exhibited wider leaf blades and more longitudinal small veins, according to the authors. Furthermore, inhibiting the expression of miR319 target genes increased cold tolerance, revealing the importance of miR319 in rice leaf morphogenesis and cold tolerance (Xu et al. 2019). Even though many stress-responsive plant miRNAs have been discovered, we still do not fully understand how these miRNAs are regulated, which makes it more difficult when multiple mature miRNAs are produced from a single locus. This is a significant barrier to understanding the functional significance of miRNAs.

19.4.6 Zinc Finger Nucleases

Zinc finger nucleases (ZFNs) are blend proteins composed of a target-specific DNA binding domain and a DNA cleavage/nuclease domain linked by a 21-amino acid linker. The mono zinc ion is bound by the zinc finger domain, which is composed of 30 amino acids in a conserved form. The sequence-independent restriction enzyme FokI from *Flavabacterium okeanokoites* is the source of the nuclease domain. It consists of two domains: a DNA binding domain that recognises the 5′-GATG-3′ sequence and the other one is a cleavage/nuclease domain that cuts downstream of the recognition region on both strands. As a result of FokI catalytic activity, cleavage in the spacer area between two ZFN units occurs, resulting in a staggered DSB with overhangs (Paschon et al. 2019). Weinthal et al. (2010) discussed the method and application of ZFNs for incorporating various sorts of point mutations, minor and large deletions and insertions.

19.4.7 TALENs

Transcription activator-like effectors nucleases function as the DNA binding domain of the protein class known as TALENs, and the DNA cleaving domain is served by the type IIS bacterial endonuclease FokI enzyme. Xanthomonas pathogenic bacteria secrete virulence proteins termed TALEs, which bind to effector binding elements (EBEs) in host gene promoters and activate gene expression. The effector molecule is composed of an N-terminal translocation domain, a C-terminal nuclear localisation signal (NLS), a transcription activation domain and a central DNA binding domain (Boch et al. 2009). The core DNA binding domain is composed of a tandem array of 12–30 identical repeats of 34 amino acids in length, with unique repeat variable diresidues (RVDs) at amino acid positions 12 and 13. RVDs and their sequences define the TALE DNA binding specificity, which can be created practically for any DNA sequence, based on this predictable one-to-one relationship. Software for designing TALEs and TALENs is widely accessible; Khan et al. have described this software and thoroughly illustrated the successful design.

19.4.8 CRISPR/Cas9

In the genome of *Escherichia coli*, CRISPR was initially identified in 1987. Jansen and his colleagues found four CRISPR-associated (Cas1–4) genes in 2002, and it was given the abbreviation CRISPR (clustered regularly interspaced short palindromic repeats). In 2005, three different research teams discovered that CRISPR spacers are derived from plasmid and phage DNA, but their biological significance remained unknown. The genomic section of invasive viruses and plasmids are inserted into bacteria as part of their defence system, claim Barrangou and his colleagues in a 2007 research (Barrangou and Marraffini 2014), to act as a memory for any infections. Class I and class II are the two main subcategories of CRISPR/Cas systems. To moderate interference with foreign nucleic acids, Class I uses many Cas proteins to form a complex known as CASCADE, whereas Class II uses a single, large Cas protein with multiple domains. The three essential stages of the CRISPR-mediated defence mechanism in bacteria are target interference, crRNA production and adaptation. Protospacers, which are spacers made of the invasive nuesthetic acid, are incorporated into the CRISPR array during the adaptation phase. The acquisition of target sequences (spacers) in various CRISPR systems depends on a brief sequence known as the protospacer adjacent motif (PAM), which is located near to a protospacer. Interference occurs during crRNA synthesis, and mature crRNAs serve as a model to selectively bind with intrusive nucleic acids (Yashika et al. 2021). In CRISPER technology, widely associated nuclease protein is used in the genetic engineering named as the Cas 9 protein creating a double strand break in DNA and then are repaired by non-homologous and homologous recombination and this technology is used to knockout or knockin the desired gene.

It is essential to make sure that the transport of the genome-engineering reagents to the proper species is achievable and that editing of the target genome is both highly specific and effective to achieve high-efficiency genome engineering in any eukaryotic cell. To create high-efficiency genome engineering technologies, reagent transport and editing specificity are significant research fields. The development of delivery platforms for genome-engineering agents, preferably for distribution into germline cells to avoid the need for tissue culture and regeneration after editing, is currently a primary emphasis for plants (Mao et al. 2016; Alok et al. 2021). Bacterial and viral vectors, as well as direct administration into various cell types, are all examples of delivery platforms. Research on specificity includes the discovery of Cas9 variants with optimal expression and single guide (sgRNA) designs that are inherently more specific than current enzymes, as well as the titration of sgRNA and Cas9 concentrations throughout the editing process. Research on editing focuses on creating efficient HDR technologies that give the user complete control over the repair procedure and the genetic result, such as the ability to make single-base substitutions, targeted gene replacements and gene fusions (Zhao et al. 2016). Upadhyay et al. (2013) demonstrated that wheat suspension cells can be used to carry out CRISPR-Cas-guided genome editing. This makes using this technology on crop plants that can be grown from cell cultures easier. They also demonstrated how simultaneous cleavage at two targeted sites within a single gene might result in the deletion of a gene fragment in wheat.

19.5 Genome Engineering in Development of Climate Resilient Crops

Climate change has become a major threat to food crops of the current world and with to meet with the demands with ever increasing population genome engineering can revolutionise the breeding of crop plants for better supply of food. From very past humans have developed different varieties of crops plants by using the different conventional breeding technology. This technology also exploit genetic variation by the means of crossbreeding to get the desired traits. This conventional technique led to the surplus increase production which was called as 'Green Revolution'. But now these techniques have become inefficient due to climate change and crops plants faces major damage every year, which cannot be solved by conventional breeding techniques. So, a new approach or tool came into existence known as genetic engineering which enable the researchers to the insights of the crop plants at the gene level and can do changes at gene level to develop new better variety, which can be climate resilient (Gao 2021). From past few decades the researchers were focused only on the weed removal, increase in yield, improvement of plant height but now the researchers are now more focused on the development of the climate resilient crop plants due to change in the earth climatic conditions. Now the researchers have started exploring at the gene level and using different techniques to edit the genome for the desired climate resilient trait using the machine learning, pangenomes and scalable gene editing techniques (Saad et al. 2022).

Crop wild relatives (CWRs) are the wild plants that can show hidden genetic diversity of the domesticated plants. These CWRs have shown to harbour different useful genes which are mostly lost by the domesticated crop plants and can be used to develop climate resilience crop. This knowledge pool of the genetic diversity has now been increasing due the advancement of the genome sequencing technology and computational biology approaches. But we are still lacking in the development of germplasm libraries of these CWRs, and it is important because the development of climate resilience plants will also be dependent on the available traits in the CWRs germplasm (Tirnaz et al. 2022). Pangenomes' is a representation of the entire set of genes of a clade including the CWRs and domesticated crop plants so, by using pangenomes, researchers can identify the structural variations and copy number in the strain. Studying of pangenomes can lead us to the understanding of the selection pressure on crop plants by human and one can also know what genes are being lost or gained (Petereit et al. 2022). So, by using the modern technology or the genetic engineering it is possible to make climate resilient plants. Some different new methods are also being used like CRISPER technology, TALENs and nucleases which are already explained above.

19.6 Status of Improved Crops with Genetic Engineering

With the advancement of technology in plant breeding by genome engineering, new varieties are being developed extensively having traits showing persistence to abiotic stress shown in Table 19.1. This technology has shown a potential to develop climate-resilient crop plants by aiding at adapt to climate change or helping in mitigating the effects of climate change (Karavolias et al. 2021). CRISPR has shown positive results in the improvement of traits of many crop plants like wheat, maize, watermelon, rice, cotton, orange, potato, tomato and sorghum (Bhat et al. 2021). Rice, which is most staple food crop globally, is most vulnerable to climate stress also. Experimental studies have shown that rice yield is mostly affected by salt and drought conditions. CRISPR-Cas9 was employed to knock out the OsRR22 gene, responsible for salt stress and eliminating this gene increased the biomass without decreasing the yield. Drought stress resistance was made possible with the reduction in the stomatal density using a cisgenic approach (Karavolias et al. 2021). Another study has shown that CRISPR-based editing of genes of phytoene desaturase (OsPDS) and mitogen-activated protein kinase (OsMPK) have also made rice tolerant to abiotic stress showing promising results (Bhat et al. 2021). CRISPR tool has also been used to develop the semi-dwarf variety of bananas showing resistance to windy climates and floods (Karavolias et al. 2021). Another improvement was also done using CRISPR technology in which a knockin of the alternate promoter was carried out before gene ARGOS8 in maize to make it drought-resistant (Karavolias et al. 2021). Other than CRISPR researchers have also used other tools and methods to develop the crop plant variety resistance to abiotic stress making them climate resilient and such techniques and method can be found in Table 19.1. The developed plant variety shows promising value addition with an increase in yield and proves that this technology can be utilised to improve crop plants' resistance.

Table 19.1 Examples of crop plants showing climate resilient traits developed using genome engineering technology.

Crop	Target gene	Technique used	Trait modified	Reference
Rice	*OsRAV2, OsDST*	Knockout using CRISPR	Salt and drought stress resistance	Yashika et al. (2021)
Corn	*ARGOS8*	Knockout using CRISPR	Drought stress resistance	Yashika et al. (2021)
Arabidopsis	*AtNHX1*	Overexpression	Salt stress resistance	Zhang et al. (2000)
Tobacco	*NADH-hydroxypyruvate reductase*	Knock-in/insertion	Resistance to heat stress	Zhang et al. (2000)
Rice, Corn	*OsDREB1, ZmDREB1A*	Overexpression	Resistance to salt, drought and cold stress	Goel and Madan (2014)
Potato	*StFLORE*	Knockout using CRISPR	Resistance to drought	Razzaq et al. (2021)
Soybean	*GmNAC06*	Knockout using CRISPR	Resistance to salt	Razzaq et al. (2021)
Tomato	*HyPRP1 domain*	Knockout using CRISPR	Resistance to salt	Razzaq et al. (2021)
Banana	*Ma04g15900, Ma06g27710*	CRISPR/cas9	Semi-drawf	Karavolias et al. (2021)
Chickpea	Transcriptome analysis	RNA sequencing	Resistance to drought, salinity, cold and heavy metal stress	Bakala et al. (2020)
Finger Millet	Transcriptome analysis	RNA sequencing	Resistance to salt stress	Bakala et al. (2020)
Tomato	*BZR1, CBF1*	CRISPR/cas9	Resistant to high and low-temperature stress	Bhat et al. (2021)
Brassica napus	*BnaA6.RGA (DELLA Protein)*	CRISPR/cas9	Resistance to drought	Bhat et al. (2021)

19.7 Problems Associated with Genetic Engineering

Crop plants are heavily genetically engineered in order to increase yield in order to fulfil the demands of the growing population or to adapt to the growing environmental stress. These modified plants are mostly resistant to biotic and abiotic stress. But as we know new technology also come with some risk or side effects so, this technology also has some potential side effects which are being ignored by researchers because of its positive benefits. To grow genetically engineered crop plants in fields they should be thoroughly studied for several generations. Experiments related to toxicity and allergy should also be done. Finally, if a new protein expression is enhanced then its toxicity, digestibility and stability experiments should also be carried out. These experiments will help to analyse any side effects of the new crop plant before commercialising for cultivation (Datta 2013). These plants can show unprecedented risk to the environment because they have not undergone any natural evolutionary process or they any natural selection. These newly cop plants can also wipe out the wild-type ones, thus posing risk to biodiversity. Studies also suggest that these can give rise to new weed varieties that can be hard to handle (Bauer-Panskus et al. 2020). Another problem that can be caused by genome editing of crops is the inactivation of genes. As in the technique, the foreign DNA sequence is added at random or the specific site but this can lead to the inactivation or can cause overexpression of the nearby genes. From recent technology like DNA sequencing, one can precisely study the genome giving better insight into the genomics of a particular crop plant but still the challenge remains to utilize that data for studying the effect of change in phenotype by changing the phenotype (Abdallah et al. 2015).

19.8 Future Aspects

Genome engineering has opened a new pathway to develop crop plants with desired traits. Identification of the trait and then transforming the trait to the other plant species at the specific loci of the chromosome is still very complex. Now, we have the genome sequence of most of the crop plants but still lagging in the annotation of genome sequence to develop more new resistant varieties. Therefore, more research is needed to identify the beneficial trait by annotating the sequence so that they can be utilised to develop new crop varieties, which can withstand the upcoming climate stress. More experimentation is needed to study the behavioural pattern of genetically modified crops to minimise any side effects like toxicity, allergy and prevention of horizontal gene transfer to prevent the development of new weed varieties. Now with the use of machine learning and artificial intelligence, new tools or techniques can be designed in the future, which can minimise the error or the side effects that can be caused by genetically modified crops.

19.9 Conclusion

Since the dawn of civilisation, humans have been superior cultivars of numerous crop species. These new varieties created by conventional breeding techniques show only traits within the same family and are sometimes not very useful. There was no approach for the interspecies gene transfer methods to develop new resistant varieties. As the genomic era came creating new technology that can identify and transfer the trait from one species to other by transferring the DNA led to the rapid advancement of the development of new plant varieties. Most of the drawbacks of conventional plant breeding were solved by genome engineering techniques using genome editing tools (ZFNs, TALENs and CRISPR), enabling fast manipulation of crops exhibiting the desired feature. As it simply requires the host's genome sequence information and is straightforward to target, a new green revolution for the production of climate-smart crops may be brought about by CRISPR/Cas. It resembles the cell's normal mutations, and screening to pick out the helpful ones is simple to execute. Now crops are being developed, which have shown potential traits of resistance to drought, salt stress as well as temperature stress, but still, there are a few drawbacks like lack of trials on genome-engineered crops, toxicity or allergy-prone, unwanted gene transfer contaminating the wild or other species and study on the stability of crop plants. As the research continues, new technology will emerge not far in the future that will also surpass current technology and eradicate the drawbacks of current technology to make new crop varieties. Man will continue to fight the shortcomings of crop production until the research will continue and will build new technology to surpass it.

References

Abdallah, N.A., Prakash, C.S., and McHughen, A. (2015). Genome editing for crop improvement: challenges and opportunities. *GM Crops & Food* 6 (4): 183–205. https://doi.org/10.1080/21645698.2015.1129937.

Alok, A., Chauhan, H., Upadhyay, S.K. et al. (2021). Compendium of plant-specific CRISPR vectors and their technical advantages. *Life* 11 (10): 1021. https://doi.org/10.3390/life11101021.

Bakala, H.S., Singh, G., and Srivastava, P. (2020). *Smart Breeding for Climate Resilient Agriculture*. IntechOpen EBooks. https://doi.org/10.5772/intechopen.94847.

Barrangou, R. and Marraffini, L.A. (2014). CRISPR-Cas systems: prokaryotes upgrade to adaptive immunity. *Molecular Cell* 54 (2): 234–244. https://doi.org/10.1016/j.molcel.2014.03.011.

Bauer-Panskus, A., Miyazaki, J., Kawall, K., and Then, C. (2020). Risk assessment of genetically engineered plants that can persist and propagate in the environment. *Environmental Sciences Europe* 32 (1): https://doi.org/10.1186/s12302-020-00301-0.

Bhat, M.A., Mir, R.A., Kumar, V. et al. (2021). Mechanistic insights of CRISPR/Cas-mediated genome editing towards enhancing abiotic stress tolerance in plants. *Physiologia Plantarum* 172 (2): 1255–1268. https://doi.org/10.1111/ppl.13359.

Blanc, E. (2012). The impact of climate change on crop yields in sub-saharan Africa. *American Journal of Climate Change* 01 (01): 1–13. https://doi.org/10.4236/ajcc.2012.11001.

Boch, J., Scholze, H., Schornack, S. et al. (2009). Breaking the code of DNA binding specificity of TAL-type III effectors. *Science* 326 (5959): 1509–1512. https://doi.org/10.1126/science.1178811.

Brant, E.J. and Budak, H. (2018). Plant small non-coding RNAs and their roles in biotic stresses. *Frontiers in Plant Science* 9: https://doi.org/10.3389/fpls.2018.01038.

Brouder, S.M. and Volenec, J.J. (2008). Impact of climate change on crop nutrient and water use efficiencies. *Physiologia Plantarum* 133 (4): 705–724. https://doi.org/10.1111/j.1399-3054.2008.01136.x.

Cardi, T. and Stewart, C.N. (2016). Progress of targeted genome modification approaches in higher plants. *Plant Cell Reports* 35 (7): 1401–1416. https://doi.org/10.1007/s00299-016-1975-1.

CGIAR (2022). Plant health – CGIAR. https://www.cgiar.org/initiative/13-plant-health-and-rapid-response-to-protect-food-and-livelihood-security/(accessed 18 February 2023).

Cingolani, P., Platts, A.E., Wang, L.Y. et al. (2012). A program for annotating and predicting the effects of single nucleotide polymorphisms, SnpEff. *Fly* 6 (2): 80–92. https://doi.org/10.4161/fly.19695.

DaMatta, F.M., Grandis, A., Arenque, B.C., and Buckeridge, M.S. (2010). Impacts of climate changes on crop physiology and food quality. *Food Research International* 43 (7): 1814–1823. https://doi.org/10.1016/j.foodres.2009.11.001.

Datta, A. (2013). Genetic engineering for improving quality and productivity of crops. *Agriculture & Food Security* 2 (1): https://doi.org/10.1186/2048-7010-2-15.

Deressa, T.T. and Hassan, R.M. (2009). Economic impact of climate change on crop production in Ethiopia: evidence from cross-section measures. *Journal of African Economies* 18 (4): 529–554. https://doi.org/10.1093/jae/ejp002.

FAO (2016). The state of food and agriculture 2016|FAO. https://www.fao.org/family-farming/detail/en/c/447860/.

Fiaz, S., Khan, S.B., Anis, G.B. et al. (2021). Application of genome engineering methods for quality improvement in important crops. In: *Genome Engineering for Crop Improvement* (ed. S.K. Upadhyay), 43–68. Wiley. https://doi.org/10.1002/9781119672425.ch3.

Gao, C. (2021). Genome engineering for crop improvement and future agriculture. *Cell* 184 (6): 1621–1635. https://doi.org/10.1016/j.cell.2021.01.005.

Goel, S. and Madan, B. (2014). *Genetic Engineering of Crop Plants for Abiotic Stress Tolerance*, 99–123. Elsevier EBooks. https://doi.org/10.1016/b978-0-12-800876-8.00005-9.

Isik, M. and Devadoss, S. (2006). An analysis of the impact of climate change on crop yields and yield variability. *Applied Economics* 38 (7): 835–844. https://doi.org/10.1080/00036840500193682.

Karavolias, N.G., Horner, W., Abugu, M.N., and Evanega, S.D. (2021). Application of gene editing for climate change in agriculture. *Frontiers in Sustainable Food Systems* 5: https://doi.org/10.3389/fsufs.2021.685801.

Kumar, M. (2016). Impact of climate change on crop yield and role of model for achieving food security. *Environmental Monitoring and Assessment* 188 (8): https://doi.org/10.1007/s10661-016-5472-3.

Mao, Y., Zhang, Z., Feng, Z. et al. (2016). Development of germ-line-specific CRISPR-Cas9 systems to improve the production of heritable gene modifications in*Arabidopsis*. *Plant Biotechnology Journal* 14 (2): 519–532. https://doi.org/10.1111/pbi.12468.

Mao, D., Xin, Y., Tan, Y. et al. (2019). Natural variation in the *HAN1* gene confers chilling tolerance in rice and allowed adaptation to a temperate climate. *Proceedings of the National Academy of Sciences of the United States of America* 116 (9): 3494–3501. https://doi.org/10.1073/pnas.1819769116.

Murovec, J., Pirc, Ž., and Yang, B.J. (2017). New variants of CRISPR RNA-guided genome editing enzymes. *Plant Biotechnology Journal* 15 (8): 917–926. https://doi.org/10.1111/pbi.12736.

Naqvi, R.Z., Siddiqui, H.L., Mahmood, M.T. et al. (2022). Smart breeding approaches in post-genomics era for developing climate-resilient food crops. *Frontiers in Plant Science* 13: https://doi.org/10.3389/fpls.2022.972164.

Paschon, D., Lussier, S., Wangzor, T. et al. (2019). Diversifying the structure of zinc finger nucleases for high-precision genome editing. *Nature Communications* 10 (1): https://doi.org/10.1038/s41467-019-08867-x.

Petereit, J., Bayer, P.E., Thomas, W. et al. (2022). Pangenomics and crop genome adaptation in a changing climate. *Plants* 11 (15): 1949. https://doi.org/10.3390/plants11151949.

Pixley, K.V., Falck-Zepeda, J.B., Paarlberg, R. et al. (2022). Genome-edited crops for improved food security of smallholder farmers. *Nature Genetics* 54 (4): 364–367. https://doi.org/10.1038/s41588-022-01046-7.

Pradhan, A.K., Naik, N., and Sahoo, K.K. (2015). RNAi mediated drought and salinity stress tolerance in plants. *American Journal of Plant Sciences* 06 (12): 1990–2008. https://doi.org/10.4236/ajps.2015.612200.

Rai, K.M. and Singh, H. (2021). Genome-editing mediated improvement of biotic tolerance in crop plants. In: *Genome Engineering for Crop Improvement* (ed. S.K. Upadhyay), 199–209. Wiley. https://doi.org/10.1002/9781119672425.ch12.

Raza, A., Razzaq, A., Mehmood, S. et al. (2019). Impact of climate change on crops adaptation and strategies to tackle its outcome: a review. *Plants* 8 (2): 34. https://doi.org/10.3390/plants8020034.

Razzaq, A., Kaur, P., Akhter, N. et al. (2021). Next-generation breeding strategies for climate-ready crops. *Frontiers in Plant Science* 12: https://doi.org/10.3389/fpls.2021.620420.

Saad, N.S.M., Neik, T.X., Thomas, W. et al. (2022). Advancing designer crops for climate resilience through an integrated genomics approach. *Current Opinion in Plant Biology* 67: 102220. https://doi.org/10.1016/j.pbi.2022.102220.

Sedeek, K.E.M., Mahas, A., and Islam, T. (2019). Plant genome engineering for targeted improvement of crop traits. *Frontiers in Plant Science* 10: https://doi.org/10.3389/fpls.2019.00114.

Shrivastava, P.V. and Kumar, R. (2015). Soil salinity: a serious environmental issue and plant growth promoting bacteria as one of the tools for its alleviation. *Saudi Journal of Biological Sciences* 22 (2): 123–131. https://doi.org/10.1016/j.sjbs.2014.12.001.

Solanke, A.U., Arora, K., Karkute, S.G., and Tomar, R.S.S. (2021). Genetic engineering for biotic stress management in rice. *Concepts and Strategies in Plant Sciences* 117–139. https://doi.org/10.1007/978-3-030-63372-1_5.

Srivastav, A., Khare, T., and Kumar, V. (2018). *Systems Biology Approach for Elucidation of Plant Responses to Salinity Stress*, 307–326. Springer International Publishing EBooks. https://doi.org/10.1007/978-3-319-90318-7_13.

Tirnaz, S., Zandberg, J., Thomas, W.J.W. et al. (2022). Application of crop wild relatives in modern breeding: an overview of resources, experimental and computational methodologies. *Frontiers in Plant Science* 13: https://doi.org/10.3389/fpls.2022.1008904.

Ullah, A., Bano, A., and Khan, N. (2021). Climate change and salinity effects on crops and chemical communication between plants and plant growth-promoting microorganisms under stress. *Frontiers in Sustainable Food Systems* 5: https://doi.org/10.3389/fsufs.2021.618092.

Upadhyay, S.K., Kumar, J., Alok, A., and Tuli, R. (2013). RNA-guided genome editing for target gene mutations in wheat. *G3: Genes, Genomes, Genetics* 3 (12): 2233–2238. https://doi.org/10.1534/g3.113.008847.

Weinthal, D.M., Tovkach, A., Zeevi, V., and Tzfira, T. (2010). Genome editing in plant cells by zinc finger nucleases. *Trends in Plant Science* 15 (6): 308–321. https://doi.org/10.1016/j.tplants.2010.03.001.

Woodrow, P., Ciarmiello, L.F., Annunziata, M.A. et al. (2017). Durum wheat seedling responses to simultaneous high light and salinity involve a fine reconfiguration of amino acids and carbohydrate metabolism. *Physiologia Plantarum* 159 (3): 290–312. https://doi.org/10.1111/ppl.12513.

Xu, J., Hou, Q., Khare, T. et al. (2019). Exploring miRNAs for developing climate-resilient crops: a perspective review. *Science of the Total Environment* 653: 91–104. https://doi.org/10.1016/j.scitotenv.2018.10.340.

Yang, L., Wu, G., and Poethig, R.S. (2011). Mutations in the GW-repeat protein SUO reveal a developmental function for microRNA-mediated translational repression in *Arabidopsis*. *Proceedings of the National Academy of Sciences of the United States of America* 109 (1): 315–320. https://doi.org/10.1073/pnas.1114673109.

Yashika, G., Ashwani, P., and Sneh, L. (2021). Raising climate resilient crops: journey from the conventional breeding to new breeding approaches. *Current Genomics* 22: https://doi.org/10.2174/1389202922666210928151247.

Zhang, J., Klueva, N., Wang, Z. et al. (2000). Genetic engineering for abiotic stress resistance in crop plants. *In Vitro Cellular & Developmental Biology – Plant* 36 (2): 108–114. https://doi.org/10.1007/s11627-000-0022-6.

Zhao, Y., Zhang, C., Liu, W. et al. (2016). An alternative strategy for targeted gene replacement in plants using a dual-sgRNA/Cas9 design. *Scientific Reports* 6 (1): https://doi.org/10.1038/srep23890.

Zörb, C., Geilfus, C., and Dietz, K. (2019). Salinity and crop yield. *Plant Biology* 21 (S1): 31–38. https://doi.org/10.1111/plb.12884.

Index

Applications of Genome Engineering in Plants, First Edition. Edited by Santosh Kumar Upadhyay.
© 2024 John Wiley & Sons Ltd. Published 2024 by John Wiley & Sons Ltd.

b

c